CELL BIOLOGY AND GENETICS

CECIE STARR/RALPH TAGGART

BIOLOGY
The Unity and Diversity of Life

TENTH EDITION

LISA STARR
Biology Illustrator

THOMSON

BROOKS/COLE

Australia • Canada • Mexico • Singapore • Spain
United Kingdom • United States

BIOLOGY PUBLISHER: Jack C. Carey

EDITOR-IN-CHIEF: Michelle Julet

DEVELOPMENTAL EDITOR: Mary Arbogast, Peggy Williams

ASSISTANT EDITOR: Suzannah Alexander

EDITORIAL ASSISTANTS: Karoliina Tuovinen, Jana Davis

MEDIA PROJECT MANAGER: Pat Waldo

TECHNOLOGY PROJECT MANAGERS: Donna Kelley, Keli Amann

MARKETING MANAGER: Ann Caven

MARKETING ASSISTANT: Sandra Perin

ADVERTISING PROJECT MANAGER: Linda Yip

SENIOR PROJECT MANAGER, EDITORIAL/PRODUCTION: Teri Hyde

PRINT/MEDIA BUYER: Karen Hunt

PERMISSIONS EDITOR: Joohee Lee

PRODUCTION SERVICE: Lachina Publishing Services, Inc.;
Grace Davidson

TEXT AND COVER DESIGN: Gary Head, Gary Head Design

ART EDITOR AND PHOTO RESEARCHER: Myrna Engler

ILLUSTRATORS: Lisa Starr, Gary Head

OFFICE SUPPORT: Brad Griffin, Verbal Clark

COVER IMAGE: *From Central America, one of the tropical rain
forests that may disappear in your lifetime.*
Kevin Schafer/Getty Images

COVER PRINTER: Phoenix Color Corp (MD)

COMPOSITOR: Preface, Inc.; Angela Harris, John Becker

FILM HOUSE: H&S Graphics; Tom Anderson

PRINTER: Quebecor/World, Versailles

For more information about our products, contact us at:
Thomson Learning Academic Resource Center
1-800-423-0563
For permission to use material from this text, contact us by:
Phone: 1-800-730-2214   Fax: 1-800-730-2215
Web: http://www.thomsonrights.com

ISBN 0-534-39745-X

Brooks/Cole—Thomson Learning
10 Davis Drive
Belmont, CA 94002
USA

Asia
Thomson Learning
60 Albert Street, #15-01
Albert Complex
Singapore 189969

Australia
Nelson Thomson Learning
102 Dodds Street
South Melbourne, Victoria 3205
Australia

Canada
Nelson Thomson Learning
1120 Birchmount Road
Toronto, Ontario M1K 5G4
Canada

Europe/Middle East/Africa
Thomson Learning
Berkshire House
168-173 High Holborn
London WC1 V7AA
United Kingdom

CONTENTS IN BRIEF

Highlighted chapters are included in CELL BIOLOGY AND GENETICS.

DETAILED CONTENTS

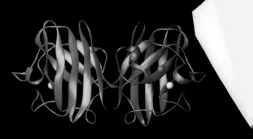

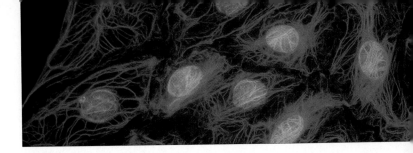

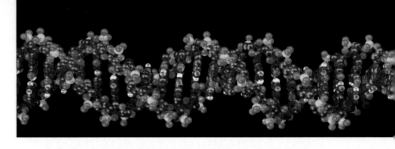

PREFACE

Successive revisions of this book span nearly thirty years and reflect feedback from many instructors and students. This new edition retains the concept spreads and other pedagogical features that are the hallmarks of the book.

CONCEPT SPREADS   Reading and absorbing textbook assignments for multiple courses in the same timeframe can overwhelm students. We make it easier for them to read about and understand biology by focusing on one concept at a time. We list key concepts on the first page of each chapter. Then we organize text, art, and evidence in support of each concept on two facing pages, at most. As shown below, each *concept spread* starts with a numbered tab and ends with a boldfaced summary of the key points. Students can preview the on-page summary before they read the concept spread. They can read it again to check whether they understand the key points before turning to the next concept.

Concept spreads also offer teachers flexibility in assigning topics to fit their course requirements. For example, those who spend less time on photosynthesis may bypass the spreads on properties of light and the chemiosmotic theory of ATP formation. They may or may not assign the Focus essay on the global impact of photosynthesis. All spreads and essays are part of a chapter story, but some offer more depth.

We incorporate headings within concept spreads to help students keep track of the hierarchy of information. Transitions between spreads help them follow the story. So does setting aside some details in optional illustrations for motivated students.

Example of a cell icon

With concept spreads, students find assigned topics fast, and they can focus on manageable amounts of information. This makes them more confident in their capacity to absorb the material. Our approach has a tangible outcome—improved test scores.

VISUALIZING CONCEPTS   We continue to develop text and art together, as an inseparable whole. Our *"read-me-first diagrams"* are a prime example of this approach. They allow visual learners to build a mental image of a concept before reading the text details about it. Figure 34.5 in the sample pages at right shows how simple descriptions walk students step by step through these preview diagrams. Many of our millions of student readers have written in to tell

us that our approach helps them far more than a reliance on "wordless" diagrams.

Many *anatomical drawings* are integrated overviews of structure and function. Students need not jump back and forth from text, to tables, then to art, and back again to visualize how an organ system is put together and what its component parts do. We also use *zoom sequences*, from macroscopic to microscopic views, to move students visually into a system or process. For example, Figures 37.19 and 37.20 start with a ballerina's biceps and move on down through levels of skeletal muscle contraction.

Icons remind students of where art fits into the story line. For instance, icons of a cell remind students where

Gold numbered tabs, such as this one, identify the start of each new concept in a chapter. Other tabs (brown) in the chapter identify Focus essays. Many essays enrich the basic text by addressing medical, environmental, and bioethical issues. Others offer detailed examples of experiments to demonstrate the power of critical thinking.

34.2

HOW ARE ACTION POTENTIALS TRIGGERED AND PROPAGATED?

Action potential propagation isn't hard to follow if you already know something about the gradients across the neural membrane. And so we now build on Section 34.1.

Approaching Threshold

When you weakly stimulate a neuron at its input zone, you disturb the ion balance across its membrane, but not much. Imagine putting a bit of pressure on the skin of a snoozing cat by gently tapping a toe on it. Tissues beneath the skin surface have receptor endings—input zones of sensory neurons. Patches of plasma membrane at these endings deform under pressure and let some ions flow across. The flow slightly changes the voltage difference across the membrane. In this case, pressure has produced a graded, local signal.

influx of ions, the cytoplasmic side of the membrane becomes less negative. This causes more gates to open and more sodium to enter. The ever increasing, inward flow of sodium is a case of **positive feedback**, whereby an event intensifies as a result of its own occurrence:

At threshold, opening of sodium gates no longer depends on the strength of the stimulus. The positive-feedback cycle is under way, and the inward-rushing sodium itself is enough to open the gated channels.

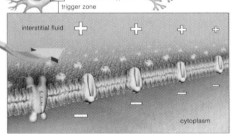
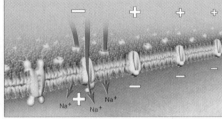

a Membrane at rest (inside negative with respect to the outside). An electrical disturbance (*yellow* arrow) spreads from an input zone to an adjacent trigger region of the membrane, which has a great number of gated sodium channels.

b A strong disturbance initiates an action potential. Sodium gates open. The sodium inflow decreases the negativity inside the neuron. The change causes more gates to open, and so on until threshold is reached and the voltage difference across the membrane reverses.

Figure 34.5   Propagation of an action potential along the axon of a motor neuron.

Graded means that signals arising at an input zone vary in magnitude. They are small to large, depending on the stimulus intensity or duration. *Local* means these signals do not spread far from the site of stimulation. Why? It takes certain kinds of ion channels to propagate a signal, and input zones simply don't have them.

When a stimulus is intense or long-lasting, graded signals spread from the input zone into an adjoining trigger zone. This patch of membrane is richly endowed with voltage-sensitive gated channels for sodium ions. *And this is where a certain amount of change in the voltage difference across the plasma membrane triggers an action potential.* The amount is the neuron's threshold level.

When these gates open, positively charged sodium ions flow into the neuron, as in Figure 34.5. With the

An All-or-Nothing Spike

Figure 34.6 shows a recording of the voltage difference across the plasma membrane before, during, and after an action potential. Notice how the membrane potential peaks once threshold is reached. All action potentials in a neuron spike to the same level above threshold as an *all-or-nothing* event. Once a positive-feedback cycle starts, nothing stops full spiking. Unless threshold is reached, the membrane disturbance subsides when the stimulation ends, and an action potential won't occur.

Each spike lasts for only a millisecond or so. Why? At the patch of membrane where the charge reversed, gated sodium channels close and shut off the sodium inflow. And about halfway into the reversal, potassium its organelles are located, as in Chapter 4. Other icons remind students of how reaction stages interconnect in a metabolic pathway, as in Chapter 7 (photosynthesis) and Chapter 8 (aerobic respiration). Others remind them of evolutionary relationships, as in Chapters 25 and 26. A multimedia icon directs students to art in the CD-ROM packaged at the back of every book. Others direct them to supplemental material on the Web and to InfoTrac® College Edition, an online database of full-length articles from 4,000 academic journals and popular sources.

Finally, we added hundreds of new *micrographs and photographs*. These are not window dressing, tacked on after the fact. They sharpen the meaning of "biodiversity" and hint at why it is worth preserving.

BALANCING CONCEPTS WITH APPLICATIONS   We draw students into each chapter with a lively or sobering application. That application gives way to a list of key concepts, an advance organizer. We attempt to maintain their interest with focus essays, which provide depth on medical, environmental, and social issues without interrupting the conceptual flow. Where we sense that the core material needs to be livened up a bit, we weave briefer applications into the text proper. On the last pages of the book, a separate Applications Index affords fast reference to our many hundreds of applications.

FOUNDATIONS FOR CRITICAL THINKING   Like all textbooks at this level, ours helps students sharpen their capacity to think critically about nature. We walk them through experiments that yielded clear evidence in favor of or against hypotheses. The main index at the back of the book lists the experiments we selected (see the entries *Experiment, examples,* and *Test, observational*).

We also selectively use chapter introductions as well as entire chapters to show productive outcomes of critical thinking. The chapter introductions to Mendelian genetics (11), DNA structure and function (13), speciation (18), immunology (39), and behavior (46) are examples. The end of each chapter has a set of *Critical Thinking* questions. Numerous *Genetics Problems* at the end of Chapters 11 and 12 help students grasp principles of inheritance.

SUPPLEMENTS   The Instructors' Examination copy for this edition lists a comprehensive package of print and multimedia supplements, including online resources that are available to qualified adopters. Please ask your local sales representative for details.

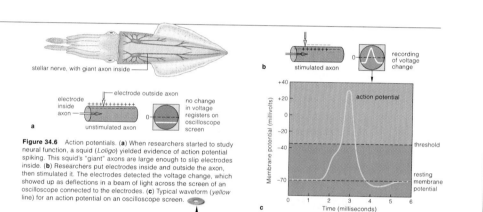

Figure 34.6 Action potentials. (**a**) When researchers started to study neural function, a squid (*Loligo*) yielded evidence of action potential spiking. This squid's "giant" axons are large enough to slip electrodes inside. (**b**) Researchers put electrodes inside and outside the axon, then stimulated it. The electrodes detected the voltage change, which showed up as deflections in a beam of light across the screen of an oscilloscope connected to the electrodes. (**c**) Typical waveform (*yellow line*) for an action potential on an oscilloscope screen.

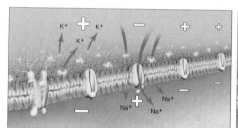

c With the reversal, sodium gates shut and potassium gates open (*red* arrows). Potassium follows its gradient out of the neuron. Voltage is restored. The disturbance triggers an action potential at the adjacent site, and so on, away from the point of stimulation.

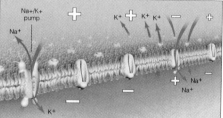

d Following each action potential, the inside of the plasma membrane becomes negative once again. However, the sodium and potassium concentration gradients are not yet fully restored. Active transport at sodium–potassium pumps restores them.

This icon signifies that we explore the concept further on our interactive CD-ROM.

This is an example of our "read-me-first" diagrams for visual learners. They are illustrated previews of the text material, complete with simple "a b c" descriptions that guide the student each step of the way.

channels opened, so potassium flows out. This restores the voltage difference at the patch but not the original gradients. Sodium–potassium pumps actively transport sodium back outside and potassium inside. Once this is done, most potassium gates close and sodium gates are in their original position—ready to be opened with the arrival of a suitable signal at the membrane patch.

The Direction of Propagation

During an action potential, the inward rush of sodium ions affects the charge distribution across the adjacent membrane patch, where an equivalent number of gated channels open. Gated channels open in the *next* patch, and the next, and so on. This positive feedback event is self-propagating and does not diminish in magnitude. You might be wondering: Do action potentials spread

back to the trigger zone? No. For a brief period after the inward rushing of sodium ions, the voltage-gated channels remain insensitive to stimulation, so sodium ions cannot move through them. This is one reason why action potentials do not spread back to the patch of membrane where they were initiated. It is why they propagate themselves away from it.

Ions cross the neural membrane through transport proteins that serve as gated or open channels. At a suitably disturbed trigger zone, sodium gates open in an all-or-nothing way, and the inward-rushing sodium causes an action potential. Sodium–potassium pumps restore the original ion gradients.

Sodium gates across the membrane are briefly inactivated after an action potential, which is one reason why an action potential is self-propagating away from a trigger zone.

Boldfaced key concepts end each tabbed section. Students can read them to preview the section's main points, then read them again to reinforce what they learn from the section.

All together, these highlighted statements are a running summary of the take-home lessons.

This icon reminds students to check out the website, which expands on the section's topic.

A MAJOR CONCEPTUAL SHIFT In the past, with each revision, we have updated details to reflect new work in biology's rapidly changing fields. This time we did more. We decided to undertake a major conceptual shift after concluding that one is long overdue in biology textbooks at this level.

The Emerging Big Picture

Biology's overriding paradigm is one of interpreting life's spectacular diversity as having evolved from simple molecular beginnings. For decades now, researchers have been chipping away at the structural secrets of biological molecules. They have discovered how different kinds are put together, how they function, and what happens when they mutate. They are casting light on how life originated, what happened during the past 3.8 billion years, and what the future may hold for humans and other organisms, individually and collectively.

This is profound stuff. *Yet introductory textbooks— including previous editions of this one—have not given students enough information to understand the remarkable connection between molecular change, evolution, and their own lives.* We rebuilt much of this tenth edition to show the connection more clearly, starting with the new models for biological molecules. The models can give students a deeper appreciation of how *structural* diversity translates into *functional* diversity.

Powerful Conceptual Tools for Students

Molecular models appear in books, newspapers, and popular magazines, but few students have a clue to what they represent. Why even ask them to memorize the fact that polypeptide chains form coils, sheets, and loops without giving clear examples of the outcomes? Why not show how different configurations form, say, membrane-spanning barrels, anchors, or jaws that grip enemy agents in the body? Why not complete the point and say that mutation may alter a configuration just enough to block or enhance a transport or anchoring or defensive function?

And why not connect that point with the paradigm? *Small changes in molecular sequences and functional domains give rise to variation in traits—the raw material of evolution.*

Once the connection becomes clear, students can apply what they learn about changes in molecular structure to many of their questions and concerns about life. Genetic disorders? Look to modifications in molecular structure that have altered how cells, organs, and the organism itself function. Events that long ago put chimpanzees and humans on their separate evolutionary paths? Look to transposons and other disruptions in shared ancestral DNA. Differences in the number of legs on a fly or centipede or human, or the number of petals on a flower? Look to small mutations in master genes that control development of the basic body plan.

And what about similarities in how, say, tulips and Tina Turner or any other organism function? Look to commonalities in molecular responses to environmental challenges.

Chapters That Develop the Conceptual Tools

Several chapters build on one another to help students develop knowledge of the molecular basis of life. The first chapter is a simple preview of the molecular basis of life's unity and diversity. Chapters 3, 5, and 6 have sections on the structure and function of enzymes and other molecules that prepare students for the important chapters on cell structure and metabolism, genetics, evolution, anatomy, and physiology.

For instance, with this background, students can sense the power of comparative molecular studies in clarifying evolutionary relationships. Among the outcomes of such studies are refined evolutionary tree diagrams and the three-domain classification system.

Chapter 28 is a new prelude to the units on anatomy and physiology. It points out that we sometimes forget how much plants and animals have in common at the molecular level, because their body plans are so different. For example, one of the essays in this chapter focuses on recurring challenges to survival of *all* individual cells in the multicelled body—specifically, the requirements for gas exchange and internal transport, for homeostasis in the internal environment, and for integration and control. This chapter also starts students thinking about how cells communicate with one another, using simple examples of signal reception, transduction, and response from plants and animals.

New Connections Essays That Reinforce the Big Picture

New to this edition are Connections essays that can help students "connect the dots" between text details and big-picture concepts. The facing page lists these essays.

Setting the Stage

Connections essay 1.4 answers a central question: How can life display both unity *and* diversity? It suggests that the theory of evolution by natural selection connects the two. Obviously, we cannot get into actual mechanisms of evolution until after the book's units on cell biology and genetics. But a simple outline of the theory can help students connect its basic premises with the topics of those units as well.

Essay 3.1 is a user-friendly introduction to the models that represent biological molecules. Most textbooks show molecules as sticks, balls, ribbons, and bubbled blobs. How many students stare blankly at them? We let them in on an unfortunately well-kept secret. Different models convey different information, such as mass, structural organization, reactivity, or function. That is why we use different kinds in different contexts. For example, the models for glycogen, hemoglobin, enzymes, and DNA tell us different things about how cells and organisms are put together and how they function.

Without this knowledge, can students extrapolate from molecular models to generic icons—to the circles and squares used to represent enzymes, ATP, and all the other biological molecules? Can they really extrapolate from generic icons alone to the concept of structure, function, and evolution? We think not.

Essay 5.2, for example, showcases some important membrane proteins. These stunning models make it easier for students to recognize that structural diversity translates into the functional diversity necessary for metabolism, gene function, integration, immunity, the formation of tissues and organs, and other tasks. They make it possible to comprehend how, say, cystic fibrosis arises from a mutation that changes the function of just one kind of membrane transporter. As another example, Essay 3.8 moves students from the molecular structure of HbA to HbS, and on to sickle-cell anemia.

Making Sense of Evolution and Biodiversity

Unless we explain continuity and change at the molecular level convincingly enough, many students will never become open to the idea of continuity and change in life itself—to the possibility that evolution is more than "just a theory."

Continuity and change—Essay 8.7 picks up on this thought. It ties together concepts from the cell biology unit into a view of how all of life connects as a result of evolution at the molecular level.

Essay 19.9 reinforces a major point—that interpreting the past scientifically requires a huge intellectual shift, from direct observation to inferences based upon molecular studies, the fossil record, and morphological comparisons. Essay 28.2 picks up on the same thought. The essay invites caution in interpreting the evolutionary history of life, because "adaptations" are not always what they seem. Its comparison of llama and camel hemoglobin underscores the point.

With tentative acceptance of evolutionary theory, students can think about applying it to their own lives. Essays 26.2, 26.11, 37.4, and 38.1 interconnect in this respect. They hint at where our amazing brains and hands came from, why we have such intricate highways for blood circulation through our body, and why we are prone to lower back pain. Essay 34.12 connects the brain's evolutionary history with its development. It shows how that connection helps explain why teenagers tend to fall asleep in class and engage in notably impulsive behavior.

Acceptance also gives broad insight into the source of biodiversity, into mass extinctions and slow recoveries. Essays 21.9, 22.1, 23.1 reinforce this point.

Connecting Our Lives to the Big Picture

Deeper understanding of how each of us connects with the sweeping story of life—that is what we hope students will take away from their introduction to biology. With this in mind, we use some essays to show how biology impacts our own lives and the choices that we make.

Essay 16.10, for example, reflects on how the ability to study and alter genomes rapidly, as with the use of DNA microarrays, is outpacing our attempts to assess its bioethical implications. What are the ramifications of human gene therapy? Cloning genetically engineered mammals? Manipulating genomes of crop plants? These are issues facing students today, and the citizens and leaders of tomorrow.

At the end of Unit V, which shows what it takes to be a plant, Essay 32.6 considers one of the costs of growing

CONNECTIONS ESSAYS

Vertical red bands down the edge of a text page flag these essays.

enough plants to feed the human population. It invites students to think about how we have become locked into using herbicides, fungicides, and pesticides on a massive scale. It invites them consider the effect of these complex molecules on nontargeted organisms, including us.

As two more examples, Essay 49.14 recounts how Rita Colwell made a sweeping connection between copepods, a bacterial life cycle, sea surface temperature changes during El Niño episodes, and horrible cholera outbreaks in Bangladesh.

The essay in our concluding chapter 50 compares life's evolution over the past 3.8 billion years with the impact of relative latecomers—humans. It asks students to think about the bioethics of mitigating the disproportionate effect of we latecomers on the world of life.

OTHER MAJOR CHANGES   Now that systematists have reached consensus, we have subsumed the kingdoms of organisms into the three-domain classification system. Section 1.3 introduces the system; Sections 19.7 and 19.8 fill in details. We took a stronger cladistic approach and did major work on the chapters of the biodiversity unit. Section 22.1 gives the rationale for doing so. The outcome is most evident in how we present the protistans (Chapter 22) and vertebrates (Chapter 26).

Again, we moved away from generic boxes and circles for molecules and present more realistic models for membranes, protein transporters, enzymes, RNAs, and other cell components. You can check out some examples of how we put them to use in Chapters 3, 5, 14, and 34.

To give an idea of what went into the new graphics, Lisa Starr rendered our molecular models from primary structural data. She even tracked down researchers who are still constructing a model of the human cardiac gap junction and worked directly with them to convert their most recent data into graphic form for this book.

We rewrote Section 13.5 (mammalian cloning) and Chapter 16 (recombinant DNA and genetic engineering) to keep up with the rapid advances in biotechnology. Section 16.10 addresses some of the bioethical issues. So do some of the detailed *Critical Thinking* questions at the end of Chapter 16. These are serious issues that should provoke discussion among students. The Chapter 15 introduction and focus essay on cancer are updated.

Given the wealth of new information to be covered in the evolution unit, we decided to condense the history of evolutionary thought into a few introductory sections for Chapter 17 (microevolution). Chapter 19 (macroevolution) is heavily rewritten and reorganized, starting with a new introduction on issues associated with measuring geologic time. There is stronger treatment of the evidence from comparative biochemistry and systematics. This chapter concludes with a reflection on how we interpret the past.

I urge our adopters to review Chapter 28. Concepts that will help students get through the two units on plant and animal anatomy/physiology are presented here. We have made so many refinements and updates—for example, on human nutrition, embryonic development, and predator–prey interactions, but we are running out of room to highlight them all. Suffice it to say that we made an honest, solid attempt to keep up with biology and to do it justice all across the board.

Writing this preview of all the changes reminded me again of how fortunate I am to be under the Wadsworth umbrella. No author can execute a huge revision every single time without a dedicated production team and enlightened management. Gary Head, Lisa Starr, Diana Starr, Suzannah Alexander, Jana Davis, Teri Hyde, Karen Hunt, Grace Davidson, Myrna Engler, Angela Harris—this is my core team, the best of the best. Pat Waldo, Donna Kelley, Keli Amann, Chris Evers, Steve Bolinger—they create our stunning multimedia package. Susan Badger, Sean Wakely, Michelle Julet, Jack Carey, Kathie Head—these are anomalies in higher education, publishers who nurture authors with intelligence, strength, and grace.

Cecie Starr, October 2002
E-mail starr@brookscole.com

ACKNOWLEDGMENTS

This book is the current version of an educational effort that started nearly three decades ago. We thank all of the students and instructors who used and commented on previous editions. We thank the individuals listed here for their significant influence on the book's development. We give special thanks to this edition's overall advisor, E. William Wischusen, who helped formulate connections to remind students of where they have been and where they are going in the book. As he pointed out, if we expect them to understand the chapters on genetics, evolution, anatomy, and physiology, we must provide them with the intellectual tools and graphics to do so. We must clearly connect chapters back to the tools that guide students through biomolecules and membranes. And we must include a new chapter on the commonalities between plant and animal systems to help students see the big picture while learning the details.

We also thank Walter Judd, who guided our overhaul of the macroevolution and biodiversity chapters, with close attention to the protistans and vertebrates. The new evolutionary tree diagram (Sections 19.8 and 22.1) and classification system (Appendix I) are his contributions.

If teachers appreciate the new clarity of thought and currency of this new edition, they might give a special nod of recognition to Bill and Walt for urging us gently to give it our best shot—and then to keep on giving more.

Major Advisor for the Tenth Edition

E. WILLIAM WISCHUSEN   *Louisiana State University*

General Advisors/Contributors

JOHN ALCOCK   *Arizona State University*
GEORGE COX   *San Diego State University*
MELANIE DEVORE   *Georgia College and State University*
DANIEL J. FAIRBANKS   *Brigham Young University*
TOM GARRISON   *Orange Coast College*
DAVID GOODIN   *The Scripps Research Institute*
PAUL E. HERTZ   *Barnard College*
JOHN D. JACKSON   *North Hennipin Community College*
WALTER JUDD   *University of Florida*
EUGENE N. KOZLOFF   *University of Washington*
KAREN E. MESSLEY   *Rock Valley College*
ELIZABETH LANDECKER-MOORE   *Rowan University*
JON REISKIND   *University of Florida*
THOMAS L. ROST   *University of California, Davis*
LAURALEE SHERWOOD   *West Virginia University*
STEPHEN L. WOLFE   *University of California, Davis*

Contributors of Influential Reviews

ALDRIDGE, DAVID   *North Carolina Agricultural/Technical State University*
ANDERSON, ROBERT C.   *Idaho State University*
ARMSTRONG, PETER   *University of California, Davis*
BAJER, ANDREW   *University of Oregon*
BAKKEN, AIMEE   *University of Washington*
BARBOUR, MICHAEL   *University of California, Davis*
BARHAM, LINDA   *Meridian Community College*

BARKWORTH, MARY *Utah State University*
BELL, ROBERT A. *University of Wisconsin, Stevens Point*
BENDER, KRISTEN *California State University, Long Beach*
BENIVENGA, STEPHEN *University of Wisconsin, Oshkosh*
BINKLEY, DAN *Colorado State University*
BORGESON, CHARLOTTE *University of Nevada*
BRENGELMANN, GEORGE *University of Washington*
BRINSON, MARK *East Carolina University*
BROWN, ARTHUR *University of Arkansas*
BUTTON, JERRY *Portland Community College*
CARTWRIGHT, PAULYN *University of Kansas*
CASE, CHRISTINE *Skyline College*
CASE, TED *University of California, San Diego*
CHRISTIANSEN, A. KENT *University of Michigan*
CLARK, DEBORAH C. *Middle Tennessee University*
COLAVITO, MARY *Santa Monica College*
CONKEY, JIM *Truckee Meadows Community College*
CROWCROFT, PETER *University of Texas at Austin*
DABNEY, MICHAEL W. *Hawaii Pacific University*
DAVIS, JERRY *University of Wisconsin, La Crosse*
DEGROOTE, DAVID K. *St. Cloud University*
DELCOMYN, FRED *University of Illinois, Urbana*
DEMMANS, DANA *Finger Lakes Community College*
DEMPSEY, JEROME *University of Wisconsin*
DENGLER, NANCY *University of California, Davis*
DENETTE, PHIL *Delgado Community College*
DENNISTON, KATHERINE *Towson State University*
DESAIX, JEAN *University of North Carolina*
DETHIER, MEGAN *University of Washington*
DEWALT, R. EDWARD *Louisiana State University*
DIBARTOLOMEIS, SUSAN *Millersville University of Pennsylvania*
DIEHL, FRED *University of Virginia*
DONALD-WHITNEY, CATHY *Collin County Community College*
DOYLE, PATRICK *Middle Tennessee State University*
DUKE, STANLEY H. *University of Wisconsin, Madison*
DYER, BETSEY *Wheaton College*
EDLIN, GORDON *University of Hawaii, Manoa*
EDWARDS, JOAN *Williams College*
ELMORE, HAROLD W. *Marshall University*
ENDLER, JOHN *University of California, Santa Barbara*
ERWIN, CINDY *City College of San Francisco*
EWALD, PAUL *Amherst College*
FALK, RICHARD *University of California, Davis*
FISHER, DAVID *University of Hawaii, Manoa*
FISHER, DONALD *Washington State University*
FLESSA, KARL *University of Arizona*
FONDACARO, JOSEPH *Hoechst Marion Roussel, Inc.*
FROEHLICH, JEFFREY *University of New Mexico*
FULCHER, THERESA *Pellissippi State Technical Community College*
GAGLIARDI, GRACE S. *Bucks County Community College*
GENUTH, SAUL M. *Mt. Sinai Medical Center*
GHOLZ, HENRY *University of Florida*
GOODMAN, H. MAURICE *University of Massachusetts Medical School*
GOSZ, JAMES *University of New Mexico*
GREGG, KATHERINE *West Virginia Wesleyan College*
GUTSCHICK, VINCENT *New Mexico State University*
HASSAN, ASLAM *Univeristy of Illinois College of Veterinary Medicine*
HELGESON, JEAN *Collin County Community College*
HUFFMAN, DAVID *Southwest Texas State University*
INEICHER, GEORGIA *Hinds Community College*
INGRAHAM, JOHN L. *University of California, Davis*
JENSEN, STEVEN *Southwest Missouri State University*
JOHNSON, LEONARD R. *University of Tennessee College of Medicine*
JOHNSTON, TIMOTHY *Murray State University*
KAREIVA, PETER *University of Washington*
KAUFMAN, JUDY *Monroe Community College*
KAYE, GORDON I. *Albany Medical College*
KAYNE, MARLENE *Trenton State College*
KENDRICK, BRYCE *University of Waterloo*
KILBURN, KERRY S. *Old Dominion University*
KILLIAN, JOELLA C. *Mary Washington College*
KIRKPATRICK, LEE A. *Glendale Community College*
KLANDORF, HILLAR *West Virginia University*
KREBS, CHARLES *University of British Columbia*

KREBS, JULIA E. *Francis Marion University*
KUTCHAI, HOWARD *University of Virginia Medical School*
LANZA, JANET *University of Arkansas, Little Rock*
LASSITER, WILLIAM *University of North Carolina*
LEVY, MATTHEW *Mt. Sinai Medical Center*
LEWIS, LARRY *Salem State College*
LITTLE, ROBERT *Medical College of Georgia*
LOHMEIER, LYNNE *Mississippi Gulf Coast Community College*
LUMSDEN, ANN *Florida State University*
LYNG, R. DOUGLAS *Indiana University, Purdue University*
MANN, ALAN *University of Pennsylvania*
MARTIN, JAMES *Reynolds Community College*
MARTIN, TERRY *Kishwaukee College*
MASON, ROY B. *Mt. San Jacinto College*
MCKEAN, HEATHER *Eastern Washington State University*
MCKEE, DOROTHY *Auburn University, Montgomery*
MCNABB, ANN *Virginia Polytechnic Institute and State University*
MENDELSON, JOSEPH R. *Utah State University*
MICKLE, JAMES *North Carolina State University*
MILLER, G. TYLER *Wilmington, North Carolina*
MILLER, GARY *University of Mississippi*
MINORKSY, PETER V. *Western Connecticut State University*
MOISES, HYLAN C. *University of Michigan Medical School*
MORENO, JORGE A. *University of Colorado, Boulder*
MORRISON-SHETLER, ALLISON *Georgia State University*
MORTON, DAVID *Frostburg State University*
MURPHY, RICHARD *University of Virginia Medical School*
MYRES, BRIAN *Cypress College*
NAPLES, VIRGINIA *Northern Illinois University*
NELSON, RILEY *University of Texas at Austin*
NORRIS, DAVID *Brigham Young University*
PECHENIK, JAN *Tufts University*
PERRY, JAMES *University of Wisconsin, Center-Fox Valley*
PETERSON, GARY *South Dakota State University*
POLCYN, DAVID M. *California State University, San Bernardino*
REESE, R. NEIL *South Dakota State University*
REID, BRUCE *Kean College of New Jersey*
RENFROE, MICHAEL *James Madison University*
RICKETT, JOHN *University of Arkansas, Little Rock*
ROSE, GREIG *West Valley College*
ROST, THOMAS *University of California, Davis*
SALISBURY, FRANK *Utah State University*
SCHAPIRO, HARRIET *San Diego State University*
SCHLESINGER, WILLIAM *Duke University*
SCHNEIDEWENT, JUDY *Milwaukee Area Technical College*
SCHNERMANN, JURGEN *University of Michigan School of Medicine*
SCHREIBER, FRED *California State University, Fresno*
SHONTZ, NANCY *Grand Valley State University*
SELLERS, LARRY *Louisiana Tech University*
SLOBODA, ROGER *Dartmouth College*
SMITH, JERRY *St. Petersburg Junior College, Clearwater Campus*
SMITH, MICHAEL E. *Valdosta State College*
SMITH, ROBERT L. *West Virginia University*
STEARNS, DONALD *Rutgers University*
STEELE, KELLY P. *Appalachian State University*
STEINERT, KATHLEEN *Bellevue Community College*
SUMMERS, GERALD *University of Missouri*
SUNDBERG, MARSHALL D. *Empire State University*
SWEET, SAMUEL *University of California, Santa Barbara*
TAYLOR, JANE *Northern Virginia Community College*
TIZARD, IAN *Texas A&M University*
TROUT, RICHARD E. *Oklahoma City Community College*
TYSER, ROBIN *University of Wisconsin, LaCrosse*
WAALAND, ROBERT *University of Washington*
WAHLERT, JOHN *City University of New York, Baruch College*
WALSH, BRUCE *University of Arizona*
WARING, RICHARD *Oregon State University*
WARNER, MARGARET R. *Purdue University*
WEBB, JACQUELINE F. *Villanova University*
WEIGL, ANN *Winston-Salem State University*
WEISS, MARK *Wayne State University*
WELKIE, GEORGE W. *Utah State University*
WHITE, EVELYN *Alabama State University*
WINICUR, SANDRA *Indiana University, South Bend*

Introduction

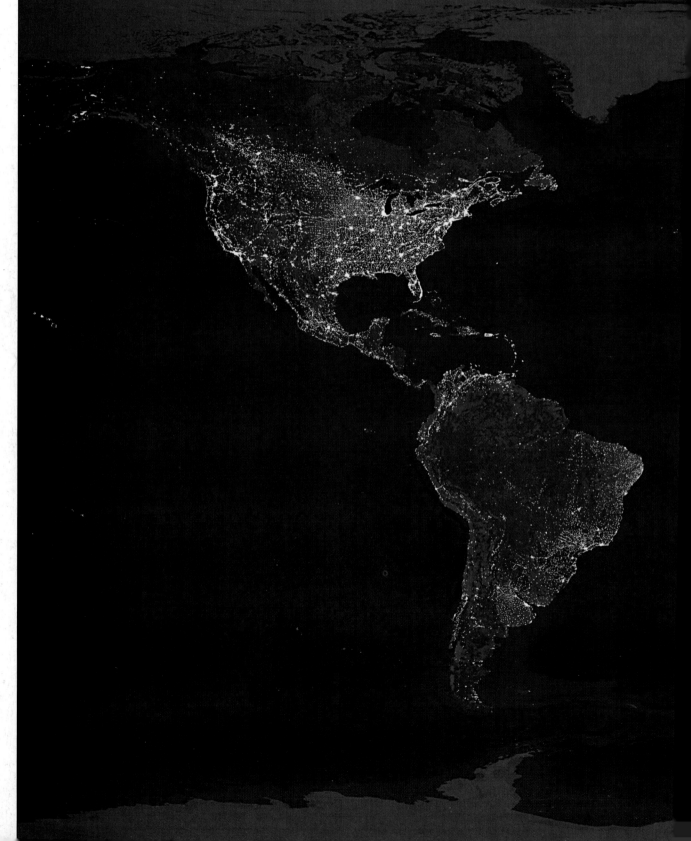

Current configurations of the Earth's oceans and land masses—the geologic stage upon which life's drama continues to unfold. This composite satellite image reveals global energy use at night by the human population. Just as biological science does, it invites you to think more deeply about the world of life—and about our impact upon it.

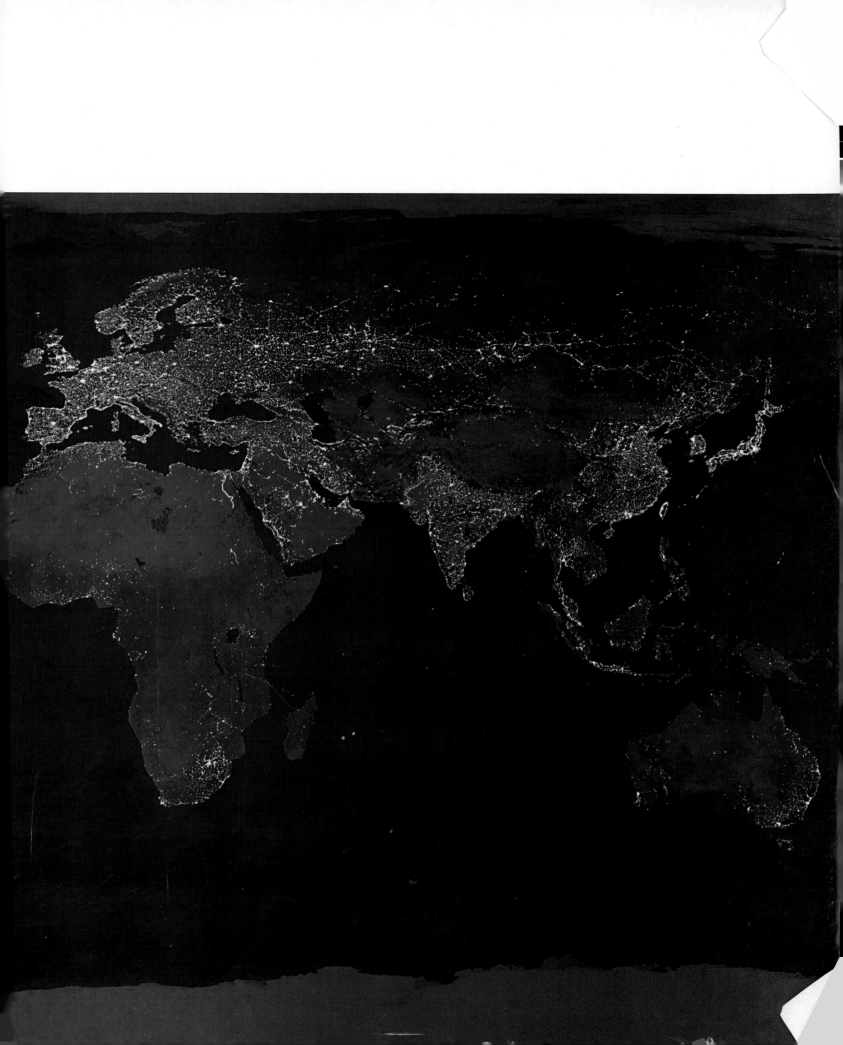

CONCEPTS AND METHODS IN BIOLOGY

Why Biology?

Leaf through a newspaper on any given Sunday and you might get an uneasy feeling that the natural world is spinning out of control. Northern forests are burning fiercely (Figure 1.1). The Arctic ice cap is melting and Greenland's glaciers are turning into lakes. Heat waves and droughts are breaking records everywhere, and maybe the whole atmosphere is warming up. The last individual of an ancient primate lineage bit the dust just as the human population neared the 6.1 billion mark. Geneticists decoded yet another garbled message in human DNA that causes another horrible disease. People are still arguing about when, exactly, the mass of embryonic cells busily dividing inside a woman's womb is "alive." In a bizarre application of genetic engineering, somebody decided to make an "artistic statement" by making a glow-in-the-dark bunny. The entire live bunny turns fluorescent green in black light.

It's enough to make you throw down the paper and go sit in a park. It's enough to make you wish you lived in the good old days when things were so much simpler.

Of course, read up on the good old days and you'll discover they weren't so good. For example, in 1918, maybe when your grandparents were alive, the Spanish flu swept around the world. Those infected often died within hours of the first symptoms. Before the pandemic subsided, *30 to 40 million* people had died. Unlike most

Figure 1.1 Biology starts with the premise that any aspect of nature—including this out-of-control forest fire in Montana—has one or more underlying causes. By giving us a way of thinking critically about life, biology helps us understand how to deal with nature, and with our place in it.

flu viruses, which mainly endanger infants, the elderly, and the very ill, the Spanish flu virus struck healthy young adults, the ones who support most of the social infrastructure. Then, as today, many thought the world was coming apart at the seams. They felt helpless and hopeless before a force of nature on the rampage.

What it boils down to is this: For a couple of million years at least, humans and their immediate ancestors have been trying to make sense of the natural world. We observe it, we come up with ideas, we test the ideas. However, the more pieces of the puzzle we fit together, the bigger the puzzle gets. We are now smart enough to know that it's almost overwhelmingly big.

You might choose to walk away from the challenge and simply let others tell you what to think. Or you might choose to develop your own understanding of the puzzle. Maybe you're interested in the pieces that affect your health, your home, the food you eat, and your offspring, should you choose to reproduce. Maybe you just find the connections among organisms and their environment fascinating. Regardless of the focus, **biology**—the scientific study of life—can help deepen your perspective on the world around you.

Start with a question that seems simple enough: *What is life?* Offhandedly, you might say you know it when you see it. Yet the question opens up a story that began at least 3.8 billion years ago.

From a biological perspective, "life" is an outcome of ancient events by which lifeless matter—atoms and molecules—became organized into the first living cells. "Life" is a way of capturing and using energy and raw materials. "Life" is a way of sensing and responding to the environment. "Life" is a capacity to reproduce. And "life" evolves, which simply means that the traits characterizing individuals of a population can change from one generation to the next.

Throughout this book, you will come across many examples of how organisms are constructed, how they function, where they live, what they do. The examples support concepts which, when taken together, convey what "life" is. *This chapter is an overview of the basic concepts.* It sets the stage for forthcoming descriptions of scientific observations, experiments, and tests that help show how you can develop, modify, and refine your views of life.

Key Concepts

1. Unity underlies the world of life, for all organisms are alike in key respects. They all consist of one or more cells made of the same kinds of substances, put together in the same basic ways. Their activities require inputs of energy, which they must get from their surroundings. All organisms sense and respond to changing conditions in their environment. They have a capacity to grow and reproduce, based on instructions contained in DNA.

2. The world of life shows immense diversity. Many millions of different kinds of organisms, or species, now inhabit the Earth. Many millions more lived in the past. Each species is unique in some of its traits—that is, in some aspects of its body plan, functioning, and behavior.

3. Theories of evolution, especially the theory of evolution by natural selection as formulated by Charles Darwin, help explain life's diversity. The theories unite all fields of biological inquiry into a single, coherent whole.

4. Biology, like other branches of science, is based on systematic observations, hypotheses, predictions, and observational and experimental tests. The external world, not internal conviction, is the testing ground for scientific theories.

DNA, ENERGY, AND LIFE

Nothing Lives Without DNA

THE MOLECULES OF LIFE Let's start this overview of life by picturing a frog busily croaking on a rock. Even without thinking about it, you know the frog is alive and the rock is not. Could you say why? After all, both consist only of protons, electrons, and neutrons, the building blocks for atoms. But atoms are building blocks for larger and more diverse bits of matter called molecules. It is at the molecular level that differences between living and nonliving things start to emerge.

You will never find a rock made of nucleic acids, proteins, carbohydrates, and lipids. In nature, only *cells* build these molecules, which you will read about later. And the signature molecule of a cell—the smallest unit that has the capacity for life—is a nucleic acid known as **DNA**. No chunk of granite or quartz has it.

DNA holds instructions for assembling a variety of proteins from smaller molecules, the amino acids. By analogy, if you follow suitable instructions and invest energy in the task, you might organize a pile of a few kinds of ceramic tiles (representing amino acids) into diverse patterns (representing proteins), as in Figure 1.2.

Enzymes are among the proteins. When these worker molecules get an energy boost, they swiftly build, split, and rearrange the molecules of life. Some work with the nucleic acids called RNAs to turn DNA instructions into proteins. Think of this as a flow of information, *from DNA to RNA to protein*. As you will see later, this molecular trinity is central to our understanding of life.

DNA AND INHERITANCE We humans tend to think we enter the world abruptly and leave it the same way. But we are more than this. We and all other organisms

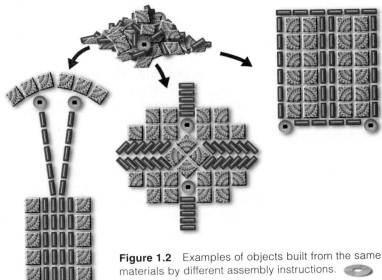

Figure 1.2 Examples of objects built from the same materials by different assembly instructions.

are part of an immense journey that began about 3.8 billion years ago with the chemical origin of the first living cells.

Under present-day conditions in nature, new cells and multicelled organisms inherit their defining traits from parents. **Inheritance** is simply the acquisition of traits through the transmission of DNA from parents to offspring. We reserve the term **reproduction** for the actual mechanisms of transmitting DNA to offspring. Why do baby storks look like storks and not pelicans? Because they inherit stork DNA, which isn't exactly the same as pelican DNA in its molecular details.

For frogs, humans, trees, and other large organisms, DNA also guides **development**. This term refers to the

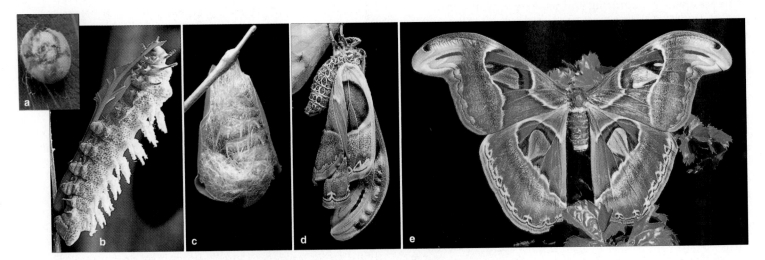

Figure 1.3 "The insect"—a series of stages of development guided largely by instructions in DNA. Different adaptive properties emerge at each stage. Shown here, a silkworm moth, from the egg (**a**), to a larval stage called a caterpillar (**b**), to a pupal stage (**c**), to the winged form of the adult (**d**,**e**).

Figure 1.4 Response to signals from pain receptors, activated by a lion cub flirting with disaster.

transformation of a new individual into a multicelled adult, typically with tissues and organs specialized for certain tasks. For example, the moth shown in Figure 1.3 is only the adult stage of a winged insect. It started out as a single cell, a fertilized egg that developed into a caterpillar. That immature larval stage ate soft leaves and grew rapidly until an internal alarm clock went off. At that time its tissues started to be remodeled into a different stage, a pupa. In time the adult, adapted to reproduce, emerged. It has special parts to make sperm or eggs; its wings have the color, pattern, and fluttering frequency that are adaptations for attracting a mate.

Like other animals, "the insect" is a series of stages, each of which must develop properly before the next begins. And instructions for forming every stage were written into moth DNA long before each time of moth reproduction—and so an ancient story of life continues.

Nothing Lives Without Energy

ENERGY DEFINED Being alive requires more than DNA. Moths and everything else in the universe also require **energy**, a capacity to do work. In their cells, work gets done as atoms give up, share, and accept electrons. It gets done as molecules are assembled, rearranged, and split apart. Energy drives these molecular events.

METABOLISM DEFINED Put it this way: Each living cell has the capacity to (1) obtain and convert energy from its surroundings and (2) use energy to maintain itself, grow, and produce more cells. We call this capacity **metabolism**. Think of one cell in a leaf that makes food by a process called photosynthesis. That cell intercepts sunlight energy and converts it to chemical energy, in the form of ATP molecules. ATP helps drive hundreds of metabolic events by transferring energy to reaction sites, where enzymes put together sugar molecules. In most cells, ATP also can form by aerobic respiration. This process can release energy that cells stored earlier, in the form of starch and other kinds of molecules.

SENSING AND RESPONDING TO ENERGY It's often said that only living things respond to the environment. Yet even a rock shows responsiveness, as when it yields to the force of gravity and tumbles down a hill or changes its shape slowly under the repeated battering of wind, rain, or tides. The difference is this: *Organisms can sense changes in their surroundings and then make compensatory, controlled responses to the changes.* How? Each organism has **receptors**, which are molecules and structures that detect stimuli. A **stimulus** is a specific form of energy that a receptor can detect. Examples are sunlight energy,

heat energy, a hormone molecule's binding energy, and the mechanical energy of a bite (Figure 1.4).

Cells adjust metabolic activities in response to signals from receptors. Each cell (and organism) can withstand only so much heat or cold. It must rid itself of harmful substances. It requires certain foods, in certain amounts. Yet temperatures do shift, harmful substances might be encountered, and food is sometimes plentiful or scarce.

Finish a snack, and sugars leave your gut and enter your blood. Blood, together with the tissue fluid that bathes your cells, makes up your internal environment. Unchecked, too much or too little sugar in the blood can cause diabetes and other problems. When the level rises, the pancreas, a glandular organ, secretes more insulin. Most of your body cells have receptors for this hormone, which stimulates cells to take up sugar. When enough cells do so, the blood sugar level returns to normal.

Organisms respond so exquisitely to energy changes that their internal operating conditions usually remain within tolerable limits. This state, called **homeostasis**, is one of the key defining features of life.

All organisms consist of one or more cells, the smallest units of life. Under present-day conditions in nature, new cells form only when existing cells reproduce.

DNA, the molecule of inheritance, encodes protein-building instructions, which RNAs help carry out. Many proteins are enzymes that speed up cellular work, which includes building all the complex molecules characteristic of life.

Cells live only for as long as they engage in metabolism. They acquire and transfer energy that is used to assemble, break down, stockpile, and dispose of materials in ways that promote survival and reproduction.

Single-celled and multicelled organisms sense and respond to environmental conditions in ways that help maintain their internal operating conditions.

ENERGY AND LIFE'S ORGANIZATION

Levels of Biological Organization

Through their individual and interactive uses of energy and materials, organisms contribute to a great pattern of biological organization. As Figure 1.5 shows, life's properties emerge when DNA and other molecules are organized as cells. A **cell** is the smallest organizational unit having a capacity to survive and reproduce on its own, given DNA instructions, building blocks, energy inputs, and suitable conditions. An obvious unit is an amoeba or another free-living cell. The definition also fits **multicelled organisms** consisting of specialized, interdependent cells, typically organized in tissues and organs. Do you question this? After all, your own cells could never live alone in nature, for body fluids must constantly bathe them, deliver substances to them, and remove their metabolic wastes. However, even isolated human cells remain alive under controlled conditions in laboratories around the world. Researchers routinely maintain isolated human cells in cultures for important experiments, as in cancer studies.

Cells and multicelled organisms usually are part of a **population**: a group of organisms of the same kind. A zebra herd is an example. The next level of organization is the **community**: all populations of all species living in the same area, such as the African savanna's bacteria, grasses, trees, zebras, lions, and so forth. The next level is the **ecosystem**: a community *together with* its physical and chemical environment. The **biosphere**, the highest level, includes all parts of the Earth's crust, waters, and atmosphere in which organisms live. Astoundingly, *this globe-spanning organization begins with the convergence of energy, certain materials, and DNA in tiny, individual cells.*

BIOSPHERE
All regions of the Earth's crust, waters, and atmosphere that sustain life

ECOSYSTEM
Community and its physical environment

COMMUNITY
Populations of all species occupying the same area

POPULATION
Group of individuals of the same kind (that is, the same species) occupying the same area

MULTICELLED ORGANISM
Individual consisting of interdependent cells typically organized in tissues, organs, and organ systems

ORGAN SYSTEM
Two or more organs interacting chemically, physically, or both in ways that contribute to organism's survival

ORGAN
Structural unit in which tissues, combined in specific amounts and patterns, perform a common task

TISSUE
Organized aggregation of cells and substances functioning together in a specialized activity

CELL
Smallest unit with the capacity to live and reproduce, independently or as part of multicelled organism

ORGANELLE
Membrane-bound internal compartment for specialized reactions (most prokaryotic cells have none)

MOLECULE
Unit of two or more bonded-together atoms of the same element or different elements

ATOM
Smallest unit of an element (a fundamental substance) that still retains the properties of that element

SUBATOMIC PARTICLE
Electron, proton, neutron, or some other fundamental unit of matter

community (all of the populations living in the same area)

populations of shrubs and trees

population of zebras

populations of grasses

ecosystem (community together with its physical environment)

multicelled organism

once-live multicelled organism that is about to revert to unorganized molecules and atoms

Figure 1.5 Levels of organization in nature.

beetle larva

Producers capture, convert, and use or store some energy from the sun.

PRODUCERS

NUTRIENT CYCLING

CONSUMERS, DECOMPOSERS

ONE-WAY FLOW OF ENERGY

Energy gets transferred from one organism to another; in time, all flows back to the environment.

Figure 1.6 Example of the one-way flow of energy and the cycling of materials through the biosphere.

(**a**) Plants of a warm, dry grassland called the African savanna capture energy from the sun and use it to build plant parts. Some of the energy ends up inside plant-eating organisms, such as this adult male elephant. He eats huge quantities of plants to maintain his eight-ton self. He produces piles of solid wastes—dung— that still hold some unused nutrients and energy. Although most organisms might not recognize it as such, elephant dung is an exploitable food source.

(**b**) And so we next have little dung beetles scrambling to the scene almost simultaneously with the uplifting of an elephant tail. Working fast, they carve fragments of moist dung into round balls, which they roll off and bury in burrows. In those balls the beetles lay eggs— a reproductive behavior that will help assure their forthcoming offspring (**c**) of a compact food supply.

Thanks to beetles, dung does not pile up and dry out into rock-hard mounds in the intense heat of the day. Instead, the surface of the land is tidied up, the beetle offspring get fed, and the leftover dung accumulates in beetle burrows—there to enrich the soil that nourishes the plants that feed (among others) the elephants.

Interdependencies Among Organisms

A great flow of energy from the sun into the world of life starts with **producers**—plants and other organisms that make their own food. Animals are **consumers**. Directly or indirectly, they use food energy that is stored in tissues of producers. For example, some energy gets transferred to zebras after they browse on the plants. It gets transferred again when a lion devours a zebra, as in Figure 1.5. And it is transferred again when fungal and bacterial decomposers extract energy from tissues and remains of lions, elephants, or any other organism. **Decomposers** break down sugars and other biological molecules to simpler materials, some of which is cycled back to the producers. In time, all of the energy that plants originally captured from the sun's rays returns to the environment, but that's another story.

For now, keep in mind that organisms connect with one another by a one-way flow of energy *through* them and a cycling of materials *among* them, as in Figure 1.6. Their interconnectedness affects the structure, size, and composition of populations and communities. It affects ecosystems, even the biosphere. Understand the extent of their interactions and you will gain insight into food shortages, cholera epidemics, acid rain, global warming, biodiversity losses, and other modern-day problems.

Nature shows levels of organization. The characteristics of life emerge at the level of single cells and extend through populations, communities, ecosystems, and the biosphere.

A one-way flow of energy through organisms and a cycling of materials among them organizes life in the biosphere. In nearly all cases, energy flow starts with sunlight energy.

IF SO MUCH UNITY, WHY SO MANY SPECIES?

So far, this overview of life has focused on its unity, on characteristics that all living things have in common. Think of it! They consist of the same lifeless materials. They stay alive by way of metabolism, ongoing energy transfers at the cellular level. They interact when using energy and raw materials. They sense and respond to the environment. They have the capacity to reproduce based on instructions in their DNA, which they inherit from individuals of a preceding generation.

Superimposed on this common heritage is immense diversity. You share the planet with many millions of kinds of organisms, or **species**. Many millions more preceded you over the past 3.8 billion years. However, their lineages vanished; they are extinct. For centuries, scholars have tried to make sense of the confounding diversity. One of them, Carolus Linneaus, promoted a classification scheme that assigns a two-part name to each newly identified species. The first part designates the **genus** (plural, genera). A genus encompasses all of the species that seem closely related because of their form, functioning, and ancestry. The second part of the name designates a particular species within a genus.

For instance, *Quercus alba* is the white oak's name. *Q. rubra* is the red oak's name. (A genus name can be abbreviated this way once it has been spelled out in a document.) Such names are passports to great worlds of information; try them out for yourself on the web.

Biologists group species that are related by descent from a common ancestor. They are still working out the groupings. For instance, scholars once recognized only two groups—animals and plants. Most biologists now favor a three-domain system of classification. The domains are called bacteria, archaea, and eukarya:

| EUBACTERIA (BACTERIA) | ARCHAEBACTERIA (ARCHAEA) | EUKARYOTES (EUKARYA) |

Within these domains are at least six groups that are still known as kingdoms—**archaebacteria**, **eubacteria**, **protistans**, **fungi**, **plants**, and **animals** (Figure 1.7).

Archaebacteria and eubacteria are all single cells of a type called *prokaryotic*. The word means they do not contain a nucleus. (In other cell types, this membrane-bound sac keeps the DNA separated from the rest of the cell's interior.) Different kinds of prokaryotic cells are producers, consumers, and decomposers. Of all the groups, theirs show the most metabolic diversity.

Archaebacteria live in hot springs, salt lakes, and other habitats as harsh as the ones that prevailed when cells originated. The eubacteria are more common and

Clockwise from top, a forest of giant kelp (*Macrocystis*), the shell of a coccolithophore, the migrating slug stage of a slime mold (*Dictyostelium discoideum*), a flagellate (*Trypanosoma brucei*) that causes African sleeping sickness, and a ciliated protozoan (*Paramecium*). Even this tiny sampling of diversity conveys why many biologists now believe that the "protistans" are not a single group but rather are many evolutionary separate lineages.

From the muck of an oxygen-free habitat, a microscopically small colony of archaebacterial cells of the genus *Methanosarcina*. Archaebacteria are evolutionarily closer to us than to true bacteria.

Figure 1.7 A few representatives of life's diversity.

widespread. Their name means true bacteria, so we will just refer to them as bacteria (singular, bacterium).

Like plants, fungi, and animals, the protistans are *eukaryotic*, meaning a nucleus houses their DNA. Most are larger and show far more internal complexity than prokaryotic cells. The grouping includes microscopic single-celled producers, consumers, and decomposers, and also giant multicelled "seaweeds." The differences among protistans are great, so this one group probably will be split into more informative ones soon.

Most fungi, such as the mushrooms sold in grocery stores, are multicelled decomposers or parasites that

Stinkhorn (*Dictyophora indusiata*) and trumpet chanterelles (*Craterellus*). Although certain fungi are parasites and cause diseases, these two species—and the vast majority of others—are decomposers. Without decomposers, communities would become buried in their own wastes.

Left: Coast redwoods (*Sequoia sempervirens*) growing in California. Like nearly all plants, redwoods are photosynthetic. *Right:* Flower of a hybrid *Gazania*. The patterns and colors of this plant's flowers guide bees to the nectar. Bees get food; the plant gets help reproducing. Like many other organisms, these two species interact in a mutually beneficial way.

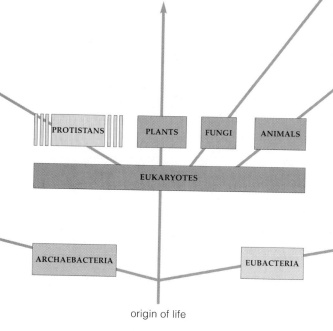

Like the ancient fishes that were its ancestors, *Basiliscus basiliscus*, a basilisk lizard, has four limbs. Unlike fishes, it is adapted to land, although it runs fast across water by stomping its feet on the surface.

PROTISTANS   PLANTS   FUNGI   ANIMALS

EUKARYOTES

ARCHAEBACTERIA   EUBACTERIA

origin of life

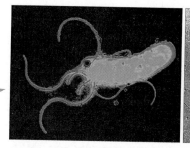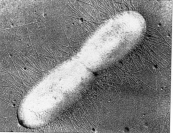

Left: Helicobacter pylori, a bacterium that can cause gastric upsets and ulcers. *Right: Escherichia coli*, a bacterial resident of the mammalian gut. It made a copy of its DNA and is now reproducing by dividing in two.

feed in a distinctive way. They secrete enzymes that digest food outside of the fungal body, then individual cells absorb the digested bits.

Plants are multicelled, photosynthetic producers. Photosynthesis means they make their own food by using sunlight energy and simple raw materials. They have cellular pipelines for water and substances that typically extend through roots, stems, and leaves.

Animals are multicelled consumers that ingest the tissues of other organisms. They include the herbivores that graze upon photosynthesizers, carnivores (meat eaters), parasites, and scavengers. All develop through a series of embryonic stages, and they actively move about during at least part of their life.

Pulling this together, can you sense what it means when someone says life has unity *and* diversity?

Although unity pervades the world of life, great diversity threads through it, for organisms differ enormously in their form, in how body parts function, and in behavior.

We group organisms into three domains—archaebacteria, eubacteria, and eukaryotes. Protistans, fungi, plants, and animals are major groups of eukaryotes.

AN EVOLUTIONARY VIEW OF DIVERSITY

How can organisms be so much alike yet also show such staggering diversity? One explanation is known as evolution by means of natural selection. All through this book, you will be looking at many examples of its predictive power. For now, simply become familiar with its basic premises. They build on the simple observation that individuals of a population vary in the details of their shared traits.

Mutation—Original Source of Variation

DNA has two striking qualities. Its instructions work to ensure that offspring will resemble their parents, yet they also permit variations in the details of most traits. Example: Having five fingers on each hand is a human trait. Yet some humans are born with six fingers on each hand instead of five. This is an outcome of a **mutation**, a heritable change in DNA. *Mutations are the original source of variations in heritable traits.*

Many mutations are harmful, for they sabotage the body's growth, development, or functioning. Yet some are harmless or beneficial. One case is a mutation that results in dark-colored instead of light-colored moths.

Moths fly at night and rest in the day, when birds that eat them are active. Birds tend to miss the light-colored moths on light-colored tree trunks. But what if people build coal-burning factories and soot-laden smoke darkens all the tree trunks? Dark moths on dark trunks aren't as conspicuous to bird predators (Figure 1.8), so they are more likely to live and then reproduce. Where soot accumulates, then, the variant (dark) form of this trait is more adaptive than the light form. An **adaptive trait** is any form of a trait that helps an individual survive and reproduce under prevailing environmental conditions.

Figure 1.9   *Facing page:* One outcome of artificial selection practices: just a few of the 300+ varieties of domesticated pigeons. Breeders started with variant forms of traits in captive populations of wild rock doves.

WILD ROCK DOVE

Evolution Defined

Now think about variant moths in terms of the whole population. At some point, a DNA mutation arose and resulted in a moth of a darker color. When the mutated individual reproduced, some of its offspring inherited its mutant DNA and the dark color. Birds saw and ate many light-colored moths, but most dark ones escaped detection and lived long enough to reproduce. So did their offspring. So did *their* offspring, and so on. The frequency of one form of the trait increased relative to the frequency of the other in the population. If the dark form becomes the most common one, people may end up referring to "the population of dark-colored moths."

Like moths, individuals of most populations have different forms of many shared traits. The relative frequencies of those different forms typically change through successive generations. When that happens, **evolution** is under way. In biology, the word means heritable change in a line of descent over time.

Natural Selection Defined

Long ago, the naturalist Charles Darwin used pigeons to explain a conceptual connection between evolution and variation in traits. Domesticated pigeons differ in size, feather color, and other traits (Figure 1.9). Pigeon breeders, Darwin knew, select certain forms of traits. If they are after black, curly-edged tail feathers, they let only individual pigeons with the most black and most curly tail feathers mate and reproduce. So "black" and "curly" become the most common tail feathers in the captive pigeon population. Lighter and less curly feathers become less common or are eliminated.

Pigeon breeding is a case of **artificial selection**, for the selection among different forms of a trait is taking place in an "artificial" environment—under contrived, manipulated conditions. Yet Darwin saw the practice as a simple model for *natural* selection, a favoring of some forms of a given trait over others in nature.

Whereas breeders are "selective agents" promoting the reproduction of some individuals but not others in captive populations, different selective agents operate across the range of variation among pigeons in the wild. Pigeon-eating peregrine falcons are such agents. Swifter or more effectively camouflaged individuals of a pigeon

Figure 1.8   An outcome of natural selection involving two forms of the same trait—body surface coloration. (**a**) Light-colored moths (*Biston betularia*) on a nonsooty tree trunk are hidden from predators. Dark ones stand out. (**b**) The dark color is more adaptive in places where soot darkens tree trunks.

population have a far better chance of escaping from falcons and living long enough to reproduce, compared to the not-so-swift or too-conspicuous individuals among them. What Darwin identified as **natural selection** is simply a result of differences in survival and reproduction among individuals of a population that vary in one or more of their shared, heritable traits.

Unless you happen to breed pigeons, the variation among pigeons may not be a burning issue in your life. So think about something closer to home. **Antibiotics** are potent secretions of certain soil-dwelling bacteria and fungi; they kill bacterial and fungal competitors for nutrients. In the 1940s, we learned to use antibiotics to kill disease-causing bacteria. Doctors prescribed these "wonder drugs" even for mild infections. In some cases antibiotics were added to toothpaste and chewing gum.

As it turned out, antibiotics are powerful agents of natural selection. Consider how one of these, called streptomycin, shuts down certain essential proteins in bacteria that are its targets. Some bacterial strains have mutations that slightly changed the molecular form of the proteins—and streptomycin no longer can bind to them. Unlike nonmutated cells, the mutants survive and reproduce. Said another way, the antibiotic acts against the bacteria susceptible to it but *favors* variant strains that resist it!

Antibiotic-resistant strains are making it difficult to treat bacterial diseases, including tuberculosis, typhoid, gonorrhea, and staph infections. As resistance evolves by selection processes, strategies for using antibiotics must also evolve to overcome new defenses. That is why drug companies try to modify parts of antibiotic molecules. Such molecular changes in the laboratory might yield more effective antibiotics. These may work only until new generations of more resistant superbugs enter the evolutionary competition for nutrients.

Later in the book, we will consider mechanisms by which populations of moths, pigeons, bacteria, and all other organisms evolve. Meanwhile, reflect on the key points about natural selection. They are the bedrock of biological inquiry into the nature of life's diversity.

1. Individuals of a population vary in form, function, and behavior. Much of the variation is heritable; it can be transmitted from parents to offspring.

2. Some forms of heritable traits are more adaptive to prevailing conditions. They improve an individual's chance of surviving and reproducing, as by helping it secure food, a mate, hiding places, and so on.

3. Natural selection is the outcome of differences in survival and reproduction among variant individuals in a given generation.

4. Natural selection leads to a better fit with prevailing environments. The adaptive forms of traits tend to become more common than other forms. Thus the population's characteristics change; it evolves.

In the evolutionary view, then, *life's diversity is the sum total of variations in traits that have accumulated in different lines of descent generation after generation, as by natural selection and other processes of change.*

Mutations in DNA introduce variations in heritable traits.

Although many mutations are harmful, some give rise to variations in form, function, or behavior that are adaptive under prevailing environmental conditions.

Natural selection is a result of differences in survival and reproduction among individuals of a population that vary in one or more heritable traits. The process helps explain evolution—change in individual lines of descent which, over time, has given rise to life's rich diversity.

THE NATURE OF BIOLOGICAL INQUIRY

The preceding sections sketched out some key concepts. Now consider approaching this or any other collection of "facts" with a critical attitude. *"Why should I accept that they have merit?"* The answer requires insight into how biologists make inferences about observations and how they test the predictive power of their inferences against actual experience in nature or the laboratory.

Observations, Hypotheses, and Tests

To get a sense of "how to do science," start by following some practices that pervade scientific research:

1. Observe some aspect of nature, carefully check what others have found out about it, then frame a question or identify a problem related to your observation.

2. Develop **hypotheses**, or educated guesses, about possible answers to questions or solutions to problems.

3. Using hypotheses as a guide, make a **prediction**—that is, a statement of what you should observe in the natural world if you were to go looking for it. This is often called the "if–then" process. (*If* gravity does not pull objects toward Earth, *then* it should be possible to observe apples falling up, not down, from a tree.)

4. Devise ways to **test** the accuracy of your predictions, as by making systematic observations, building models, and conducting experiments. **Models** are theoretical, detailed descriptions or analogies that help us visualize something that has not yet been directly observed.

5. If tests do not confirm the prediction, check to see what might have gone wrong. For example, maybe you overlooked some factor that influenced the test results. Or maybe the hypothesis is not a good one.

6. Repeat the tests or devise new ones—the more the better, for hypotheses that withstand many tests are likely to have a higher probability of being useful.

7. Objectively analyze and report the test results and the conclusions drawn from them.

You might hear someone refer to these practices as "the scientific method," as if all scientists march to the drumbeat of an absolute, fixed procedure. They do not. Many observe, describe, and report on some subject, then leave the hypothesizing to others. Some are lucky; they stumble onto information they aren't even looking for. (Of course, chance does favor a mind that's already prepared to know what the information means.) It is not one single method scientists have in common. It's a critical attitude about being shown rather than told, and taking a logical approach to problem solving.

Logic encompasses thought patterns by which an individual draws a conclusion that does not contradict evidence used to support it. Lick a cut lemon and you notice it's mouth-puckeringly sour. Lick ten more. Each

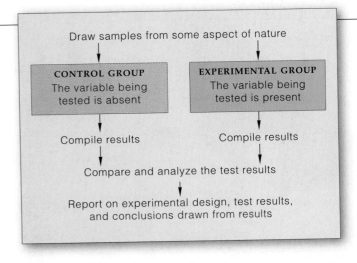

Figure 1.10 Generalized sequence of steps involved in an experimental test of a prediction based on a hypothesis.

time you notice the same thing, and you conclude that all lemons are mouth-puckeringly sour. You correlated one specific (lemon) with another (sour). By this pattern of thinking, called **inductive logic**, an individual derives a general statement from specific observations.

Express the generalization in "if–then" terms, and you have a hypothesis: "If you lick any lemon, then you will experience a sour taste." By this pattern of thinking, **deductive logic**, an individual makes inferences about specific consequences or specific predictions that must follow from a hypothesis.

You decide to test the hypothesis by sampling lemons. One variety, the Meyer, is relatively mellow. You also recall that some people can't taste anything. So you modify the hypothesis: "If most people lick any lemon *except* the Meyer lemon, they will experience a sour taste." Suppose you sample all known lemon varieties in the world and conclude the modified hypothesis is a good one. You'll never *prove* it; there may be lemon trees in places we don't even know about. You *can* say the hypothesis has a high probability of not being wrong.

Comprehensive observations are a logical means to test the predictions that flow from hypotheses. So are **experiments**. These tests simplify observation in nature or the laboratory by manipulating and controlling the conditions under which observations are made. When suitably designed, observational and experimental tests allow you to predict that something will happen if a hypothesis isn't wrong (or won't happen if it *is* wrong). Figure 1.10 gives a general idea of the steps involved.

AN ASSUMPTION OF CAUSE AND EFFECT Experiments start from this premise: *Any aspect of nature has one or more underlying causes.* With this premise, science is very different from faith in the supernatural ("beyond nature"). Scientific experiments deal with potentially falsifiable hypotheses, which means they are tested in the natural world in ways that might disprove them.

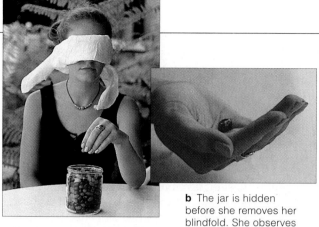

b The jar is hidden before she removes her blindfold. She observes a single green jellybean in her hand and assumes the jar holds only green jellybeans.

a Natalie, blindfolded, randomly plucks a jellybean from a jar of 120 green and 280 black jellybeans. That's a ratio of 30 to 70 percent.

c Still blindfolded, Natalie randomly picks 50 jellybeans from the jar and ends up with 10 green and 40 black ones.

d The larger sample leads her to assume one-fifth of the jar's jellybeans are green and four-fifths are black (a ratio of 20 to 80). Her larger sample more closely approximates the jar's green-to-black ratio. The more times Natalie repeats the sampling, the greater the chance she will come close to knowing the actual ratio.

Figure 1.11 A simple demonstration of sampling error.

EXPERIMENTAL DESIGN To get conclusive test results, experimenters rely on certain practices. They refine test designs by searching the literature for information that may relate to their inquiry. They design experiments to test one prediction of a hypothesis at a time. Whenever possible, they set up a **control group**: a standard for comparison with one or more experimental groups. A control group should be identical with experimental groups *except* for the one variable being investigated. **Variables** are specific aspects of objects or events that may differ or change over time and among individuals.

Section 1.6 shows the design of one experiment. As you will see, the experimenters directly manipulated a single variable in an attempt to support or disprove a prediction. They also tried to hold constant any other variables that could influence the results.

SAMPLING ERROR Rarely can experimenters observe *all* the individuals of a group. Instead, they must use subsets or samples of populations, events, and other aspects of nature. They must avoid **sampling error**, or risking tests with subsets that are not representative of the whole. Generally, distortion in test results is less likely when the samplings are large and when they are repeated (Figure 1.11).

About the Word "Theory"

Suppose no one has disproved a hypothesis after years of rigorous tests. Suppose scientists use it to interpret more data or observations, which could involve more hypotheses. When a hypothesis meets these criteria, it may become accepted as a **scientific theory**.

You may hear someone apply the word "theory" to a speculative idea, as in the expression "It's only a theory." But a scientific theory differs from speculation for this reason: *Researchers have tested its predictive power many times and in many ways in the natural world and have yet to find evidence that disproves it.* This is why the theory of natural selection is respected. We use it successfully to explain diverse issues, such as how life originated, how plant toxins relate to plant-eating animals, what the sexual advantage of feather color might be, why certain cancers run in families, and why antibiotics are losing effectiveness. Together with the record of Earth history, the theory has even helped explain how life evolved.

Maybe an extensively tested theory is as close to the truth as scientists can get with available evidence. For instance, after more than a century of many thousands of different tests, Darwin's theory stands, with only minor modification. We can't prove it holds under all possible conditions; that would take an infinite number of tests. As for any theory, we can only say *it has a high probability of being a good one.* Even so, biologists keep their eyes open for any new information and new tests that might disprove its premises.

This last point gets us back to the value of thinking critically. Scientists must keep asking themselves: *Will observations or experiments show that a hypothesis is false?* They expect one another to put aside pride or bias by testing ideas even in ways that may prove them wrong. When someone doesn't or won't do this, others will— for science is a competitive yet cooperative community. Ideally, individuals share ideas, knowing it's as useful to expose errors as to applaud insights. They can and often do change their mind when shown contradictory evidence. This is a strength of science, not a weakness.

A scientific approach to studying nature is based on asking questions, formulating hypotheses, making predictions, devising tests, and objectively reporting the results.

A scientific theory is a longstanding hypothesis, supported by scientific tests, that explains the cause or causes of a broad range of related phenomena. It remains open to tests, revision, and tentative acceptance or rejection.

The Power of Experimental Tests

BIOLOGICAL THERAPY EXPERIMENTS Worry about the new strains of pathogenic (disease-causing) bacteria that resist antibiotics, and you won't be alone. Bruce Levin of Emory University and Jim Bull of the University of Texas have been looking for alternatives to antibiotic therapy. They studied literature on a *biological therapy* that enlists bacteriophages—"bacteria eaters"—to fight infections. Bacteriophages, a class of viruses, attack a narrow range of bacterial strains. When they contact a target, they inject genetic material into it. What happens next is a hostile takeover: The cell's metabolic machinery makes new viral particles. It dies after viral enzymes rupture its outer membrane. Virus particles escape and infect new cells.

The biologists wondered, as others had decades ago, whether injections of bacteriophages could help people fight bacterial infections. The discovery of antibiotics had diverted attention from that idea, at least in the West. Now the bacteria eaters were looking good again. Levin

and Bull focused on duplicating earlier experiments by researchers H. Williams Smith and Michael Huggins.

They chose 018:K1:H7, a harmful strain of *Escherichia coli*. Bacteriophages utilized in laboratory work usually ignore 018:K1:H7. However, like a harmless strain that lives in the intestines of humans and other mammals, it exits the body in feces. So Levin and Bull looked for *E. coli* killers in samples from a sewage treatment plant in Atlanta. They isolated two kinds and named them *H* (Hero, an effective killer) and *W* (the less effective Wimp).

With graduate student Terry DeRouin and technician Nina Moore Walker, they grew a population of 018:K1:H7 in a culture flask. They used a strain of laboratory mice, all females of the same age. In one experimental group, fifteen mice were injected with 018:K1:H7 in one thigh and over 10^7 particles of *H* bacteriophage in the other. All survived. Fifteen mice of a control group received 018:K1:H7 injections only. None survived. Figure 1.12*a*

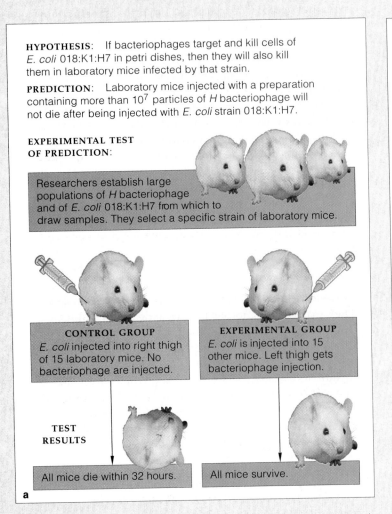

HYPOTHESIS: If bacteriophages target and kill cells of *E. coli* 018:K1:H7 in petri dishes, then they will also kill them in laboratory mice infected by that strain.

PREDICTION: Laboratory mice injected with a preparation containing more than 10^7 particles of *H* bacteriophage will not die after being injected with *E. coli* strain 018:K1:H7.

EXPERIMENTAL TEST OF PREDICTION:

Researchers establish large populations of *H* bacteriophage and of *E. coli* 018:K1:H7 from which to draw samples. They select a specific strain of laboratory mice.

CONTROL GROUP
E. coli injected into right thigh of 15 laboratory mice. No bacteriophage are injected.

EXPERIMENTAL GROUP
E. coli is injected into 15 other mice. Left thigh gets bacteriophage injection.

TEST RESULTS

All mice die within 32 hours.

All mice survive.

a

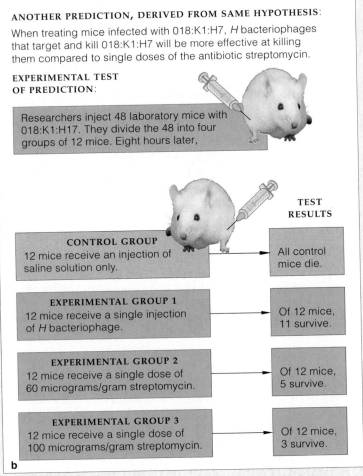

ANOTHER PREDICTION, DERIVED FROM SAME HYPOTHESIS:

When treating mice infected with 018:K1:H7, *H* bacteriophages that target and kill 018:K1:H7 will be more effective at killing them compared to single doses of the antibiotic streptomycin.

EXPERIMENTAL TEST OF PREDICTION:

Researchers inject 48 laboratory mice with 018:K1:H17. They divide the 48 into four groups of 12 mice. Eight hours later,

TEST RESULTS

CONTROL GROUP
12 mice receive an injection of saline solution only.

All control mice die.

EXPERIMENTAL GROUP 1
12 mice receive a single injection of *H* bacteriophage.

Of 12 mice, 11 survive.

EXPERIMENTAL GROUP 2
12 mice receive a single dose of 60 micrograms/gram streptomycin.

Of 12 mice, 5 survive.

EXPERIMENTAL GROUP 3
12 mice receive a single dose of 100 micrograms/gram streptomycin.

Of 12 mice, 3 survive.

b

Figure 1.12 Two experiments to compare the effectiveness of bacteriophage injections as opposed to antibiotics for treating a bacterial infection.

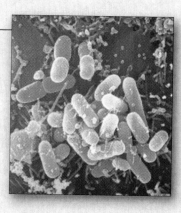

Figure 1.13 Some cells of a pathogenic strain of *Escherichia coli* (pink) adhering to a human intestinal cell in tissue culture.

outlines the experimental test. Figure 1.13 shows an example of the target pathogens.

Figure 1.12*b* outlines a second experimental test of another prediction derived from the same hypothesis. The results reinforced Smith and Huggins's conclusion that certain bacteriophages are as good as or better than antibiotics at stopping certain bacterial infections.

Each type of bacteriophage targets one or at most a few strains of bacterial species. It now takes too long to discover which bacterium is causing a severe infection, so the right bacteriophage might not be enlisted until it's too late. New procedures may speed up identification.

IDENTIFYING IMPORTANT VARIABLES In nature, many factors can change the course of an infection. To give examples, genetic differences among individuals can lead to differences in immune responses to infection. Other factors are the individual's health, nutritional habits, and age. Some pathogens are deadlier than others, and so on.

That's why Levin and Bull simplified and controlled the variables. Variables, remember, are specific aspects of objects or events that may differ among individuals and over time. All cells in their experiments descended from the same parent cell. They received the same nutrients at the same temperature in a flask. They could expect them to respond the same way to an attack. They were the same age and sex, and they were raised under identical laboratory conditions. Each mouse in each experimental group was injected with the same amount of antibiotic or bacteriophages. Each mouse in a control group got the same injection of saline solution. The focus was on *one variable*—a specific bacteriophage versus a specific antibiotic—in a simple, controlled, artificial situation.

BIAS IN REPORTING RESULTS Whether intentionally or not, experimenters run a risk of interpreting their data in terms of what they want to prove or dismiss. A few have even been known to fake measurements or nudge findings to reinforce their bias. That is why science emphasizes reporting test results in *quantitative* terms—with actual counts or some other precise form. Doing so will allow other experimenters to check or test the results readily and systematically, as Levin and Bull did.

THE LIMITS OF SCIENCE

The call for objective testing strengthens the theories that emerge from scientific studies. It also puts limits on the kinds of studies that can be carried out. Beyond the realm of science, some events remain unexplained. Why do we exist, for what purpose? Why does any one of us have to die at a particular moment? Such questions lead to *subjective* answers. These come from within, as an outcome of all the experiences and mental connections that shape human consciousness. Because people differ vastly in this regard, subjective answers do not readily lend themselves to scientific analysis and experiments.

This is not to say subjective answers are without value. No human society can function for long unless its members share a commitment to certain standards for making judgments, even subjective ones. Moral, aesthetic, philosophical, and economic standards vary from one society to the next. But they all guide people in deciding what is important and good, and what is not. All attempt to give meaning to what we do.

Every so often, scientists stir up controversy when they explain something that was thought to be beyond natural explanation—as belonging to the supernatural. This is often the case when a society's moral codes are interwoven with religious narratives. Exploring a long-standing view of the natural world from the scientific point of view might be misinterpreted as questioning morality, even though the two are not the same thing.

As one example, centuries ago in Europe, Nicolaus Copernicus studied the planets and concluded the Earth circles the sun. Today this seems obvious enough. Back then, it was heresy. The prevailing belief was that the Creator made the Earth—and, by extension, humans—the immovable center of the universe. Later a respected scholar, Galileo Galilei, studied the Copernican model of the solar system, thought it was a good one, and said so. He was forced to retract his statement publicly, on his knees, and put the Earth back as the fixed center of things. (Word has it that when he stood up he muttered, "Even so, it *does* move.") Later still, Darwin's theory of evolution ran up against the same prevailing belief.

Today, as then, society has sets of standards. Those standards might be questioned when some new, natural explanation runs counter to supernatural beliefs. This doesn't mean that scientists who raise the questions are less moral, less lawful, less sensitive, or less caring than anyone else. It simply means one more standard guides their work: *The external world, not internal conviction, must be the testing ground for scientific beliefs.*

Systematic observations, hypotheses, predictions, tests. In all these ways, science differs from systems of belief that are based on faith, force, or simple consensus.

SUMMARY   *Gold* indicates text section

1. There is unity in the living world, for all organisms share these characteristics: They consist of one or more cells. They are assembled from the same kinds of atoms and molecules according to the same laws of energy. They engage in metabolism, and they also sense and respond to specific aspects of the environment. Their DNA gives them a capacity to survive and reproduce. Table 1.1 lists these characteristics. *1.1*

2. The hierarchy of biological organization includes cells, multicelled organisms, populations, communities, ecosystems, and the biosphere. *1.2*

3. Many millions of species (kinds of organisms) exist. Many millions more existed in the past but are extinct. Classification schemes are ever more inclusive groupings of species, starting with genus and proceeding through family, order, class, and phylum, to kingdom. *1.3*

4. Life's diversity arises through mutations: heritable changes in the structure of DNA molecules. The changes are the basis for variation in heritable traits. These are traits that parents bestow on their offspring, including most details of body form and function. *1.4*

5. Darwin's theory of evolution by natural selection is central to biological inquiry. Its key premises are: *1.4*

   a. Individuals of a population differ in the details of their shared heritable traits. Variant forms of traits may affect the ability to survive and reproduce.

   b. Natural selection is the outcome of differences in survival and reproduction among individuals that differ in one or more traits. Adaptive forms of a trait tend to become more common; less adaptive ones become less common or vanish. Thus the defining traits may change over successive generations; the population may evolve.

6. The methods of scientific inquiry are diverse. These terms are important aspects of the methods: *1.5*

   a. Theory: Explanation of a broad range of related phenomena supported by numerous tests. An example is Darwin's theory of evolution by natural selection.

   b. Hypothesis: A proposed explanation of a specific phenomenon. Sometimes called an educated guess.

   c. Prediction: A claim about what can be expected in nature, based on premises of a theory or hypothesis.

   d. Test: An attempt to produce actual observations that match predicted or expected observations.

   e. Conclusion: A statement about whether to accept, modify, or reject a theory or hypothesis, based on tests of predictions that were derived from it.

7. Logic is a pattern of thought by which an individual draws a conclusion that does not contradict evidence used to support the conclusion. Inductive logic means individuals derive a general statement from specific observations. Deductive logic means individuals make inferences about particular consequences or predictions

Table 1.1 *Summary of Life's Key Characteristics*

SHARED CHARACTERISTICS THAT REFLECT LIFE'S UNITY

1. Organisms consist of one or more cells.

2. Organisms are constructed of the same kinds of atoms and molecules according to the same laws of energy.

3. Organisms engage in metabolism; they acquire and use energy and materials to survive and reproduce.

4. Organisms sense and make controlled responses to their internal and external environments.

5. Heritable instructions encoded in DNA give organisms their capacity to grow and reproduce. DNA instructions also guide the development of complex multicelled organisms.

6. Characteristics that define a population of organisms can change through the generations; the population can evolve.

FOUNDATIONS FOR LIFE'S DIVERSITY

1. Mutations (heritable changes in the structure of DNA) give rise to variation in heritable traits, including most details of body form, functioning, and behavior.

2. Diversity is the sum total of variations that accumulated in different lines of descent over the past 3.8 billion years, as by natural selection and other processes of evolution.

that must follow from a hypothesis. Such a pattern of thinking is often expressed in "if–then" terms. *1.5*

8. Predictions that flow from hypotheses can be tested by comprehensive observations or by experiments. *1.5*

9. Experimental tests simplify observations in nature or the laboratory because observations are made under precisely manipulated and controlled conditions. The tests are based on this premise: Any aspect of nature has one or more underlying causes, whether obvious or not. Hypotheses are scientific only when they may be tested in ways that might disprove them. *1.5, 1.6*

10. A control group is a standard against which one or more experimental groups (test groups) are compared. Ideally, it is the same as each experimental group in all variables except the one being investigated. *1.5, 1.6*

11. A variable is a specific aspect of an object or event that might differ over time and between individuals. Experimenters directly manipulate the variable they are studying to support or disprove a prediction. *1.5, 1.6*

12. Test results might be distorted by sampling error: chance differences between a population, event, or some other aspect of nature and the samples chosen to represent it. Distortion is less likely when samplings are large and when they are repeated. *1.5, 1.6*

13. Scientific theories arise from systematic observations, hypotheses, predictions, and experimental testing. The external world is the testing ground for theories. *1.7*

Review Questions

For this chapter and subsequent chapters, *italics* after a review question identify the section where you can find answers. They include section numbers and *CI* (for *Chapter Introduction*).

1. Why is it difficult to formulate a simple definition of life? *CI*

2. Name the molecule of inheritance in cells. *1.1*

3. Write out simple definitions of the following terms: *1.1*
   a. cell   b. metabolism   c. energy   d. ATP

4. How do organisms sense changes in their surroundings? *1.1*

5. Study Figure 1.5. Then, on your own, arrange and define the levels of biological organization. *1.2*

6. Study Figure 1.6. Then, on your own, make a sketch of the one-way flow of energy and the cycling of materials through the biosphere. To the side of the sketch, write out definitions of producers, consumers, and decomposers. *1.2*

7. List the shared characteristics of life. *CI, 1.3*

8. What are the two parts of the scientific name for each kind of organism? *1.3*

9. List the six kingdoms of species as outlined in this chapter. Also list some of their general characteristics. *1.3*

10. Define mutation and adaptive trait. Explain the connection between mutation and the diversity of life. *1.4*

11. Write brief definitions of evolution, artificial selection, and natural selection. *1.4*

12. Define and distinguish between: *1.5*
   a. hypothesis and prediction
   b. observational test and experimental test
   c. inductive and deductive logic
   d. speculation and scientific theory

13. With respect to experimental tests, define variable, control group, and experimental group. *1.5*

14. What does sampling error mean? *1.5*

Self-Quiz   ANSWERS IN APPENDIX III

1. _____ is the capacity of cells to extract energy from sources in their environment, and to transform and use energy to grow, maintain themselves, and reproduce.

2. _____ is a state in which the internal environment is being maintained within tolerable limits.

3. The _____ is the smallest unit of life.

4. If a form of a trait improves chances for surviving and reproducing in a given environment, it is a(n) _____ trait.

5. The capacity to evolve is based on variations in heritable traits, which originally arise through _____ .

6. You have some number of traits that also were present in your great-great-great-great-grandmothers and great-great-great-great-grandfathers. This is an example of _____ .
   a. metabolism   c. a control group
   b. homeostasis   d. inheritance

7. DNA molecules _____ .
   a. contain instructions for traits
   b. undergo mutation
   c. are transmitted from parents to offspring
   d. all of the above

8. For many years in a row, a dairy farmer allowed his best milk-producing cows but not the poor producers to mate. Over many generations, milk production increased. This outcome is an example of _____ .
   a. natural selection   c. evolution
   b. artificial selection   d. both b and c

9. A control group is _____ .
   a. a standard against which experimental groups are compared
   b. identical to experimental groups except for one variable
   c. a standard with several variables against which an experimental group is compared
   d. both a and b

10. A specific aspect of an object or event that may change over time or change among individuals is a _____ .
   a. control group   c. variable
   b. experimental group   d. sampling error

11. The fewer the number of individuals from a population that are chosen at random for an experimental group, _____ .
   a. the greater the chance of sampling error
   b. the smaller the chance of sampling error
   c. the less likely that differences among them will distort the test results

12. Match the terms with the most suitable descriptions.
   ____ adaptive   a. statement of what you should find in
        trait   nature if you were to go looking for it
   ____ natural   b. proposed explanation; educated guess
        selection   c. improves chances to survive and
   ____ scientific   reproduce in prevailing environment
        theory   d. related set of hypotheses that form a
   ____ hypothesis   broadly useful, testable explanation
   ____ prediction   e. outcome of differences in survival and
                        reproduction among individuals that
                        differ in details of one or more traits

Critical Thinking

1. Some spiders (*Dolomedes*) that feed on insects around ponds occasionally capture tadpoles and small fishes, as in Figure 1.14, and isn't that fun to think about? While they are immature, the female spiders confine themselves to a small patch of vegetation next to the pond. When sexually mature, they mate and store sperm that fertilize the eggs. Only then do they move out and occupy larger areas around the pond. Develop hypotheses to explain what might cause the spiders to live in different places at different times. Design an experiment to test each hypothesis.

Figure 1.14 A spider (*Dolomedes*) and its prey: a tiny minnow. The spider delivered paralyzing venom as well as digestive enzymes into its captive and is now sucking predigested juices from it. Such spiders can move about below the water's surface. Hairs on their body trap oxygen (for aerobic respiration) during hunting expeditions.

2. A scientific theory about some aspect of nature rests upon inductive logic. The assumption is that, because an outcome of some event has been observed to happen with great regularity, it will happen again. However, we can't know this for certain, because there is no way to account for all possible variables that may affect the outcome. To illustrate this point, Garvin McCain and Erwin Segal offer a parable:

Once there was a highly intelligent turkey. The turkey lived in a pen, attended by a kind, thoughtful master, and it had nothing to do but reflect upon the world's wonders and regularities. It observed some major regularities. Morning always began with the sky getting light, followed by the clop, clop, clop of its master's friendly footsteps, which was followed by the appearance of delicious food. Other things varied—sometimes the morning was warm and sometimes cold—but food always followed footsteps. The sequence of events was so predictable that it eventually became the basis of the turkey's theory about the goodness of the world.

One morning, after more than a hundred confirmations of the goodness theory, the turkey listened for the clop, clop, clop, heard it, and had its head chopped off.

The turkey learned the hard way that explanations about the world only have a high or low probability of not being wrong. Today, some people take this uncertainty to mean that "facts are irrelevant—facts change." If that is so, should we just stop doing scientific research? Why or why not?

3. Witnesses in a court of law are asked to "swear to tell the truth, the whole truth, and nothing but the truth." What are some of the problems inherent in the question? Can you think of a better alternative?

4. Many popular magazines publish an astounding number of articles on diet, exercise, and other health-related topics. Some authors recommend a specific diet or dietary supplement. What kinds of evidence do you think the articles should include so that you can decide whether to accept the recommendations?

5. Although scientific information is often used when making a decision, it can't tell an individual what's "right" or "wrong." Give an example of this from your own experience.

6. As the old saying goes, Everybody complains about the weather, but nobody does anything about it. Maybe you can start thinking about how we humans might change at least one aspect of local weather conditions. Two researchers at Arizona State University found that, at least for their investigation of the northeast coast of North America, it really does rain more on weekends, when we'd rather be out having fun!

As R. Cerveny and R. Balling, Jr. reasoned, if the amount of rain falling each day of the week is a random event, then rules of probability should apply. (*Probability* means the chance that each possible outcome of an event will occur is proportional to the number of ways it can be reached.) Thus, each day of the week should get one-seventh (14.3 percent) of the total rainfall for the week. But it turns out Monday is driest (13.1 percent). Days of the week are wetter, and Saturday is the wettest of all (with 16 percent of the total). Sunday is a bit above average.

What causes this effect? As the researchers hypothesized, if *air pollution* is greater on weekdays than weekends, then human activities might influence regional patterns of rainfall. Monday through Friday, coal-burning factories and gas-powered vehicles release quantities of tiny particles into the air. The particles can promote updrafts and act as "platforms" for the formation of water droplets. Air pollution must build up during the week and extend into Saturday. Over the weekend, fewer factories operate and fewer commuters are on the road. Therefore, by Monday, the air must be cleaner—and drier.

What kind of evidence do you suppose Cerveny and Balling gathered to test the hypothesis? Jot down a few ideas and check them against the researchers' article in the 6 August 1998 issue of *Nature*. Whether or not you continue in biology, this exercise will be good practice in searching through scientific literature for actual data that you can evaluate on topics of interest to you.

7. Scientists devised experiments to shed light on whether different fish species of the same genus compete in their natural habitat. They constructed twelve ponds, identical in chemical composition and physical characteristics. Then they released the following individuals in each pond:

Ponds 1, 2, 3:	Species A	300 individuals each pond
Ponds 4, 5, 6:	Species B	300 individuals each pond
Ponds 7, 8, 9:	Species C	300 individuals each pond
Ponds 10, 11, 12:	Species A, B, C	300 individuals of each species in each pond

Does this experimental design take into consideration all factors that can affect the outcome? If not, how would you modify it?

Selected Key Terms

For this chapter and subsequent chapters, these are the **boldface** terms that appear in the text, in the sections indicated here by *italic* numbers (or *CI*, short for Chapter Introduction). As a study aid, make a list of the terms, write a definition for each, then check it against the one in the text. You will use these terms later on. Becoming familiar with each one will help give you a foundation for understanding the material in later chapters.

adaptive trait *1.4*	ecosystem *1.2*	mutation *1.4*
animal *1.3*	energy *1.1*	natural selection *1.4*
antibiotic *1.4*	eubacteria *1.3*	plant *1.3*
archaebacterium *1.3*	evolution *1.4*	population *1.2*
artificial selection *1.4*	experiment *1.5*	prediction *1.5*
biology *CI*	fungus *1.3*	producer *1.2*
biosphere *1.2*	genus *1.3*	protistan *1.3*
cell *1.2*	homeostasis *1.1*	receptor *1.1*
community *1.2*	hypothesis *1.5*	reproduction *1.1*
consumer *1.2*	inductive logic *1.5*	sampling error *1.5*
control group *1.5*	inheritance *1.1*	species *1.3*
decomposer *1.2*	metabolism *1.1*	stimulus *1.1*
deductive logic *1.5*	model *1.5*	test, scientific *1.5*
development *1.1*	multicelled	theory, scientific *1.5*
DNA *1.1*	organism *1.2*	variable *1.5*

Readings

Carey, S. 1994. *A Beginner's Guide to the Scientific Method.* Belmont, California: Wadsworth. Paperback.

McCain, G., and E. Segal. 1988. *The Game of Science.* Fifth edition. Pacific Grove, California: Brooks/Cole. Paperback.

Moore, J. 1993. *Science as a Way of Knowing.* Cambridge, Massachusetts: Harvard University Press.

On-Line readings at Student Guide for InfoTrac: www.brookscole.com/biology

I Principles of Cellular Life

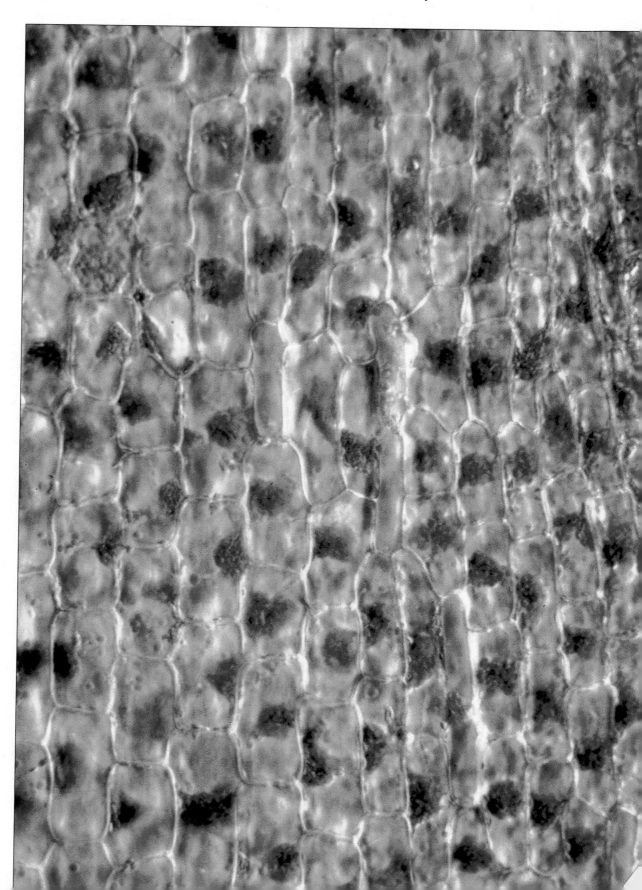

Living cells of one kind of plant (Elodea), observed with the aid of a microscope. Each of these rectangular cells contains efficient chemical factories called chloroplasts (the round, green structures inside).

CHEMICAL FOUNDATIONS FOR CELLS

How Much Are You Worth?

Hollywood thinks actress Julia Roberts is worth $20 million per picture, the Texas Rangers think shortstop Alex Rodriguez is worth $252 million per decade, and the United States thinks the average teacher isn't worth much more than $41,820 per year. From a biochemical perspective, though, how much is a human body really worth?

Think of it! Your body is a collection of elements. An **element** is a fundamental form of matter that has mass and takes up space, and at least here on Earth it can't be broken down into something else (except in physics laboratories). Break a chunk of copper into smaller and smaller bits and the smallest bit you end up with, even a lone atom, is still copper. Each solid, liquid, or gaseous substance consists of characteristic proportions of one or more such elements.

Ninety-two elements occur naturally on Earth. Four kinds—oxygen, hydrogen, carbon, and nitrogen —are the most abundant ones in your body, as they are in all other living organisms. Others in your body include phosphorus, potassium, sulfur, calcium, and sodium. Also present are *trace* elements, meaning that each represents less than 0.01 percent of body weight.

Figure 2.1*a* lists each element in your body. It also lists the cost per kilogram for that element. It all adds up to a paltry $118.63 for someone who weighs seventy kilograms, or 154.3 pounds.

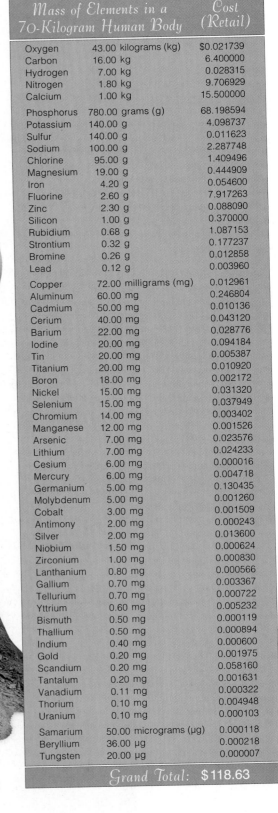

Figure 2.1 (**a**) How much is a human being worth, chemically speaking? Shown here, the mass of each element in a seventy-kilogram human body and its commercial cost. Prices are based on the forms of elements most likely to be found in nature (for example, the hydrogen and oxygen atoms in molecules of water rather than in purified form). Note that the market price for gold changes monthly. The cost for uranium, no longer available through chemical suppliers, was obtained from E-Bay.

(**b**) Proportions of some elements in a human body and in the fruit of a pumpkin compared to proportions of elements making up the Earth's crust. How are the proportions similar? How do they differ?

Mass of Elements in a 70-Kilogram Human Body		Cost (Retail)
Oxygen	43.00 kilograms (kg)	$0.021739
Carbon	16.00 kg	6.400000
Hydrogen	7.00 kg	0.028315
Nitrogen	1.80 kg	9.706929
Calcium	1.00 kg	15.500000
Phosphorus	780.00 grams (g)	68.198594
Potassium	140.00 g	4.098737
Sulfur	140.00 g	0.011623
Sodium	100.00 g	2.287748
Chlorine	95.00 g	1.409496
Magnesium	19.00 g	0.444909
Iron	4.20 g	0.054600
Fluorine	2.60 g	7.917263
Zinc	2.30 g	0.088090
Silicon	1.00 g	0.370000
Rubidium	0.68 g	1.087153
Strontium	0.32 g	0.177237
Bromine	0.26 g	0.012858
Lead	0.12 g	0.003960
Copper	72.00 milligrams (mg)	0.012961
Aluminum	60.00 mg	0.246804
Cadmium	50.00 mg	0.010136
Cerium	40.00 mg	0.043120
Barium	22.00 mg	0.028776
Iodine	20.00 mg	0.094184
Tin	20.00 mg	0.005387
Titanium	20.00 mg	0.010920
Boron	18.00 mg	0.002172
Nickel	15.00 mg	0.031320
Selenium	15.00 mg	0.037949
Chromium	14.00 mg	0.003402
Manganese	12.00 mg	0.001526
Arsenic	7.00 mg	0.023576
Lithium	7.00 mg	0.024233
Cesium	6.00 mg	0.000016
Mercury	6.00 mg	0.004718
Germanium	5.00 mg	0.130435
Molybdenum	5.00 mg	0.001260
Cobalt	3.00 mg	0.001509
Antimony	2.00 mg	0.000243
Silver	2.00 mg	0.013600
Niobium	1.50 mg	0.000624
Zirconium	1.00 mg	0.000830
Lanthanum	0.80 mg	0.000566
Gallium	0.70 mg	0.003367
Tellurium	0.70 mg	0.000722
Yttrium	0.60 mg	0.005232
Bismuth	0.50 mg	0.000119
Thallium	0.50 mg	0.000894
Indium	0.40 mg	0.000600
Gold	0.20 mg	0.001975
Scandium	0.20 mg	0.058160
Tantalum	0.20 mg	0.001631
Vanadium	0.11 mg	0.000322
Thorium	0.10 mg	0.004948
Uranium	0.10 mg	0.000103
Samarium	50.00 micrograms (µg)	0.000118
Beryllium	36.00 µg	0.000218
Tungsten	20.00 µg	0.000007
Grand Total:		$118.63

a

All you have to do is briefly observe Julia or Alex or any teacher and you know that a human body is far more than a list of its ingredients. Even so, that list is a starting point for the body's distinct characteristics. For example, compare its elemental proportions with those of, say, the Earth's crust or a pumpkin (Figure 2.1*b*). Although it has many of the same elements as dirt and pumpkins, the proportions differ, and the kinds and amounts are integrated as a unique, highly organized, dynamic form.

Another point: That listing of ingredients is like a snapshot, frozen in time. Your 43 kilograms of oxygen atoms are bound to other atoms in body water or in molecules that cells use as building blocks or energy —but this isn't your lifetime supply. At any moment, uncounted numbers of oxygen atoms are leaving your body, as in the carbon dioxide being exhaled from your lungs, and new ones from air, food, and water are replacing them. Like all other organisms, you must continually take up sufficient amounts of oxygen and the other elements listed to build and maintain cells— and, ultimately, yourself. If you were to inhale merely a few breaths of oxygen-free gas, you would fall into a coma within five seconds. Deprived of oxygen for two to four minutes more, you would be dead.

Remember this example if someone tries to tell you "chemistry" isn't important. *You* are chemical, and so is every living and nonliving thing in the universe. Hamburgers and pasta, refrigerators and jet fuel, sickle-cell anemia and other genetic disorders, health and disease, pesticides and agriculture, acid rain and old-growth forests, global warming, nerve gas in the hands of terrorists—you name it, and "chemistry" has something to say about it. Study it well.

Key Concepts

1. All substances consist of one or more elements, most notably oxygen, carbon, hydrogen, and nitrogen. Each element consists of atoms, which are the smallest units of matter that still display the element's properties. The atoms making up each element consist of protons, electrons and, except for the hydrogen atom, neutrons.

2. The atoms that make up each kind of element have the same number of protons and electrons. They may differ slightly from one another in their number of neutrons. Variant forms of an element's atoms are called isotopes.

3. Atoms have no overall electric charge unless they become ionized—that is, unless they lose electrons or acquire more of them. An ion is an atom or molecule that has lost or gained one or more electrons and now carries an electric charge.

4. Whether a given atom will interact with other atoms depends on how many electrons it has and how they are arranged. When energetic interactions unite two or more atoms, this produces a chemical bond.

5. The molecular organization and activities of living things arise largely from ionic, covalent, and hydrogen bonds between atoms.

6. Life probably originated in water and is exquisitely adapted to its properties. Foremost among its properties are temperature-stabilizing effects, cohesiveness, and a capacity to dissolve or repel a variety of substances.

7. All organisms depend on the controlled formation, use, and disposal of hydrogen ions (H^+). The pH scale is a measure of the concentration of hydrogen ions in solutions.

Earth's Crust		Human		Pumpkin	
Oxygen	46.6%	Oxygen	65%	Oxygen	85%
Silicon	27.7	Carbon	18	Hydrogen	10.7
Aluminum	8.1	Hydrogen	10	Carbon	3.3
Iron	5.0	Nitrogen	3	Potassium	0.34
Calcium	3.6	Calcium	2	Nitrogen	0.16
Sodium	2.8	Phosphorus	1.1	Phosphorus	0.05
Potassium	2.1	Potassium	0.35	Calcium	0.02
Magnesium	1.5	Sulfur	0.25	Magnesium	0.01
		Sodium	0.15	Iron	0.008
		Chlorine	0.15	Sodium	0.001
		Magnesium	0.05	Zinc	0.0002
		Iron	0.004	Copper	0.0001
		Iodine	0.004		

b

REGARDING THE ATOMS

Structure of Atoms

What are the smallest particles that retain the properties of an element? **Atoms**. A line of about a million of them would fit in the period ending this sentence. Physicists have split atoms into more than 100 kinds of smaller parts. But the only subatomic particles we will need to consider are called **protons**, **electrons**, and **neutrons**.

All atoms have one or more protons, which carry a positive electric charge (p^+). Except for hydrogen, atoms also have one or more neutrons, which are uncharged. Protons and neutrons make up an atom's core region, or atomic nucleus (Figure 2.2). Zipping about the nucleus and occupying most of the volume of an atom are one or more electrons, which carry a negative charge (e^-). Each atom normally has as many electrons as protons. In this configuration, the atom carries no net charge.

Each element has a unique *atomic* number, based on the number of protons in its atoms. The hydrogen atom has one proton, so the number is 1. For the carbon atom, with six protons, it is 6 (Table 2.1). Each element has a *mass* number, or the combined number of protons and neutrons in the atomic nucleus. The carbon atom, with six protons and six neutrons, has a mass number of 12.

Why bother with atomic and mass numbers? They give you a sense of whether and how substances will interact. *This information can help you predict how a given substance might behave in individual cells, in multicelled organisms, and in the environment, under many conditions.*

Isotopes—Variant Forms of Atoms

Although all atoms of an element have the same number of protons, they do not all have the same number of

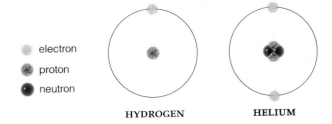

Figure 2.2 The hydrogen atom and helium atom according to one model for atomic structure. The model is highly simplified; the nucleus of these two representative atoms would actually be an invisible speck at the scale employed here.

neutrons. For each element, atoms that vary in neutron number are called **isotopes**. Most naturally occurring elements have two or more common isotopes. Carbon has three, nitrogen has two, and so on. A superscript number to the left of a symbol for an element signifies the isotope. Carbon's three isotopes are ^{12}C or carbon 12 (the most common form, with six protons and six neutrons), ^{13}C (six protons, seven neutrons), and ^{14}C (six protons, eight neutrons).

All isotopes of an element interact chemically with other atoms in the same way. This means that cells can use any isotope of an element for metabolic activities.

What about radioactive isotopes (radioisotopes)? A physicist, Henri Becquerel, discovered them in 1896 after he put a heavily wrapped rock on top of an unexposed photographic plate inside a desk drawer. The rock held energy-emitting isotopes of uranium. A few days after the plate was exposed to the emissions, a faint image of the rock appeared on it. Marie Curie, a coworker, gave the name "radioactivity" to this chemical behavior.

As we now know, a **radioisotope** is an isotope that has an unstable nucleus and becomes more stabilized by spontaneously emitting energy and particles. The process of radioactive decay transforms a radioisotope into an atom of a different element at a known rate. For instance, we know that over a predictable time span, carbon 14 becomes nitrogen 14 (Section 19.2).

Table 2.1	*Atomic Number and Most Common Mass Number of Selected Elements*		
Element	Symbol	Atomic Number	Mass Number
Hydrogen	H	1	1
Carbon	C	6	12
Nitrogen	N	7	14
Oxygen	O	8	16
Sodium	Na	11	23
Magnesium	Mg	12	24
Phosphorus	P	15	31
Sulfur	S	16	32
Chlorine	Cl	17	35
Potassium	K	19	39
Calcium	Ca	20	40
Iron	Fe	26	56
Iodine	I	53	127

Elements are forms of matter that occupy space, have mass, and cannot be degraded to something else by ordinary means.

Atoms, the smallest particles unique to each element, have one or more positively charged protons, negatively charged electrons, and (except for hydrogen) neutrons.

Most elements have two or more isotopes, which are atoms that differ in the number of neutrons. A radioisotope has an unstable nucleus. Radioactive decay transforms it into another element at a predictable and characteristic rate.

Using Radioisotopes to Track Chemicals and Save Lives

Radioisotopes are often used as **tracers**, or substances with an isotope attached to them. Researchers insert them into a cell, a multicelled body, an ecosystem, or any other system and use them like shipping labels. Devices detect the emissions from radioactive tracers and follow them through a pathway or pinpoint its destination.

For example, Melvin Calvin and other botanists used radioactive tracers to track the steps of photosynthesis. As they knew, all isotopes of an element have the same number of electrons, so all interact with other atoms the same way. Plants, they hypothesized, use any carbon isotope when they build carbohydrates. By putting plant cells in a medium enriched with a tracer (^{14}C rather than ^{12}C), they tracked the uptake of carbon through each reaction step in the formation of sugars and starches.

Botanists use a phosphorus 32 tracer to identify how plants take up and use soil nutrients and fertilizers. They apply such findings to work on improving crop yields.

Radioisotopes are tools for medical diagnosis as well as for research. When clinicians analyze, say, the human thyroid, they use a type that swiftly decays into a harmless element. The thyroid is the only gland of ours that uses iodine to build thyroid hormones, which greatly influence body growth and functions. If a patient's symptoms point to abnormal thyroid output, clinicians may inject a trace amount of iodine 123 into the patient's blood. They would then use a photographic imaging device to scan the gland. Figure 2.3 shows examples of the type of images they get.

PET (for *Positron-Emission Tomography*) also uses radioisotopes to form images of body tissues. Clinicians attach a tracer to glucose or another molecule. They inject the labeled glucose into the patient, who is moved into a PET scanner (Figure 2.4*a*). All body cells use glucose, so they take up the labeled kind, too. Uptake is greater in some tissues than others, depending on their metabolic activity at the time of examination. Laboratory devices detect the radioactive emissions and use them to form an image. Such images may reveal variations or abnormalities in metabolic activity, as in Figure 2.4*d*.

Radioisotopes energetic enough to kill cells still can be useful. In *radiation therapy*, radioisotopes stop or impair the activity of abnormal body cells. To give an example, emissions from a source of radium 226 or cobalt 60 may destroy small, localized cancers. As another example, the emissions from plutonium 238 drive artificial pacemakers that smooth out irregular heartbeats. The radioisotope is sealed in a case so its emissions won't damage body cells.

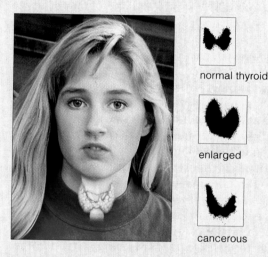

normal thyroid

enlarged

cancerous

Figure 2.3   Location of the human thyroid relative to the trachea (windpipe), and scans of the thyroid gland from three patients.

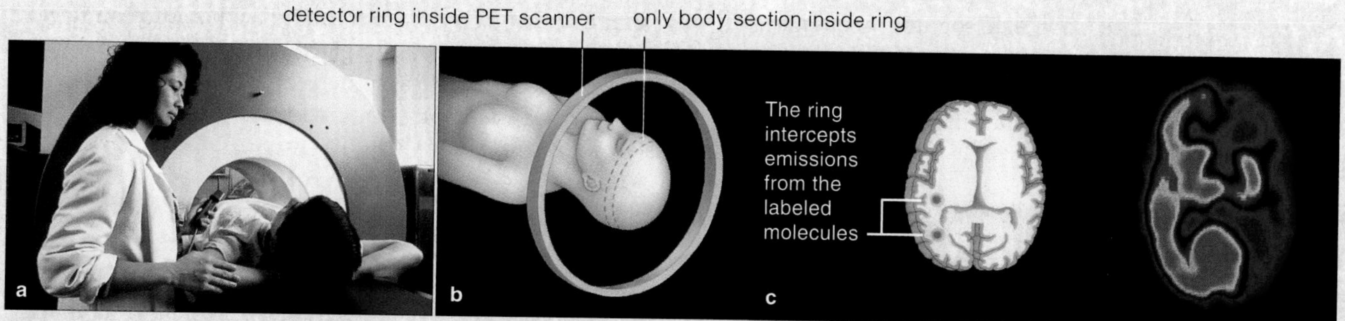

---

detector ring inside PET scanner   only body section inside ring

The ring intercepts emissions from the labeled molecules

Figure 2.4   (**a**) Patient moving into a PET scanner. (**b,c**) Inside, a ring of detectors intercepts radioactive emissions from labeled molecules that were injected into the patient. Computers analyze and color-code the number of emissions from each location in the scanned body region.

d PET scan (*red* indicates areas of greatest activity; *blue*, of least activity)

(**d**) Different colors in a brain scan signify differences in metabolic activity. Cells of this brain's left half absorbed and used the labeled molecules at expected rates. However, cells in the right half showed little activity. The patient was diagnosed as having a neurological disorder.

WHAT HAPPENS WHEN ATOM BONDS WITH ATOM?

Electrons and Energy Levels

Cells stay alive because energy inherent in all electrons makes things happen (Figure 2.5). Countless atoms in cells acquire, share, and donate electrons. Atoms of some elements do this easily; others don't. What determines whether an atom will interact with others? *The outcome depends on the number and arrangement of its electrons.*

Tinker with magnets and you can get a sense of the attractive force between unlike charges (+ −) as well as the repulsive force between like charges (++ or − −). Electrons carry a negative charge. In atoms, the positive charge of protons attracts electrons, but they repel each other. They respond to the pulls and pushes by moving in different orbitals. An orbital is a volume of space around the atomic nucleus in which electrons are likely to be at any instant (Figure 2.6). Visualize two children circling a cookie jar, not yet expert in the art of sharing. Each is drawn to the cookies but dreads being hit on the head by the other. They bob about on opposite sides of the jar to avoid each other. They never occupy the same space at the same time. In this analogy, the orbital corresponds to the perimeter in which the children are moving around the cookie jar.

Each orbital can house one or at most two electrons. Atoms differ in how many occupied orbitals they have because they differ in their number of electrons.

Hydrogen is the simplest atom. It has one electron in a spherical orbital closest to the nucleus. The orbital corresponds to the *lowest available energy level.* In other atoms, two electrons fill the first orbital, and two more electrons occupy a second spherical orbital around the first. The larger atoms have more electrons in orbitals farther from the nucleus, at *higher energy levels.*

An electron at any energy level has a specific amount of energy. On its own, it cannot move to a different level. But

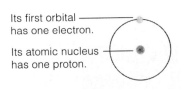

Its first orbital has one electron.

Its atomic nucleus has one proton.

a Shell model of an H atom

Figure 2.5 How a lone electron can make things happen. Electrons were confined in a bubble of liquid helium, then fluctuating sound waves popped the bubble. An electron escaped and made this white-centered red flash.

what if an atom absorbs extra energy from the sun? If the energy input is just the right amount, the electron will become excited enough to reach a higher energy level. Left to itself, it will quickly return to the lower energy level by emitting the extra energy. Most often, its emission is in the form of light. You will read about this event in Sections 7.3 and 7.4, on photosynthesis.

Think "Shells"

The **shell model** is a simple way to show how electrons are distributed in an atom. By this model, a "shell" encloses all orbitals available to electrons at the same energy level (Figure 2.7). The first, spherical orbital is in the first shell. Enclosing the first shell is a second shell at a higher energy level. The second shell has four more available orbitals. More orbitals fit in a third shell, a fourth shell, and so on through large, complex atoms of the heavier elements listed in Appendix VI.

From Atoms to Molecules

How can you predict whether two atoms will interact? Check for electron vacancies in their outermost shell. Vacancies mean an atom might give up, gain, or share electrons under suitable conditions, which of course will change the distribution or number of its electrons. You can use the shell model to picture what goes on here.

Figure 2.7 shows the electron distribution for some atoms. Electrons are assigned to an energy level (a shell). Count the vacancies—one or more unfilled orbitals in

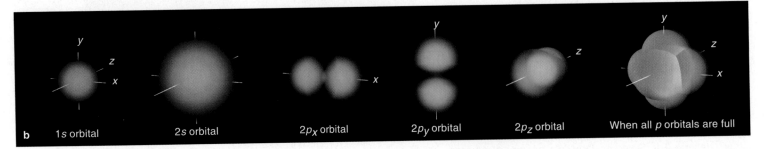

| 1s orbital | 2s orbital | $2p_x$ orbital | $2p_y$ orbital | $2p_z$ orbital | When all p orbitals are full |

b

Figure 2.6 Electron arrangements in atoms with respect to three axes. Each axis (x, y, or z) is perpendicular to the other two. (**a**) A hydrogen atom's electron occupies a spherical 1s orbital, at the lowest energy level. (**b**) Every atom has a 1s orbital occupied by one or two electrons. Other orbitals include the 2s and p orbitals at the second energy level. Each can hold two electrons.

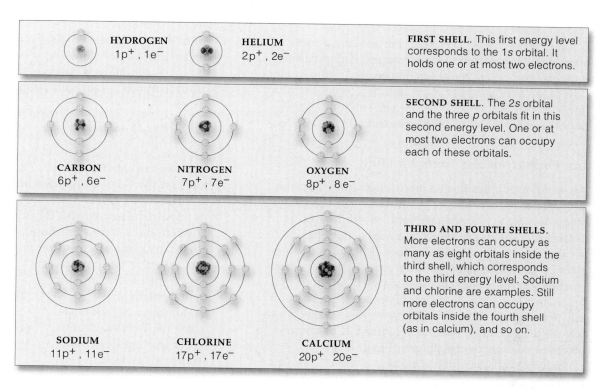

Figure 2.7 Examples of the shell model for the distribution of electrons in atoms. Successive shells correspond to higher energy levels.

Hydrogen, carbon, and other atoms with electron vacancies (unfilled orbitals) inside their outermost shell tend to give up, accept, or share electrons. Helium and other atoms with no electron vacancies in their outermost shell are inert; they show little if any tendency to interact with other atoms.

HYDROGEN
$1p^+$, $1e^-$

HELIUM
$2p^+$, $2e^-$

FIRST SHELL. This first energy level corresponds to the 1s orbital. It holds one or at most two electrons.

CARBON
$6p^+$, $6e^-$

NITROGEN
$7p^+$, $7e^-$

OXYGEN
$8p^+$, $8e^-$

SECOND SHELL. The 2s orbital and the three p orbitals fit in this second energy level. One or at most two electrons can occupy each of these orbitals.

SODIUM
$11p^+$, $11e^-$

CHLORINE
$17p^+$, $17e^-$

CALCIUM
$20p^+$ $20e^-$

THIRD AND FOURTH SHELLS. More electrons can occupy as many as eight orbitals inside the third shell, which corresponds to the third energy level. Sodium and chlorine are examples. Still more electrons can occupy orbitals inside the fourth shell (as in calcium), and so on.

Figure 2.8 Chemical bookkeeping. We use symbols for elements when writing *formulas*, which identify the composition of compounds. For instance, water has the formula H_2O. The subscript number tells you two hydrogen (H) atoms are present for each oxygen (O) atom. Use such symbols and formulas when writing *chemical equations:* representations of the reactions among atoms and molecules. Substances entering a reaction (reactants) are to the left of the reaction arrow, and products are to the right, as shown by this chemical equation for photosynthesis.

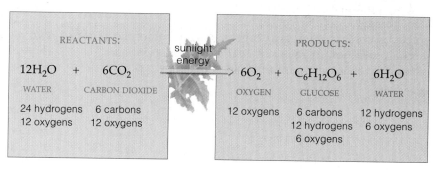

REACTANTS:

$12H_2O$ + $6CO_2$

WATER   CARBON DIOXIDE

24 hydrogens   6 carbons
12 oxygens   12 oxygens

sunlight energy

PRODUCTS:

$6O_2$ + $C_6H_{12}O_6$ + $6H_2O$

OXYGEN   GLUCOSE   WATER

12 oxygens   6 carbons   12 hydrogens
     12 hydrogens   6 oxygens
     6 oxygens

each atom's outermost shell. As you can see, helium is one of the atoms with no vacancies. It is an *inert* atom; it shows little tendency to enter chemical reactions.

Now look again at Figure 2.1, which lists the most abundant elements that make up a typical organism. These elements include hydrogen, oxygen, carbon, and nitrogen. The atoms of all four elements have electron vacancies. *Such atoms tend to eliminate their vacancies by forming bonds with other atoms.*

Each **chemical bond** is a union between the electron structures of atoms. Take a moment to study Figure 2.8. It summarizes a few conventions we use when talking about metabolic reactions. A **molecule** forms when two or more atoms join up. Some molecules consist of only one element. Molecular nitrogen (N_2), with two nitrogen atoms, is like this. So is molecular oxygen (O_2).

The molecules of **compounds** consist of two or more different elements in proportions that never vary. Water is an example. Each water molecule has one oxygen atom and two hydrogen atoms. The water molecules in

rainclouds, the ocean, a Siberian lake, your bathtub, the petals of a leaf or flower, or anywhere else always have twice as many hydrogen as oxygen atoms.

In a **mixture**, two or more elements or compounds intermingle in proportions that can and usually do vary. Swirling water with the sugar sucrose (a compound of carbon, hydrogen, and oxygen) gives you a mixture.

Electrons occupy orbitals, or volumes of space around an atom's nucleus. By a simplified model, orbitals are arranged as a series of shells enclosing the nucleus. Successive shells correspond to levels of energy, which become higher with distance from the nucleus.

One or at most two electrons occupy any orbital. Atoms with unfilled orbitals in their outermost shell tend to interact with other atoms; those with no vacancies do not.

In molecules of an element, all atoms are the same kind. In molecules of a compound, atoms of two or more elements are bonded together, in unvarying proportions.

IMPORTANT BONDS IN BIOLOGICAL MOLECULES

Eat your peas! Drink your milk! For as long as you've been alive, you've been eating complex carbohydrates, proteins, and other "biological molecules." Only living cells make and use these molecules, which consist of a few kinds of atoms joined by a few kinds of bonds—mainly ionic, covalent, and hydrogen bonds.

Ion Formation and Ionic Bonding

An atom, recall, has just as many electrons as protons, so it carries no net charge. That balance can change for atoms having a vacancy—an unfilled orbital—in their outermost shell. For example, a chlorine atom has such a vacancy and can acquire another electron. A sodium atom has a lone electron in its outermost shell, and that electron can be pulled or knocked out. An atom that has either lost or gained one or more electrons is an **ion**. The balance between its protons and its electrons has shifted, so now the atom is ionized; it has become positively or negatively charged (Figure 2.9*a*).

In living cells, neighboring atoms commonly accept or donate electrons among themselves. When one loses an electron and another grabs it, both become ionized. Depending on cellular conditions, the two ions might not separate; they might stay together as a result of the mutual attraction of opposite charges. An association of two ions that have opposing charges is known as an **ionic bond**. You see one outcome of ionic bonding in Figure 2.9*b*, which shows a portion of a crystal of table salt, or NaCl. In such crystals, sodium ions (Na^+) and chloride ions (Cl^-) interact through ionic bonds.

Covalent Bonding

Suppose two atoms, each with an unpaired electron in its outermost shell, meet up. Each exerts an attractive force on the other's unpaired electron but not enough to yank it away. Each becomes more stable by *sharing* its unpaired electron with the other. A sharing of a pair of electrons is a **covalent bond**. Example: A hydrogen atom partly fills the electron vacancy in its outermost shell when it has bonded covalently with another hydrogen atom.

Two hydrogen atoms, each with a single proton, share two electrons

MOLECULAR HYDROGEN

In structural formulas, a line between two atoms is a *single* covalent bond. It means an electron pair, as in molecular hydrogen (H_2, or H—H). In a *double* covalent bond, two atoms share two electron pairs, as in molecular oxygen (O=O). In a *triple* covalent bond, two atoms share three pairs, as in molecular nitrogen (N≡N). These examples are gaseous molecules. Each time you breathe in, a stupendous number of H_2, O_2, and N_2 molecules flow into your lungs.

Covalent bonds can be nonpolar or polar. Atoms exert the same pull on electrons and share them equally in a *nonpolar* covalent bond. "Nonpolar" means there is no difference in charge between the two ends of the bond—that is, at its two poles. Molecular hydrogen has a nonpolar covalent bond. Its two H atoms, each with one proton, attract the shared electrons equally.

SODIUM ATOM
11 p$^+$,
11 e$^-$

electron transfer

CHLORINE ATOM
17 p$^+$,
17 e$^-$

SODIUM ION
11 p$^+$,
10 e$^-$

CHLORIDE ION
17 p$^+$,
18 e$^-$

+   −

a Formation of a sodium ion and a chloride ion

Figure 2.9 **(a)** Ionization by way of an electron transfer. In this case, a sodium atom donates the single electron in its outermost shell to a chlorine atom, which has an unfilled orbital in *its* outermost shell. A sodium ion (Na^+) and a chloride ion (Cl^-) are the outcome of this interaction. **(b)** In each crystal of table salt, or NaCl, many sodium and chloride ions remain together owing to the mutual attraction of opposite charges. Their interaction is a case of ionic bonding.

1 mm

Na^+ Cl^- Na^+
Cl^- Na^+ Cl^-
Na^+ Cl^- Na^+

b Crystals of sodium chloride (NaCl)

In a *polar* covalent bond, atoms of different elements (which have different numbers of protons) don't exert the same pull on shared electrons. The more attractive atom ends up with a slight negative charge. We say it's "electronegative." However, its effect is balanced out by the other atom, which ends up with a slight positive charge. Thus, taken together, the two atoms of a polar covalent bond have no *net* charge.

Consider a water molecule. It has two polar covalent bonds: H—O—H. The electrons are less attracted to the two hydrogens than to the oxygen, which has more

Figure 2.10 Like skydivers who briefly clasp hands to form an orderly pattern, weak attractions within and between molecules can form and break easily.

protons. The molecule is slightly negative at one end and positive at the other, but has no *net* charge overall. As you will see, its polarity attracts polar substances.

Hydrogen Bonding

The patterns of electron sharing in covalent bonds hold atoms together in specific arrangements in molecules. Some of the patterns also give rise to weak attractions and repulsions between charged functional groups of molecules, and between molecules. Like interacting skydivers, these interactions can form and break easily (Figure 2.10). Even so, they have important roles in the structure and functioning of biological molecules.

For example, a **hydrogen bond** is a weak attraction between an electronegative atom (such as an oxygen or nitrogen atom taking part in a polar covalent bond) and a hydrogen atom taking part in a second polar covalent bond. Hydrogen's slight positive charge weakly attracts the atom with the slight negative charge (Figure 2.11).

Hydrogen bonds often form between different parts of a molecule that twists and folds back on itself. Such bonds are pervasive among proteins (Section 3.7). They commonly form between two or more molecules, as when they hold the two strands of a DNA molecule together (Figure 2.11*d*). Although individually easy to break, hydrogen bonds collectively stabilize the DNA. Extensive hydrogen bonding between individual water molecules also contributes to water's life-sustaining properties, which is the topic of the next section.

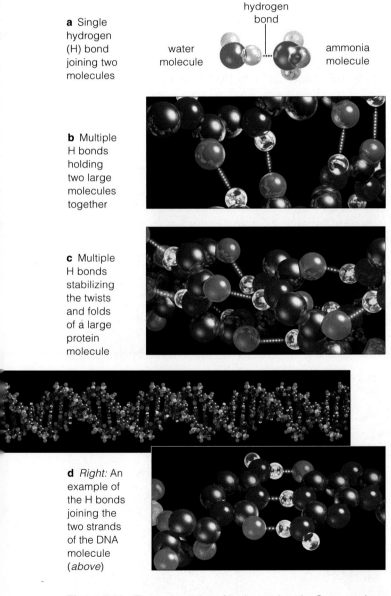

a Single hydrogen (H) bond joining two molecules

water molecule — hydrogen bond — ammonia molecule

b Multiple H bonds holding two large molecules together

c Multiple H bonds stabilizing the twists and folds of a large protein molecule

d *Right:* An example of the H bonds joining the two strands of the DNA molecule (*above*)

Figure 2.11 Three examples of hydrogen bonds. Compared to a covalent bond, a hydrogen bond is easier to break. Collectively, extensive hydrogen bonding has a major role in the structure and behavior of water, DNA, proteins, and many other substances, as you will see in later chapters.

In an ionic bond, two ions of opposite charge attract each other and stay together. Ions form when atoms gain or lose electrons and so acquire a net positive or negative charge.

In a covalent bond, atoms share a pair of electrons. When atoms share the electrons equally, the bond is nonpolar. When the sharing is not equal, the bond itself is polar—slightly positive at one end, slightly negative at the other.

In a hydrogen bond, a covalently bound atom showing a slight negative charge weakly interacts with a covalently bound hydrogen atom showing a slight positive charge.

PROPERTIES OF WATER

No sprint through basic chemistry is complete unless it leads us to the collection of molecules called water. Life originated in water. Organisms still live in it, or they cart water around with them in cells and tissue spaces. Many metabolic reactions require water as a reactant. Cell shape and internal structure depend on it. These topics will repeatedly occupy our attention in the book, so you may find it useful to become familiar with the following points about water's properties.

Polarity of the Water Molecule

A water molecule, remember, has no net charge, but the charge it does carry is unevenly distributed. Its electron arrangements and bond angles make its oxygen "end" a bit negative and its hydrogen end a bit positive, as in Figure 2.12*a*. Because of the polarity resulting from this charge distribution, one water molecule attracts and hydrogen-bonds with others (Figure 2.12*b,c*).

Sugars and other polar molecules easily hydrogen-bond with a water molecule; its polarity attracts them.

They are **hydrophilic** (water-loving) **substances**. That same polarity repels oils and other nonpolar molecules, which are **hydrophobic** (water-dreading) **substances**. Shake a bottle full of water and salad oil, then set it on a table. New hydrogen bonds soon replace those broken when you shook the bottle. As water molecules reunite, they push oil molecules aside, forcing them to cluster as oil droplets or as an oily film at the water's surface.

A thin, oily membrane separates water inside a cell from water outside it. Membrane organization, hence life itself, starts with both hydrophilic and hydrophobic interactions, as Sections 4.1 and 5.1 describe.

Water's Temperature-Stabilizing Effects

Cells consist mainly of water. They release a lot of heat energy by metabolism. If it weren't for water's hydrogen bonds, they might cook in their own juices. To see why, start with these facts: Each molecule vibrates nonstop, and its motion increases as it absorbs heat. **Temperature** is a measure of molecular motion. Compared with most

slight negative
charge at this end

The + and – balance each other; the whole molecule carries no net charge, overall

a   slight positive charge at this end

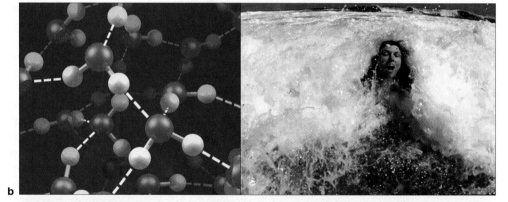

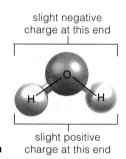

Figure 2.12 Water, a substance vital for life. (**a**) A water molecule's polarity. (**b**) Hydrogen bonding pattern among water molecules in liquid water. Dashed lines signify H bonds.

(**c**) The hydrogen bonding pattern of ice. Below 0°C, each water molecule hydrogen-bonds to four others in a three-dimensional lattice. Molecules are farther apart than they would be in liquid water at room temperature, when molecular motion is greater and not as many bonds form. Ice floats because it has fewer molecules than the same volume of liquid water; its lattice is less dense.

The Arctic Ocean's ice cap is melting. At current rates of melting, it will be gone in fifty years. So will polar bears. Already their seal-hunting season is shorter; the bears are thinner and producing fewer cubs.

b

c

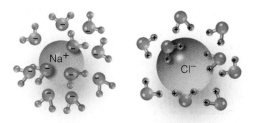

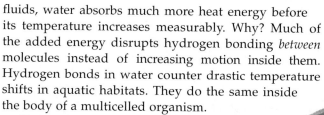

Figure 2.13 Water's cohesion. (**a**) A water strider uses its fine, water-resistant leg hairs and water's high surface tension to scoot on water. (**b**) Cohesion, and evaporation from leaves, pulls water from roots to leaves.

Figure 2.14 Two spheres of hydration.

fluids, water absorbs much more heat energy before its temperature increases measurably. Why? Much of the added energy disrupts hydrogen bonding *between* molecules instead of increasing motion inside them. Hydrogen bonds in water counter drastic temperature shifts in aquatic habitats. They do the same inside the body of a multicelled organism.

Even when liquid water's temperature is not shifting much, hydrogen bonds constantly break but form again just as fast. By contrast, large energy inputs increase molecular motion so much that hydrogen bonds stay broken, so individual molecules at the water's surface escape into air. By this process, called **evaporation**, heat energy converts liquid water to the gaseous state. The energy input overcomes the attractive forces between molecules of liquid water, molecules break free, and surface temperature of the water is lowered.

Evaporative water loss is one of the main ways in which you, and some other mammals, cool off when you sweat on hot, dry days. Sweat is about 99 percent water, and it evaporates from your skin.

Below 0°C, hydrogen bonds resist breaking and lock water molecules in the latticelike bonding pattern of ice (Figure 2.12c). Ice is less dense than water. During winter freezes, ice sheets may form near the surface of ponds, lakes, and streams. The ice blanket "insulates" liquid water beneath it and helps protect many fishes, frogs, and other aquatic organisms against freezing.

Water's Cohesion

Life also depends on water's cohesion. **Cohesion** means something has a capacity to resist rupturing when placed under tension—that is, stretched—as by the weight of a bug's legs (Figure 2.13a). Think of a lake or some other body of liquid water. Uncountable hydrogen bondings exert a continual inward pull on molecules of water at or near the surface. The bonding causes a high surface tension. You can see evidence of this when, say, flying insects splat against water and float on it.

Cohesion operates inside organisms, as well. Plants, for example, require nutrient-laden water for growth and metabolism. Largely because of cohesion, narrow columns of liquid water move up through the pipelines of vascular tissues. Such tissues extend from roots to all leaves and all other growing parts of plants. On sunny days, water evaporates from leaves as molecules break free and diffuse into the air, as in Figure 2.13b. Hydrogen bonding pulls up more molecules into leaf cells as replacements, in ways you'll read about in Section 30.3.

Water's Solvent Properties

Finally, water is an excellent solvent, for ions and polar molecules easily dissolve in it. A dissolved substance is known as a **solute**. In general, we say that a substance is *dissolved* after water molecules cluster around ions or molecules of it and thereby keep them dispersed in fluid. Such clusters are called "spheres of hydration." Spheres of hydration form around solutes in cellular fluid, the sap of maple trees, blood, fluid in your gut, and every other fluid associated with life.

Watch this happen when you pour some table salt (NaCl) into a cup of water. After a while, the crystals of salt separate into Na^+ and Cl^-. Each Na^+ attracts the negative end of some water molecules at the same time that Cl^- attracts the positive end of others (Figure 2.14). The spheres of hydration formed in this way keep the ions dispersed in the fluid.

A water molecule has no net charge, but it is slightly polar owing to an uneven distribution of charge. Its oxygen "end" is a bit negative and its hydrogen "end" is a bit positive.

Polarity allows water molecules to hydrogen-bond to each other and to other polar (hydrophilic) substances. Water molecules tend to repel nonpolar (hydrophobic) substances.

Water has temperature-stabilizing effects, internal cohesion, and a capacity to dissolve many substances. These properties influence the structure and functioning of organisms.

ACIDS, BASES, AND BUFFERS

A great variety of ions dissolved in the fluids inside and outside cells influence cell structure and functioning. Among the most influential are **hydrogen ions**, or H^+. Hydrogen ions are the same thing as free (unbound) protons. They have far-reaching effects largely because they are chemically active and there are so many of them.

The pH Scale

At any instant in liquid water, some water molecules are breaking apart into hydrogen ions and hydroxide ions (OH^-). This ionization of water is the basis of the **pH scale**, as in Figure 2.15. Biologists use this scale when measuring the H^+ concentration of seawater, tree sap, blood, and other fluids. Pure water (not rainwater or tapwater) always has just as many H^+ as OH^- ions. This condition, which also may occur in other fluids, signifies neutrality on the pH scale.

Neutrality is calculated to be 7. Each change by one unit from neutrality corresponds to a tenfold increase or decrease in H^+ concentration. Most commonly, the pH scale ranges from 0 (the highest H^+ concentration) to 14 (the lowest). *The greater the H^+ concentration, the lower the pH).* An easy way to sense that range is to dip your tongue in baking soda (pH 9), water (7), then lemon juice (2.3).

How Do Acids Differ From Bases?

When they dissolve in water, substances called **acids** *donate* protons (H^+) to the water solution. By contrast, substances called **bases** *accept* H^+ when dissolved in water. *Acidic* solutions, such as lemon juice, gastric fluid, and coffee, release H^+; their pH is below 7. *Basic* solutions, which include seawater, baking soda, and egg white, take up or combine with H^+. Basic solutions are also called "alkaline" fluids, with a pH above 7.

The fluid inside most human cells is about 7 on the pH scale. Most fluids bathing these cells have values between 7.3 and 7.5. This also is the case for the fluid portion of human blood. Seawater is more alkaline than the body fluids of organisms that live in it.

Most acids are weak or strong. Weak acids, such as carbonic acid (H_2CO_3), are reluctant H^+ donors. But strong acids readily give up H^+ when they dissociate in water. Examples are hydrochloric acid (HCl), nitric acid (HNO_3), and sulfuric acid (H_2SO_4).

Sniff and eat spicy food. Swallowing sends it onward to gastric fluid in your stomach. The meal stimulates cells of the stomach's lining to secrete a strong acid, HCl, that dissociates into H^+ and Cl^-. The ions increase the H^+ concentration, thereby making the gastric fluid more acidic. Increased acidity activates enzymes that digest the food's proteins and attack most bacteria that may have lurked in or on the food. When people eat too much spicy food, they may get an *acid stomach* and reach for an antacid, such as milk of magnesia. This is a base. When it dissolves, it releases magnesium ions and OH^- to neutralize the acid. OH^- combines with excess H^+ in gastric fluid, which reduces the H^+ concentration and thereby calms things down.

High concentrations of strong acids or bases also disrupt the external environment and pose dangers to life. Read the labels on bottles of ammonia, drain cleaner, or other products commonly stored in households. Many cause severe *chemical burns*. So does the sulfuric acid in car batteries.

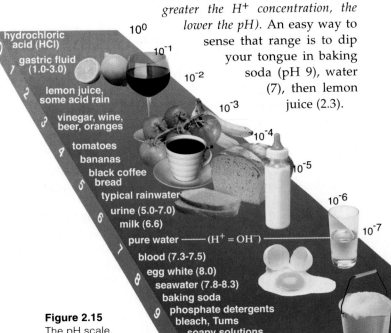

hydrochloric acid (HCl)
gastric fluid (1.0–3.0)
lemon juice, some acid rain
vinegar, wine, beer, oranges
tomatoes
bananas
black coffee
bread
typical rainwater
urine (5.0–7.0)
milk (6.6)
pure water ——— ($H^+ = OH^-$) ———
blood (7.3–7.5)
egg white (8.0)
seawater (7.8–8.3)
baking soda
phosphate detergents
bleach, Tums
soapy solutions, milk of magnesia
household ammonia (10.5–11.9)
hair remover
oven cleaner
sodium hydroxide (NaOH)

10^0
10^{-1}
10^{-2}
10^{-3}
10^{-4}
10^{-5}
10^{-6}
10^{-7}
10^{-8}
10^{-9}
10^{-10}
10^{-11}
10^{-12}
10^{-13}
10^{-14}

0 1 2 3 4 5 6 7 8 9 10 11 12 13 14

Figure 2.15
The pH scale, which represents concentrations of hydrogen ions as measured in 1 liter of a solution. Also shown are the approximate pH values for some solutions. The pH scale ranges from 0 (most acidic) to 14 (most basic). A change of 1 on the scale means a tenfold change in H^+ concentration.

The burning of fossil fuels—as by power plants and cars—and nitrogen fertilizers release strong acid that alter the pH of rain (Figure 2.16). Certain regions are sensitive to this *acid rain* because of their soil type and vegetation cover. The modified chemistry of habitats in such regions can harm organisms. We will return to this topic in Section 50.1.

Buffers Against Shifts in pH

Metabolic reactions are sensitive to even slight shifts in pH, for H^+ and OH^- can combine with many different molecules and alter their functions. Normally, control mechanisms respond to shifts in pH, as they do when HCl enters the stomach in response to a meal. Many of the controls involve buffer systems.

A **buffer system** is a partnership between a weak acid and the weak base that forms as the acid dissolves in water. The two work as a pair to counter slight shifts in pH. Remember, the OH^- level rises when any strong base enters a fluid. But a weak acid can neutralize part of the incoming OH^- by donating H^+. The weak acid's partner (a weak base) forms as a result of its action. Later, if a strong acid floods in, the weak base will accept H^+ from it and become the buffer system's acid.

Bear in mind, the action of a buffer system cannot make new hydrogen ions or eliminate ones that already are present. It can only bind or release them.

In all complex, multicelled organisms, diverse buffer systems operate in the internal environment—in blood and tissue fluids. For example, metabolic reactions in lungs and kidneys help control the acid–base balance of this environment, at levels suitable for life (Chapters 40 and 41). For now, simply think of what happens when the blood level of H^+ decreases and the blood is not as acidic as it should be. At such times, carbonic acid that is dissolved in blood releases H^+ and so becomes the partner base, bicarbonate:

$$H_2O + CO_2 \longrightarrow H_2CO_3 \longrightarrow HCO_3^- + H^+$$

WATER CARBON DIOXIDE   CARBONIC ACID   BICARBONATE

When blood becomes more acidic, more H^+ becomes bound to the base, thus forming the partner acid:

$$HCO_3^- + H^+ \longrightarrow H_2CO_3 \longrightarrow H_2O + CO_2$$

BICARBONATE   CARBONIC ACID   WATER CARBON DIOXIDE

Any buffer system can mop up only so much extra acid or base. Once its capacity is exceeded, a system's pH destabilizes quickly. Uncontrolled shifts in pH have bad effects. If blood's pH (7.3–7.5) declines even to 7,

Figure 2.16 Sulfur dioxide emissions from a coal-burning power plant. Airborne pollutants such as sulfur dioxide dissolve in water vapor to form acidic solutions. They are a major component of acid rain.

this invites a *coma*, a sometimes irreversible state of unconsciousness. An increase to 7.8 can lead to *tetany*, a potentially lethal condition in which skeletal muscles contract uncontrollably. With *acidosis*, carbon dioxide builds up in blood, too much carbonic acid forms, and blood pH plummets. With *alkalosis*, a rise in blood pH cannot be corrected. Both conditions can be lethal.

Salts

Salts are compounds that release ions *other than* H^+ and OH^- in solution. Salts and water often form when an acid and a base interact. Depending on a solution's pH, salts are able to dissolve easily. Consider how sodium chloride forms:

$$HCl \text{ (acid)} + NaOH \text{ (base)} \longrightarrow NaCl \text{ (salt)} + H_2O$$

HYDROCHLORIC ACID   SODIUM HYDROXIDE   SODIUM CHLORIDE

$$Na^+ \quad Cl^- \text{ (ionization)}$$

Many salts dissolve into ions that play key roles in cells. For example, nerve cell activity depends on ions of sodium, potassium, and calcium. Your muscles contract with the help of calcium ions. And water absorption by plant cells depends largely on potassium ions.

Hydrogen ions (H^+) and other ions dissolved in the fluids inside and outside cells affect cell function.

When dissolved in water, acidic substances release H^+, and basic (alkaline) substances accept them. Certain acid–base interactions help maintain the pH value of a fluid—that is, its H^+ concentration—by a buffering action.

A buffer system counters slight shifts in pH by releasing hydrogen ions when their concentration is too low or by combining with them when the concentration is too high.

Salts are compounds that release ions other than H^+ and OH^-, and many of those ions have key roles in cell functions.

SUMMARY

1. Chemistry helps us understand the nature of all the substances that make up cells, organisms, and the Earth, its waters, and the atmosphere. Each substance consists of one or more elements. Of the ninety-two naturally occurring elements, the most common in organisms are oxygen, carbon, hydrogen, and nitrogen. Organisms also have lesser amounts of many other elements, such as calcium, phosphorus, potassium, and sulfur. *CI*

2. Table 2.2 summarizes some key chemical terms that you will encounter throughout this book. *CI, 2.1–2.6*

3. Elements consist of atoms. An atom has one or more positively charged protons plus an equal number of negatively charged electrons and (except for hydrogen atoms) one or more uncharged neutrons. The protons and neutrons occupy the core region, or the atomic nucleus. Most elements have isotopes, which are two or more forms of atoms that have the same number of protons but different numbers of neutrons. *2.1*

4. Whether an atom interacts with others depends on the number and arrangement of its electrons, which occupy orbitals (volumes of space) in a series of shells around the atomic nucleus. When an atom has unfilled orbitals in its outermost shell, it tends to bond with atoms of other elements. *2.3*

5. An atom may lose or gain one or more electrons and thus become an ion, which has an overall positive or negative charge. *2.4*

6. Generally, a chemical bond is a union between the electron structures of atoms. *2.3*

   a. In an ionic bond, a positive ion and negative ion stay together because of the mutual attraction of their opposite charges. *2.4*

   b. Atoms often share one or more pairs of electrons in single, double, and triple covalent bonds. Electron sharing is equal in nonpolar covalent bonds, and it is unequal in polar covalent bonds. The interacting atoms have no net charge overall; the bond is slightly negative at one end and slightly positive at the other. *2.4*

   c. In a hydrogen bond, one covalently bonded atom (such as oxygen) that has a slight negative charge is weakly attracted to the slight positive charge of a hydrogen atom that is taking part in a different polar covalent bond. *2.4*

7. Polar covalent bonds join together three atoms in a water molecule (two hydrogens and one oxygen). The polarity of a water molecule invites hydrogen bonding between water molecules. Such hydrogen bonding is the basis of liquid water's ability to resist temperature changes more than other fluids do, display internal cohesion, and easily dissolve polar or ionic substances. These properties greatly affect the metabolic activity, structure, shape, and internal organization of cells. *2.5*

8. By the pH scale, a solution is assigned a number that reflects its H^+ concentration. A typical range is from 0 (highest concentration) to 14 (lowest). At pH 7, the H^+ and OH^- concentrations are equal. Acids release H^+ in water, and bases combine with them. Buffer systems help maintain the pH values of blood, tissue fluids, and the fluid inside cells. *2.6*

Table 2.2	*Summary of Important Players in the Chemical Basis of Life*
ELEMENT	Fundamental form of matter that occupies space, has mass, and cannot be broken apart into a different form of matter by ordinary physical or chemical means.
ATOM	Smallest unit of an element that still retains the characteristic properties of that element.
Proton (p^+)	Positively charged particle of the atomic nucleus. All atoms of an element have the same number of protons, which is the atomic number. A proton without an electron zipping around it is a hydrogen ion (H^+).
Electron (e^-)	Negatively charged particle that can occupy a volume of space (orbital) around an atomic nucleus. Electrons can be shared or transferred among atoms.
Neutron	Uncharged particle of the nucleus of all atoms except hydrogen. For a given element, the mass number is the number of protons and neutrons in the nucleus.
MOLECULE	Unit of matter in which two or more atoms of the same element, or different ones, are bonded together.
Compound	Molecule composed of two or more different elements in unvarying proportions. Water is an example.
Mixture	Intermingling of two or more elements in proportions that can and usually do vary.
ISOTOPE	One of two or more forms of an element's atoms that differ in their number of neutrons.
Radioisotope	Unstable isotope, having an unbalanced number of protons and neutrons, that emits particles and energy.
Tracer	Molecule of a substance to which a radioisotope is attached. Together with tracking devices, it is used to identify movement or destination of the substance in a metabolic pathway, the body, or some other system.
ION	Atom that has gained or lost one or more electrons, thus becoming negatively or positively charged.
SOLUTE	Any molecule or ion dissolved in some solvent.
Hydrophilic substance	Polar molecule or molecular region that can readily dissolve in water.
Hydrophobic substance	Nonpolar molecule or molecular region that strongly resists dissolving in water.
ACID	Substance that donates H^+ when dissolved in water.
BASE	Substance that accepts H^+ when dissolved in water.
	Compound that releases ions other than H^+ or OH^- when dissolved in water.

Review Questions

1. What is an element? Name four elements (and their symbols) that make up more than 95 percent of the body weight of all living organisms. *CI, 2.1*

2. Define atom, isotope, and radioisotope. *2.1*

3. How many electrons can occupy each orbital around an atomic nucleus? Using the shell model, explain how the orbitals available to electrons are distributed in an atom. *2.3*

4. Define molecule, compound, and mixture. *2.3*

5. Distinguish between:
   a. ionic and hydrogen bonds *2.4*
   b. polar and nonpolar covalent bonds *2.4*
   c. hydrophilic and hydrophobic interactions *2.5*

6. If a water molecule has no net charge, then why does it attract polar molecules and repel nonpolar ones? *2.5*

7. Label the atoms in each water molecule in the sketch at right. Indicate which parts of each molecule carry a slight positive charge (+) and which carry a slight negative charge (−). *2.5*

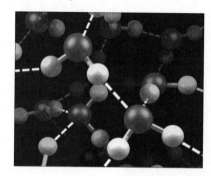

8. Define acid and base. Then describe the behavior of a weak acid in solutions having a high or low pH value. *2.6*

Self-Quiz   ANSWERS IN APPENDIX III

1. Electrons carry a _____ charge.
   a. positive   b. negative   c. zero

2. Atoms share electrons unequally in a(n) _____ bond.
   a. ionic   c. polar covalent
   b. nonpolar covalent   d. hydrogen

3. An individual water molecule shows _____ .
   a. polarity   d. spherelike hydration
   b. hydrogen-bonding capacity e. a and b
   c. notable heat resistance   f. all of the above

4. In liquid water, spheres of hydration form around _____ .
   a. nonpolar molecules   d. solvents
   b. polar molecules   e. b and c
   c. ions   f. all of the above

5. Hydrogen ions (H⁺) are _____ .
   a. the basis of pH values   d. dissolved in blood
   b. unbound protons   e. both a and b
   c. targets of certain buffers f. all of the above

6. When dissolved in water, a(n) _____ donates H⁺; however, a(n) _____ accepts H⁺.

7. Match the terms with the most suitable descriptions.
   _____ trace element   a. weak acid and its partner base
   _____ buffer system   work as a pair to counter pH shifts
   _____ chemical bond   b. union between electron structures
   _____ temperature   of two atoms
                          c. less than 0.01% of body weight
                          d. measure of molecular motion
                             in some defined region

Critical Thinking

1. An ionic compound forms when calcium combines with chlorine. Referring to Figure 2.8, give the compound's formula. (*Hint:* Be sure the outermost shell of each atom is filled.)

2. David, an inquisitive three-year-old, poked his fingers into warm water inside a metal pan on the stove, then touched the pan and got a nasty burn. Explain why water in a metal pan heats up far more slowly than the pan itself.

3. When molecules absorb microwaves, which are a form of electromagnetic radiation, they move more rapidly. Explain why a microwave oven can heat foods.

4. From what you know about cohesion, devise a hypothesis to explain why water forms droplets.

5. Edward is trying to study a chemical reaction that an enzyme catalyzes (speeds up). H⁺ forms in the reaction, but the enzyme is destroyed at low pH. What can he include in his reaction mix to protect the enzyme while he studies the reaction?

6. Through interactions with water and ions, a soluble protein becomes dispersed in cellular fluid. An electrically charged cushion around it makes this happen. Using the sketch at right as a guide, explain the chemical interactions through which this cushion forms. Start with the major bonds in the protein molecule.

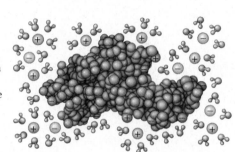

7. Reactants and products often are expressed in moles. A *mole* is a certain number of atoms or molecules of any substance, just as "a dozen" can refer to any twelve cats, roses, and so forth. Molar weight, in grams, equals the total atomic weight of all atoms making up that substance. For example, the atomic weight of carbon is 12, so one mole of carbon weighs 12 grams. A mole of oxygen (atomic weight 16) weighs 16 grams.

   State why a mole of water (H_2O) weighs 18 grams, and why a mole of glucose ($C_6H_{12}O_6$) weighs 180 grams.

Selected Key Terms

acid *2.6*	evaporation *2.5*	molecule *2.3*
atom *2.1*	hydrogen bond *2.4*	neutron *2.1*
base *2.6*	hydrogen ion (H⁺) *2.6*	pH scale *2.6*
buffer system *2.6*	hydrophilic	proton *2.1*
chemical bond *2.3*	substance *2.5*	radioisotope *2.1*
cohesion *2.5*	hydrophobic	salt *2.6*
compound *2.3*	substance *2.5*	shell model *2.3*
covalent bond *2.4*	ion *2.4*	solute *2.5*
electron *2.1*	ionic bond *2.4*	temperature *2.5*
element *CI*	isotope *2.1*	tracer *2.2*
	mixture *2.3*	

Readings

Ritter, P. 1996. *Biochemistry: A Foundation.* Pacific Grove, California: Brooks/Cole. Good, easy-to-read introduction to biochemistry, with plenty of human-interest applications.

CARBON COMPOUNDS IN CELLS

Carbon, Carbon, In the Sky—Are You Swinging Low and High?

Maybe you've already heard that the past decade was the warmest ever recorded. Maybe you know that a long-term rise in the lower atmosphere's temperature—a **global warming**—is under way. It is something you might want to worry about. In 2001, for example, rare thunderstorms pelted Alaska and 27 inches of rainfall in twenty-four hours washed homes away in Hawaii.

If, as predicted, the atmospheric temperature spikes even 2.5 degrees higher in this century, the sea level may rise enough to permanently flood coastal cities. Changes in rainfall patterns over broad regions may give new meaning to the expressions "crop failure" and "water wars." The already bad flooding and mudslides in the tropics and California will get worse, interiors of the continents will become drier, and deserts will expand.

And what about organisms other than ourselves? Think about the trees of those scenic forests in the high mountains of the Pacific Northwest (Figure 3.1). The trees don't do much during the winter, because the water required for growth is locked away in the form of snow and ice. The trees entered dormancy during autumn's cool, dry days; their metabolic activities idled, and growth ceased. Only water already inside them kept their living cells from dying. Only when spring arrives do the trees break dormancy. As temperatures rise and snow melts, mineral-laden water becomes available. And now photosynthetic cells take carbon dioxide from the air inside leaves and make sugars, starches, and other carbon-based compounds.

Figure 3.1 Above, a forest of conifers beneath the first snows of winter on Silver Star Mountain. The form, function, and very survival of these trees—and of people, bears, and all other organisms—start with the carbon atom and its diverse molecular partners in organic compounds. Even the gasoline and other fossil fuels that we humans have come to depend on had their beginning in carbon atoms used millions of years ago, by the growing trees of ancient forests.

Look closely and you will find the same connection between air temperature and photosynthesis all over the planet. Ultimately it affects you and all other kinds of organisms that can or can't produce their own food.

Here is something else to think about. The carbon dioxide concentration in the air declines in spring and summer when photosynthesizers take up stupendous amounts of it. It rises in autumn, when photosynthesis declines and decomposers go to work. The wastes and remains of photosynthesizers feed decomposers and fan their population growth. Collectively, decomposers release huge amounts of carbon dioxide as a metabolic by-product.

Now researchers find that plants of the northern forests and the great plains of Canada, the United States, and elsewhere are breaking dormancy earlier than they did two decades ago. Also, the atmospheric concentration of carbon dioxide is swinging more than before—as much as 20 percent more in Hawaii and a whopping 40 percent in Alaska. Global warming is probably promoting longer growing seasons, hence more uptake of carbon dioxide and the wider swings.

We don't know why the air is warming up, but a long-term rise in carbon dioxide levels in the air is accompanying it. We do know that human populations burn huge amounts of gas, coal, and other carbon-rich fossil fuels for energy. The carbon dioxide that burning releases actually may be part of the problem, as you will read in Section 48.9.

The point is this: *Carbon permeates the world of life—from the energy-requiring activities and structural organization of cells, to physical and chemical conditions that span the globe and influence ecosystems everywhere.*

With this chapter, we turn to life-giving properties that emerge from the molecular structure of carbon-rich compounds. Study the chapter well, including the summary in Section 3.10. It will serve as a foundation for understanding how diverse organisms put together such compounds, how they use them, and how the effects of all those uses ripple through the biosphere.

Key Concepts

1. Organic compounds are molecules containing carbon and at least one hydrogen atom. We define cells partly by their capacity to assemble the organic compounds called carbohydrates, lipids, proteins, and nucleic acids. These are the molecules of life.

2. Cells assemble biological molecules from pools of smaller organic compounds, including simple sugars, fatty acids, amino acids, and nucleotides.

3. Glucose and other simple sugars are carbohydrates. So are oligosaccharides, short chains of covalently bonded sugar units. The most complex carbohydrates are the polysaccharides, many of which consist of hundreds or thousands of sugar units.

4. Lipids are greasy or oily compounds that show little tendency to dissolve in water but dissolve easily in nonpolar compounds, such as other lipids. Neutral fats (triglycerides), phospholipids, waxes, and sterols are important lipids.

5. Cells use carbohydrates and lipids as building blocks and as major sources of energy.

6. Proteins show the greatest diversity in structure and function. Many serve as construction materials in cells. Many others are enzymes, a class of molecules that hugely increase the rate of specific metabolic reactions. Other proteins transport substances in cells or across membranes, contribute to cell movements, trigger changes in cell activities, and defend the body against injury and disease.

7. ATP and other kinds of nucleotides have essential roles in metabolism. DNA and RNA are strandlike nucleic acids assembled from nucleotide units. They are the basis of inheritance and reproduction.

THE MOLECULES OF LIFE—FROM STRUCTURE TO FUNCTION

What Is An Organic Compound?

Under present-day conditions on Earth, *only living cells synthesize complex carbohydrates, lipids, proteins, and nucleic acids.* These are the molecules that typify life. Different classes of biological molecules are the cell's instant energy sources, structural materials, metabolic workers, cell-to-cell signals, and libraries of hereditary information. As you will see, their three-dimensional shape influences how each functions—and that shape is influenced by how a molecule's atoms are arranged and how electric charge is distributed among them.

Molecules of life are **organic compounds**, which we define as containing the element carbon and at least one hydrogen atom. The term is a holdover from a time when chemists thought "organic" substances were the ones they got from animals and vegetables, as opposed to "inorganic" substances they got from minerals. The term persists even though scientists now make organic compounds in laboratories. It persists even though we now have reason to believe that organic compounds were present on Earth before organisms were.

The **hydrocarbons** consist only of hydrogen atoms covalently bonded to carbon. Gasoline and other fossil fuels are examples. Like other organic compounds, they have a specific number of atoms arranged in specific ways. Also, organic compounds have one or more **functional groups**, which are particular atoms or clusters of atoms covalently bonded to carbon.

In this book we adhere to a standardized color code for the main atoms of organic compounds:

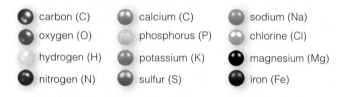

carbon (C)   calcium (C)   sodium (Na)
oxygen (O)   phosphorus (P)   chlorine (Cl)
hydrogen (H)   potassium (K)   magnesium (Mg)
nitrogen (N)   sulfur (S)   iron (Fe)

It All Starts With Carbon's Bonding Behavior

Living organisms consist mainly of oxygen, hydrogen, and carbon (Figure 2.1). The oxygen and hydrogen are mainly in the form of water. Put aside the water, and carbon makes up more than half of what's left.

Carbon's importance in life arises from its versatile bonding behavior. *Each carbon atom can covalently bond with as many as four other atoms.* Such bonds, in which two atoms share one, two, or three pairs of electrons, are relatively stable. They commonly join carbon atoms together as backbones to which hydrogen, oxygen, and other elements are attached. Such bonds are the start of the three-dimensional shapes of organic compounds.

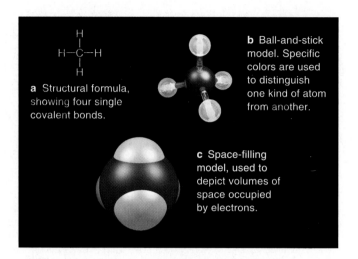

Figure 3.2 Molecular models for methane (CH_4), the simplest organic compound.

Ways of Representing Organic Compounds

Methane is the simplest organic compound of all. This colorless, odorless gas is abundant in the atmosphere, marine sediments, termite colonies, stagnant swamps, and stockyards. Its four hydrogen atoms are covalently bonded to a carbon atom (CH_4). Figure 3.2 shows ways to represent it. A ball-and-stick model is used to depict bond angles and convey how the molecule's mass is distributed (in atomic nuclei). The space-filling model is better for conveying a molecule's size and surfaces.

Now let's use the ball-and-stick model to depict an organic compound with six covalently bonded carbon atoms from which hydrogen *and* oxygen atoms project:

ball-and-stick model for the linear structure of glucose

In cells, this type of carbon backbone sometimes forms chains. But most of the time it coils back on itself, and its two ends connect to form a ring structure:

six-carbon ring structure of glucose that usually forms inside cells

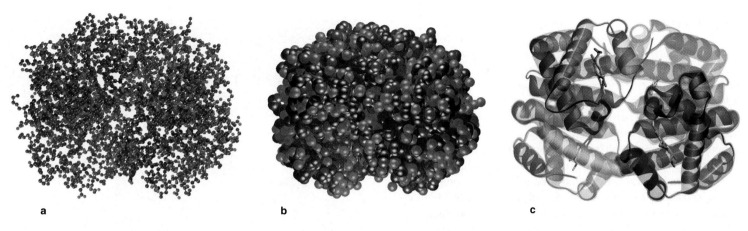

a   b   c

Figure 3.3 Visualizing the structure of hemoglobin, the oxygen-transporting molecule in red blood cells. (**a**) Ball-and-stick model, (**b**) space-filling model, and (**c**) ribbon model. Unlike the color coding for atoms, colors used for ribbon models and simple icons for complex molecules vary, depending on the context.

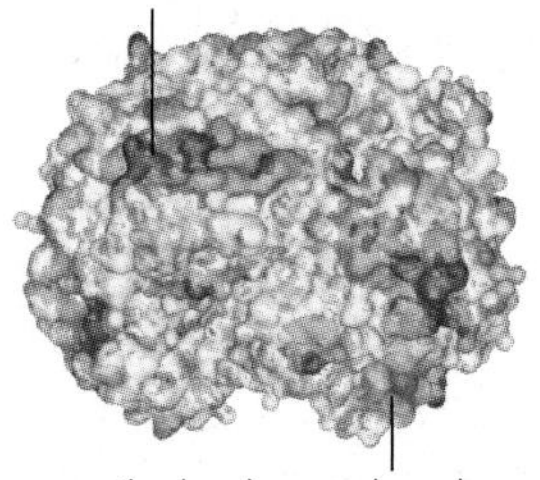

positively charged region

Figure 3.4 Model for the electrostatic potential energy at the surface of part of a hemoglobin molecule.

negatively charged region

We often depict carbon ring structures in simpler ways. A flat structural model may show the carbons but not other atoms bonded to it. If an icon for the ring shows no atoms at all, it is understood that one carbon atom occupies each "corner" of the ring:

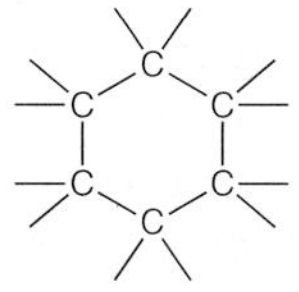

simplified structural formula for a six-carbon ring

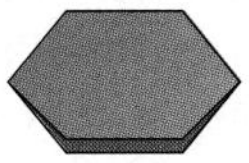

icon for a six-carbon ring

Figure 3.3 shows different ways of representing a much larger molecule: the protein hemoglobin. Like all all other vertebrates, your life depends on hemoglobin, which transports oxygen to tissue neighborhoods of all living cells in your body. The ball-and-stick and the space-filling models give you an idea of this molecule's mass and structural complexity. But neither tells you much about its oxygen-transporting function.

Now look at Figure 3.3*c*. This ribbon model shows how the hemoglobin molecule actually consists of four chains. As you will read later, each chain is a string of subunits known as amino acids. Certain regions of each chain are straight or folded, others are coiled. For now, it is enough to know that the three-dimensional shape of hemoglobin includes four pockets, each containing a small cluster of atoms known as a heme group (coded *red-orange* in this figure). Each heme group is able to bind and then release an oxygen molecule in response to differences in local tissue conditions.

More sophisticated models are now in use. Some computer models, for instance, show local differences in electric charge across molecular surfaces. Areas color-coded, say, *red* on the surface of one molecule

would be attractive to a *blue* surface on another part of the same molecule or on a different one (Figure 3.4).

Ultimately, such insights into the three-dimensional structure of molecules help us to understand how cells and multicelled organisms function. For instance, virus particles can infect a cell when they dock at specific proteins at the plasma membrane. Like Lego blocks, the proteins have ridges, clefts, and charged regions at their surface that are complementary to ridges, clefts, and charged regions of a protein at the surface of the virus particle. Design a drug molecule that matches up with a viral protein, figure out how to deliver enough copies of it into a patient, and a lot of virus particles may bind to the decoys instead of infecting body cells.

Throughout this book, you will be coming across different kinds of molecular models. In each case, the model selected gives you a glimpse into the structure and function of the molecule being described.

Carbohydrates, lipids, proteins, and nucleic acids are the main biological molecules, the organic compounds that only living cells assemble under present-day conditions in nature.

Organic compounds have diverse, three-dimensional shapes and functions that start with their carbon backbone and the bonding arrangements that arise from it.

Insights into the structure of molecules ultimately help us understand how cells, and multicelled organisms, function.

OVERVIEW OF FUNCTIONAL GROUPS

As you just read in the preceding section, functional groups are lone atoms or clusters of atoms covalently bonded to carbon atoms of organic compounds. Each has specific chemical and physical properties that are consistent from one molecule to the next. Compared to hydrocarbon regions, they are more reactive. Important features of carbohydrates, lipids, proteins, and nucleic acids arise from the number, kind, and arrangement of functional groups, such as those shown in Figure 3.5.

For example, sugars in your diet belong to a class of organic compounds called **alcohols**, which have one or more *hydroxyl* groups (—OH). Small alcohols dissolve fast because water molecules hydrogen-bond with these groups. Larger ones don't; they have long hydrocarbon

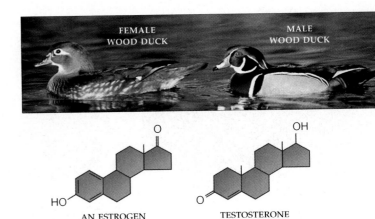

Figure 3.6 Observable differences in traits between the male and female wood duck (*Aix sponsa*). Two sex hormones govern the development of feather color and other traits that help males and females recognize each other and so influence reproductive success. Both hormones—testosterone and one of the estrogens—have the same carbon ring structure. They differ in the position of functional groups attached to the ring.

chains. As you will see shortly, enzyme action can split molecules and join them together at hydroxyl groups. By contrast, hydrocarbons are water insoluble. Chains of them occur in fatty acids, which is why lipids with fatty acid tails resist dissolving in water.

Carbonyl groups are highly reactive and prone to electron transfers. They are building blocks of fats and carbohydrates. *Carboxyl* groups are present in amino acids, fatty acids, and other important molecules. The *phosphate* group has oxygen atoms that form covalent bonds. It dictates ATP's energy-carrying function, and it also combines with sugars to form the backbones of DNA and RNA. As you will see, the *sulfhydryl* group, a component of the amino acid cysteine, helps stabilize the structure of many proteins.

How much can one functional group do? Consider a seemingly minor difference in the functional groups of two structurally similar sex hormones (Figure 3.6). Early on, an embryo of a wood duck, human, or any other vertebrate is neither male nor female; it just has a set of tubes and ducts that can develop either way. If an embryo starts making testosterone, that hormone will drive development of the tubes and ducts into male sex organs and, later, govern the emergence of male traits such as feather color or hairiness. Female sex organs form only in the *absence* of testosterone. Estrogens will guide the development of distinctively female traits.

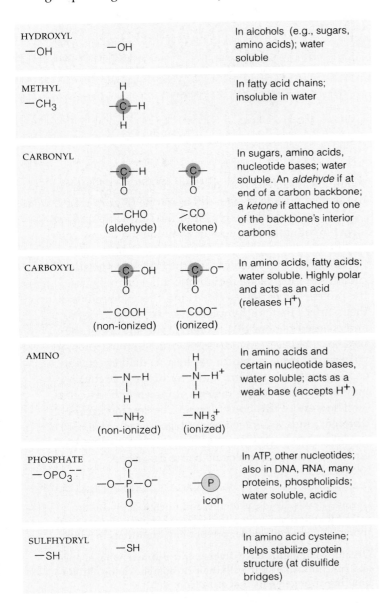

Figure 3.5 Common functional groups in biological molecules, with examples of their occurrences.

Functional groups covalently bonded to carbon backbones add enormously to the structural and functional diversity of organic compounds, cells, and multicelled organisms.

HOW DO CELLS BUILD ORGANIC COMPOUNDS?

Four Families of Building Blocks

Think back on the trees of the world's great forests, as described in the chapter introduction. These producers take carbon (from carbon dioxide), water, and sunlight (for energy) and transform them first into small organic compounds. Simple sugars, fatty acids, amino acids, and nucleotides are the four major families of these small compounds. Each family includes many kinds of molecules that contain two to thirty-six carbon atoms, at the most.

Cells maintain and replenish their pools of small organic compounds, which collectively account for only about 10 percent of the organic material inside a cell. They continually withdraw some molecules as sources of energy. They withdraw others for use as individual subunits, or **monomers**, of larger molecules that they require for their structure and functioning. These larger molecules, called **polymers**, consist of three to millions of subunits that may or may not be identical. In turn, when large molecules are broken down, their released monomers may be used at once for energy, or they may rejoin the cellular pools as free molecules.

Five Categories of Reactions

So how do cells actually do the construction work? It will take more than one chapter to sketch out answers (and best guesses) to the question. At this point, simply become aware that reactions by which a cell assembles, rearranges, and splits apart organic compounds require more than energy inputs. They also require **enzymes**, a class of proteins that make metabolic reactions proceed much faster than they would on their own. Different kinds of enzymes mediate different reactions. In later chapters, you will come across specific examples of five categories of reactions:

1. *Functional-group transfer.* One molecule gives up a functional group, which another molecule accepts.

2. *Electron transfer.* One or more electrons stripped from one molecule are donated to another molecule.

3. *Rearrangement.* A juggling of its internal bonds converts one type of organic compound into another.

4. *Condensation.* Through covalent bonding, two molecules combine to form a larger molecule.

5. *Cleavage.* A molecule splits into two smaller ones.

To get a sense of these cell activities, think of what happens in a **condensation reaction**: Enzymes split off an —OH group from one molecule and an H atom from another, and a covalent bond forms between both at

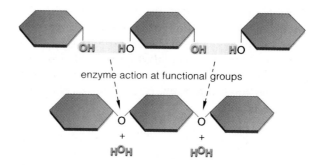

a Two condensation reactions. Enzymes remove an —OH group and H atom from two molecules, which covalently bond as a larger molecule. Two water molecules form.

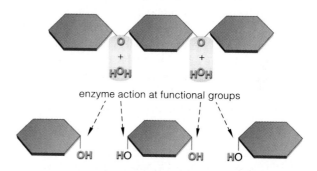

b Hydrolysis, a water-requiring cleavage reaction. Enzyme action splits a molecule into three parts, then attaches an —OH group and an H atom derived from a water molecule to each exposed site.

Figure 3.7 Examples of the types of metabolic reactions by which most biological molecules are put together, rearranged, and broken apart.

their exposed sites. The discarded atoms often form water, or H_2O (Figure 3.7a). Starch and other polymers form by way of repeated condensation reactions.

Another example: A type of cleavage reaction called **hydrolysis** is like condensation in reverse (Figure 3.7b). Enzymes split molecules at specific groups, then attach one —OH group and a hydrogen atom derived from a water molecule to the exposed sites. With hydrolysis, cells can cleave polymers into smaller molecules when these are required for building blocks or for energy.

Cells build the large molecules of life mainly from four families of small organic compounds called simple sugars, fatty acids, amino acids, and nucleotides.

Cells continually assemble, rearrange, and degrade both small and large organic compounds. They do so mainly by enzyme-mediated reactions involving the transfer of functional groups or electrons, rearrangement of internal bonds, and a combining or splitting of molecules.

3.4

CARBOHYDRATES—THE MOST ABUNDANT MOLECULES OF LIFE

Which biological molecules are most abundant? The **carbohydrates**. Most carbohydrates consist of carbon, hydrogen, and oxygen in a 1:2:1 ratio $(CH_2O)_n$. Cells use them as structural materials, transportable forms of energy, or storage forms of energy. There are three main classes, called the **monosaccharides**, **oligosaccharides**, and **polysaccharides**.

The Simple Sugars

"Saccharide" comes from a Greek word meaning sugar. A *mono*saccharide (meaning one sugar monomer) is the simplest carbohydrate. It has at least two —OH groups bonded to the carbon backbone plus an aldehyde or a ketone group. Most dissolve easily in water. Common types have a backbone of five or six carbon atoms that tends to form a ring structure when dissolved in cells or body fluids. Ribose and deoxyribose, the sugar unit of RNA and DNA, respectively, have five carbon atoms. Glucose has six (Figure 3.8a). Most organisms use it as their main energy source. Glucose also is a precursor

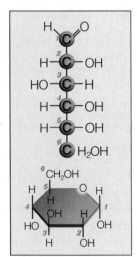
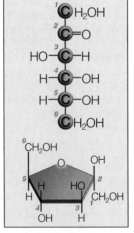

a Structure of glucose   **b** Structure of fructose

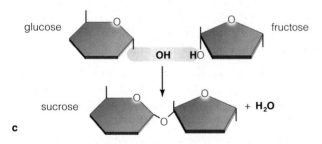

c

Figure 3.8 Straight-chain and ring forms of (**a**) glucose and (**b**) fructose. For reference purposes, carbon atoms of simple sugars are numbered in sequence, starting at the end closest to the molecule's aldehyde or ketone group. (**c**) Condensation of two monosaccharides into a disaccharide.

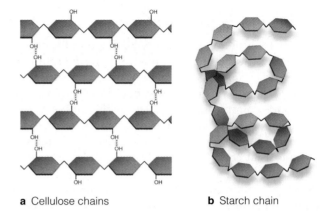

a Cellulose chains   **b** Starch chain

Figure 3.9 Bonding patterns for glucose units in cellulose and in starch. (**a**) In cellulose, bonds form between glucose chains. This pattern stabilizes the chains and allows them to become tightly bundled. (**b**) In amylose, a form of starch, bonds form between units within a chain, which coils up.

(parent molecule) of many compounds and a building block for larger carbohydrates. Vitamin C (a well-known sugar acid) and glycerol (an alcohol that has three —OH groups) are compounds made from sugar monomers.

Short-Chain Carbohydrates

Unlike the simple sugars, an *oligo*saccharide is a short chain of covalently bonded sugar monomers. (*Oligo–* means "a few.") The ones called *di*saccharides consist of only two sugar units. Lactose, sucrose, and maltose are examples. Lactose, a sugar in milk, has one glucose and one galactose unit. Sucrose, the most plentiful sugar in nature, has a glucose and a fructose unit (Figure 3.8c). Plants convert larger carbohydrates to sucrose, which is easily transported through leaves, stems, and roots. Table sugar is sucrose crystallized from sugarcane and sugarbeets. The backbone of some proteins and other large molecules often bears oligosaccharide side chains.

Complex Carbohydrates

The "complex" carbohydrates, or *poly*saccharides, are straight or branched chains of many sugar monomers (often hundreds or thousands) of the same or different types. Cellulose, starch, and glycogen, the most common polysaccharides, consist only of glucose—yet they have very different properties. Why? The answer starts with differences in covalent bonding patterns between their glucose units, which are joined together in chains.

In cellulose, many glucose chains stretch out side by side and hydrogen-bond to one another at —OH groups (Figure 3.9a). This bonding arrangement stabilizes the chains into a tightly bundled pattern that resists being digested, at least by most enzymes. Fibers of cellulose

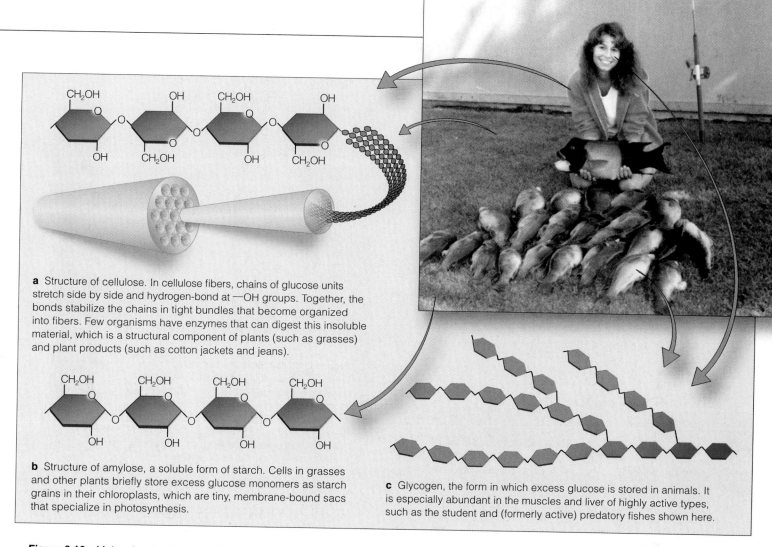

a Structure of cellulose. In cellulose fibers, chains of glucose units stretch side by side and hydrogen-bond at —OH groups. Together, the bonds stabilize the chains in tight bundles that become organized into fibers. Few organisms have enzymes that can digest this insoluble material, which is a structural component of plants (such as grasses) and plant products (such as cotton jackets and jeans).

b Structure of amylose, a soluble form of starch. Cells in grasses and other plants briefly store excess glucose monomers as starch grains in their chloroplasts, which are tiny, membrane-bound sacs that specialize in photosynthesis.

c Glycogen, the form in which excess glucose is stored in animals. It is especially abundant in the muscles and liver of highly active types, such as the student and (formerly active) predatory fishes shown here.

Figure 3.10 Molecular structure of cellulose, starch, and glycogen, and their typical locations in a few organisms. All three carbohydrates consist only of glucose units.

are a structural component of plant cell walls (Figure 3.10*a*). Like steel rods in reinforced concrete, the fibers are tough, insoluble, and resistant to weight loads and mechanical stress.

In starch, the pattern of covalent bonding puts each glucose unit at an angle relative to the next unit in line. The chain ends up coiling like a spiral staircase (Figure 3.9*b*). The coils are easily digestible. In starches that have branched chains, they are even more so. Many —OH groups project outward from the coiled chains, and this makes them readily accessible to enzymes. Plants store their photosynthetically produced sugars as large starch molecules (Figure 3.10*b*). Enzymes easily hydrolyze the starch to glucose units.

In animals, glycogen is the sugar-storage equivalent of starch in plants. Muscle and liver cells store a lot of it. When the level of sugar in blood declines, liver cells degrade glycogen, so glucose is released and enters the blood. Exercise strenuously but briefly and your muscle cells tap glycogen for a burst of energy. Figure 3.10*c* shows a few of glycogen's many branchings.

Figure 3.11 Scanning electron micrograph of a tick. Its body covering is a protective cuticle reinforced with chitin.

A different polysaccharide, chitin, is distinctive in that it has nitrogen-containing groups attached to its glucose monomers. Chitin is a material that strengthens external skeletons and other hard body parts of many animals, including crabs, earthworms, insects, and ticks (Figure 3.11). Chitin also is a structural material that strengthens the cell walls of many kinds of fungi.

The simple sugars (such as glucose), oligosaccharides, and polysaccharides (such as starch) are carbohydrates. Every cell requires carbohydrates as structural materials, stored forms of energy, and transportable packets of energy.

GREASY, OILY—MUST BE LIPIDS

If something is greasy or oily to the touch, you can bet it is a lipid or has lipid components. **Lipids** are nonpolar hydrocarbons. They resist dissolving in water but easily dissolve in nonpolar substances (think butter into warm cream sauce). Cells use different lipids as energy stores, as structural materials (for example, in cell membranes and surface coatings), and as signaling molecules. Let's look first at the kinds having fatty acid components—fats, phospholipids, and waxes. Then we will consider the sterols, each with a backbone of four carbon rings.

Fats and Fatty Acids

Lipids known as **fats** have one, two, or three fatty acids attached to glycerol. Each **fatty acid** has a backbone of as many as thirty-six carbon atoms, a carboxyl group (—COOH) at one end, and hydrogen atoms occupying

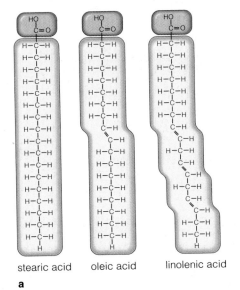

stearic acid   oleic acid   linolenic acid

a

Figure 3.12 (**a**) Structural formulas for three fatty acids. In stearic acid, the carbon backbone is fully saturated with hydrogen atoms. Oleic acid, with a double bond in its backbone, is an unsaturated fatty acid. Linolenic acid, with its three double bonds, is a polyunsaturated fatty acid. (**b**) Space-filling model for stearic acid.

b

most or all of the remaining bonding sites. Most stretch out like a flexible tail. Tails that are *unsaturated* have one or more double bonds. *Saturated* tails have single bonds only. Figure 3.12 shows examples.

Most animal fats have many saturated fatty acids, which pack together by weak interactions. They remain solid at temperatures that keep most plant fats liquid, as "vegetable oils." Similar packing interactions in plant fats aren't as stable because of rigid kinks in their fatty acid tails. That is why vegetable oils flow freely.

Butter, lard, vegetable oils, and other natural fats are mostly **triglycerides**. These "neutral" fats have three fatty acid tails attached to a glycerol unit (Figure 3.13). Triglycerides are the body's most abundant lipids and its richest energy source. Gram for gram, they yield more than twice as much energy when degraded, compared to complex carbohydrates such as starches. Quantities of triglycerides are stored as droplets in the cells of body fat (adipose tissue) in every vertebrate. A thick layer of triglycerides under the skin helps penguins and some other animals resist cold (Figure 3.13b). It insulates the body against extreme temperatures of icy habitats.

glycerol

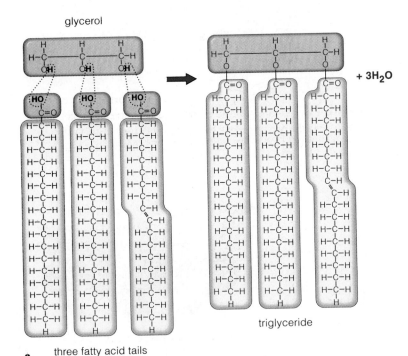

+ 3H₂O

triglyceride

a   three fatty acid tails

b

Figure 3.13 (**a**) Condensation of fatty acids and a glycerol molecule into a triglyceride. (**b**) Triglyceride-protected penguins taking the plunge.

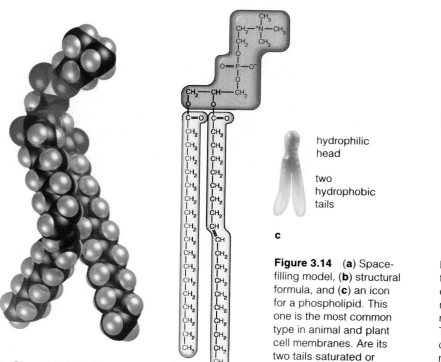

a Phosphatidylcholine

hydrophilic head

two hydrophobic tails

c

Figure 3.14 (**a**) Space-filling model, (**b**) structural formula, and (**c**) an icon for a phospholipid. This one is the most common type in animal and plant cell membranes. Are its two tails saturated or unsaturated?

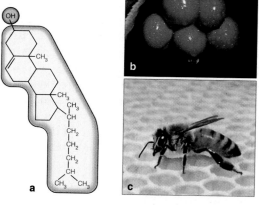

a

b

c

Figure 3.15 (**a**) Structural formula for cholesterol, the major sterol of animal tissues. Your liver produces enough cholesterol for your body. A fat-rich diet may result in high cholesterol levels in the blood, which may contribute to the clogging of small arteries. Two waxy materials: (**b**) The waxy, water-repelling cuticle of cherries. (**c**) Honeycomb, a structural material of a firm, water-repellent, waxy secretion called beeswax.

One fatty acid is a precursor of eicosanoids: local signaling molecules that include prostaglandins. As you will see, they have roles in contraction, defense, message flow in the nervous system, and other key processes.

Phospholipids

A **phospholipid** has a glycerol backbone, two fatty acid tails, and a hydrophilic "head" with a phosphate group and another polar group (Figure 3.14). Phospholipids are the main materials of cell membranes, which have two layers of lipids. Heads of one layer are dissolved in the cell's fluid interior, and heads of the other layer are dissolved in the surroundings. Sandwiched between the two are all the fatty acid tails, which are hydrophobic.

Sterols and Their Derivatives

Sterols are among the many lipids with no fatty acids. Sterols differ in the number, position, and type of their functional groups, but all have a rigid backbone of four fused-together carbon rings, as below. Eukaryotic cells have sterols in their membranes. Cholesterol is the most common type in tissues of animals. Figure 3.15a shows its structure. Also, cholesterol gets remodeled into compounds including vitamin D (necessary for good bones and teeth), steroids, and bile salts. Steroids include sex hormones of the sort shown

sterol backbone

in Figure 3.6. These hormones govern the formation of gametes and development of secondary sexual traits, such as feather color and hair distribution. Bile salts have roles in the digestion of fats in the small intestine.

Waxes

Waxes have long-chain fatty acids tightly packed and linked to long-chain alcohols or carbon rings. All have a firm consistency; all repel water. The cuticle covering aboveground plant parts consists mostly of waxes and another lipid, cutin (Figure 3.15b). It restricts water loss and fends off certain parasites. Waxy secretions protect, lubricate, and impart pliability to skin and to hair. Birds secrete waxes, fatty acids, and fats from preen glands to waterproof feathers. Bees use beeswax to construct honeycomb, which houses not only honey but new bee generations (Figure 3.15c). Tiny organisms drifting in the seas even use waxes as their main energy source.

Being largely hydrocarbon, lipids can dissolve in nonpolar substances, but they resist dissolving in water.

Triglycerides, or neutral fats, have a glycerol head and three fatty acid tails. They are the body's major energy reservoirs. Phospholipids are the main components of cell membranes.

Sterols such as cholesterol serve as membrane components and precursors of steroid hormones and other compounds. Waxes are firm yet pliable components of water-repelling and lubricating substances.

A STRING OF AMINO ACIDS: PROTEIN PRIMARY STRUCTURE

Of all large biological molecules, the **proteins** are the most diverse. The ones called enzymes make reactions happen faster than they would on their own. Structural proteins are the stuff of spider webs, butterfly wings, feathers, bone, and many other body parts and products. Some proteins transport substances across membranes and through fluids. Nutritious proteins abound in eggs and seeds. Protein hormones are signals for change in cell activities. Many proteins act as weapons against pathogens. Amazingly, cells build thousands of diverse proteins from only twenty kinds of amino acids!

Amino Acid Structure

An **amino acid** is a small organic compound consisting of an amino group (which is basic), a carboxyl group (which is an acid), a hydrogen atom, and one or more atoms called its R group. Figures 3.16 and 3.17 show the structure of some amino acids you will encounter later on in the book. Generally, their components are all bonded covalently to the same carbon atom.

Polypeptide Chain Formation

When a cell synthesizes a protein, enzymes join amino acids, one after the other, by peptide bonds. This type of covalent bond forms between the amino group ($-NH_3^+$) of one amino acid and the carboxyl group ($-COO^-$) of the next amino acid in line.

When peptide bonds join three or more amino acids, we have a **polypeptide chain**. In such chains, the carbon backbone has nitrogen atoms positioned in this regular pattern: $-N-C-C-N-C-C-$.

For each kind of protein, different amino acids are selected one at a time from the twenty kinds available.

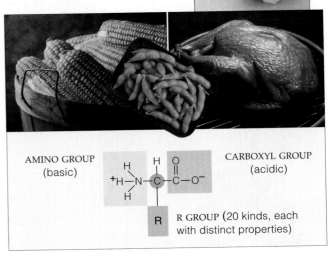

AMINO GROUP (basic)   **CARBOXYL GROUP** (acidic)

R GROUP (20 kinds, each with distinct properties)

Figure 3.16 Generalized structural formula for amino acids, along with soybeans and a few other dietary sources of these small organic compounds.

Instructions encoded in the cell's DNA determine the selection. Overall, the protein's amino acid sequence is unique and is its *primary* structure (Figure 3.18).

Thousands of different kinds of proteins occur in nature. Many are *fibrous* proteins that have polypeptide chains arranged as strands or sheets. They contribute to a cell's shape and organization. *Globular* proteins have one or more chains folded in compact, rounded shapes. Most enzymes are globular proteins. So are proteins that help cells, and cell parts move.

Regardless of the type of protein, its shape and its function arise from the primary structure—that is, from information encoded in its amino acid sequence. That

Figure 3.17 Structural formulas and ball-and-stick models for a sampling of the twenty common amino acids in cells. Appendix VI shows all twenty.

Green boxes indicate R groups, which are side chains that include functional groups. Each type of side chain makes a contribution to the chemical and physical properties of each amino acid.

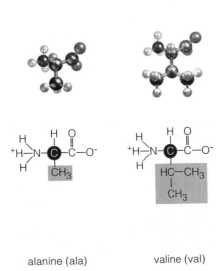
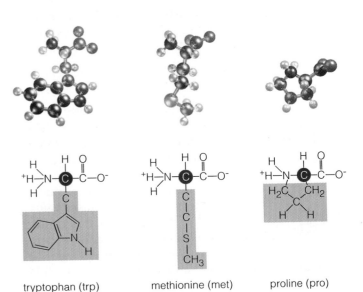

alanine (ala)　　valine (val)　　tryptophan (trp)　　methionine (met)　　proline (pro)

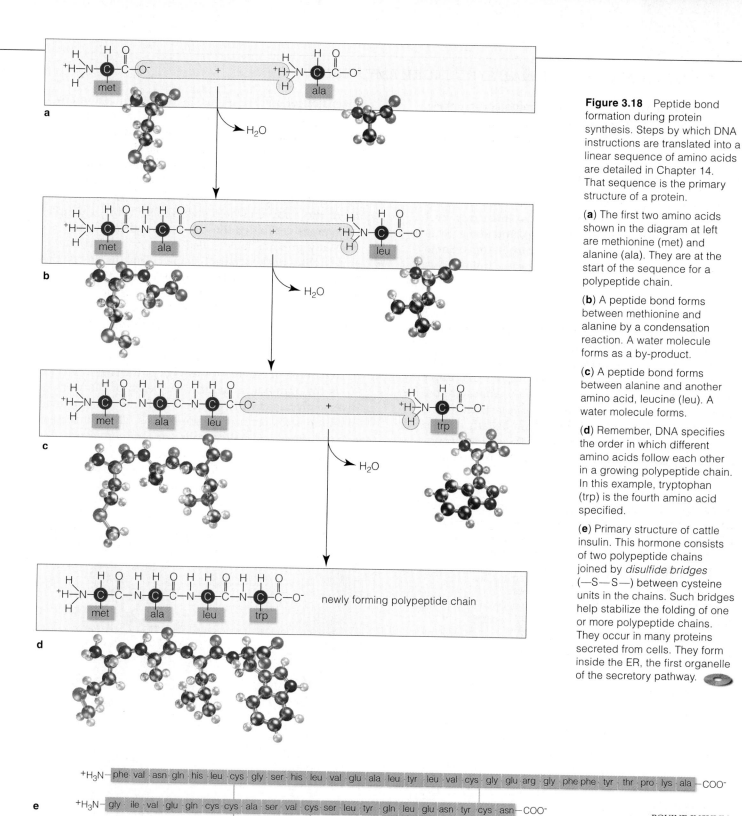

Figure 3.18 Peptide bond formation during protein synthesis. Steps by which DNA instructions are translated into a linear sequence of amino acids are detailed in Chapter 14. That sequence is the primary structure of a protein.

(**a**) The first two amino acids shown in the diagram at left are methionine (met) and alanine (ala). They are at the start of the sequence for a polypeptide chain.

(**b**) A peptide bond forms between methionine and alanine by a condensation reaction. A water molecule forms as a by-product.

(**c**) A peptide bond forms between alanine and another amino acid, leucine (leu). A water molecule forms.

(**d**) Remember, DNA specifies the order in which different amino acids follow each other in a growing polypeptide chain. In this example, tryptophan (trp) is the fourth amino acid specified.

(**e**) Primary structure of cattle insulin. This hormone consists of two polypeptide chains joined by *disulfide bridges* (—S—S—) between cysteine units in the chains. Such bridges help stabilize the folding of one or more polypeptide chains. They occur in many proteins secreted from cells. They form inside the ER, the first organelle of the secretory pathway.

newly forming polypeptide chain

$^+H_3N-$ phe val asn gln his leu cys gly ser his leu val glu ala leu tyr leu val cys gly glu arg gly phe phe tyr thr pro lys ala $-COO^-$

$^+H_3N-$ gly ile val glu gln cys cys ala ser val cys ser leu tyr gln leu glu asn tyr cys asn $-COO^-$

BOVINE INSULIN

sequence influences which parts of a polypeptide chain will coil, bend, or interact with other chains. The type and arrangement of atoms in the coiled, stretched out, and folded parts dictate whether the protein will function as, say, an enzyme, a transporter, a receptor, or even an unlucky target for a bacterium or virus.

Each protein consists of one or more polypeptide chains of amino acids. The amino acid sequence (which kind of amino acid follows another in the chain) is unique for each kind of protein and gives rise to its unique structure, chemical behavior, and function.

HOW DOES A PROTEIN'S FINAL STRUCTURE EMERGE?

Now that you have a sense of how amino acids make a polypeptide chain, take a look at examples of the ways in which the sequence dictates a protein's final shape.

Second and Third Levels of Protein Structure

Picture the primary structure as a set of rigid playing cards joined by sticks (covalent bonds) that can swivel a bit. Each "card" is a peptide group, and some of its atoms favor bond formation with nearby atoms. Some swivels favor hydrogen bonding between amino acids along the length of a polypeptide chain. Others put R groups in positions that invite bond formation.

Certain sequences of amino acids favor a pattern of bonding that causes part of the polypeptide chain to coil and twist into a helix, a bit like a spiral staircase (Figure 3.19a). Other sequences give rise to sheetlike regions (Figure 3.19b) and to looped regions that differ in length and how they twist. These outcomes are the dominant features of a protein's *secondary* structure.

Adjacent coils or strands in a chain usually end up close together in the protein, with their side groups packed together. These regions now fold up into one to

dozens of compact domains. By definition, a **domain** is a polypeptide chain, or part of it, that has become self-organized as a structurally stable, functional unit. We call domain formation the protein's *tertiary* structure.

Figure 3.19b shows adjacent, strandlike regions of one kind of polypeptide chain with looped connecting regions. Hydrogen bonds and other interactions bring about the array and help stabilize it. In this example, the regions get folded into a barrel-shaped domain.

Many proteins have coiled domains around one or more tightly packed, barrel-shaped domains. Inside the barrel's interior are hydrophobic R groups that exclude water and shield R groups of looped regions, which serve in enzyme action and other tasks. These proteins differ from ones that have coils flanking a sheetlike domain (Figure 3.20a). The sheetlike proteins are quite diverse in structure. They carry out their specific tasks at some type of crevice at the edge of the sheet.

Is the amino acid sequence for each of the world's diverse proteins totally unique? No. Some combinations of amino acids recur in most proteins. Those that give rise to barrels and helical coils are common, especially among enzymes and transporters.

Figure 3.19 Emergence of secondary and tertiary structure. (**a**) Extensive hydrogen bonding (dotted lines) along a polypeptide chain or a local region of it can result in a helical structure. (**b**) Hydrogen bonding between strandlike regions of a chain can result in a sheetlike structure. The helical or sheetlike regions are a protein's secondary structure. Adjacent helixes and sheets pack together in structurally stable, functional units (domains), the protein's tertiary structure.

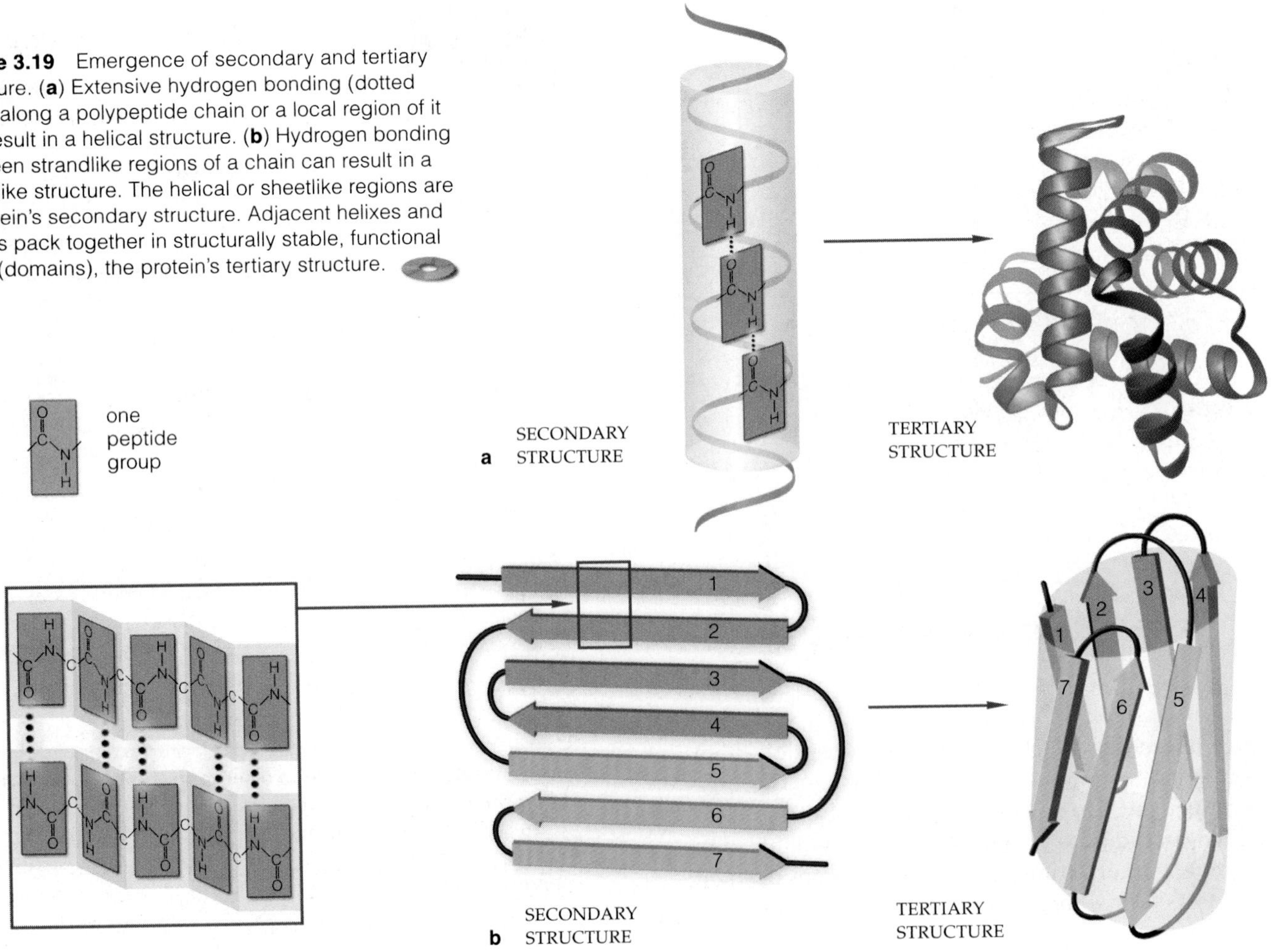

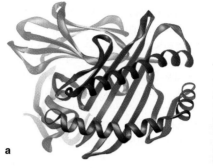

a

where the molecule binds and displays "enemies" (*arrow*)

one of the chains spans the plasma membrane and anchors the molecule

b

Figure 3.20 A protein that helps your body defend itself against bacteria and other foreign agents. (**a**) HLA-A2 quaternary structure, which consists of two polypeptide chains that are like jaws. (**b**) The presumed location of this protein relative to the outer surface of the plasma membrane.

Fourth Level of Protein Structure

Many final proteins consist of two or more polypeptide chains held together by hydrogen bonds and sometimes covalent bonds. Such proteins have *quaternary* structure.

Proteins called **HLAs** (human leukocyte antigens) are like this. All body cells have HLAs at their surface, but white blood cells, or leukocytes, bristle with them. HLAs function in recognizing "self"—the body's own cells—from "nonself." They bind and, later on, display fragments of nonself proteins, thus making the body's defending cells aware of an invader. We know of more than 400 slightly different versions of HLAs. Humans differ in terms of which versions occur on their cells.

Figure 3.20 shows the quaternary structure of the version designated HLA-A2. Notice, in Figure 3.20*a*, how it consists of two polypeptide chains. Part of one chain is a sheetlike domain with two helixes arching above it. These helixes are like jaws. They are able to grip a bit of a captured and digested "enemy agent" and offer it up to the body's defending cells. Receptors on the defending cells recognize and latch on to HLA molecules that are displaying foreign fragments. Such bindings trigger an immune response.

Now look at Figure 3.20*b*. Another part of the same chain continues across the plasma membrane of a body cell and ties in with fibers in the cytoplasm to anchor the whole molecule.

Protein structure doesn't stop here. Short, linear, or branched oligosaccharides get attached to many new polypeptide chains. Most proteins at the cell surface and secreted from cells are such *glyco*proteins. Lipids get attached to other proteins. One such *lipo*protein forms as proteins in blood combine with cholesterol, triglycerides, and phospholipids absorbed from the gut after a meal. Such attachments add to protein diversity.

Denaturation—How to Undo the Structure

Breaking weak bonds of a protein or any other large molecule disrupts its three-dimensional shape, an event called **denaturation**. For example, weak hydrogen bonds are sensitive to increases and decreases in temperature and pH. If the temperature or pH exceeds a protein's range of tolerance, its polypeptide chains will unwind or change shape, and the protein will lose its function. Consider the protein albumin, concentrated in the "egg white" of uncooked chicken eggs. When you cook eggs, the heat does not disrupt the strong covalent bonds of albumin's primary structure. But it destroys weaker bonds contributing to the three-dimensional shape. For some proteins, denaturation might be reversed when normal conditions are restored—but albumin isn't one of them. There is no way to uncook a cooked egg.

A protein's primary structure is the sequence of different kinds of amino acids along a polypeptide chain.

A protein has secondary structure. Local regions along the length of a polypeptide chain twist and fold into helical coils, sheetlike arrays of strands, and loops.

A protein may have tertiary structure. A polypeptide chain or parts of it becomes organized into domains: structurally stable, compact units that may have distinct functions.

A protein with quaternary structure consists of two or more polypeptide chains joined by hydrogen bonds. Covalent bonds, disulfide bridges, and other interactions stabilize it.

WHY IS PROTEIN STRUCTURE SO IMPORTANT?

Just One Wrong Amino Acid...

Earlier sections showed you elegant models for protein structure. Cells usually are good at construction tasks and turn out protein molecules that are just what their DNA specifies. But sometimes mistakes are made in the assembly process. Sometimes nature sabotages the specifications, as by ultraviolet radiation attacks. If even one amino acid is replaced by the wrong kind, the substitution may have far-reaching consequences.

Consider such a change in hemoglobin's structure. This protein has a quaternary structure; it consists of four polypeptide chains called globin. During protein synthesis, each globin molecule becomes tightly folded. The folding produces a small pocket that is chemically attractive to an iron-containing heme group, the part of hemoglobin that transports oxygen (Figure 3.21a). As you read this, each mature red blood cell in your body is transporting about a billion oxygen molecules, all bound to 250 million or so hemoglobin molecules.

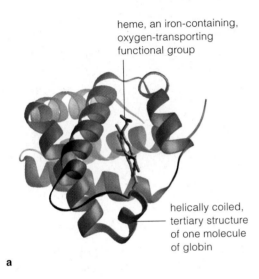

heme, an iron-containing, oxygen-transporting functional group

helically coiled, tertiary structure of one molecule of globin

a

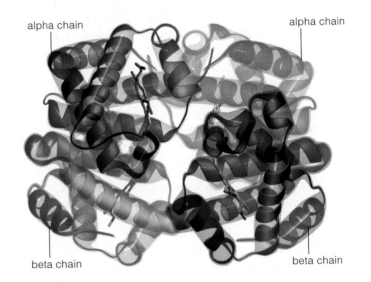

alpha chain   alpha chain

beta chain   beta chain

b

Figure 3.21 *Above:* (**a**) A globin molecule. This coiled polypeptide chain associates with heme, an iron-containing functional group that strongly attracts oxygen.

(**b**) Quaternary structure of hemoglobin, an oxygen-transporting pigment in red blood cells. It consists of four globin molecules and four heme groups. Two of the chains, designated alpha, differ a bit from the other two (beta chains) in amino acid sequence. To help you visualize all four chains, helical regions are enclosed in transparent tubes.

Right: (**c**) Normal sequence of amino acids at the start of a beta chain for hemoglobin.

(**d**) A single amino acid substitution results in the abnormal beta chain found in HbS molecules. During protein synthesis, valine was added instead of glutamate at the sixth position of the growing polypeptide chain. The ball and stick models for the original amino acid and the substitution are shown in the diagrams.

Glutamate has an overall negative charge; valine has no net charge. This difference in charge gives rise to a water-repelling, sticky patch on the HbS molecule.

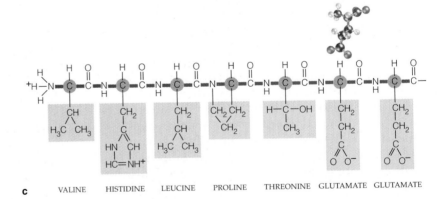

c   VALINE   HISTIDINE   LEUCINE   PROLINE   THREONINE   GLUTAMATE   GLUTAMATE

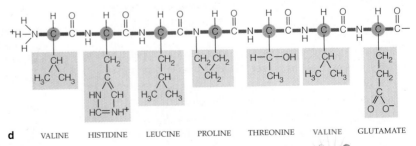

d   VALINE   HISTIDINE   LEUCINE   PROLINE   THREONINE   VALINE   GLUTAMATE

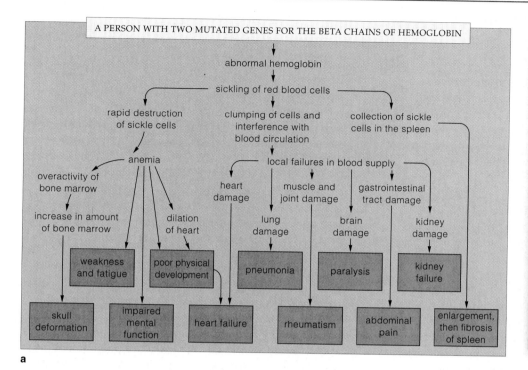

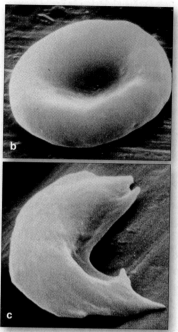

Figure 3.22 (**a**) Diagram tracking the wide range of symptoms characteristic of sickle-cell anemia, a genetic disorder. (**b**) Red blood cell from a person affected by the disorder. This scanning electron micrograph shows the surface appearance of a red blood cell from an affected individual when the blood is adequately oxygenated. (**c**) This scanning electron micrograph shows the sickle shape of a red blood cell carrying HbS when the concentration of oxygen in blood is low.

The globin in hemoglobin has two slightly different forms: alpha and beta (Figure 3.21b). There is a type of mutation that changes DNA's instructions for the beta chain. It specifies valine instead of glutamate at the sixth position shown in Figure 3.21c. Glutamate has an overall negative charge; valine, however, has no net charge. The difference puts a "sticky," water-repelling patch on beta chains. Hemoglobin that incorporates the altered chain is called HbS (instead of HbA).

Sickle-Shaped Cells and a Serious Disorder

People inherit two genes for the beta chain, one from each parent. (A "gene" is a unit of information about a heritable trait.) If one is normal but the other calls for HbS, people usually do not run into trouble. Their red blood cells can still make enough normal hemoglobin molecules to compensate for the abnormal ones. But in someone who inherited two mutated genes, red blood cells can only make HbS—the outcome being a severe genetic disorder called *sickle-cell anemia*.

Humans, like most organisms, must take in a lot of oxygen; cells use it to get energy, by aerobic respiration. Oxygen flows into our lungs, then diffuses into blood. There it binds to hemoglobin, which travels through arteries, arterioles, then the small-diameter, thin-walled capillaries that thread through all tissues. Most of the oxygen diffuses into tissues, then cells. Cellular uptake lowers its concentration in blood. Its decline is greatest at high altitudes and during strenuous activity.

Where oxygen concentrations are lowest, abnormal hemoglobin molecules are attracted to one another's sticky patches. They get stuck in rod-shaped clumps. The rods distort red blood cells into a sickle shape, as in Figure 3.22. (A sickle is a farm tool that has a long, crescent-shaped blade.) Thus distorted, cells rupture easily, and the remnants clog and rupture capillaries. Their rapid destruction leads to oxygen-starved cells in affected tissues. Clumping also causes local failures in the circulatory system's capacity to deliver oxygen and carry away carbon dioxide and other metabolic wastes. In time, ongoing expression of the mutant gene may damage tissues and organs throughout the body.

Hemoglobin, hormones, enzymes, receptors—these are the kinds of proteins necessary for your survival. Understand the structure and functions of proteins in general, and you are on your way to understanding life in its richly normal and abnormal expressions.

The molecular structure of proteins dictates how they can perform their functions. How proteins function dictates the quality of life, sometimes life or death itself.

NUCLEOTIDES AND NUCLEIC ACIDS

The Diverse Roles of Nucleotides

The small organic compounds called **nucleotides** each consist of a sugar, at least one phosphate group, and a nitrogen-containing base. The sugar is either ribose or deoxyribose. Although both sugars have a five-carbon ring structure, ribose has an oxygen atom attached to carbon 2 and deoxyribose does not. The bases have a single or double carbon ring structure. Cells put these small compounds to different uses.

The nucleotide **ATP** (adenosine triphosphate) has a string of three phosphate groups attached to its sugar component (Figure 3.23*a*). ATP can readily transfer a phosphate group to many other molecules inside cells, which makes the acceptor molecules energized enough to enter a reaction. ATP's phosphate-group transfers are one of the most important aspects of metabolism.

Other nucleotides have different metabolic roles. Certain kinds are subunits of **coenzymes**, or enzyme helpers, which accept hydrogen atoms and electrons stripped from molecules at one reaction site and then transfer them to different sites in the cell. Figure 3.23*b* shows the structure of NAD$^+$ (short for nicotinamide adenine dinucleotide), a player in aerobic respiration. FAD (flavin adenine dinucleotide) is another kind.

Still other nucleotides serve as chemical messengers between cells and between one part of the cytoplasm and another. Later on in the book, you will be reading about one of these messengers, cAMP (cyclic adenosine monophosphate).

Last but not least, nucleotides also serve as building blocks for the larger molecules called nucleic acids.

Regarding DNA and the RNAs

Besides performing tasks for metabolic reactions, five kinds of nucleotides have vital roles in the storage and retrieval of heritable information in all cells. They are monomers for single- and double-stranded molecules classified as **nucleic acids**. In such strands, a covalent bond connects the sugar component of one nucleotide with the phosphate group of the next nucleotide in the sequence (Figure 3.24).

All cells start out life and then maintain themselves by way of instructions they inherited in some number of double-stranded molecules of deoxyribonucleic acid, or **DNA**. This nucleic acid consists of four kinds of nucleotides. Figure 3.24 gives their structural formulas. As you can see, the four differ only in their component base, which is adenine, guanine, thymine, or cytosine.

Figure 3.25 shows how hydrogen bonds between bases join the two strands together along the length of a DNA molecule. Think of each "base pair" as one rung of a ladder, and the two sugar–phosphate backbones as the ladder's two posts. The ladder twists and turns in a regular pattern, forming a double helix.

The sequence of bases in DNA encodes heritable information about how to synthesize all of the proteins that give each new cell the potential to grow, maintain itself, and reproduce. The particular bases in at least some parts of the sequence are unique to each species.

Like DNA, the **RNAs** (ribonucleic acids) consist of four kinds of nucleotide monomers. Unlike DNA, the bases are adenine, guanine, cytosine, and uracil. Also unlike DNA, RNA molecules are usually single strands

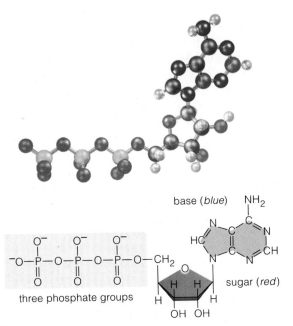

base (*blue*)

three phosphate groups

sugar (*red*)

a

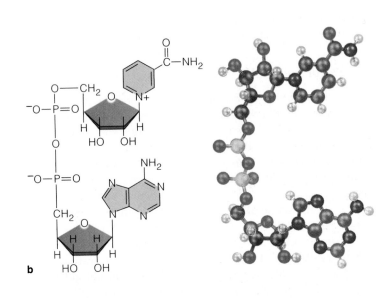

b

Figure 3.23 Ball-and-stick models and structural formulas for (**a**) ATP and (**b**) NAD$^+$. Both nucleotides have central roles in cell metabolism.

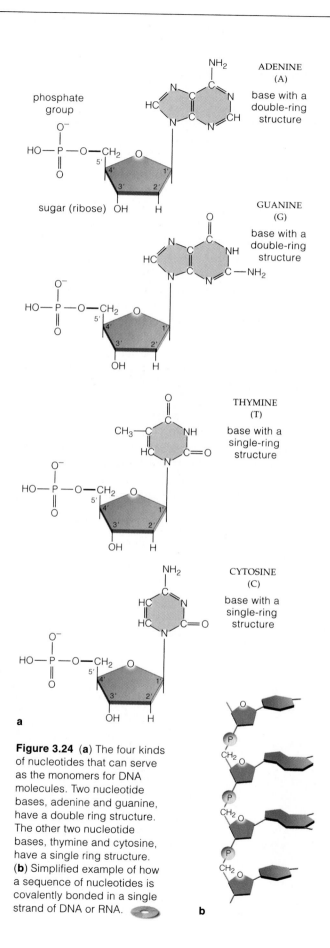

ADENINE
(A)

base with a
double-ring
structure

phosphate
group

sugar (ribose)

GUANINE
(G)

base with a
double-ring
structure

THYMINE
(T)

base with a
single-ring
structure

CYTOSINE
(C)

base with a
single-ring
structure

a

Figure 3.24 (**a**) The four kinds of nucleotides that can serve as the monomers for DNA molecules. Two nucleotide bases, adenine and guanine, have a double ring structure. The other two nucleotide bases, thymine and cytosine, have a single ring structure. (**b**) Simplified example of how a sequence of nucleotides is covalently bonded in a single strand of DNA or RNA.

b

Figure 3.25 Models for a DNA molecule, with its two strands of nucleotides hydrogen-bonded at their bases and twisted into a double helix.

covalent
bonding
in carbon
backbone

hydrogen bonding
between bases

of nucleotides. Certain RNAs encode protein-building instructions, and act as a messenger to help translate the code into proteins. Other RNAs carry out the translation. Ever since cells first appeared on Earth, RNAs have been key to the processes by which genetic information is used to build proteins.

The small organic compounds called nucleotides serve as energy carriers, chemical messengers, and subunits for coenzymes and for nucleic acids. The nucleic acids consist of nucleotides joined one after another by covalent bonds.

The nucleic acid DNA consists of two nucleotide strands, joined by hydrogen bonds, and twisted as a double helix. Encoded in its nucleotide sequence is heritable information about all proteins a cell requires to survive and reproduce.

Different RNAs are single-stranded nucleic acids with roles in the processes by which a cell retrieves and uses genetic information in DNA to build proteins.

SUMMARY

1. Organic compounds consist of carbon and at least one hydrogen atom. Their carbon atoms are often bonded covalently to form a linear or ring-shaped backbone, and functional groups are attached to the backbone. *3.1, 3.2*

2. In nature, only living cells assemble large organic compounds called biological molecules (Table 3.1). Cells build these complex carbohydrates, lipids, proteins, and nucleic acids from simple sugars, fatty acids, amino acids, nucleotides, and other small organic compounds. *3.1*

3. Cells assemble, rearrange, and break down organic compounds by different enzyme-mediated reactions, as when they transfer functional groups or electrons. *3.3*

4. Cells use simple sugars and oligosaccharides for energy and building blocks. They use complex carbohydrates as structural materials and as energy storage forms. They use the lipids (especially triglycerides) as energy storage forms, as structural components of cell membranes, and as precursors of other compounds. *3.4–3.5*

5. Proteins are the most diverse organic compounds. They function in enzyme activity, structural support, signaling, shape changes, transport, movements, and defense against disease. Nucleotides are energy carriers (especially ATP), coenzymes, and subunits for strands of the nucleic acids DNA and RNA, which interact and form the basis of inheritance and reproduction. *3.6–3.8*

6. An amino acid sequence, unique for each kind of polypeptide chain, is the start of each protein's unique structure, chemical behavior, and function. *3.6–3.9*

    a. Local helical, strandlike, and looped regions form along the chain and are a protein's secondary structure. The local regions self-organize into a tertiary structure: compact, structurally stable, functional units (domains).

    b. Many proteins have quaternary structure. Two or more polypeptide chains are joined and stabilized by hydrogen bonds and other interactions. *3.9*

Table 3.1 *Summary of the Main Organic Compounds in Living Things*

Category	Main Subcategories	Some Examples and Their Functions	
CARBOHYDRATES . . . contain an aldehyde or a ketone group, and one or more hydroxyl groups	**Monosaccharides** (simple sugars)	Glucose	Energy source
	Oligosaccharides	Sucrose (a disaccharide)	Most common form of sugar; the form transported through plants
	Polysaccharides (complex carbohydrates)	Starch, glycogen Cellulose	Energy storage Structural roles
LIPIDS . . . are mainly hydrocarbon; generally do not dissolve in water but do dissolve in nonpolar substances, such as other lipids	**Lipids with fatty acids** *Glycerides:* Glycerol backbone with one, two, or three fatty acid tails	Fats (e.g., butter), oils (e.g., corn oil)	Energy storage
	Phospholipids: Glycerol backbone, phosphate group, one other polar group, and (often) two fatty acids	Phosphatidylcholine	Key component of cell membranes
	Waxes: Alcohol with long-chain fatty acid tails	Waxes in cutin	Conservation of water in plants
	Lipids with no fatty acids *Sterols:* four carbon rings; the number, position, and type of functional groups differ among sterols	Cholesterol	Component of animal cell membranes; precursor of many steroids and vitamin D
PROTEINS . . . are one or more polypeptide chains, each with as many as several thousand covalently linked amino acids	**Fibrous proteins** Long strands or sheets of polypeptide chains; often tough, water-insoluble	Keratin Collagen	Structural component of hair, nails Structural component of bone
	Globular proteins One or more polypeptide chains folded into globular shapes; many roles in cell activities	Enzymes Hemoglobin Insulin Antibodies	Great increase in rates of reactions Oxygen transport Control of glucose metabolism Tissue defense
NUCLEIC ACIDS (AND NUCLEOTIDES) . . . are chains of units (or individual units) that each consist of a five-carbon sugar, phosphate, and a nitrogen-containing base	**Adenosine phosphates**	ATP cAMP (Section 36.2)	Energy carrier Messenger in hormone regulation
	Nucleotide coenzymes	NAD^+, $NADP^+$, FAD	Transfer of electrons, protons (H^+), from one reaction site to another
	Nucleic acids Chains of thousands to millions of nucleotides	DNA, RNAs	Storage, transmission, translation of genetic information

Review Questions

1. Define organic compound. Name the type of chemical bond that predominates in the backbone of such a compound. *3.1*

2. Define hydrocarbon. Also define functional group. *3.1, 3.2*

3. Name the molecules of life and the families of small organic compounds from which they are built. Do they break apart most easily at their hydrocarbon portion or at functional groups? *3.2*

4. Define condensation reaction and hydrolysis. *3.3*

5. Select a carbohydrate, lipid, protein, or nucleic acid. How do its functional groups and bonds between carbon atoms in its backbone contribute to its final shape and function? *3.4–3.9*

6. Which item listed includes all of the other items listed? *3.5*
   a. triglyceride   c. wax   e. lipid
   b. fatty acid   d. sterol   f. phospholipid

7. Explain how hemoglobin's three-dimensional shape arises, starting with the primary structure of its four chains. *3.9*

Self-Quiz   ANSWERS IN APPENDIX III

1. Each carbon atom can share pairs of electrons with as many as _____ other atoms.
   a. one   b. two   c. three   d. four

2. Hydrolysis is a (an) _____ reaction.
   a. functional group transfer   d. condensation
   b. electron transfer   e. cleavage
   c. rearrangement   f. both b and d

3. _____ is a simple sugar (monosaccharide).
   a. Glucose   c. Ribose   e. both a and b
   b. Sucrose   d. Chitin   f. both a and c

4. In unsaturated fats, fatty acid tails have one or more _____ .
   a. single covalent bonds   b. double covalent bonds

5. _____ are to proteins as _____ are to nucleic acids.
   a. Sugars; lipids   c. Amino acids; hydrogen bonds
   b. Sugars; proteins   d. Amino acids; nucleotides

6. A denatured protein or DNA molecule has lost its _____ .
   a. hydrogen bonds   c. function
   b. shape   d. all of the above

7. Nucleotides occur in _____ .
   a. ATP   b. DNA   c. RNA   d. all are correct

8. Match each molecule with the most suitable description.
   _____ long sequence of amino acids   a. carbohydrate
   _____ the main energy carrier   b. phospholipid
   _____ glycerol, fatty acids, phosphate   c. protein
   _____ two strands of nucleotides   d. DNA
   _____ one or more sugar monomers   e. ATP

Critical Thinking

1. About 2,000 years ago, Plutarch suspected that the oracle at Delphi was high on a sweet-smelling gas when she made her prophecies (Figure 3.26). Recently, geologists discovered two intersecting faults beneath her temple. They analyzed sediments under the temple's foundation and found trapped molecules of methane and ethane. They also found these gases and another one, the sweet-smelling ethylene, bubbling in a natural spring near the temple. Apparently the oracle sniffed gases from oil-rich deposits, which collected in the temple's sunken floor after escaping through fissures that opened when the faults slipped.

Figure 3.26 Temple of Apollo at Delphi. To ancient Greeks, prophecies made by the temple's oracle were the voice of Apollo. A geologist now points out that the oracle made her prophecies while in a hydrocarbon-induced trance.

Reflect on the assumption of cause and effect (Section 1.5). Do you find it irritating or enlightening that science has given us a natural explanation for these "supernatural" prophecies?

2. It seems there are "good" and "bad" unsaturated fats. The double bonds of both put a bend in their fatty acid tails. But the bend in *cis* fatty acids keeps the whole tail aligned in the same direction. The bend in *trans* fatty acids makes it zigzag:

| *Cis* fatty acid | *Trans* fatty acid |

Some *trans* fatty acids occur naturally in beef. Most form by processes that solidify vegetable oils for margarine, shortening, and the like. These substances are widely used in prepared foods (such as cookies) and in french fries and other fast-food products. *Trans* fatty acids are linked to high levels of LDL, a "bad" form of cholesterol that can set the stage for a heart attack. Speculate on why your body might have an easier time dealing with *cis* fatty acids than *trans* fatty acids.

3. In the following list, identify which is the carbohydrate, the fatty acid, the amino acid, and the polypeptide:
   a. $^+NH_3$—CHR—COO$^-$ c. (glycine)$_{20}$
   b. $C_6H_{12}O_6$ d. $CH_3(CH_2)_{16}COOH$

4. A clerk in a health-food store tells you that certain "natural" vitamin C tablets extracted from rose hips are better for you than synthetic vitamin C tablets. Given your understanding of the structure of organic compounds, what would be your response? Design an experiment to test whether the vitamins differ.

Selected Key Terms

alcohol *3.2*	fat *3.5*	oligosaccharide *3.4*
amino acid *3.6*	fatty acid *3.5*	organic compound *3.1*
ATP *3.9*	functional group *3.1*	phospholipid *3.5*
carbohydrate *3.4*	global warming *CI*	polymer *3.3*
coenzyme *3.9*	HLA *3.7*	polypeptide chain *3.6*
condensation	hydrocarbon *3.1*	polysaccharide *3.4*
reaction *3.3*	hydrolysis *3.3*	protein *3.6*
denaturation *3.7*	lipid *3.5*	RNA *3.9*
DNA *3.9*	monomer *3.3*	sterol *3.5*
domain *3.7*	monosaccharide *3.4*	triglyceride *3.5*
enzyme *3.3*	nucleic acid *3.9*	wax *3.5*
	nucleotide *3.9*	

Readings

Alberts, B., et al. 2002. *Molecular Biology of the Cell.* Fourth edition. New York: Garland. Big book, easy to read.

On-Line readings at Student Guide for InfoTrac:
www.brookscole.com/biology

4

CELL STRUCTURE AND FUNCTION

Animalcules and Cells Fill'd With Juices

Early in the seventeenth century, a scholar by the name of Galileo Galilei put two glass lenses inside a cylinder. With this instrument he happened to look at an insect, and later he described the stunning geometric patterns of its tiny eyes. Thus Galileo, who wasn't a biologist, was one of the first to record a biological observation made through a microscope. The study of the cellular basis of life was about to begin. First in Italy, then in France and England, scholars set out to explore a world whose existence had not even been suspected.

At midcentury Robert Hooke, Curator of Instruments for the Royal Society of England, was at the forefront of those studies. When Hooke first turned a microscope to thinly sliced cork from a mature tree, he observed tiny compartments (Figure 4.1c). He gave them the Latin name *cellulae*, meaning small rooms—hence the origin of the biological term "cell." They

actually were the interconnecting walls of dead plant cells, which is what cork is made of, but Hooke didn't think of them as being dead because neither he nor anyone else at the time knew cells could be alive. In other plant tissues, he observed cells "fill'd with juices" but didn't have a clue to what they represented.

Given the simplicity of their instruments, it is just amazing that the pioneers in microscopy observed as much as they did. Antony van Leeuwenhoek, a Dutch shopkeeper, had exceptional skill in constructing lenses and possibly the keenest vision of all (Figure 4.1a). By the late 1600s, he was discovering natural wonders everywhere, including "many very small animalcules, the motions of which were very pleasing to behold," in scrapings of tartar from his teeth. Elsewhere he observed protistans, sperm, and even a bacterium—an organism so small it would not be seen again for another two centuries!

c

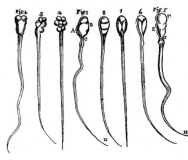

d

Figure 4.1 Early glimpses into the world of cells. (**a**) Antony van Leeuwenhoek, with microscope in hand. (**b**) Robert Hooke's compound microscope and (**c**) his drawing of cell walls from cork tissue. (**d**) One of van Leeuwenhoek's early sketches of sperm cells. (**e**) Cartoon evidence of the startling impact of microscopic observations on nineteenth-century London.

In the 1820s, improvements in lenses brought cells into sharper focus. Robert Brown, a botanist, noticed an opaque spot in a variety of cells and called it a nucleus. In 1838 another botanist, Matthias Schleiden, wondered if the nucleus had something to do with development. As he hypothesized, each plant cell might develop as an independent unit even though it's part of the plant.

By 1839, after years of studying animal tissues, the zoologist Theodor Schwann had this to say: Animals as well as plants consist of cells and cell products—and even though the cells are part of a whole organism, to some extent they have an individual life of their own.

A decade later a question remained: Where do cells come from? Rudolf Virchow, a physiologist, completed his own studies of a cell's growth and reproduction—that is, its division into two daughter cells. Every cell, he reasoned, comes from a cell that already exists.

And so, by the middle of the nineteenth century, microscopic analysis had yielded three generalizations, which together constitute the **cell theory**. *First, every organism is composed of one or more cells. Second, the cell is the smallest unit having the properties of life. Third, the continuity of life arises directly from the growth and division of single cells.* All three insights still hold true.

This chapter is not meant for memorization. Read it simply to gain an overview of current understandings of cell structure and function. In later chapters, you can refer back to it as a road map through the details. With its images from microscopy, this chapter and others in the book can transport you into spectacular worlds of juice-fill'd cells and animalcules.

MONSTER SOUP commonly called THAMES WATER

Key Concepts

1. All organisms consist of one or more cells. The cell is the smallest unit that still retains the characteristics of life. Since the time of life's origins, each new cell has descended from a cell that is already alive. These are the three generalizations of the cell theory.

2. All cells have a plasma membrane. This outermost, double-layered membrane separates a cell's interior from its surroundings, although it selectively allows substances to cross it.

3. All cells contain cytoplasm, an organized internal region where energy conversions, protein synthesis, movements of cell parts, and other required activities proceed. In eukaryotic cells only, the DNA is enclosed in a membrane-bound nucleus. In prokaryotic cells (archaebacteria and eubacteria), the DNA simply is concentrated in part of the cell interior.

4. The plasma membrane and internal cell membranes consist mainly of lipids and proteins. The lipids are organized as two adjacent layers. This bilayer gives a membrane its basic structure and prevents water-soluble substances from freely crossing it. Proteins embedded in the bilayer or positioned at its surfaces carry out many functions, such as transporting substances across the membrane.

5. Diverse organelles, including the nucleus, divide the interior of eukaryotic cells into membrane-bound, functional compartments. Prokaryotic cells do not have comparable organelles.

6. When cells are growing, they increase faster in volume than in surface area. This physical constraint on growth influences their size and shape.

7. Microscopes modify light rays or accelerated beams of electrons in ways that allow us to form images of incredibly small specimens. They are the foundation for our current understanding of cell structure and function.

BASIC ASPECTS OF CELL STRUCTURE AND FUNCTION

Inside your body and at its moist surfaces, trillions of cells live in interdependency. In northern forests, pollen grains—four-celled structures—escape from pine trees. In scummy pondwater, a single-celled amoeba moves on its own. For humans, pines, amoebas, and all other organisms, the **cell** is the smallest unit that lives on its own or has the potential to do so (Figure 4.2). Each cell is structurally organized for metabolism. It senses and responds to its environment. And heritable instructions in its DNA give it the potential to reproduce.

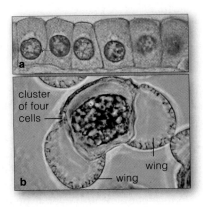

Figure 4.2 (**a**) A row of cells at the surface of a tiny tube inside a human kidney. Each cell has a nucleus (the sphere stained darker red). (**b**) A winged pollen grain. One of its cells will give rise to a sperm that may meet up with an egg and help start a new pine tree.

Structural Organization of Cells

Cells differ hugely in size, shape, and activities, as you might gather by comparing a tiny bacterium with one of your relatively giant liver cells. Yet they are alike in three respects: They all start out life with a plasma membrane, a region of DNA, and a region of cytoplasm:

1. **Plasma membrane**. This thin, outermost membrane maintains the cell as a distinct entity. By doing so, it allows metabolic events to proceed apart from random events in the environment. This membrane does not totally isolate the cell interior. Substances and signals continually move across it in highly controlled ways.

2. **Nucleus** or **nucleoid**. Depending on the species, the DNA occupies a membrane-bound sac (nucleus) in the cell or simply a region of the cell interior (nucleoid).

3. **Cytoplasm**. Cytoplasm is everything between the plasma membrane and the region of DNA. It consists of a semifluid matrix and other components, such as **ribosomes** (structures on which proteins are built).

This chapter introduces two very different kinds of cells. **Eukaryotic cells** have organelles—tiny sacs with one or more outer membranes, in the cytoplasm. One sac, the nucleus, is their defining feature. **Prokaryotic cells** don't have a nucleus; nothing intervenes between their DNA and cytoplasm. The only prokaryotic cells are archaebacteria and eubacteria. All other organisms —from amoebas to peach trees to puffball mushrooms to whales and zebras—consist of eukaryotic cells.

Organization of Cell Membranes

All cell membranes have the same structural framework of two sheets of lipid molecules. Figure 4.3 shows this **lipid bilayer** arrangement for the plasma membrane, the continuous, oily boundary that prevents the free passage of water-soluble substances into and out of all cells. In eukaryotic cells, other membranes divide the cytoplasm into functional zones in which substances are synthesized, processed, stockpiled, or degraded.

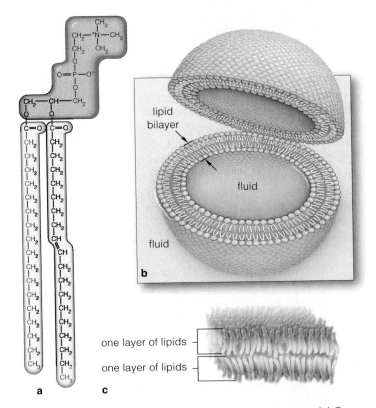

Figure 4.3 Lipid bilayer organization of cell membranes. (**a**) One of the phospholipids, the most abundant molecules of membranes. (**b,c**) These and other lipids are arranged into two layers. Their hydrophobic tails are sandwiched between the hydrophilic heads. In cells, the heads are dissolved in cytoplasm on one side of the bilayer and in extracellular fluid on the other side.

Diverse proteins embedded in the lipid bilayer or positioned at one of its surfaces carry out most of the membrane functions (Figure 4.4). For example, some proteins act as a channel for water-soluble substances. Others pump substances across the bilayer. Still others serve as receptors, which can latch onto hormones and other types of signaling molecules that trigger changes in cell activities. You will read more about membrane proteins in the chapter to follow.

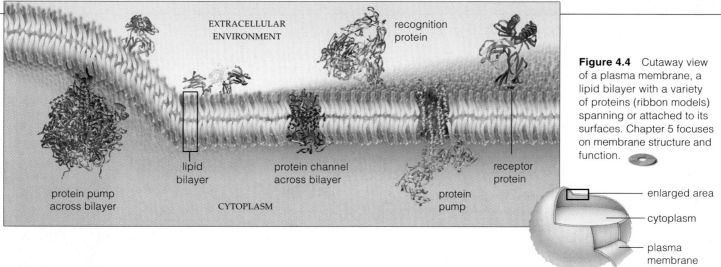

Figure 4.4 Cutaway view of a plasma membrane, a lipid bilayer with a variety of proteins (ribbon models) spanning or attached to its surfaces. Chapter 5 focuses on membrane structure and function.

enlarged area

cytoplasm

plasma membrane

Why Aren't All Cells Big?

You may be wondering how small cells really are. Can any be observed with the unaided human eye? Just a few, such as the "yolks" of bird eggs, cells in the red part of watermelons, and fish eggs. Eggs can get large because they are metabolically inert at maturity; most metabolically active cells are too tiny to be seen except with microscopes. To give you a sense of *how* tiny, a red blood cell is about 8 millionths of one meter across, so about 2,000 of them would fit across your thumbnail!

Why aren't all cells big? A physical relationship called the **surface-to-volume ratio** constrains increases in cell size. By this relationship, any object's volume increases with the cube of the diameter, but the surface area increases only with the square.

Apply this to a round cell. As Figure 4.5 shows, *if a cell expands in diameter during growth, then its volume will increase faster than its surface area will*. Suppose you make a round cell grow four times wider than normal. Its volume increases 64 times (4^3). But the surface area increases only 16 times (4^2). Now each unit of plasma membrane must serve four times as much cytoplasm as before. Moreover, beyond a certain point, the inward flow of nutrients and the outward flow of wastes won't be fast enough, so you'll end up with a dead cell.

A large, round cell also would have trouble moving materials *through* its cytoplasm. In small cells, random, tiny motions of molecules easily distribute materials. If a cell isn't small, you usually can expect it to be long and thin or have outfoldings and infoldings that increase its surface relative to its volume. *The smaller or narrower or more frilly-surfaced the cell, the more efficiently materials cross its surface and become distributed through the interior.*

We also see evidence of surface-to-volume constraints on body plans of multicelled species. For example, cells attach end to end in strandlike algae; each one interacts directly with its surroundings. Muscle cells are thin, but each is as long as the muscle of which it is part.

Figure 4.5 Example of the surface-to-volume ratio. This physical relationship between increases in volume and in surface area imposes restrictions on the size and the shape of cells. This includes frog eggs 2–3 mm across. They are among the largest of all animal cells. Compare Figure 4.8 in Section 4.2.

diameter (cm):	0.5	1.0	1.5
surface area (cm^2):	0.79	3.14	7.07
volume (cm^3):	0.06	0.52	1.77
surface-to-volume ratio:	13.17:1	6.04:1	3.99:1

All cells have an outermost plasma membrane, an internal region of cytoplasm, and an internal region of DNA. Only eukaryotic cells have a nucleus and other organelles.

A lipid bilayer, consisting of a two-layer arrangement of lipid molecules, gives a cell membrane its overall structure. Proteins embedded in the bilayer or positioned at one of its surfaces carry out diverse membrane functions.

Metabolic activity is largely a function of cell volume and surface area. As cells grow, their volume increases faster than their surface area does. The surface-to-volume ratio constrains increases in size. It also influences cell shape and the body plans of multicelled organisms.

Microscopes—Gateways to Cells

Modern microscopes are gateways to astounding worlds. Some even afford glimpses into the structure of molecules. Different kinds use wavelengths of light or accelerated electrons (Figures 4.6 and 4.7). The micrographs in Figures 4.8 and 4.9 only hint at the incredible details now being observed. A **micrograph** is simply a photograph of an image that came into view with the help of a microscope.

LIGHT MICROSCOPES Picture a series of waves moving across an ocean. Each **wavelength** is the distance from one wave's peak to the peak of the wave behind it. Light also travels as waves from sources such as the sun and illuminated specimens. In a *compound light microscope*, two or more sets of glass lenses bend light emanating from a cell or some other specimen in ways that form an enlarged image of it (Figure 4.6a,b).

A living cell must be small or thin enough for light to pass through. It would help if cell parts differed in color and density from the surroundings, but most are nearly colorless and appear uniformly dense. To get around this problem, microscopists stain cells (expose them to dyes that react with some parts of a specimen but not others). Staining may alter and kill cells. Dead cells break down

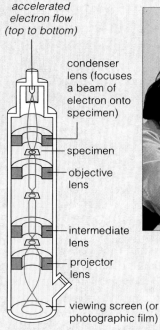

accelerated electron flow (top to bottom)

condenser lens (focuses a beam of electron onto specimen)

specimen

objective lens

intermediate lens

projector lens

viewing screen (or photographic film)

Figure 4.7 Generalized diagram of an electron microscope. You can get an idea of the diameter of the lenses from the photograph above.

rapidly, which is why technicians typically preserve or pickle a cell before staining it.

Suppose you use the best glass lens system. When you magnify the diameter of a specimen by 2,000 times or more, you discover that cell parts appear larger but are not clearer. To see why, think about the distance between the two crests of a wavelength of red or violet light. The distance is about 750 nanometers for red light and 400 nanometers for violet (wavelengths of all other colors fall in between). If a cell structure is less than one-half of a wavelength long, it will not be able to disturb enough of the rays of light streaming past to become visible.

ELECTRON MICROSCOPES Resolution of fine details is better with the help of electrons. Electrons, recall, are particles of matter. But they also behave like waves. In electron microscopy, streams of electrons are accelerated to wavelengths of about 0.005 nanometer—about 100,000 times shorter than wavelengths of visible light. Electrons cannot pass through glass lenses, but a magnetic field can bend them from their path and focus them.

In a *transmission electron microscope*, a magnetic field is the "lens." Accelerated electrons are directed through a specimen, focused into an image, and magnified. With *scanning electron microscopes*, a narrow beam of electrons moves back and forth across a specimen to which a thin coat of metal has been applied. The metal responds by emitting electrons. A detector tied to electronic circuitry converts energy of the emitted electrons into an image of the specimen's surface on a television screen. Most of the scanning images have fantastic depth (Figure 4.9d).

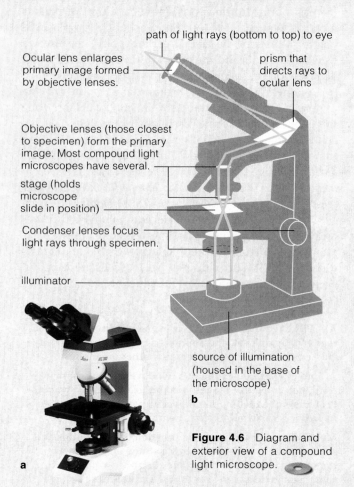

path of light rays (bottom to top) to eye

Ocular lens enlarges primary image formed by objective lenses.

prism that directs rays to ocular lens

Objective lenses (those closest to specimen) form the primary image. Most compound light microscopes have several.

stage (holds microscope slide in position)

Condenser lenses focus light rays through specimen.

illuminator

source of illumination (housed in the base of the microscope)

b

a

Figure 4.6 Diagram and exterior view of a compound light microscope.

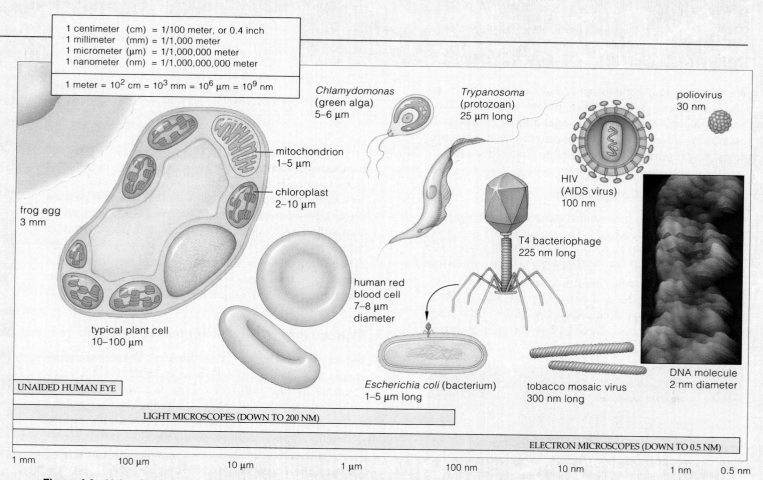

1 centimeter (cm) = 1/100 meter, or 0.4 inch
1 millimeter (mm) = 1/1,000 meter
1 micrometer (μm) = 1/1,000,000 meter
1 nanometer (nm) = 1/1,000,000,000 meter

$1 \text{ meter} = 10^2 \text{ cm} = 10^3 \text{ mm} = 10^6 \text{ μm} = 10^9 \text{ nm}$

Chlamydomonas (green alga) 5–6 μm

Trypanosoma (protozoan) 25 μm long

poliovirus 30 nm

mitochondrion 1–5 μm

chloroplast 2–10 μm

HIV (AIDS virus) 100 nm

frog egg 3 mm

T4 bacteriophage 225 nm long

human red blood cell 7–8 μm diameter

typical plant cell 10–100 μm

Escherichia coli (bacterium) 1–5 μm long

tobacco mosaic virus 300 nm long

DNA molecule 2 nm diameter

UNAIDED HUMAN EYE

LIGHT MICROSCOPES (DOWN TO 200 NM)

ELECTRON MICROSCOPES (DOWN TO 0.5 NM)

1 mm   100 μm   10 μm   1 μm   100 nm   10 nm   1 nm   0.5 nm

Figure 4.8 Units of measure used in microscopy. The photomicrograph of DNA was obtained with a *scanning tunneling microscope*, which yields magnifications up to 100 million times. The scope's needlelike probe has a single atom at its tip. When voltage is applied between the tip and an atom at a specimen's surface, electrons "tunnel" from the probe to the specimen. As the tip moves over a specimen's contours, a computer analyzes the tunneling motion and makes a three-dimensional view of the surface atoms.

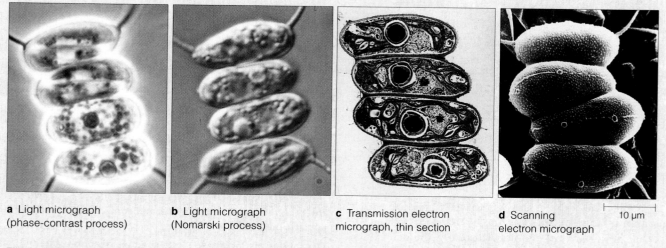

a Light micrograph (phase-contrast process)

b Light micrograph (Nomarski process)

c Transmission electron micrograph, thin section

d Scanning electron micrograph

10 μm

Figure 4.9 How different microscopes reveal different aspects of the same organism—a green alga (*Scenedesmus*). Images of all four specimens are at the same magnification. Phase-contrast and Nomarski processes (**a,b**) create optical contrasts without staining the cells. Both processes enhance the usefulness of light micrographs. As for other micrographs in the book, a horizontal bar below the micrograph, as in (**d**), provides you with a visual reference for size. A micrometer (μm) is 1/1,000,000 of a meter. Using the scale bar, can you estimate the length and width of *Scenedesmus*?

DEFINING FEATURES OF EUKARYOTIC CELLS

We turn now to organelles and other structural features that are typical of the cells of plants, animals, fungi, and protistans. We define an **organelle** as an internal, membrane-bound sac or compartment that serves one or more specialized functions inside eukaryotic cells.

Major Cellular Components

Study micrographs of a typical eukaryotic cell, such as the ones in the preceding section, and you'll probably notice that the nucleus is one of the most conspicuous features. Remember, any cell that starts out life with a nucleus is a eukaryotic cell; it houses a "true nucleus." Many other organelles and structures also are typical of these cells, although the numbers and kinds differ from one cell type to the next. Table 4.1 lists common features.

Table 4.1 *Common Features of Eukaryotic Cells*

ORGANELLES AND THEIR MAIN FUNCTIONS

Nucleus	Localizing the cell's DNA
Endoplasmic reticulum	Routing and modifying newly formed polypeptide chains; also, synthesizing lipids
Golgi body	Modifying polypeptide chains into mature proteins; sorting and shipping proteins and lipids for secretion or for use inside the cell
Various vesicles	Transporting or storing a variety of substances; digesting substances and structures in the cell; other functions
Mitochondria	Producing many ATP molecules in highly efficient fashion

NON-MEMBRANOUS STRUCTURES AND THEIR FUNCTIONS

Ribosomes	Assembling polypeptide chains
Cytoskeleton	Imparting overall shape and internal organization to the cell; moving the cell and its internal structures

Think about Table 4.1 and you might find yourself asking: What is the advantage of partitioning the cell interior with such organelles? *The compartmentalization allows a large number of activities to occur simultaneously in very limited space.* Consider a photosynthetic cell in a leaf. It can put together starch molecules by one set of reactions and break them apart by another set. Yet the cell would gain nothing if the synthesis and breakdown reactions proceeded at the exact same time on the same starch molecule. Without membranes of organelles, the

Figure 4.10 *Facing page:* Generalized sketches showing some of the features of (**a**) a typical photosynthetic plant cell and (**b**) a typical animal cell.

balance of different chemical activities that helps keep eukaryotic cells alive would spiral out of control.

Organelle membranes have another function besides physically separating incompatible reactions. They allow compatible and interconnected reactions to proceed at different times. For instance, a plant's photosynthetic cells produce starch molecules in an organelle called a chloroplast, then store and later release starch for use in different reactions inside the same organelle.

Which Organelles Are Typical of Plants?

Figure 4.10*a* can start you thinking about organelles inside a typical plant cell. Bear in mind, calling a cell "typical" is like calling a cactus or a water lily or an apple tree a "typical" plant. As is the case for animal cells, variations on the basic plan are mind-boggling. With this qualification in mind, the illustration shows some common locations of organelles and structures you are likely to observe in many plant cells.

Which Organelles Are Typical of Animals?

Now start thinking about organelles of a typical animal cell, such as the one in Figure 4.10*b*. Like plant cells, it has a nucleus, mitochondria, and the other common features listed in Table 4.1. *These structural similarities correlate with functions that are necessary for survival, regardless of the cell type.* We will return to this concept throughout the book.

Comparing Figure 4.10*a* with 4.10*b* also will give you an initial idea of how plant and animal cells differ in their structure. For example, you will never observe an animal cell surrounded by a cell wall. (You might see many different kinds of fungal and protistan cells with one, however.) What other differences can you identify?

Eukaryotic cells have a number of organelles. These are internal, membrane-bound sacs and compartments with specific metabolic functions.

Organelles physically separate chemical reactions, many of which are incompatible. They also separate different reactions in time, as when certain molecules are assembled, stored, then used later in other reaction sequences.

All eukaryotic cells contain certain organelles (such as the nucleus) and structures (such as ribosomes) that perform functions essential for survival. Specialized cells also may incorporate additional kinds of organelles and structures.

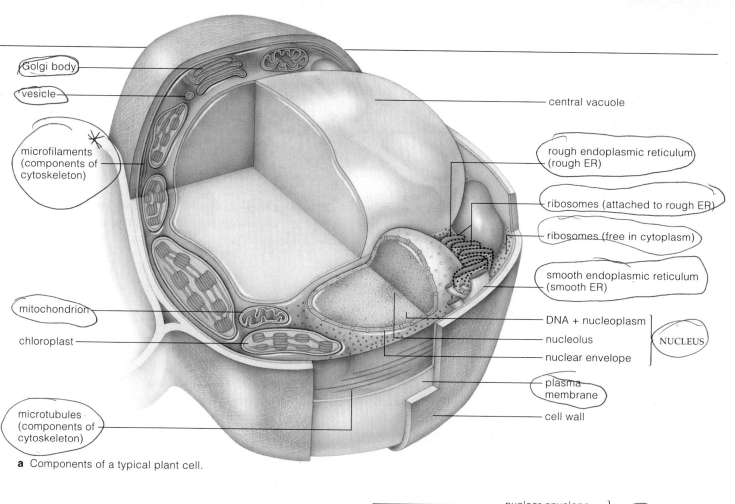

Golgi body

vesicle

microfilaments
(components of
cytoskeleton)

mitochondrion

chloroplast

microtubules
(components of
cytoskeleton)

central vacuole

rough endoplasmic reticulum
(rough ER)

ribosomes (attached to rough ER)

ribosomes (free in cytoplasm)

smooth endoplasmic reticulum
(smooth ER)

DNA + nucleoplasm

nucleolus   NUCLEUS

nuclear envelope

plasma
membrane

cell wall

a Components of a typical plant cell.

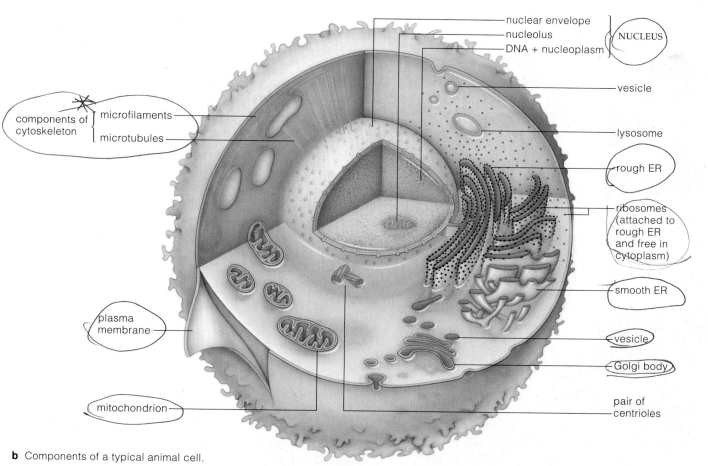

nuclear envelope
nucleolus   NUCLEUS
DNA + nucleoplasm

vesicle

lysosome

rough ER

ribosomes
(attached to
rough ER
and free in
cytoplasm)

smooth ER

vesicle

Golgi body

pair of
centrioles

components of
cytoskeleton
  microfilaments
  microtubules

plasma
membrane

mitochondrion

b Components of a typical animal cell.

THE NUCLEUS

Constructing, operating, and reproducing cells simply cannot be done without carbohydrates, lipids, proteins, and nucleic acids. It takes a class of proteins—enzymes —to build and use these molecules. Said another way, a cell's structure and function begin with proteins. *And instructions for building proteins are located in DNA.*

Unlike bacteria, eukaryotic cells have their genetic instructions distributed among several to many DNA molecules of different lengths. Each human body cell, for instance, has forty-six DNA molecules. End to end,

the molecules would extend outward for a meter or so. Similarly, stretch a frog cell's twenty-six DNA molecules end to end and they would extend ten meters. We have very little idea what frogs are doing with so much of it, but that's a lot of DNA!

Eukaryotic cells, recall, protect DNA in a nucleus. Figure 4.11 shows the structure of this organelle. The nucleus has two functions. *First*, it physically separates all the DNA molecules from the cytoplasm's complex metabolic machinery. (Or, as one professor put it, the nucleus keeps chromosomes in and mitochondria out.) Localizing DNA in a sac makes it easier for a parent cell to copy and sort out hereditary instructions before the cell divides. This is important, because each of the two daughter cells that form later requires a full set of DNA molecules. *Second*, the nuclear membranes serve as a boundary where cells control the movement of substances and signals to and from the cytoplasm. Table 4.2 lists the components of the nucleus.

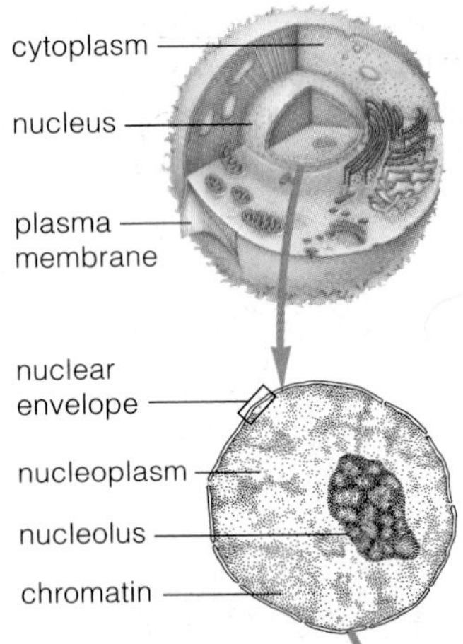

Table 4.2	*Components of the Nucleus*
Nuclear envelope	Pore-riddled double-membrane system that selectively controls the passage of various substances into and out of the nucleus
Nucleoplasm	Fluid interior portion of the nucleus
Nucleolus	Dense cluster of RNA and proteins that will be assembled into subunits of ribosomes
Chromosome	One DNA molecule and many proteins that are intimately associated with it
Chromatin	Total collection of all DNA molecules and their associated proteins in the nucleus

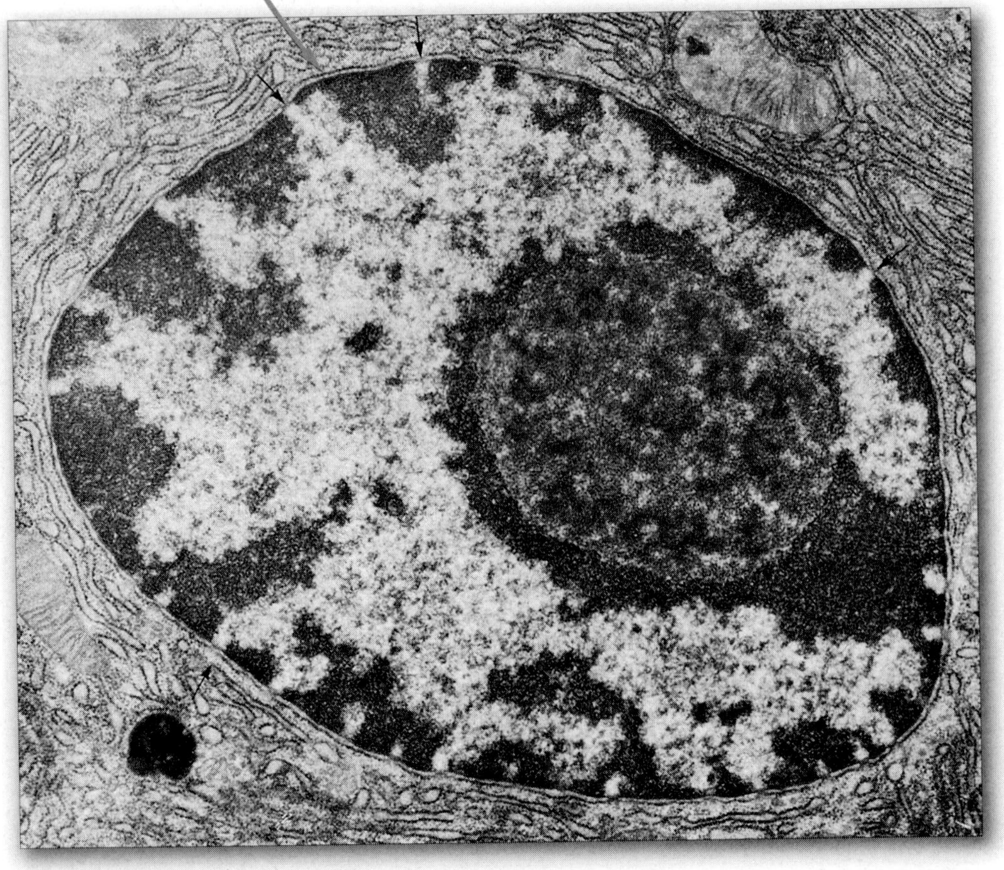

Nuclear Envelope

Unlike cells, a nucleus has two outer membranes, one wrapped around the other. Its double-membrane system is the **nuclear envelope**. It has not one but two lipid bilayers in which many molecules of proteins are embedded (Figure 4.12). The nuclear envelope surrounds the nucleoplasm, the fluid portion of a nucleus. The innermost surface contains attachment sites for strandlike proteins that anchor DNA to the envelope and also help keep it organized. On the outer surface is a profusion of ribosomes. Proteins are

Figure 4.11 Pancreatic cell nucleus. The small arrows on this transmission electron micrograph point to pores where control systems operate to restrict or permit the passage of specific substances across the nuclear envelope.

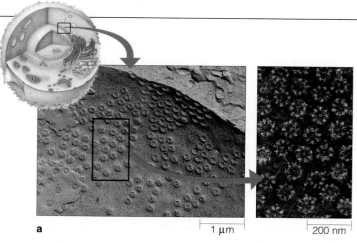

a

1 μm   200 nm

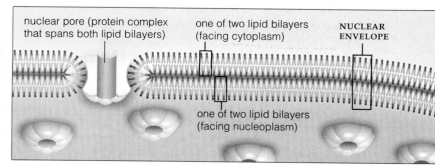

nuclear pore (protein complex that spans both lipid bilayers)

one of two lipid bilayers (facing cytoplasm)

NUCLEAR ENVELOPE

one of two lipid bilayers (facing nucleoplasm)

b

Figure 4.12 (**a**) Part of the outer surface of a nuclear envelope. *Left:* This specimen was fractured to show the intimate layering of its two lipid bilayers. *Right:* Nuclear pores. Each pore across the envelope is an organized cluster of membrane proteins. It permits the selective transport of substances into and out of the nucleus. (**b**) Sketch of the nuclear envelope's structure.

assembled on membrane-bound ribosomes or on other ribosomes positioned in the cytoplasm.

Like all cell membranes, a nuclear envelope's lipid bilayers prevent water-soluble substances from moving freely into and out of a nucleus. However, many pores, each a cluster of proteins, span both bilayers. Small, water-soluble molecules and ions can cross the nuclear envelope at the pores. As you will see later in the book, large molecules (including the ribosomal subunits) also cross through the pores, in highly controlled ways.

Nucleolus

Take another look at the nucleus in Figure 4.11. Inside is a somewhat spherical, dense mass of material. What is it? When eukaryotic cells grow, one or more of these masses form in the nucleus. Each is a **nucleolus** (plural, nucleoli), a construction site where certain RNAs and proteins are combined to make ribosomal subunits. The subunits move through nuclear pores to the cytoplasm. There, two subunits converge as an intact ribosome, where amino acids are assembled into proteins.

Chromosomes

When eukaryotic cells aren't dividing, their individual DNA molecules do not show up in micrographs, which look grainy (Figure 4.11). As cells get ready to divide, however, they duplicate their DNA molecules so each daughter cell will receive all of the required hereditary instructions. Then, the DNA no longer looks grainy; each molecule folds into a compact structure.

Early microscopists bestowed the name *chromatin* on the seemingly grainy substance and *chromosomes* on

the condensed structures. We now define **chromatin** as the cell's collection of DNA, together with all proteins associated with it (Table 4.2). Each **chromosome** is one DNA molecule and its associated proteins, regardless of whether it is in threadlike or condensed form:

one chromosome (one threadlike DNA molecule + proteins; not duplicated)

one chromosome (threadlike but now duplicated; two DNA molecules + proteins)

one chromosome (duplicated and also condensed)

In other words, the appearance of "the chromosome" changes over the life of a eukaryotic cell! In chapters to come, you will look at different aspects of chromosomes, so you may find it useful to remember this point.

What Happens to the Proteins Specified by DNA?

Outside the cell nucleus, new polypeptide chains for proteins are assembled on ribosomes. What happens to them? Many are stockpiled in the cytoplasm or are used at once. Others enter the endomembrane system. As described in the next section, endoplasmic reticulum, Golgi bodies, and diverse vesicles make up this system.

Thanks to DNA's instructions, many proteins take on specific, final forms in the endomembrane system. Also in this system, lipids are assembled and packaged by enzymes and other proteins (these, too, were built according to DNA's instructions). As you will see next, vesicles deliver the proteins and lipids to specific sites within the cell or to the plasma membrane, for export.

The nucleus, an organelle with two outer membranes, keeps the DNA molecules of eukaryotic cells separated from the metabolic machinery of the cytoplasm.

The localization makes it easier to organize the DNA and to copy it before a parent cell divides into daughter cells.

Pores across the nuclear envelope help control the passage of many substances between the nucleus and cytoplasm.

THE ENDOMEMBRANE SYSTEM

The **endomembrane system** is a series of functionally connected organelles in which lipids are assembled and new polypeptide chains are modified. Its products are sorted and shipped to different destinations. Figure 4.13 shows how its organelles—the ER, Golgi bodies, and vesicles—interconnect with one another.

Endoplasmic Reticulum

The functions of the endomembrane system begin with **endoplasmic reticulum**, or **ER**. In animal cells, the ER is continuous with the nuclear envelope, and it extends through the cytoplasm. Its membranes appear rough or smooth, depending on whether ribosomes are attached to the membrane facing the cytoplasm.

We typically observe *rough* ER arranged into stacks of flattened sacs with many ribosomes attached (Figure 4.14*a*). Every new polypeptide chain is synthesized on ribosomes. But only the newly forming chains having a built-in signal can enter the space within rough ER or become incorporated into ER membranes. (That signal is a sequence of fifteen to twenty specific amino acids.) Once the chains are in rough ER, enzymes may attach oligosaccharides and other side chains to them. Many specialized cells secrete the final proteins. Rough ER is abundant in such cells. For example, in your pancreas, ER-rich gland cells make and secrete enzymes that end up in the small intestine and help digest your meals.

Smooth ER is free of ribosomes and curves through cytoplasm like connecting pipes (Figure 4.14*b*). Many cells assemble most lipids inside the pipes. Smooth ER is well developed in seeds. In liver cells, certain drugs as well as toxic metabolic wastes are inactivated in it. Sarcoplasmic reticulum, a type of smooth ER in skeletal muscle cells, functions in controlling contraction.

Golgi Bodies

In **Golgi bodies**, enzymes put the finishing touches on proteins and lipids, sort them out, then package them inside vesicles for shipment to specific locations. For example, an enzyme in one Golgi region might attach a phosphate group to a new protein, thereby giving it a mailing tag to its proper destination.

Commonly, a Golgi body looks rather like a stack of pancakes; it is composed of a series of flattened, membrane-bound sacs (Figure 4.15). In functional terms, the last portion of a Golgi body corresponds to the top pancake. Here, vesicles form as patches of the membrane bulge out, then break away into the cytoplasm.

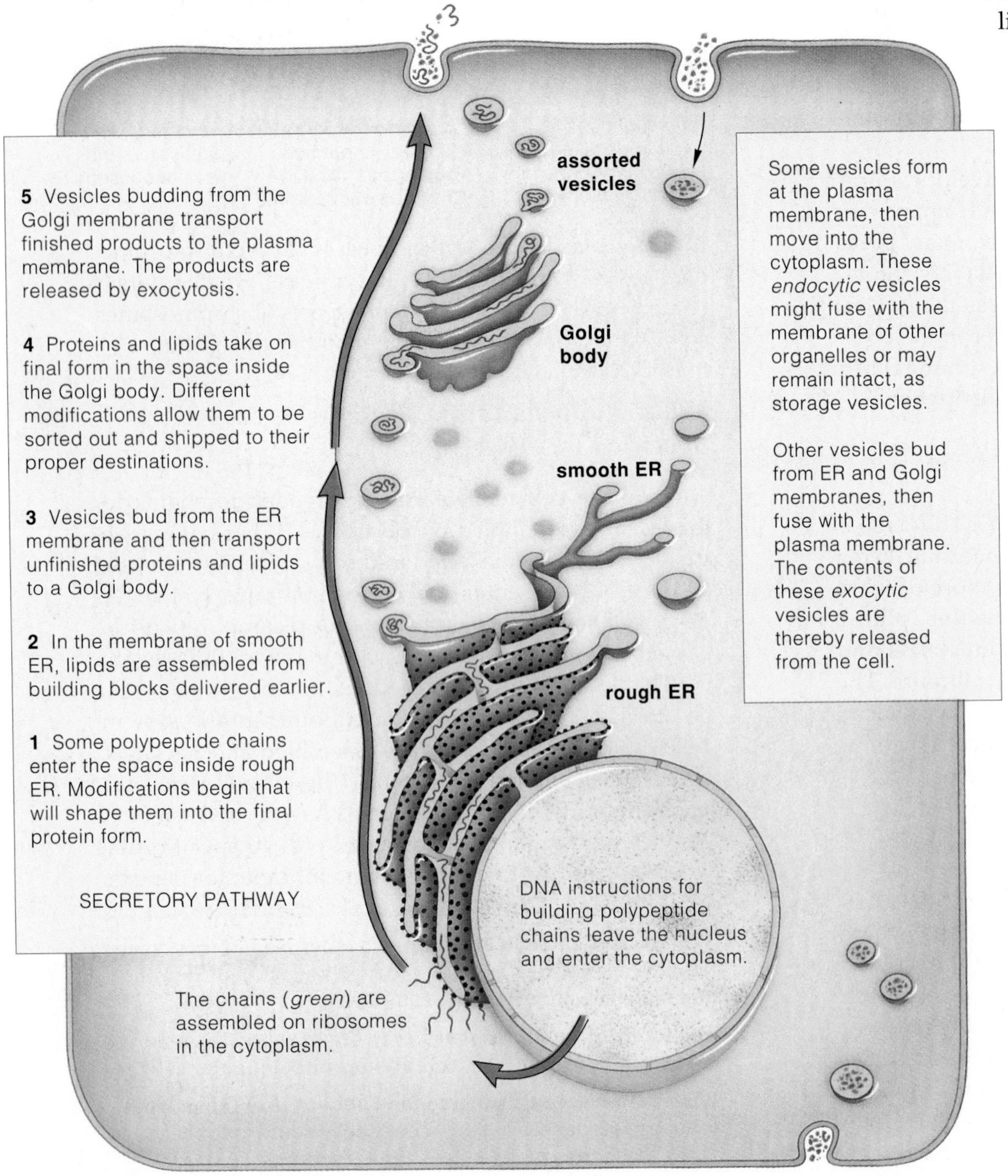

5 Vesicles budding from the Golgi membrane transport finished products to the plasma membrane. The products are released by exocytosis.

4 Proteins and lipids take on final form in the space inside the Golgi body. Different modifications allow them to be sorted out and shipped to their proper destinations.

3 Vesicles bud from the ER membrane and then transport unfinished proteins and lipids to a Golgi body.

2 In the membrane of smooth ER, lipids are assembled from building blocks delivered earlier.

1 Some polypeptide chains enter the space inside rough ER. Modifications begin that will shape them into the final protein form.

SECRETORY PATHWAY

assorted vesicles

Golgi body

smooth ER

rough ER

Some vesicles form at the plasma membrane, then move into the cytoplasm. These *endocytic* vesicles might fuse with the membrane of other organelles or may remain intact, as storage vesicles.

Other vesicles bud from ER and Golgi membranes, then fuse with the plasma membrane. The contents of these *exocytic* vesicles are thereby released from the cell.

DNA instructions for building polypeptide chains leave the nucleus and enter the cytoplasm.

The chains (*green*) are assembled on ribosomes in the cytoplasm.

Figure 4.13 Endomembrane system, a membrane system in the cytoplasm that synthesizes, modifies, packages, and ships proteins and lipids. *Green* arrows highlight a secretory pathway by which some proteins and lipids are packaged and released from many types of cells, including gland cells that secrete mucus, sweat, and digestive enzymes.

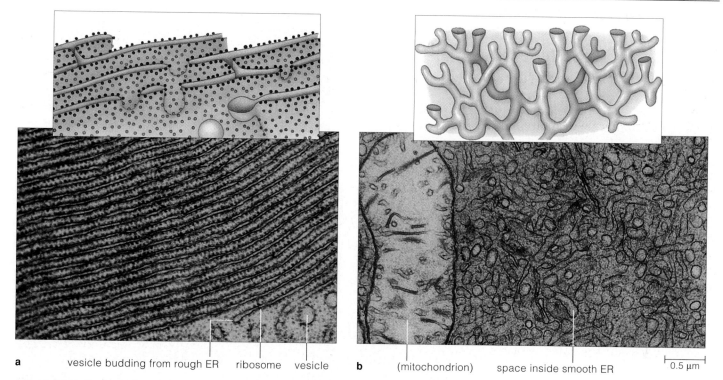

a vesicle budding from rough ER ribosome vesicle b (mitochondrion) space inside smooth ER 0.5 µm

Figure 4.14 Transmission electron micrographs and sketches of endoplasmic reticulum. (**a**) Many ribosomes dot the flattened surfaces of rough ER that face the cytoplasm. (**b**) This section reveals the diameters of many interconnected, pipelike regions of smooth ER.

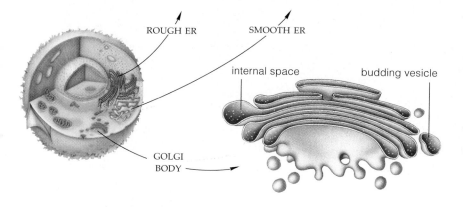

ROUGH ER SMOOTH ER

internal space budding vesicle

GOLGI BODY

Figure 4.15 Sketch and micrograph of a Golgi body from an animal cell. 0.25 µm

A Variety of Vesicles

Vesicles are tiny, membranous sacs that move through the cytoplasm or take up positions in it. The lysosome, a common type, buds from Golgi membranes of animal cells and some fungal cells. **Lysosomes** are organelles of intracellular digestion. They hold enzymes that digest complex carbohydrates, proteins, nucleic acids, and some lipids. Often they fuse with vesicles formed earlier from patches of plasma membrane that surrounded bacteria, molecules, and other items that docked at membrane receptors. Lysosomes also digest entire cells and cell parts. For example, when a tadpole is developing into an adult frog, its tail gradually disappears. Lysosomal enzymes are responding to developmental signals and are helping to destroy cells that make up the tail.

Peroxisomes, another type, are sacs of enzymes that break down fatty acids and amino acids. A product of the reactions, hydrogen peroxide, is toxic, as the Chapter 5 introduction describes. But enzyme action converts it to water and oxygen or uses it to break down alcohol and other toxins. Drink alcohol, and the peroxisomes of liver and kidney cells will degrade nearly half of it.

In the ER and Golgi bodies of the endomembrane system, many proteins take on final form and lipids are assembled.

Lipids, proteins (such as enzymes), and other items become packaged in vesicles destined for export, storage, membrane building, intracellular digestion, and other cell activities.

MITOCHONDRIA

Recall, from Section 3.9, that ATP is an energy carrier. At many different reaction sites, it delivers energy—in the form of phosphate groups—that drives nearly all cell activities. Many ATP molecules form when organic compounds are fully broken down to carbon dioxide and water in a **mitochondrion** (plural, mitochondria).

Only eukaryotic cells contain these organelles. Figure 4.16 gives an idea of their structure. ATP-forming reactions that proceed in mitochondria extract more energy from organic compounds than can be done by any other means. They cannot run to completion without plenty of oxygen. Like all other land-dwelling vertebrates, every time you breathe in, you take in oxygen mainly for the mitochondria in cells—in your case, many trillions of cells.

Each mitochondrion has a double-membrane system. The outer one faces the cytoplasm. The inner one usually is folded repeatedly.

What's the point of such a complex system? It creates two compartments. Enzymes and other proteins of electron transport systems stockpile hydrogen ions in the outer compartment. Ions then flow to the inner compartment in controlled ways. Energy inherent in the flow drives ATP formation, as Chapter 8 describes. Oxygen binds to hydrogen and to the spent electrons, forming the end product, water. Its removal keeps the transfer systems clear for operation.

All eukaryotic cells contain one or more mitochondria. You may find only one in a single-celled yeast. You might find one thousand or more in cells that demand a lot of energy, such as those of muscles. Liver cells, too, contain a profusion of mitochondria. This alone tells you that the liver is a particularly active, energy-demanding organ.

Mitochondria are a lot like bacteria in their size and biochemistry. They have their own DNA and some ribosomes. They divide independently. Did they evolve from ancient prokaryotic cells through "endosymbiosis"? By this idea, a predatory, amoeba-like cell engulfed other cells, which escaped digestion. The engulfed cells, then their descendants, reproduced in the host. In time they became permanent, protected residents. And some of the structures and functions that they formerly needed to maintain an independent life were no longer essential and were lost. In this way

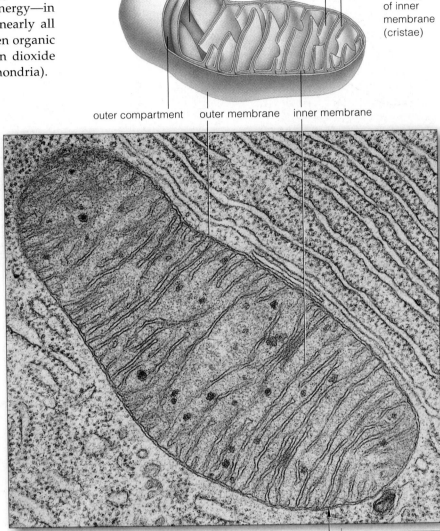

Figure 4.16 Sketch and transmission electron micrograph, thin section, of a typical mitochondrion. This organelle specializes in the production of large quantities of ATP, the main energy-carrying molecule between different reaction sites in cells. The process that produces ATP in mitochondria can't proceed without free oxygen.

inner compartment — repeated foldings of inner membrane (cristae)

outer compartment — outer membrane — inner membrane

25 µm

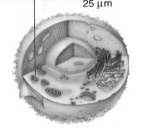

the descendants evolved into mitochondria. We return to this idea of endosymbiotic origins in Section 20.4.

The organelles called mitochondria are the ATP-producing powerhouses of all eukaryotic cells.

Energy-releasing reactions proceed at the compartmented, internal membrane system of mitochondria. The reactions, which require oxygen, produce far more ATP than can be produced by any other cellular reactions.

SPECIALIZED PLANT ORGANELLES

Chloroplasts and Other Plastids

Many plant cells contain plastids, a general category of organelles that specialize in photosynthesis or storage. Three types are common in different tissues of plants: chloroplasts, chromoplasts, and amyloplasts.

Of all eukaryotic cells, only the photosynthetic ones have **chloroplasts**. These organelles use sunlight energy to drive the formation of ATP and NADPH. These compounds are then used to assemble sugars and some other organic compounds.

Most chloroplasts have oval or disklike shapes. The semifluid interior, the stroma, is enclosed in two outer membrane layers. In the stroma is an inner membrane, called the thylakoid membrane, that forms a single sac, or compartment. That compartment is folded into interconnecting, flattened disks that are often arranged as stacks, as in Figure 4.17. Botanists call each stack a granum (plural, grana).

The first stage of photosynthesis starts where many light-trapping pigments, enzymes, and other proteins are embedded in the thylakoid membrane. It ends with formation of ATP and NADPH. The ATP and NADPH are used at sites in the stroma where sugars, then starch and other organic compounds, are built from carbon dioxide and water. Clusters of new molecules of starch (starch grains) may briefly accumulate in the stroma.

The most abundant photosynthetic pigments are chlorophylls that reflect or transmit green light. Others include carotenoids, which reflect or transmit yellow, orange, and red light. The relative abundances of such pigments contribute to the colors of plant parts.

In many ways, chloroplasts are like photosynthetic bacteria. Like mitochondria, they may have evolved by endosymbiosis in the manner outlined in Sections 4.6 and 20.4.

Chromoplasts lack chlorophylls, but they have an abundance of carotenoids. These pigments are the basis of red-to-yellow colors of many flowers, autumn leaves, ripening fruits, and carrots and other roots. The colors commonly attract animals that pollinate the plants or disperse seeds.

Amyloplasts lack pigments. Often they store starch grains and are abundant in cells of stems, potato tubers (underground stems), and seeds.

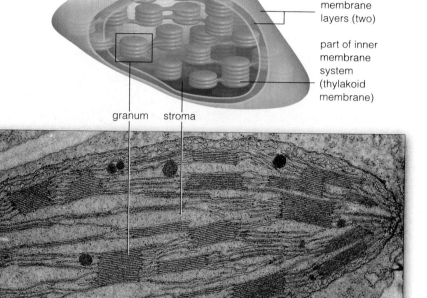

granum stroma

outermost membrane layers (two)

part of inner membrane system (thylakoid membrane)

0.5 µm

Figure 4.17 Generalized sketch of a chloroplast, the key defining feature of every photosynthetic eukaryotic cell. The transmission electron micrograph shows a thin section of a chloroplast from corn (*Zea mays*).

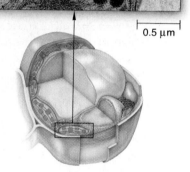

Central Vacuole

Many mature, living plant cells have a **central vacuole** (Figure 4.10a). This fluid-filled organelle stores amino acids, sugars, ions, and toxic wastes, and helps the cell grow. A central vacuole expands during growth. Fluid pressure builds up inside the cell and forces its pliable wall to enlarge. The cell surface area increases, and so does the rate at which water and other substances can be absorbed across the plasma membrane.

In most cases, the central vacuole increases so much in volume that it takes up 50 to 90 percent of the cell's interior. The cytoplasm ends up as a very narrow zone between the central vacuole and plasma membrane.

Photosynthetic eukaryotic cells have chloroplasts and other plastids that function in food production and storage.

Many plant cells have a central vacuole. When this storage vacuole enlarges during growth, cells are forced to enlarge, and this increases the surface area available for absorption.

4.8

SUMMARY OF TYPICAL FEATURES OF EUKARYOTIC CELLS

Before leaving the organelles, take a moment to study Figures 4.18 and 4.19 to put them together into a big picture of cell structure and function.

Figure 4.18 Transmission electron micrograph of a photosynthetic cell from a blade of Timothy grass (*Phleum pratense*), in cross-section. The sketches highlight key organelles.

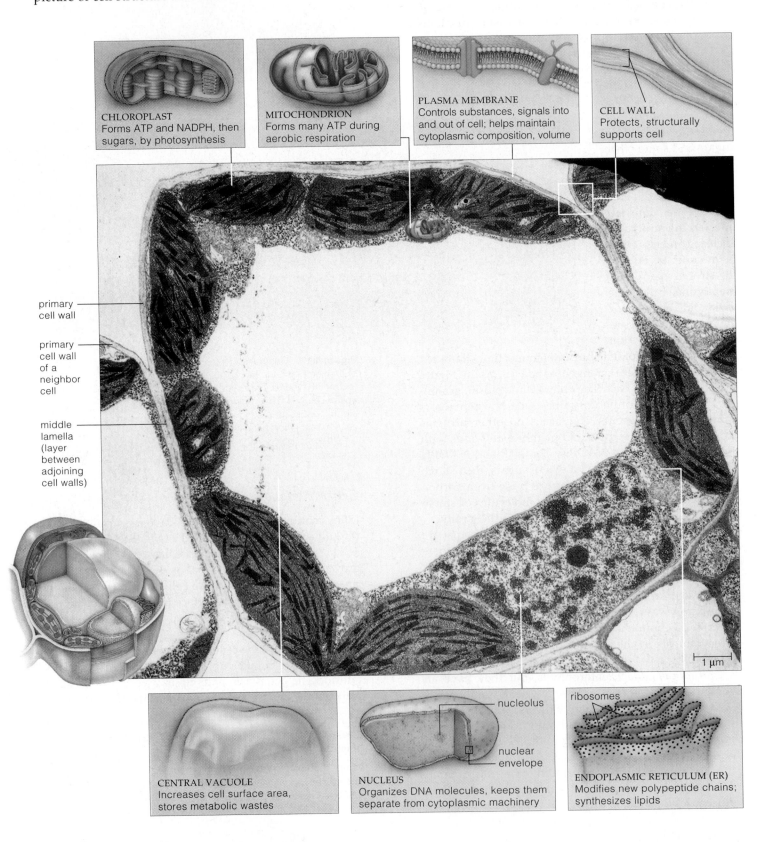

CHLOROPLAST
Forms ATP and NADPH, then sugars, by photosynthesis

MITOCHONDRION
Forms many ATP during aerobic respiration

PLASMA MEMBRANE
Controls substances, signals into and out of cell; helps maintain cytoplasmic composition, volume

CELL WALL
Protects, structurally supports cell

primary cell wall

primary cell wall of a neighbor cell

middle lamella (layer between adjoining cell walls)

1 µm

CENTRAL VACUOLE
Increases cell surface area, stores metabolic wastes

NUCLEUS
Organizes DNA molecules, keeps them separate from cytoplasmic machinery

nucleolus

nuclear envelope

ribosomes

ENDOPLASMIC RETICULUM (ER)
Modifies new polypeptide chains; synthesizes lipids

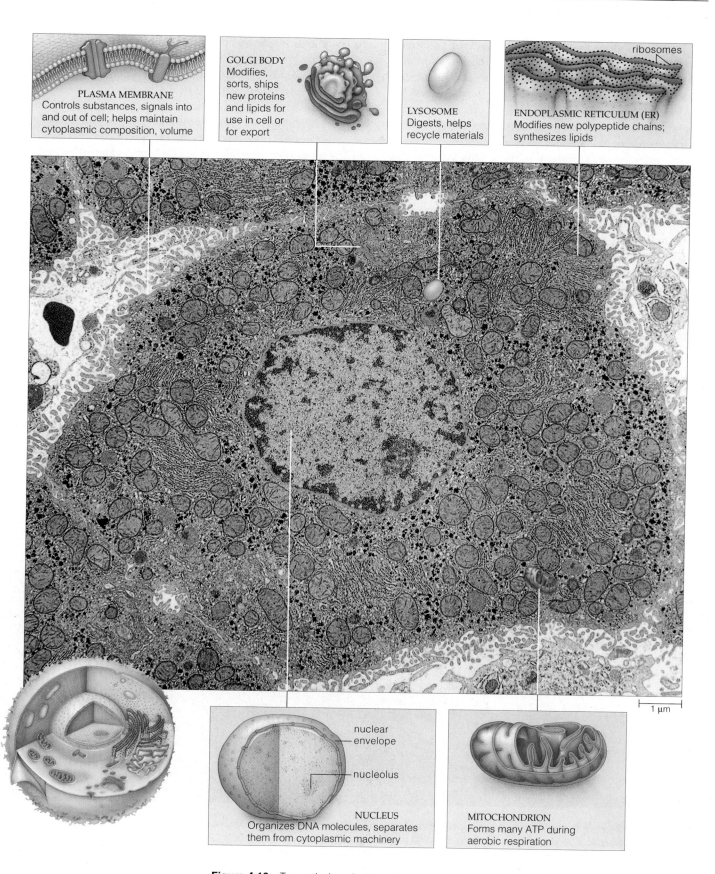

PLASMA MEMBRANE Controls substances, signals into and out of cell; helps maintain cytoplasmic composition, volume

GOLGI BODY Modifies, sorts, ships new proteins and lipids for use in cell or for export

LYSOSOME Digests, helps recycle materials

ENDOPLASMIC RETICULUM (ER) Modifies new polypeptide chains; synthesizes lipids

ribosomes

1 μm

nuclear envelope

nucleolus

NUCLEUS Organizes DNA molecules, separates them from cytoplasmic machinery

MITOCHONDRION Forms many ATP during aerobic respiration

Figure 4.19 Transmission electron micrograph of an animal cell, cross-section. The sketches highlight key organelles. The specimen is a cell from the liver of a rat.

4.9

EVEN YOUR CELLS HAVE A SKELETON

What pops into mind when you hear the word *skeleton*? An organized system of bones that includes a skull and rib cage? That is only one kind of skeleton in nature. If you use the general meaning of the word—*something that forms a structural framework*—then eukaryotic cells, too, have a skeleton. A complex, interconnected system of protein filaments extends between their nucleus and plasma membrane. Different parts of this system, the **cytoskeleton**, reinforce, organize, and move internal cell parts and often function in motility. Many parts are permanent, and others form only at certain times in the life of a cell. Figure 4.20 shows examples from one of the main types of cells making up your body.

Microtubules and **microfilaments** are two classes of cytoskeletal elements with roles in most eukaryotic cell movements. Certain animal cells also have ropelike **intermediate filaments** that impart strength. Although prokaryotic cells do not contain a cytoskeleton, some do have reinforcing, actin-like protein filaments.

Microtubules—The Big Ones

Microtubules function mainly in internal organization and in sustained, directional movements that put cell structures and organelles in new locations. Many of them become assembled in a spindle apparatus, which moves chromosomes in specific ways before the cell divides. Other microtubules help shunt chloroplasts, vesicles, and other organelles about.

A microtubule is a straight, hollow cylinder which, at twenty-five nanometers across, is the largest of all cytoskeletal elements in plant and animal cells (Figure 4.21*a*). Cells build it from tubulin monomers. Tubulin consists of two chemically distinct polypeptide chains, each folded into a globular shape. When the cylinder is being assembled, all of its monomers are oriented in the same direction. This assembly pattern puts slightly different chemical properties at opposite ends of the cylinder. Like bricks being stacked up into a new wall, monomers join the cylinder's *plus* (fast-growing) end, which at first grows freely through the cytoplasm.

In animal cells, microtubules usually grow out in all directions from the centrosome, a type of microtubule organizing center (MTOC). Microtubules form at these sites of dense material, and their *minus* (slow-growing) ends remain anchored in them. But microtubules are inherently unstable, and they can abruptly fall apart.

Cells control which microtubules disassemble and which ones remain intact by selectively capping some of them with other proteins. For instance, caps might be added only to microtubules lengthening in a specific direction—say, the ones that happen to be reinforcing the advancing end of a white blood cell on the prowl. Microtubules are not needed at the trailing end of the cell; they are not capped, and so they fall apart.

Microtubules are targets in the ongoing chemical warfare between plants and the animals that eat them. For instance, the autumn crocus (*Colchicum autumnale*) in Figure 4.22*a* makes colchicine. This poison blocks the assembly and promotes disassembly of microtubules in animal cells. It does not hurt the plants, which have an evolved insensitivity to it. (Colchicine cannot bind well to the type of tubulin monomers that plants produce.) Similarly, taxol is a poison made by the western yew

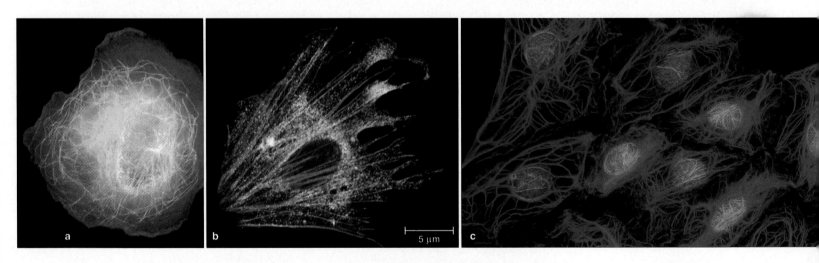

5 µm

Figure 4.20 Cells stained to reveal their cytoskeleton. (**a**) Microtubules (*gold*) and distribution of actin filaments (*red*) in a pancreatic epithelial cell. This cell secretes bicarbonate, which functions in digestion by neutralizing stomach acid. The *blue* region is DNA. (**b**) Actin filaments (*red*) and accessory proteins (*green* and *blue*) help new epithelial cells such as this one migrate to proper locations in tissues while human embryos are developing. (**c**) Keratin intermediate filaments (*red*) of kangaroo rat cells in culture. The *blue*-stained organelle in each cell is the nucleus.

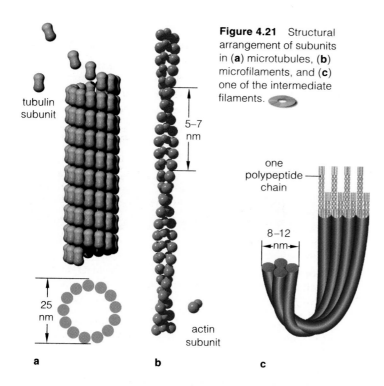

tubulin
subunit

5–7
nm

25
nm

one
polypeptide
chain

8–12
nm

actin
subunit

a

b

c

Figure 4.21 Structural arrangement of subunits in (**a**) microtubules, (**b**) microfilaments, and (**c**) one of the intermediate filaments.

Figure 4.22 Two sources of microtubule poisons: (**a**) Autumn crocus, *Colchicum autumnale*. (**b**) Western yew, *Taxus brevifolia*.

(*Taxus brevifolia*), shown in Figure 4.22*b*. Taxol also can block cell division in animal cells. Doctors now use a synthetic version of taxol to stop the uncontrolled cell divisions that give rise to some kinds of tumors.

Microfilaments—The Thin Ones

Microfilaments, at five to seven nanometers across, are the thinnest cytoskeletal elements. Each is composed of two helically twisted polypeptide chains. The chains are assembled from actin monomers (Figure 4.21*b*).

Like microtubules, microfilaments can be assembled and disassembled in controlled ways. They are often arranged as bundles or networks. For instance, the **cell cortex** is a mesh of cytoskeletal elements beneath the plasma membrane. It has bundles, crosslinked arrays, and gel-like arrays of microfilaments. Some reinforce a cell's shape. Some reconfigure the surface, as when the bundle around an animal cell's midsection contracts to pinch the cell in two. Some anchor membrane proteins. Others serve in muscle contraction. When crosslinked ones loosen in some large cells, *cytoplasmic streaming* occurs. Local gel-like regions become more fluid, and they flow vigorously. The flow redistributes substances and cell components through the cell interior.

Myosin and Other Accessory Proteins

The genes for tubulin and actin apparently have been highly conserved over time; all eukaryotic cells make similar forms of these monomers for microtubules and microfilaments. In spite of the uniformity, both kinds of cytoskeletal elements play diverse roles. How? Other proteins associate with them to execute different tasks. For example, as you will see, kinesin and myosin are two kinds of *motor* proteins. They use ATP energy to move on cytoskeletal tracks and help move some cell component to new locations. The *crosslinking* proteins splice together microfilaments as the cell cortex.

Intermediate Filaments

Intermediate filaments are the most stable elements of some cytoskeletons. They are between eight and twelve nanometers wide (Figure 4.21*c*). The six known groups strengthen and help maintain the shape of cells or cell parts. For example, lamins help form a basketlike mesh that reinforces the nucleus. They anchor many adjacent actin and myosin filaments into units of contraction inside muscle cells. Desmins and vimentins help hold the contractile units in position. Different cytokeratins reinforce cells that make nails, claws, horns, and hairs.

In certain animal cells only, intermediate filaments are present in the cytoplasm. Because they have only one or at most two kinds of these filaments, researchers can use them to identify the type of cell. Such typing is a useful tool in diagnosing the tissue origin of different forms of cancer cells. Intermediate filaments also may reinforce the nuclear envelope of all eukaryotic cells.

A cytoskeleton is the basis of the shape, internal structure, and movement of eukaryotic cells. It consists of the same kinds of protein filaments in all eukaryotic cells. Accessory proteins associate with these filaments and extend their range of functions.

Microtubules, the largest cytoskeletal element, are cylinders made of tubulin monomers. They help spatially organize the cell interior and have roles in moving cell components.

Microfilaments, the thinnest cytoskeletal element, consist of two polypeptide chains of actin monomers, helically coiled together. They form flexible, linear bundles and networks that reinforce or restructure the cell surface.

Intermediate filaments help strengthen and maintain the shape of cells or cell parts. Some are present only in the cytoplasm of certain types of animal cells. Others probably help reinforce the nuclear envelope of all eukaryotic cells.

HOW DO CELLS MOVE?

Chugging Along With Motor Proteins

Think of the bustle at a train station during the busiest holiday season, and you get an idea of what is going on inside a cell. Instead of people, cells move vesicles, chromosomes, and other components from one place or position to another. For tracks, they use microtubules and microfilaments. Instead of train engines, they use **motor proteins** which, when repeatedly energized by ATP hydrolysis, move cell components in a sustained, directional fashion. Generally, motor proteins called *kinesins* and *dyneins* move along microtubules; *myosins* move along microfilaments (Figures 4.23 and 4.24).

Consider the kinesins, a diverse group of freight engines. Some use microtubules to move chromosomes during cell division. Like dyneins, some can make one microtubule slide over another. Others line up ER and Golgi bodies with microtubules that extend from the centrosome to the cell periphery. Still others chug along tracks in nerve cells that extend from your spine to your toes. A number of these engines are organized in series, with each moving its freight partway along the track before transferring it to the next engine in line. It takes them a day or two to move vesicles filled with signaling molecules to the end of the line. In plants, kinesins can put chloroplasts in new light-intercepting positions as the angle of the sun changes overhead.

How do they do it? Here we leave the train, so to speak, in favor of a recent model for kinesin "walking." A kinesin molecule has two helical domains that can bind a specific structure or organelle at one end. The other end has two globular domains, analogous to feet, that can bind to a microtubular track (Figure 4.24). One foot moves forward when ATP energizes it. Then the other foot does the same and moves ahead, past the first foot. Through repeats of this molecular action, kinesin chugs along the microtubule, dragging freight with it. Typically it finishes a 1,000-nanometer journey in a few seconds, then lets go of the microtubule.

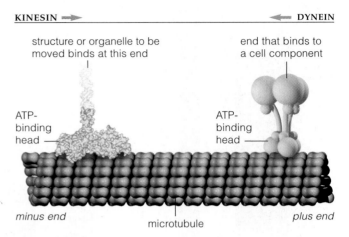

KINESIN ➡️   ⬅️ DYNEIN

structure or organelle to be moved binds at this end

end that binds to a cell component

ATP-binding head

ATP-binding head

minus end   microtubule   plus end

Figure 4.23 Kinesin, one of the motor proteins. Most kinesins move toward the plus end of a microtubule (or microfilament). Most dyneins move toward the minus end. If a microtubule is anchored near the center of a cell, kinesins move components away from it; dyneins move them toward it. These proteins are not drawn to scale. Each "foot" is about the size of one tubulin subunit.

Myosins work in much the same way. They move structures on microfilaments or slide one microfilament over another. For example, each muscle cell is divided lengthwise into many contractile units (Chapter 37). Each unit has microfilaments arranged in parallel with bundles of myosin. Myosin is shaped like a golf club, with a shaft and head. A head energized by ATP binds a microfilament, thrusts it in some direction, and lets go. Repeated power strokes by many myosin molecules make microfilaments slide toward each unit's center. A muscle cell shortens (contracts) as all the units shorten.

Cilia, Flagella, and False Feet

Besides moving internal components, many cells move their body or some part of it through the environment. Some types do this with **flagella** (singular, flagellum)

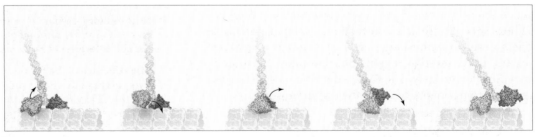

Figure 4.24 (**a**) Ribbon model for kinesin. (**b**) One model for how the kinesin molecule's two heads repeatedly move along a microtubule, driven by phosphate-group transfers from ATP.

a

b

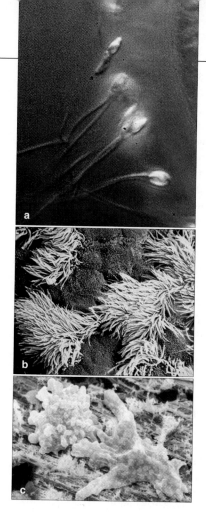

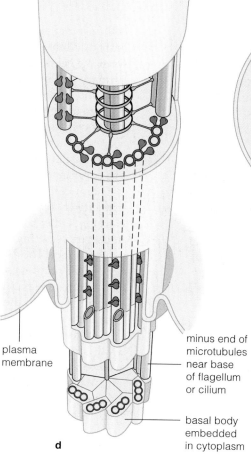

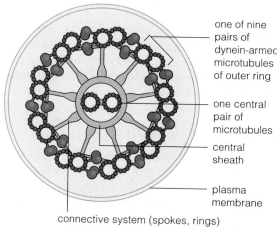

one of nine pairs of dynein-armed microtubules of outer ring

one central pair of microtubules

central sheath

plasma membrane

connective system (spokes, rings)

plasma membrane

minus end of microtubules near base of flagellum or cilium

basal body embedded in cytoplasm

d

Figure 4.25 A few motile structures. (**a**) Human sperm. Each sperm has a flagellum that has one function: moving its DNA-packed head to an egg. (**b**) From the lining of an airway to the lungs, the free surface of ciliated cells (*yellow*) and mucus-secreting cells (*brown*). (**c**) Pseudopods of a macrophage (*light blue*) about to contact another cell. (**d**) Internal organization of flagella and cilia. Both structures have a 9 + 2 array: a ring of nine pairs of microtubules around two pairs at the core. All are connected by spokes and linking elements that restrict the range of sliding.

or **cilia** (cilium). Compared to cilia, flagella usually are longer and less profuse on a cell. Sperm and other free-living, motile cells use them as whiplike tails to swim, as in Figure 4.25*a*. In many multicelled species, ciliated cells stir air or fluid. Many thousands line the airways in your chest (Figure 4.25*b*). Their coordinated motion directs mucus-trapped particles away from the lungs.

Both motile structures have an internal, cylindrical array of nine pairs of microtubules plus a central pair. Protein spokes and links connect to and stabilize this "9+2 array," which arises from a barrel-shaped MTOC known as a **centriole**. The centriole remains positioned below the finished array as a **basal body** (Figure 4.25*d*).

Both flagella and cilia beat by a sliding mechanism that begins at the minus end of microtubules. Inside an unbent flagellum (or cilium), all microtubule doublets extend the same distance into the tip. When the motile structure bends, doublets on the side bending the most are being displaced farthest from the tip.

The motor protein dynein forms many short arms down the length of each microtubule pair in the outer ring. When energized by ATP, all the arms bind the pair in front of them, tilt in a short, downward stroke, then let go. As the bound pair slides down, its arms

bind and force the doublet in front of it to slide down a bit also, and so on around the ring. The microtubules cannot slide far. The system of spokes and links stops them, but it has enough slack built in to permit each microtubule to *bend* a bit. Thus a sliding motion caused by motor proteins is converted to a bending motion.

A final example: Amoebas, macrophages, and many other free-living cells form **pseudopods** ("false feet"). These temporary, irregular lobes project from the body and are used for locomotion and prey capture (Figure 4.25*c*). Microfilaments rapidly elongate in the lobe and push it forward. Motor proteins may chug along these tracks, dragging the plasma membrane with them. The animal cell in Figure 4.20*b* also was moving forward with the help of sheetlike extensions of microfilaments.

Cell contractions and migrations, chromosome movements, and all other forms of cell motility arise at organized arrays of microtubules, microfilaments, and accessory proteins.

When energized by ATP, motor proteins move steadily, in specific directions, along tracks of microtubules and microfilaments. They deliver specific cell components that are bound to them to new locations or into new positions.

CELL SURFACE SPECIALIZATIONS

Our survey of eukaryotic cells concludes with a look at cell walls and some other specialized surface structures. Many are architectural marvels constructed from cell secretions. Others are clustered membrane proteins that structurally and functionally connect neighboring cells.

Eukaryotic Cell Walls

Single-celled eukaryotic species are directly exposed to their surroundings. Many have a **cell wall**, a structural component that wraps continuously around the plasma membrane. A cell wall protects and physically supports its owner. The wall is porous, so water and solutes can easily move to and from the plasma membrane. A great variety of protistans, as in Figure 4.26, have a cell wall. So do plant cells and many types of fungal cells.

For instance, young cells in actively growing parts of plants secrete gluelike polysaccharides (such as pectin), glycoproteins, and cellulose. The

Figure 4.26 From a freshwater habitat, one of the single-celled, walled protistans (*Ceratium*, a dinoflagellate).

cellulose molecules form ropelike strands that become embedded in the gluey matrix. All secretions combined form the cell's **primary wall** (Figure 4.27). Being sticky, primary walls cement abutting cells together. They are thin and pliable, which permits the cell to continue to enlarge under the pressure of incoming water.

At cell surfaces exposed to air, waxes and other cell secretions accumulate. The deposits form a cuticle. This semitransparent, protective surface cover helps restrict evaporative water loss from plants (Figure 4.28a).

Plant cells that produce only a thin wall retain the capacity to divide or to change shape as they grow and develop. Other plant cells stop enlarging as they mature and start secreting material on the primary wall's inner surface. The deposits form a rigid, **secondary wall** that reinforces cell shape (Figure 4.27e). Whereas cellulose makes up less than 25 percent of the cell's primary wall, the secondary wall deposits are more extensive and contribute more to its structural support.

In woody plants, up to 25 percent of the secondary wall is made of lignin. Lignin consists of a six-carbon ring to which a three-carbon chain and an oxygen atom are attached. It makes plant parts more waterproof, less inviting to plant-attacking organisms, and stronger.

Figure 4.27 Examples of cell walls in (**a**) flax plants. (**b**) Primary cell wall of young cells in flower petals. Cell secretions form the middle lamella, a layer between the walls of adjoining cells. The layer is thickest in adjoining corners. Plasmodesmata, membrane-lined channels across the adjacent walls, connect the cytoplasm of neighboring cells. (**c,d**) Part of the lustrous fibers in a cell from a flax stem. We make linen from such fibers, which are three times stronger than cotton fibers. (**e**) In flax fibers, as in many other types of plant cells, more layers become deposited inside the primary wall. The layers stiffen the wall and help maintain its shape. Later, the cell dies, leaving the stiffened walls behind. This also happens in water-conducting pipelines that thread through most plant tissues. Interconnected, stiffened walls of dead cells form the tubes.

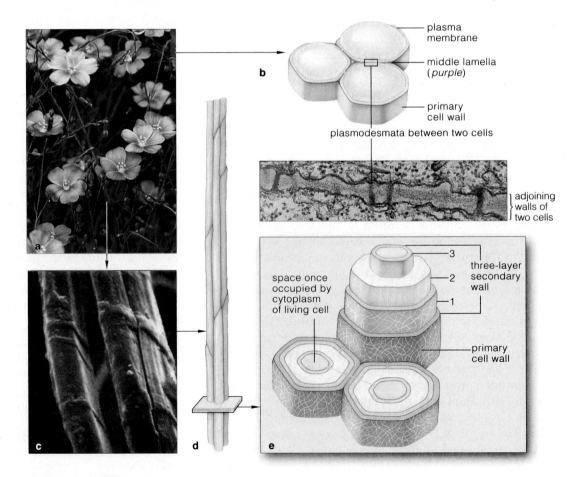

plasma membrane

middle lamella (*purple*)

primary cell wall

plasmodesmata between two cells

adjoining walls of two cells

space once occupied by cytoplasm of living cell

three-layer secondary wall

primary cell wall

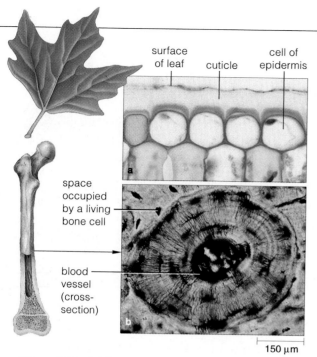

surface
of leaf   cuticle   cell of
epidermis

a

space
occupied
by a living
bone cell

blood
vessel
(cross-
section)

b

150 µm

Figure 4.28 (**a**) Section through a plant cuticle, an outer surface layer of cell secretions. (**b**) Cross-section through compact bone tissue, stained for microscopy.

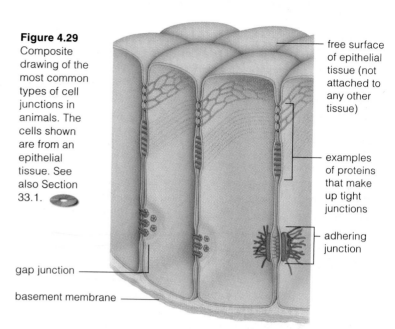

Figure 4.29 Composite drawing of the most common types of cell junctions in animals. The cells shown are from an epithelial tissue. See also Section 33.1.

free surface of epithelial tissue (not attached to any other tissue)

examples of proteins that make up tight junctions

adhering junction

gap junction

basement membrane

Matrixes Between Animal Cells

Animal cells have no walls. But intervening between many of these cells are matrixes made of cell secretions and materials taken up from the surroundings. Think of cartilage at the knobby ends of your leg bones. Cartilage consists of scattered cells plus collagen or elastin fibers formed from their secretions, all embedded in a ground substance of polysaccharides (Section 33.2). Or think of bone tissue, in which an extensive matrix separates the living bone cells (Figure 4.28*b* and Section 37.5).

Cell Junctions

Even when a wall or some other structure imprisons a cell in its own secretions, the only contact that cell has with the outside world is through its plasma membrane. In multicelled species, membrane components project into adjoining cells and into the surrounding medium. Among the components are **cell junctions** where a cell sends and receives signals and materials, and where it recognizes and cements itself to cells of the same type.

In plants, for instance, many tiny channels cross the adjacent primary walls of living cells and interconnect their cytoplasm. Figure 4.27*b* shows a few. Each channel is a plasmodesma (plural, plasmodesmata). The plasma membranes of adjoining cells merged during growth. Now they line the channels, so there is an uninterrupted flow of substances between the cells. That is how living cells in the plant body engage in quick exchanges.

In most animal tissues, three categories of cell-to-cell junctions are common (Figure 4.29). *Tight* junctions link the cells of most body tissues, including epithelia that line outer surfaces, internal cavities, and organs. They seal abutting cells so water-soluble substances can't pass between them. That is why gastric fluid cannot leak across the stomach's lining and damage internal tissues. *Adhering* junctions join cells in tissues of the skin, heart, and other organs subjected to stretching. Gap junctions link the cytoplasm of neighboring cells. They are open channels for a rapid flow of signals and substances.

Cell Communication

Also at the plasma membrane are diverse receptors for signals from other cells. **Cell communication** refers to ways in which one cell signals another to change its activities. For example, certain hormones dock at body cells as a signal to start dividing. That external signal is essential for growth and development of all complex organisms. Throughout this book you will come across specific examples of cell communication. Chapters 15, 28, 32, 34, 36, 39, and 43 especially provide details of the kinds of signaling mechanisms used to coordinate activities of all living cells in the multicelled body.

A variety of protistan, plant, and fungal cells have a porous but protective wall that surrounds the plasma membrane. The cells themselves secrete wall-forming materials.

The combined secretions of many cells form the cuticle that covers the surface of plants and certain animals, as well as the extracellular matrixes of animal tissues.

In multicelled organisms, diverse proteins at the plasma membrane function as physical links, cytoplasmic bridges, and receptors for communication signals between cells.

4.12

PROKARYOTIC CELLS

We turn now to prokaryotic cells. Unlike the cells you've considered so far, a nucleus does not enclose their DNA. *Prokaryotic* means "before the nucleus." Biologists chose the word to remind us that this type of cell appeared on Earth before the nucleus evolved in the forerunners of eukaryotic cells.

These are the smallest known cells. Most are not much more than a micrometer wide; even rod-shaped species are only a few micrometers long. In structural terms, they are the simplest cells to think about. Most have a semirigid or rigid cell wall that surrounds the plasma membrane, supports the cell, and imparts shape to it (Figure 4.30a). As you will see later, their wall differs from walls of eukaryotic cells. Dissolved substances can move freely to and from the plasma membrane because the wall is permeable. Commonly, sticky polysaccharides enclose a wall and help cells attach to interesting surfaces, such as river rocks, teeth, and the vagina. Many pathogenic (disease-causing) eubacteria have a capsule made of thick, jellylike polysaccharides. It surrounds and protects the wall.

As in eukaryotic cells, the plasma membrane of prokaryotic cells mediates the flow of substances to and from the cytoplasm. It, too, has proteins that serve as channels, transporters, and receptors for diverse signals and substances. It, too, incorporates built-in machinery for metabolic reactions, such as the breakdown of energy-rich compounds. In most of the photosynthetic prokaryotic cells, organized arrays of proteins in the plasma membrane harness light energy, then convert it to chemical energy of ATP.

Prokaryotic cells are so small they have only a tiny volume of cytoplasm. But they have many ribosomes on which polypeptide chains are built. The cytoplasm is distinct from that of eukaryotic cells; it is continuous with an irregularly shaped region of DNA. Membranes don't enclose this region, which is the nucleoid (Figure 4.30b). A circular DNA molecule, also called a bacterial chromosome, occupies the region. Given their tiny size and internal simplicity, the cells don't need a complex cytoskeleton, but actin-like protein filaments do form a scaffold that helps shape the sturdier cell wall.

Projecting out from the surface of many prokaryotic cells are one or more bacterial flagella. These threadlike motile structures differ from eukaryotic flagella; they do not have a 9 + 2 array of microtubules. Bacterial flagella help a cell move rapidly through fluid environments. Other surface projections include pili (singular, pilus), the main protein filaments that help many types of cells attach to various surfaces, even to one another.

We recognize two major groups of prokaryotic cells: Archaebacteria and Eubacteria (Sections 19.7 and 21.3).

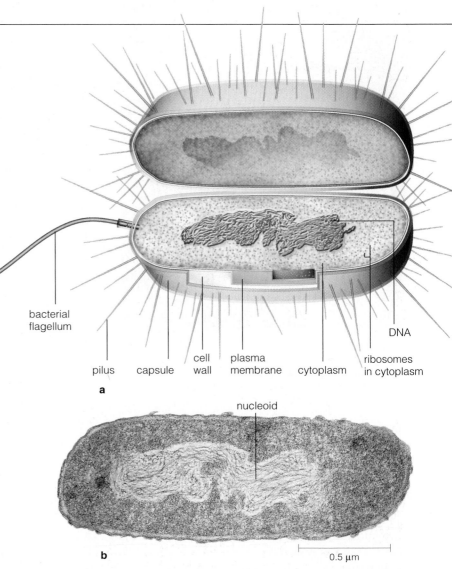

bacterial flagellum

pilus   capsule   cell wall   plasma membrane   cytoplasm   ribosomes in cytoplasm   DNA

a

nucleoid

b

0.5 µm

Figure 4.30 (**a**) Generalized sketch of a typical prokaryotic body plan. (**b**) Micrograph of the bacterium *Escherichia coli*.

The gut of humans and other mammals holds a huge population of a normally harmless strain of *E. coli*. Some harmful strains can contaminate meat sold commercially. One strain also can taint cider. Cooking meat thoroughly or pasteurizing cider kills these pathogens. Where this was not done, people who ate the meat or drank the cider became sick. Some died.

Facing page: (**c**) Researchers manipulated this *E. coli* cell to release its single, circular molecule of DNA. (**d**) Cells of various bacterial species are shaped like balls, rods, or corkscrews. Ball-shaped cells of *Nostoc*, a photosynthetic eubacterium, stick together in a thick, jellylike sheath of their own secretions. Chapter 21 gives other examples. (**e**) Like this *Pseudomonas marginalis* cell, many species have one or more bacterial flagella that propel the cell body in fluid environments.

Together, the species in these groupings are the most metabolically diverse organisms of all. Different kinds have managed to exploit energy and raw materials in just about every kind of environment, ranging from hot springs and snow fields to hot, dry deserts. In addition, ancient prokaryotic cells gave rise to all the protistans,

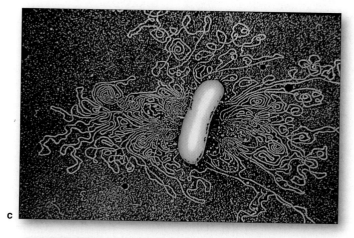
c

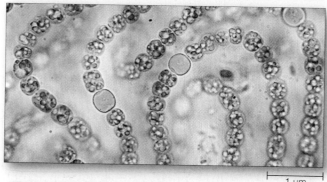
d

1 µm

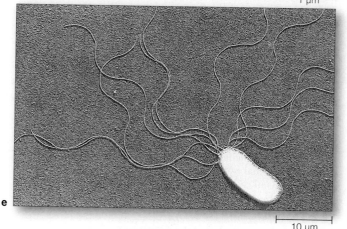
e

10 µm

plants, fungi, and animals ever to appear on Earth. The evolution, structure, and functions of these remarkable cells are topics of later chapters.

Archaebacteria and eubacteria are prokaryotic cells; their DNA is not housed inside a nucleus. Most species have a cell wall around the plasma membrane. Generally, they do not have organelles comparable to those of eukaryotic cells.

These are the simplest cells, but as a group they show the most metabolic diversity. Their metabolic activities proceed at the plasma membrane and within the cytoplasm.

SUMMARY   *Gold* indicates text section

1. Three generalizations constitute the cell theory: *CI*
   a. All living things are composed of one or more cells.
   b. The cell is the smallest entity that still retains the properties of life. That is, it lives independently or has a heritable, built-in capacity to do so.
   c. New cells arise only from cells that already exist.

2. All cells start out life with a plasma membrane, a region of DNA, and cytoplasm. The plasma membrane (a thin, outer membrane) maintains a cell as a distinct entity and lets metabolic activities proceed apart from random events in the environment. Many substances and signals cross it in controlled ways. The cytoplasm consists of ribosomes, structural elements, fluids, and (only in eukaryotic cells) organelles between the region of DNA and the plasma membrane. *4.1*

3. Membranes of lipids and proteins are essential for cell structure and function. The lipids form two layers, with their hydrophobic tails sandwiched between their hydrophilic heads (which are dissolved in fluid). This bilayer gives the membrane its basic structure. Water-soluble substances cannot cross it. Proteins embedded in the bilayer or attached to it have diverse roles. Many are channels or pumps where water-soluble substances cross the bilayer, others are receptors, and so on. *4.1*

4. The surface-to-volume ratio constrains cell size, cell shape, and the body plans of multicelled organisms. *4.1*

5. Modern microscopes harness wavelengths of light or focused beams of electrons to reveal cell structures. *4.2*

6. Membranes divide the cytoplasm of eukaryotic cells into functional compartments (organelles). Prokaryotic cells do not have comparable organelles. *4.3, 4.12*

7. Organelle membranes separate metabolic reactions inside the cytoplasm and let different kinds proceed in orderly fashion. Many similar reactions proceed at the plasma membrane of prokaryotic cells. *4.3*
   a. The nuclear envelope functionally separates DNA from the metabolic machinery of the cytoplasm. *4.4*
   b. The endomembrane system includes the ER, Golgi bodies, and vesicles. Many new proteins are modified into final form and lipids are assembled in this system. Finished products are packaged and then shipped off to destinations inside or outside the cell. *4.5*
   c. Mitochondria specialize in producing numerous ATP molecules by oxygen-requiring reactions. *4.6*
   d. Chloroplasts specialize in capturing energy from the sun and using it to build organic compounds. *4.7*

8. The cytoskeleton of eukaryotic cells functions in cell shape, internal organization, and movement. Many cells have walls and other surface specializations. *4.9, 4.10*

9. Table 4.3 on the next page summarizes the defining features of prokaryotic and eukaryotic cells.

Table 4.3 *Summary of Typical Components of Prokaryotic and Eukaryotic Cells*

Cell Component	Archaebacteria, Function	PROKARYOTIC Eubacteria	EUKARYOTIC Protistans	Fungi	Plants	Animals
Cell wall	Protection, structural support	✓*	✓*	✓	✓	None
Plasma membrane	Control of substances moving into and out of cell	✓	✓	✓	✓	✓
Nucleus	Physical separation and organization of DNA	None	✓	✓	✓	✓
DNA	Encoding of hereditary information	✓	✓	✓	✓	✓
RNA	Transcription, translation of DNA messages into polypeptide chains of specific proteins	✓	✓	✓	✓	✓
Nucleolus	Assembly of subunits of ribosomes	None	✓	✓	✓	✓
Ribosome	Protein synthesis	✓	✓	✓	✓	✓
Endoplasmic reticulum (ER)	Initial modification of many of the newly forming polypeptide chains of proteins; lipid synthesis	None	✓	✓	✓	✓
Golgi body	Final modification of proteins, lipids; sorting and packaging them for use inside cell or for export	None	✓	✓	✓	✓
Lysosome	Intracellular digestion	None	✓	✓*	✓*	✓
Mitochondrion	ATP formation	**	✓	✓	✓	✓
Photosynthetic pigments	Light–energy conversion	✓*	✓*	None	✓	None
Chloroplast	Photosynthesis; some starch storage	None	✓*	None	✓	None
Central vacuole	Increasing cell surface area; storage	None	None	✓*	✓	None
Bacterial flagellum	Locomotion through fluid surroundings	✓*	None	None	None	None
Flagellum or cilium with 9 + 2 microtubular array	Locomotion through or motion within fluid surroundings	None	✓*	✓*	✓*	✓
Complex cytoskeleton	Cell shape; internal organization; basis of cell movement and, in many cells, locomotion	Rudimentary***	✓*	✓*	✓*	✓

  * Known to be present in cells of at least some groups.

  ** Many groups use oxygen-requiring (aerobic) pathways of ATP formation, but mitochondria are not involved.

*** Protein filaments form a simple scaffold that helps support cell wall in at least some species.

Review Questions

1. State the three key points of the cell theory. *CI*

2. Suppose you wish to observe the three-dimensional surface of an insect's eye. Would you see more details with the aid of a compound light microscope, transmission electron microscope, or scanning electron microscope? *4.2*

3. Label the organelles in the two diagrams below of a plant cell and an animal cell. What are the main differences between the two kinds of cells? In what ways are they similar? *4.3*

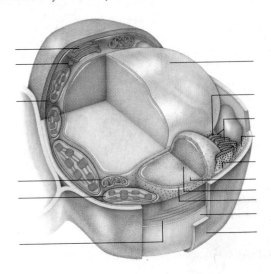

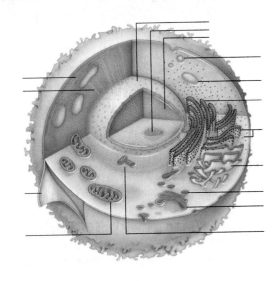

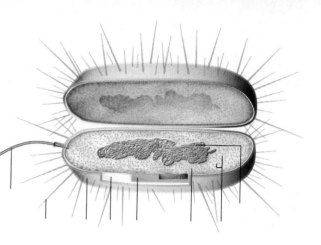

4. Describe three features that all cells have in common. After reviewing Table 4.3, write a paragraph on the key differences between eukaryotic and prokaryotic cells. *4.1, 4.3, 4.12*

5. Briefly characterize the structure and function of the cell nucleus, nuclear envelope, and nucleolus. *4.4*

6. Distinguish between chromosome and chromatin. Do chromosomes always maintain the same appearance during the life of a cell? *4.4*

7. Name three kinds of organelles that are components of the cell's endomembrane system. Briefly describe the function of each component. *4.5*

8. Is this statement true or false: Plant cells have chloroplasts, but not mitochondria. Explain your answer. *4.3, 4.6, 4.7*

9. What are the functions of the central vacuole? *4.7*

10. Define cytoskeleton. How does it aid in cell functions? *4.9*

11. What gives rise to the 9 + 2 array of cilia and flagella? *4.10*

12. Which of the following kinds of organisms typically have walled cells: bacteria, protistans, fungi, plants, animals? *4.11*

13. State the function and describe the general characteristics of cell walls. For example, are the walls permeable? Why or why not? *4.11*

14. As they grow and develop, many kinds of plant cells form a secondary wall. Is this wall deposited inside or outside the surface of the primary wall that formed earlier? *4.11*

15. In multicelled organisms, coordinated interactions depend on physical linkages and chemical communications between cells. What types of junctions are present between neighboring animal cells? Plant cells? *4.11*

16. Label the parts of the bacterial cell diagram shown at the upper right. *4.12*

Self-Quiz   ANSWERS IN APPENDIX III

1. Cell membranes consist mainly of a _____ .
   a. carbohydrate bilayer and proteins
   b. protein bilayer and phospholipids
   c. lipid bilayer and proteins
   d. none of the above

2. Organelles _____ .
   a. are membrane-bound compartments
   b. are typical of eukaryotic cells, not prokaryotic cells
   c. separate chemical reactions in time and space
   d. all of the above are features of the organelles

3. Cells of many protistans, plants, and fungi, but not animals, commonly have _____ .
   a. mitochondria   c. ribosomes
   b. a plasma membrane   d. a cell wall

4. Is this statement true or false: The plasma membrane is the outermost component of all cells. Explain your answer.

5. Unlike eukaryotic cells, prokaryotic cells _____ .
   a. lack a plasma membrane   c. do not have a nucleus
   b. have RNA, not DNA   d. all of the above

6. Match each cell component with its function.
   ____ mitochondrion   a. synthesis of polypeptide chains
   ____ chloroplast   b. initial modification of new
   ____ ribosome   polypeptide chains
   ____ rough ER   c. final modification of proteins;
   ____ Golgi body   sorting, shipping tasks
                         d. photosynthesis
                         e. formation of many ATP

Critical Thinking

1. In *The Immunity Syndrome*, a classic Star Trek episode, a stupendous amoeba engulfs a starship. Spock uses a shuttle-craft to plant explosives inside the amoeba to blow it up before it reproduces and destroys the universe. What is wrong with this scenario? Similarly, why is it likely that you will never encounter even a two-ton amoeba on a sidewalk?

2. Your professor shows you an electron micrograph of a cell with large numbers of mitochondria and Golgi bodies. You notice that this particular cell also contains a great deal of rough endoplasmic reticulum. What kinds of cellular activities would require such an abundance of the three kinds of organelles?

3. The 1 in 30,000 individuals affected by a very rare disorder, *immotile cilia syndrome*, cannot clear mucus from airways that lead to their lungs. Bacteria thrive in the mucus that collects in the airways, and their metabolic by-products damage tissues. Chronic inflammation follows and contributes to the damage.
   Researchers using video microscopy discovered that the cilia of affected people beat poorly or not at all, even when their surface appearance is normal. Speculate on which components of the cilium may be contributing to the disorder. How do you suppose they became abnormal?

Selected Key Terms

basal body *4.10*	endomembrane	nuclear envelope *4.4*
cell *4.1*	system *4.5*	nucleoid *4.1*
cell communication	ER (endoplasmic	nucleolus *4.4*
4.11	reticulum) *4.5*	nucleus *4.1, 4.4*
cell cortex *4.9*	eukaryotic cell *4.1*	organelle *4.3*
cell junction *4.11*	flagellum *4.10*	peroxisome *4.5*
cell theory *CI*	Golgi body *4.5*	plasma membrane *4.1*
cell wall *4.11*	intermediate	primary wall *4.11*
central vacuole *4.7*	filament *4.9*	prokaryotic cell *4.1, 4.12*
centriole *4.10*	lipid bilayer *4.1*	pseudopod *4.10*
chloroplast *4.7*	lysosome *4.5*	ribosome *4.1*
chromatin *4.4*	microfilament *4.9*	secondary wall *4.11*
chromosome *4.4*	micrograph *4.2*	surface-to-volume
cilium *4.10*	microtubule *4.9*	ratio *4.1*
cytoplasm *4.1*	mitochondrion *4.6*	vesicle *4.5*
cytoskeleton *4.9*	motor protein *4.10*	wavelength *4.2*

Readings

Alberts, B., et al., 2002. *Molecular Biology of the Cell*. Fourth edition. New York: Garland. Clear look at the cytoskeleton.

Wolfe, S., 1993. *Molecular and Cellular Biology*. Belmont, California: Brooks–Cole. Good introduction to the kinds of experiments and research that are revealing details of cell structure and function.

On-Line readings at Student Guide for InfoTrac:
www.brookscole.com/biology

A CLOSER LOOK AT CELL MEMBRANES

One Bad Transporter and Cystic Fibrosis

As small as it may be, a cell is a living thing engaged in risky business. Think of the cells in your body and how they must respond to something as ordinary as water. At all times, they must move water and solutes in one direction or the other across their plasma membrane. The membrane crossings must be exceedingly selective, for they help keep conditions favorable for survival of

individual cells and, ultimately, you. If all goes well, cells take in and send out the right amounts—not too little, not too much. But who is to say that life always goes well?

CFTR is one type of protein channel across the plasma membrane of epithelial cells. Sheets of these cells line sweat glands, airways and sinuses, and ducts in the digestive and reproductive systems. Chloride ions move through them, and water follows to form a thin film on the free surface of the linings. Mucus, which lubricates tissues and helps prevent infection, slides freely on the watery film.

Sometimes the gene coding for CFTR is mutated. Not enough chloride and water cross to the free surface of the linings, so the film does not form. In its absence, mucus gets dried out, thick, and difficult to displace.

For example, thickened mucus clogs ducts from the pancreas, so digestive enzymes cannot reach the small intestine where most food is digested and absorbed. Weight loss follows. Sweat glands secrete too much salt and alter the body's water–salt balance, which affects the heart and other organs. Males become sterile.

What happens in airways to the lungs? Ciliated cells normally sweep away bacteria and other particles that become trapped in the mucus (Section 4.10). Not now. **Biofilms** form. These consist of microbial populations anchored to the lining by stiff, sticky polysaccharides of their own making. Biofilms strongly resist the body's defenses and antibiotics. *Pseudomonas aeruginosa* is the most efficient of the colonizers. It causes low-grade infections that may persist for years. Most patients die because their lungs fail. Most expect to live no longer than thirty years. At present there is no cure.

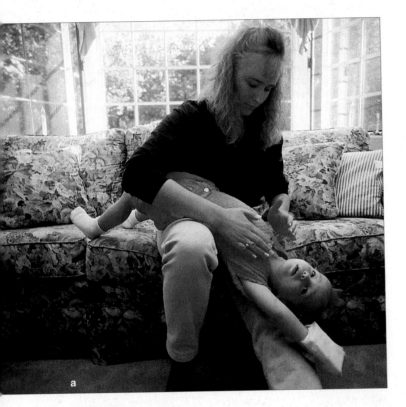

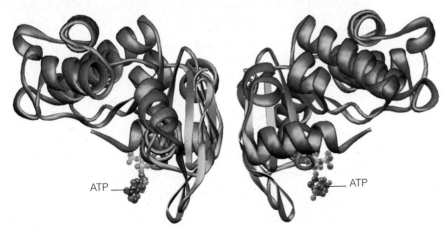

ATP   ATP

b

Figure 5.1 (a) Child affected by cystic fibrosis, or CF, who each day endures chest and back thumps, and repositionings to dislodge the thick mucus that collects in airways to the lungs. The symptoms can vary from one affected individual to the next, partly because the abnormal protein that causes CF has mutated in hundreds of different ways. Environmental factors and a person's genetic makeup also affect the outcome.

(b) Partial ribbon model for one of the ABC transporters, which include CFTR. The two parts shown are the ATP-driven motors that widen an ion channel across the plasma membrane. Other molecular details of CFTR, including the structure of its two membrane-spanning domains, are being worked out.

The mutant CFTR protein that sets these events in motion is the molecular basis of *cystic fibrosis* (CF), now the most common fatal genetic disorder in the United States. More than 10 million people carry the mutant gene. However, CF develops only when they have inherited two copies of it, one from each parent. This happens in about 1 of every 3,300 live births.

CFTR belongs to the family of **ABC transporters**, membrane proteins in all prokaryotic and eukaryotic species. Some, like CFTR, are channels through which hydrophobic substances cross the membrane. Others pump substances across. By their action, some types even control what other membrane proteins are doing.

In all but 10 percent of CF patients, loss of a single amino acid during protein synthesis causes the fatal disorder. Before a new CFTR protein is shipped to the plasma membrane, it is supposed to be modified in the endomembrane system you read about in Chapter 4. Copies of the mutant protein do indeed enter the ER, but enzymes destroy 99 percent of them before they even reach Golgi bodies. Thus few chloride channels reach their normal destinations.

CFTR may also contribute to sinus problems of an estimated 30 million people in the United States alone. *Sinusitis* is a chronic inflammation of the lining of cavities in the skull, around the nose. For one study at Johns Hopkins University, researchers took blood samples from 147 sinusitis patients and from a control group of 147 unaffected people. They focused on only sixteen of the 500+ known mutated forms of the CFTR gene. It turns out that ten of the sinusitis patients inherited a single copy of one of the sixteen mutated CFTR genes, the only ones considered for this study.

By this example, you can see how problems among a huge number of individuals can arise when cells are denied working copies of just one of the proteins that span the plasma membrane. Think about it, and you also may come to appreciate even more the precision with which normal cells use their proteins.

For eukaryotic cells, that precision extends also to the membranes of those internal compartments called organelles. Aerobic respiration, photosynthesis, and many other processes require the selective movement of substances across internal boundary layers, too. Gain insight into the structure and functioning of cell membranes, and you gain insight into survival at life's most fundamental level.

Key Concepts

1. All cell membranes consist primarily of lipids, organized as a double layer. The lipid bilayer gives the membrane its primary structure and prevents haphazard movement of water-soluble substances across it. Diverse proteins embedded in the bilayer or associated with one of its surfaces carry out most membrane functions.

2. The plasma membrane incorporates many receptor proteins, which receive chemical signals that trigger changes in cell activities. It incorporates recognition proteins, which identify a cell as being of a certain type. Spanning all cell membranes are transport proteins through which specific water-soluble substances can cross the bilayer.

3. A concentration gradient is a difference in the number per unit volume of molecules (or ions) of a substance between two regions. The molecules tend to show a net movement down such a gradient, to the region where they are less concentrated. This behavior is called diffusion.

4. Metabolism depends on concentration gradients that drive the directional movements of substances. Cells have mechanisms for increasing or decreasing water and solute concentrations across the plasma membrane and internal cell membranes.

5. With passive transport, a solute crosses a membrane simply by diffusing through a channel in the interior of a transport protein. With active transport, a different kind of transport protein pumps a solute across the membrane, against its concentration gradient. Active transport requires an energy input, typically from ATP.

6. By a molecular behavior called osmosis, water diffuses across any selectively permeable membrane to a region where its concentration is lower.

7. Larger packets of material move across the plasma membrane by processes of endocytosis and exocytosis.

MEMBRANE STRUCTURE AND FUNCTION

Revisiting the Lipid Bilayer

Imagine you are peering at a living cell with the aid of a microscope. What do you suppose would happen if you punctured it with an extremely fine needle? You may be surprised to find that its cytoplasm won't ooze out. Its plasma membrane will flow over the puncture site and seal it! Like the cytoplasmic and extracellular fluids bathing it, a cell membrane is not a solid, static wall; it has a fluid quality.

How does a fluid membrane remain distinct from its fluid surroundings? Think back on what you know about phospholipids, its most abundant components. Each **phospholipid**, recall, has a phosphate-containing head and two fatty acid tails attached to a glycerol backbone (Figure 5.2a). The head is hydrophilic, which means it readily dissolves in water. Its two tails are hydrophobic; water repels them. When you immerse many phospholipid molecules in water, they interact with water molecules and with one another until they spontaneously cluster as a sheet or film at the water's surface. Their interactions might even force them to become positioned as two layers, with all fatty acid tails sandwiched between all hydrophilic heads. This **lipid bilayer** arrangement is the structural basis of cell membranes (Section 4.1 and Figure 5.2c).

A lipid bilayer arrangement minimizes the number of hydrophobic groups exposed to water. A puncture in a cell membrane invites sealing behavior because it is energetically unfavorable; too many hydrophobic groups become exposed to the surroundings.

Few cells get jabbed by fine needles. But the self-sealing behavior of membrane phospholipids is good for more than damage control. Among other things, it helps vesicles form. When, say, a patch of ER or Golgi membrane is budding off at one end, its phospholipids are being repelled by water in the cytoplasm around it. The water "pushes" the phospholipids together, the bud rounds off into a vesicle, and the rupture seals.

What Is the Fluid Mosaic Model?

Figure 5.3 shows a small patch of cell membrane that corresponds to the **fluid mosaic model**. By this model, a cell membrane has a mixed composition—a *mosaic*—of phospholipids, glycolipids, sterols, and proteins. The phospholipid heads differ, and their tails differ in length and saturation. (Unsaturated fatty acids have one or more double bonds in their backbone and fully saturated ones have none.) Glycolipids are structurally similar to them, but their heads also have one or more sugars attached to them. In animal cell membranes, cholesterol is the most abundant sterol (Figure 5.2b). Plant cell membranes incorporate phytosterols.

Also by this model, a membrane is *fluid* as a result of the motions and interactions of its component parts. Most of the phospholipids are free to drift sideways. They also spin about their long axis and flex their tails, which keeps adjacent molecules from packing together into a solid layer. The short or kinked (unsaturated) hydrophobic tails contribute to the membrane fluidity. As you will read later, some membrane proteins also are free to drift laterally through the fluid bilayer.

Figure 5.2 (**a**) Structural formula for phosphatidylcholine. This is a phospholipid and one of the most common components of animal cell membranes. *Orange* indicates its hydrophilic head; *yellow* indicates its hydrophobic tails.

(**b**) Structural formula for the main sterol in animal tissues: cholesterol. Phytosterols are its equivalent in plant tissues.

(**c**) Spontaneous organization of lipid molecules into two layers (a bilayer structure). When immersed in liquid water, their hydrophobic tails become sandwiched between their hydrophobic heads, which dissolve in the water.

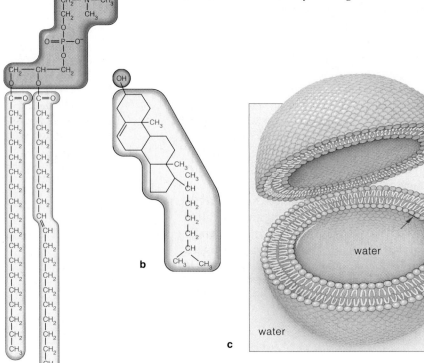

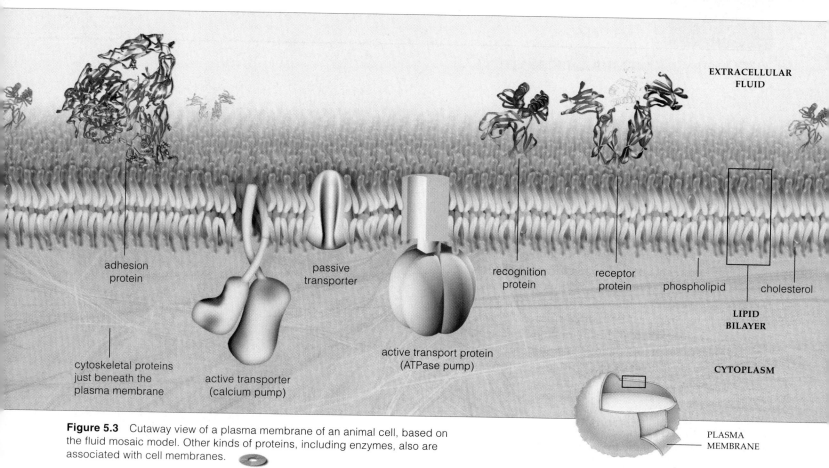

EXTRACELLULAR FLUID

adhesion protein

passive transporter

recognition protein

receptor protein

phospholipid

cholesterol

LIPID BILAYER

cytoskeletal proteins just beneath the plasma membrane

active transporter (calcium pump)

active transport protein (ATPase pump)

CYTOPLASM

PLASMA MEMBRANE

Figure 5.3 Cutaway view of a plasma membrane of an animal cell, based on the fluid mosaic model. Other kinds of proteins, including enzymes, also are associated with cell membranes.

The fluid mosaic model is a good starting point for exploring membranes. But bear in mind, membranes differ in which molecules they incorporate and how their molecules are arranged. Even the two surfaces of the same bilayer differ. For example, oligosaccharides and other carbohydrates are attached to many protein and lipid components of a plasma membrane, but only on its outward-facing surface (Figure 5.3). The number

and kinds of carbohydrates differ from one species to the next. They differ even among different cells of the same individual.

How do membrane surfaces become so distinct? Their characteristics start inside the ER, where many polypeptide chains are modified before being shipped to their final destinations (Section 4.5). Proteins that will become part of the plasma membrane are shipped off in vesicles that fuse with a Golgi body. There they are further modified, then shipped off in other vesicles that fuse with the plasma membrane. Figure 5.4 shows how fusion releases the proteins to the outside face of the membrane.

The next section introduces the kinds of membrane proteins you will be encountering in many chapters. Here again, you will see how the function of a protein correlates with its molecular structure.

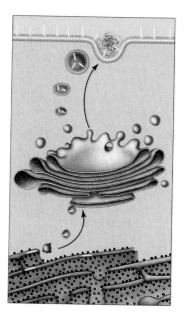

Figure 5.4 Source of the asymmetrical distribution of proteins, carbohydrates, and lipids in membranes. Proteins that function at the cell surface start out as new polypeptide chains that pass through the ER and Golgi bodies. There, they become modified. Many are then shipped to the plasma membrane inside vesicles. A vesicle's membrane fuses with the plasma membrane. As it does, proteins inside automatically become oriented correctly in the plasma membrane.

A cell membrane includes two layers composed mainly of lipids, phospholipids especially. This lipid bilayer is the structural foundation for the membrane and also serves as a barrier to water-soluble substances.

Hydrophobic parts of membrane lipids are sandwiched between the hydrophilic parts, which are dissolved either in cytoplasmic fluid or in extracellular fluid.

Many different proteins having different functions are embedded in the membrane or positioned at its surfaces.

A GALLERY OF MEMBRANE PROTEINS

Where Are the Proteins Positioned?

Cells make contact with their surroundings through lipids and proteins of their plasma membrane. Here, they receive and send out a variety of substances and signals. For multicelled organisms, this is where cells that defend the body encounter evidence of an attack. This is where cells adhere to one another in tissues or to the extracellular matrix. Proteins have roles in most of these tasks, and here again we see how molecular structure translates into specific functional roles.

Integral proteins interact with hydrophobic parts of the bilayer's phospholipids, and they are not easy to remove. Most span the bilayer, with their hydrophilic domains extending past both of its surfaces. *Peripheral* proteins are positioned at the membrane surface, not in the bilayer. Weak interactions, including hydrogen bonds, allow them to associate with integral proteins and with polar heads of the membrane lipids.

What Are Their Functions?

Figure 5.5 introduces the main categories of membrane proteins and their functions. All span the bilayer. In multicelled organisms, **adhesion proteins** help cells of the same type locate, stick together, and remain in the proper tissues. After tissues form, some adhesion sites become part of cell junctions, as Section 4.11 describes.

The **communication proteins** form channels that match up across the plasma membranes of two cells; they let signals and substances flow rapidly between their cytoplasm. **Receptor proteins** bind extracellular substances, such as hormones, that trigger changes in cell activities. For example, enzymes that crank up machinery for cell growth and division are switched on when a specific hormone binds with receptors for it. Different cells have many different combinations of receptors. **Recognition proteins** are like molecular fingerprints; they identify each cell as belonging to a particular tissue or individual. **Transport proteins** passively let solutes cross the membrane through their interior or they actively pump them through.

Other proteins, including enzymes, are embedded in cell membranes. As you will see, some enzymes also are associated with other kinds of proteins and are necessary for their functioning.

Membrane proteins include transporters, which passively or actively move water-soluble substances across the lipid bilayer. They also include diverse enzymes.

The plasma membrane, especially of multicelled species, has many receptors as well as proteins that function in cell-to-cell adhesion, cell communication, and self-recognition.

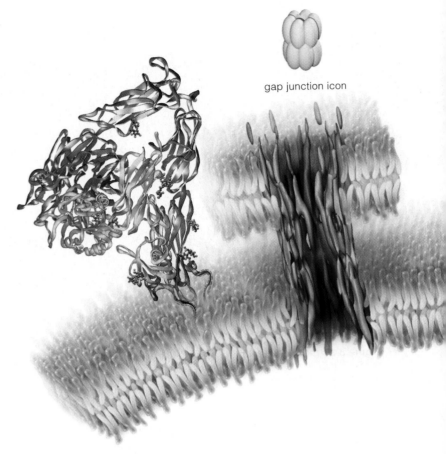

gap junction icon

ADHESION PROTEINS

These proteins are embedded within the plasma membrane. They help one cell adhere to another or to a protein, such as collagen, that is part of an extracellular matrix.

Integrins, including the one shown here, also relay signals across the membrane. Other examples: The cadherins of one cell bind with identical cadherins in adjoining cells. The selectins join together cells by adhering to specific polysaccharides. They are abundant in endothelium, the special lining of blood vessels and the heart.

CELL-TO-CELL COMMUNICATION PROTEINS

These proteins match up with identical proteins in the plasma membrane of an adjoining cell and form a channel, which directly connects their cytoplasm. Chemical and electrical signals flow through the channel.

This type, a cardiac gap junction, is abundant in your heart muscle cells. Signals flow so fast from cell to cell that they all contract together, as a functional unit. This model shows only the parts of the gap junction assembly with greatest density; details are being worked out.

Figure 5.5 Major categories and functions of membrane proteins. You will encounter specific examples of these proteins in this book, especially those with the simple icons above them.

recognition protein icon

passive transporter icon

active transporter icon
(P type)

active transporter icon
(F type)

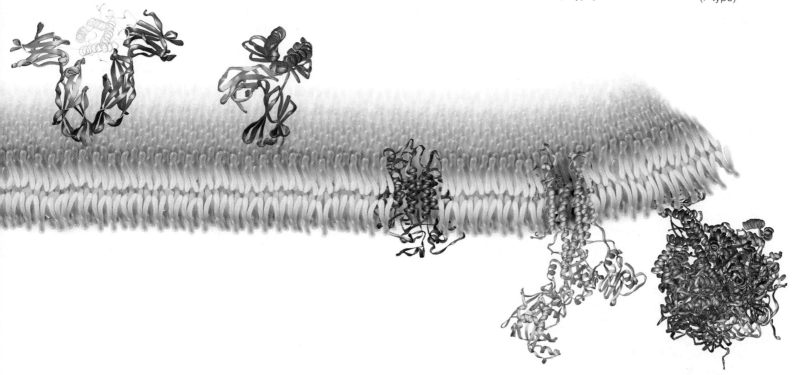

RECEPTOR PROTEINS

Receptors embedded in a membrane are docks for diverse hormones and other signals. They cause target cells to change their activities. Targets are any cells with receptors for the signaling molecule.

A signal might make a cell synthesize a certain protein, speed or block a metabolic reaction, secrete some substance, or get ready to divide.

Shown above, a receptor for somatotropin (also called growth hormone). Copies of this receptor occur on most cells of the vertebrate body. A somatotropin molecule (*gray*) bound to it triggers cell divisions that bring about increases in size during growth and development.

RECOGNITION PROTEINS

Each plasma membrane has glycoproteins (and glycolipids) with sugar side chains projecting above it. These identify a cell as *nonself* (foreign) or *self* (belonging to one's own body or a tissue).

Some function in tissue defense. Remember the HLA proteins (Section 3.7)? Foreign fragments bound to HLA incite immune cells to battle. Still other recognition proteins help cells of the same type find and attach to one another as tissues are forming.

Shown here, one of the glycophorins, an identity tag on red blood cells. *See also* Section 11.4 on ABO blood groups.

PASSIVE TRANSPORTERS

Passive transporters are channels that allow specific solutes to move through them without requiring an energy input. The solute simply diffuses across the membrane, following concentration or electric gradients (Section 5.4).

Diverse animal cells and plant vacuoles have plenty of open channels for water (aquaporins). Other channels change shape when the solute passes through them. The example shown is GluT1, which binds glucose and helps it cross a red blood cell's plasma membrane. A chloride–bicarbonate cotransporter passively lets chloride and bicarbonate ions cross a membrane in opposite directions at the same time.

Other kinds are *ion-selective channels* with molecular gates. Some gates open or close fast when a small molecule binds to them or when the distribution of electric charge across a membrane changes. Nerve and muscle cells have gated channels for sodium, calcium, potassium, and chloride ions.

ACTIVE TRANSPORTERS

Active transport proteins pump a specific solute across the membrane to the side where it is more concentrated and less likely to move on its own. Some pumps are cotransporters that let one kind of solute flow passively "downhill" while pumping a different kind "uphill."

ATP-dependent transporters are known as ATPase pumps. Most *P type* ATPases consist of one polypeptide chain. The calcium pump shown here (*blue*) is one example. The sodium–potassium pump is another.

V type ATPases pump hydrogen ions (H^+). They activate hydrolytic enzymes in fungal, plant, and animal cells.

F type ATPases in prokaryotic, plant, and animal cells pump H^+ across membranes. They can also run in reverse, with a flow of H^+ through them driving the synthesis of ATP. Hence their more precise name, the ATP synthases (Chapters 7 and 8).

Do Membrane Proteins Stay Put?

Some time ago, researchers figured out how to split a plasma membrane right down the middle of its bilayer. They found that proteins were not spread like a coat on the bilayer, as some had hypothesized, but rather that a great many proteins were embedded in it:

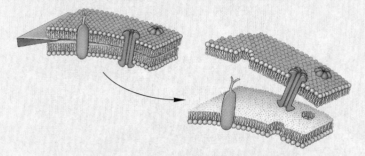

Were those proteins rigidly positioned in the membrane? No one knew until researchers designed an ingenious experiment. They induced an isolated human cell and an isolated mouse cell to fuse. The plasma membranes of the cells from the two species merged to form a single, continuous membrane in the new, hybrid cell. Most of the membrane proteins had become mixed together in less than an hour (Figure 5.6).

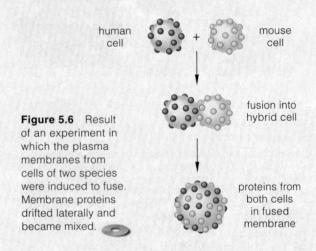

Figure 5.6 Result of an experiment in which the plasma membranes from cells of two species were induced to fuse. Membrane proteins drifted laterally and became mixed.

human cell + mouse cell

fusion into hybrid cell

proteins from both cells in fused membrane

We now know that many membrane proteins are free to move laterally through the lipid bilayer, but others do indeed stay put. Some kinds cluster together and do not move relative to one another. The protein receptors for acetylcholine, which function in information flow through nervous systems, are like this. Other kinds are tethered to cytoskeletal elements that limit their lateral movements. That glycophorin shown in Section 5.2 is attached to a mesh of crosslinked spectrin molecules. So is a type of transport protein that exchanges chloride ions for bicarbonate ions across the plasma membrane.

THINK DIFFUSION

What determines whether a substance will move one way or another across the lipid bilayer or through a membrane protein? Part of the answer has to do with something called diffusion. Think about the water at a bilayer's surfaces. Plenty of substances are dissolved in it, but the kinds and amounts are not the same near the two surfaces. The membrane itself set up the difference and is now maintaining it. How? Each cell membrane shows **selective permeability**: Its molecular structure lets some substances but not others cross it in certain ways, at certain times.

As one example, the lipid chains of the bilayer are largely nonpolar. As a result, molecular oxygen, carbon dioxide, and other small, nonpolar molecules can freely cross it. Although water molecules are polar, some slip through gaps that open up when the hydrocarbon tails flex and bend (Figure 5.7).

By contrast, the bilayer is not permeable to large, polar molecules such as glucose, nor is it permeable to ions (Figure 5.7). Membrane proteins help them across. Sometimes water molecules in which these substances are dissolved cross with them.

The membrane barriers and crossings are essential, because metabolism depends on the cell's capacity to increase, decrease, and maintain concentrations of the molecules and ions for specific reactions.

What Is A Concentration Gradient?

Now picture molecules or ions of some substance near a cell membrane. They move constantly and randomly collide and bounce off one another. They collide more frequently in places where they are more concentrated. When the concentration in one region is not the same as in an adjoining region, this condition is a *gradient*. Pulling this all together, a **concentration gradient** is a difference in the number per unit volume of ions or molecules of a substance between adjoining regions.

In the absence of other forces, any substance moves from a region where it is more concentrated to a region where it is less concentrated. *At biological temperatures, the thermal energy of the molecules drives this movement.* Although these molecules collide randomly and careen back and forth millions of times per second, their *net* movement is away from the place where they are most concentrated.

Diffusion is the name for the net movement of like molecules or ions down a concentration gradient. It is a factor in how substances move across membranes and through cytoplasmic fluid. In multicelled species, it also moves substances between regions and between the body and its environment. For instance, if oxygen builds up in leaf cells, it may diffuse into air in the leaf, then into air outside where its concentration is lower.

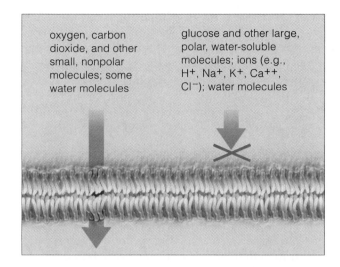

oxygen, carbon dioxide, and other small, nonpolar molecules; some water molecules

glucose and other large, polar, water-soluble molecules; ions (e.g., H^+, Na^+, K^+, Ca^{++}, Cl^-); water molecules

Figure 5.7 Selective permeability of cell membranes. Small, nonpolar molecules and some water molecules cross the lipid bilayer. Ions and large, polar, water-soluble molecules and water that is dissolving them cannot cross it. Transport proteins actively or passively help them across.

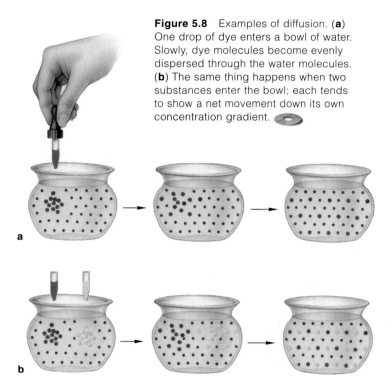

Figure 5.8 Examples of diffusion. (**a**) One drop of dye enters a bowl of water. Slowly, dye molecules become evenly dispersed through the water molecules. (**b**) The same thing happens when two substances enter the bowl; each tends to show a net movement down its own concentration gradient.

a

b

Like other substances, oxygen tends to diffuse in a direction established by its *own* concentration gradient, not gradients of other substances dissolved in the same fluid. You can observe the outcome of this tendency by squeezing a drop of dye into water. The dye molecules diffuse to the region where they are less concentrated. At the same time, water molecules move to the region where *they* are less concentrated (Figure 5.8).

What Determines Diffusion Rates?

Several factors influence the rates of movement down a concentration gradient. Among these are the gradient's steepness, molecular size, temperature, and electric or pressure gradients that may be present.

Diffusion is faster when gradients are steep. Why? Far more molecules move out from a region of greater concentration compared to the number moving in. As the gradient decreases, so does the difference in how many are moving either way. The individual molecules would still be in motion if the gradient were to vanish. But the total number moving one way or the other in a given interval would stay much the same. When the net distribution of molecules is nearly uniform in two adjoining regions, we call this *dynamic equilibrium*.

What about temperature? More heat energy (that is, higher temperature) causes molecules to move faster and collide more often. That is why diffusion is more rapid than in a cooler adjoining region. What about molecular size? Generally, smaller molecules flow down their concentration gradient faster than large ones do.

An electric gradient also can influence the rate and direction of diffusion. An **electric gradient** is simply a difference in electric charge between adjoining regions. Example: Many ions are dissolved in the fluids bathing a membrane, each contributing to the electric charge of one fluid or the other. Opposite charges attract. So the fluid with the more negative charge, overall, exerts the greatest pull on positively charged substances, such as sodium ions. Many events, including information flow through nervous systems, involve the combined force of electric and concentration gradients.

Finally, as you will see shortly, diffusion also may be affected by a **pressure gradient**, which is a difference in the pressure being exerted in adjoining regions.

Molecules or ions of a substance constantly collide because of their inherent energy of motion. The collisions result in diffusion, a net outward movement of a substance from one region into an adjoining region where it is less concentrated.

A concentration gradient is a form of energy. It can drive the directional movement of a substance across a cell membrane. The steepness of the gradient, temperature, molecular size, electric gradients, and pressure gradients influence the rates of diffusion.

Metabolic reactions rely on chemical energy inherent in concentrated amounts of molecules and ions. Cells have mechanisms to increase or decrease those concentrations across the plasma membrane and internal cell membranes.

TYPES OF CROSSING MECHANISMS

Before getting into the actual mechanisms that move substances across membranes, study the overview in Figure 5.9. These are the mechanisms that help supply cells and organelles with raw materials and get rid of wastes. Collectively, they help maintain the volume and pH of cells or organelles within functional ranges.

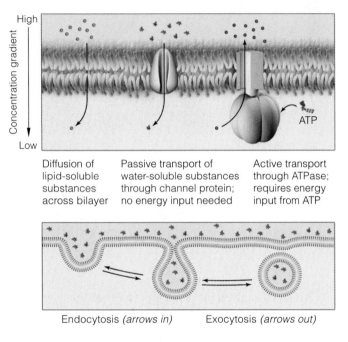

Diffusion of lipid-soluble substances across bilayer

Passive transport of water-soluble substances through channel protein; no energy input needed

Active transport through ATPase; requires energy input from ATP

ATP

Endocytosis *(arrows in)*   Exocytosis *(arrows out)*

Figure 5.9 Membrane crossing mechanisms.

Again, small, nonpolar molecules such as oxygen diffuse across a lipid bilayer. Polar molecules and ions cross by diffusing through transport proteins that span the bilayer. The passive transporters simply permit the substance to follow its concentration gradient across. This is called *passive transport*, or "facilitated" diffusion.

ATPase pumps engage in *active transport*. They, too, let polar substances cross the membrane through their interior, but the net direction of movement is against the concentration gradient. Unlike diffusion across the bilayer or passive transport, active transport requires an input of energy to counter a concentration gradient.

Other mechanisms move substances in bulk across the plasma membrane. *Exocytosis* involves fusion of the plasma membrane and a membrane-bound vesicle that formed inside the cytoplasm. *Endocytosis* involves an inward sinking of a patch of plasma membrane, which seals back on itself to form a vesicle in the cytoplasm.

Different mechanisms work with and against concentration gradients to move substances across the lipid bilayer itself or through proteins that span the bilayer.

HOW DO THE TRANSPORTERS WORK?

All transporters let water-soluble molecules and ions diffuse through some kind of channel or tunnel inside them. When the solute enters a channel opening at one side of the membrane, it weakly binds to the protein. The protein's shape changes, making the opening close behind the bound solute and another open in front of it. This exposes the solute to the fluid that bathes the membrane's other surface. There, the solute binding site reverts to its former state and releases the cargo.

Passive Transport

Passive transport is the name for unassisted diffusion of a specific solute through a transport protein. Which way it moves across the membrane depends only on its concentration gradient, electric gradient, or both.

Energetically speaking, the membrane crossing costs only what the cell has already spent to produce and maintain the gradients. Passive transport itself adds no more to the energy cost. The *net* direction of movement

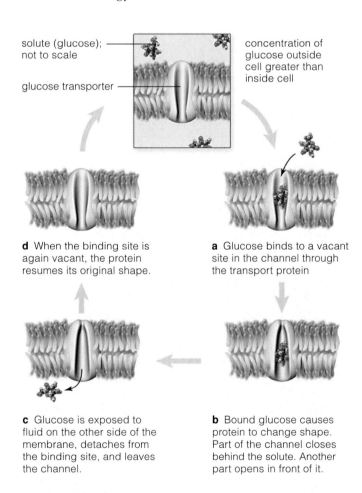

solute (glucose); not to scale

glucose transporter

concentration of glucose outside cell greater than inside cell

d When the binding site is again vacant, the protein resumes its original shape.

a Glucose binds to a vacant site in the channel through the transport protein

c Glucose is exposed to fluid on the other side of the membrane, detaches from the binding site, and leaves the channel.

b Bound glucose causes protein to change shape. Part of the channel closes behind the solute. Another part opens in front of it.

Figure 5.10 Passive transport, using a glucose transporter as the model. Glucose moves both ways across a membrane. But the *net* movement will be down the concentration gradient until concentrations are equal on both sides of the membrane.

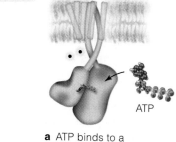
higher concentration of calcium outside cell

lower concentration of calcium inside cell

Figure 5.11 Active transport by a calcium pump. This partial model shows the channel across the membrane. ATP transfers a phosphate group to a calcium pump. Reversible changes in this transporter's shape can result in greater net movement of solute particles *against* the concentration gradient.

in a given interval depends on how many molecules or ions of a specific solute are randomly contracting the transporters (Figure 5.10). Encounters are simply more frequent on the side of the membrane where the solute concentration is greatest. So the solute's *net* movement is in the direction where it is less concentrated.

If nothing else were going on, the passive transport would continue until concentrations were equal across the membrane. But other processes affect the outcome.

For instance, the glucose transporters (Figure 5.10) passively move glucose from blood into the cell, where it is used for biosynthesis and instant energy. When the blood glucose level is high, cells maintain the gradient even when uptake is rapid. How? As fast as glucose molecules are diffusing into cells, others are entering metabolic reactions. In other words, when the cells use glucose, they are maintaining a concentration gradient that favors uptake of *more* glucose.

Active Transport

Only in a dead cell have solute concentrations become equal on both sides of membranes. Living cells never stop expending energy to pump solutes into and out from their interior. In **active transport**, energy-driven protein motors help move a specific solute across the cell membrane, *against* its concentration gradient.

Reflect on the active transporters in Figure 5.5. As one of them binds ATP, a channel through its interior opens up. Only a particular kind of solute can enter the channel and bind reversibly to molecular groups that line it. Binding causes the transporter to accept a phosphate group from the ATP. That phosphate-group transfer changes the transporter's shape. And with that change, the solute is released and enters the fluid on the other side of the cell membrane.

Figure 5.11 shows one **calcium pump**. This active transporter helps keep the concentration of calcium in a cell at least a thousand times lower than it is outside. A different protein, the **sodium–potassium pump,** is a cotransporter. When ATP activates it, this pump binds sodium ions (Na⁺) from one side of the membrane and releases them on the other side. The release facilitates binding of potassium ions (K⁺) at a different site inside the tunnel through the transporter, which reverts to its original shape after K⁺ is released to the other side.

Through operation of such active transport systems, concentration and electric gradients can be maintained across membranes. The gradients are vital to many cell

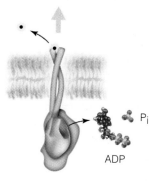
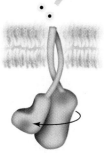

e The shape returns to its resting position.

a ATP binds to a calcium pump. ATP

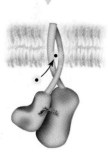

d Shape change permits calcium release at opposite membrane surface. Phosphate group and ADP are released. Pᵢ ADP

b Calcium enters tunnel through pump, binds to functional groups inside.

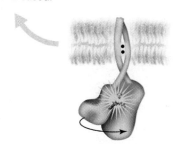

c ATP transfers a phosphate group to pump. Energy input will cause pump's shape to change.

activities and physiological processes, including muscle contraction and nerve cell (neuron) function. We return to the mechanisms of active transport in later chapters.

All transport proteins bind solutes on one side of a cell membrane and reversibly change shape. The charge shunts the solute through some type of tunnel in their interior.

In passive transport, a solute diffuses through a transporter, and its net movement is down its concentration gradient.

In active transport, the net diffusion of one type of solute is uphill, against its concentration gradient. The transporter must be activated by an energy input from ATP to counter the energy inherent in the gradient.

WHICH WAY WILL WATER MOVE?

By far, more water diffuses across cell membranes than any other substance, so the main factors that influence its directional movement deserve special attention.

Osmosis

Think about something as simple as a running faucet or as stupendous as Niagara Falls. Either way, moving water is providing you with an example of **bulk flow** —the mass movement of one or more substances in response to pressure, gravity, or some other external force. Bulk flow accounts for some water movement in plants and animals. A beating heart generates the fluid pressure that pumps blood (mostly water) through the body. Sap flows in the conducting tissues that thread through maple trees, and this, too, is bulk flow.

What about the movement of water into and out of cells or organelles? If the concentration of water is not equal across a cell membrane, **osmosis** tends to occur. Water molecules will tend to diffuse down the water concentration gradient and thus cross that membrane, which is selectively permeable.

The concentrations of solutes at the two sides of selectively permeable membranes influence osmosis. This type of membrane lets the

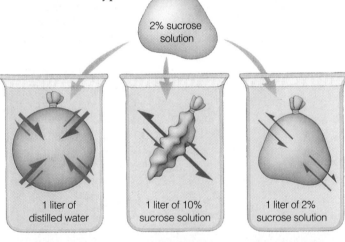
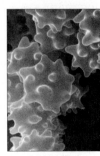

2% sucrose solution

| 1 liter of distilled water | 1 liter of 10% sucrose solution | 1 liter of 2% sucrose solution |

a

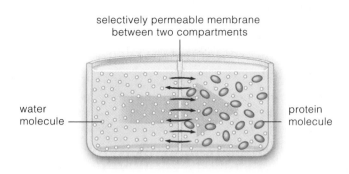

selectively permeable membrane between two compartments

water molecule

protein molecule

Figure 5.12 Demonstration of how a solute concentration gradient influences osmotic movement. Start with a container divided by a membrane that water but not proteins can cross. Pour water into the left compartment. Pour the same volume of a protein-rich solution into the right compartment. There, proteins occupy some of the space. The net diffusion of water in this example is from left to right (large *gray* arrow).

small, polar water molecules cross but restricts passage of large, polar molecules. The side with the most solute particles ends up with the lower concentration of water (Figure 5.12). To illustrate this, dissolve some glucose in water. Compared to an equivalent volume of water, the glucose solution has fewer water molecules. Why? Because each glucose molecule is occupying space that was formerly occupied by water molecules.

It is primarily the *total number* of molecules or ions, not the type of solute, that dictates the concentration of water. Dissolve an amount of an amino acid or urea in a liter of water, and the water concentration changes about as much as it did in the glucose solution. Add some sodium chloride (NaCl) to a liter of water, and it dissociates into equal numbers of sodium ions and chloride ions. This results in twice as many particles of solute as there were in the glucose solution, and so the concentration of water will decrease proportionately.

Effects of Tonicity

Suppose you decide to test the statement that water tends to move toward a region where solutes are *more* concentrated. You make three sacs from a membrane that water but not sucrose can cross (Figure 5.13). You

HYPOTONIC CONDITIONS	HYPERTONIC CONDITIONS	ISOTONIC CONDITIONS
Water diffuses into red blood cells, which swell up	*Water diffuses out of the cells, which shrink*	*No net movement of water, no change in cell size or shape*

b

Figure 5.13 Effect of tonicity on water movement. (**a**) Arrow widths indicate the direction and the relative amounts of flow. (**b**) The micrograph by each sketch shows shapes that a human red blood cell assumes when you put it in fluids of higher, lower, or equal solute concentrations. Solutions inside and outside this type of cell normally are balanced. A red blood cell has no built-in mechanisms to help it adjust to drastic changes in solute levels in its fluid surroundings.

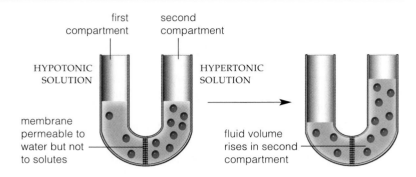

Figure 5.14 Increase in fluid volume owing to osmosis. In time, the net diffusion across a membrane separating two compartments is equal, but the fluid volume in compartment 2 is greater because the membrane is impermeable to solutes.

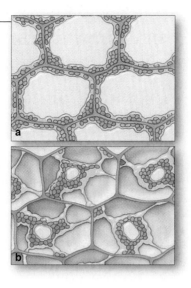

Figure 5.15 (a) Young plant cells. (b) Example of what plasmolysis, an osmotically induced loss of internal fluid pressure, does to the cells. The cytoplasm and central vacuole both shrink, and the plasma membrane moves away from the cell wall.

fill each sac with a solution that is 2 percent sucrose. You immerse one sac in a liter of distilled water (which has no solutes). You immerse another in a solution that is 10 percent sucrose. You immerse the third sac in a solution that is 2 percent sucrose. Each time, tonicity dictates the extent and direction of water movement.

Tonicity refers to the relative solute concentrations of two fluids. When two fluids on opposing sides of a membrane differ in solute concentration, the **hypotonic solution** is the one with fewer solutes. The one that has more is the **hypertonic solution**. Water tends to diffuse from hypotonic to hypertonic fluids. **Isotonic solutions** have the same solute concentrations, so water shows no net osmotic movement from one to the other.

Normally, the fluids inside and outside your cells are isotonic. When any tissue fluid becomes drastically hypotonic, too much water diffuses into cells, and they burst. When fluid becomes too hypertonic, an outward diffusion of water shrivels them.

Most cells have built-in mechanisms that adjust to shifts in tonicity. Red blood cells do not. (Figure 5.13 shows what happened to them in an observational test of tonicity differences.) That's why severely dehydrated patients get infusions of a solution isotonic with blood.

Effects of Fluid Pressure

Animal cells avoid bursting by engaging in selective transport of solutes across the plasma membrane. Cells of plants and many protistans, fungi, and bacteria do the same with the help of pressure on their cell walls.

Pressure differences as well as solute concentrations influence the osmotic movement of water. Take a look at Figure 5.14. It shows how water continues to move across a membrane between a hypotonic solution and a hypertonic solution until the solute concentration is the same on both sides. As you can see, the *volume* of the formerly hypertonic solution has increased (because its solutes cannot diffuse out).

Hydrostatic pressure is pressure that any volume of fluid exerts against a wall, membrane, or some other structure enclosing it. (In plants, this is called the *turgor* pressure.) The greater the fluid's solute concentration, the greater will be the hydrostatic pressure it exerts.

Living cells cannot increase in volume indefinitely (Section 4.1). At some point, fluid pressure that builds up inside the cell counters water's inward diffusion. That point is the **osmotic pressure**, the amount of force preventing further increase in a solution's volume.

Think of a young plant cell, with its pliable primary wall. Water diffuses into the cell as it grows. This puts more fluid pressure on its wall and makes it expand, so the cell volume can increase. Further expansion of the wall (and the cell) ends when internal fluid pressure develops enough to counterbalance the water uptake.

Plant cells are vulnerable to water losses, which can occur when soil dries out or becomes too salty. Water stops diffusing in and starts diffusing out, so internal fluid pressure falls. An osmotically induced shrinkage of cytoplasm is called plasmolysis (Figure 5.15). Plants adjust somewhat to the loss of pressure, as when they actively take up potassium ions against a concentration gradient by the mechanisms outlined in Section 5.6.

As you will read in Chapters 38 and 42, hydrostatic and osmotic pressure also influence the distribution of water in the tissue fluids and cells of animals.

Osmosis is a net diffusion of water between two solutions that differ in solute concentration and that are separated by a selectively permeable membrane. The greater the number of molecules and ions dissolved in a solution, the lower its water concentration will be.

Water tends to move osmotically to regions of greater solute concentration (from hypotonic to hypertonic solutions). There is no net diffusion between isotonic solutions.

The fluid pressure that a solution exerts against a membrane or wall also influences the osmotic movement of water.

MEMBRANE TRAFFIC TO AND FROM THE CELL SURFACE

Exocytosis and Endocytosis

Transport proteins can move only small molecules and ions into or out of cells. When it comes to taking in or expelling large molecules or particles, cells use vesicles that form through exocytosis and endocytosis.

By **exocytosis**, a vesicle moves to the cell surface, and the protein-studded lipid bilayer of its membrane fuses with the plasma membrane. While this exocytic vesicle is losing its identity, its contents are released to the surroundings (Figure 5.16a). In three pathways of **endocytosis**, a cell takes in substances near its surface.

A small patch of plasma membrane balloons inward and pinches off. The result, an endocytic vesicle, then transports its contents to some organelle or stores them in the cytoplasm (Figures 4.13 and 5.16b).

By *receptor-mediated* endocytosis, the first pathway, membrane receptors chemically recognize and bind to a specific substance—for example, a hormone, vitamin, or mineral. The receptors become concentrated in tiny pits that form in the plasma membrane (Figure 5.17). Each pit looks like a woven basket on its cytoplasmic side. The basket consists of protein filaments (clathrin) interlocked into stable, geometric patterns. When the pit sinks into the cytoplasm, the basket closes back on itself and becomes the vesicle's structural framework.

The second pathway, *bulk-phase* endocytosis, is less selective. An endocytic vesicle forms around a small volume of extracellular fluid regardless of what kinds of substances happen to be dissolved in it. Bulk-phase endocytosis operates at a fairly constant rate in nearly all eukaryotic cells. By continually pulling patches of plasma membrane into the cytoplasm, this pathway compensates for membrane that steadily arrives from the cytoplasm in the form of exocytic vesicles.

The third pathway is **phagocytosis** (meaning cell eating). By this active form of endocytosis, a cell engulfs microbes, particles, and cellular debris. Amoebas and some other protistans get food this way. In multicelled species, macrophages and some other white blood cells phagocytize pathogenic viruses or bacteria, cancerous body cells, and other threats to health.

Phagocytosis, too, is mediated by receptors. First, a target binds with receptors bristling from a phagocytic

plasma membrane

exocyosis
(from cytoplasm)

a

endocyosis
(into cytoplasm)

b

Figure 5.16 (a) Exocytosis and (b) endocytosis.

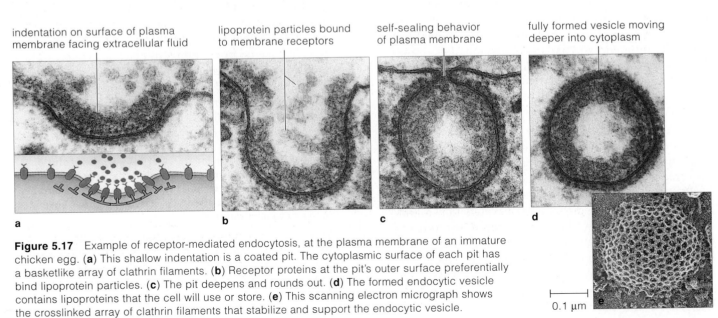

indentation on surface of plasma membrane facing extracellular fluid

lipoprotein particles bound to membrane receptors

self-sealing behavior of plasma membrane

fully formed vesicle moving deeper into cytoplasm

a **b** **c** **d**

0.1 μm **e**

Figure 5.17 Example of receptor-mediated endocytosis, at the plasma membrane of an immature chicken egg. (**a**) This shallow indentation is a coated pit. The cytoplasmic surface of each pit has a basketlike array of clathrin filaments. (**b**) Receptor proteins at the pit's outer surface preferentially bind lipoprotein particles. (**c**) The pit deepens and rounds out. (**d**) The formed endocytic vesicle contains lipoproteins that the cell will use or store. (**e**) This scanning electron micrograph shows the crosslinked array of clathrin filaments that stabilize and support the endocytic vesicle.

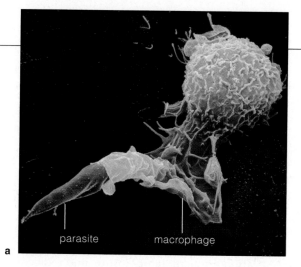

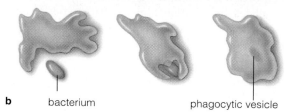

b   bacterium   phagocytic vesicle

Figure 5.18   Phagocytosis. (**a**) Scanning electron micrograph showing a macrophage in the act of engulfing *Leishmania mexicana*. This parasitic protozoan causes a potentially fatal disease called leishmaniasis. Bites from infected sandflies transmit the parasite to humans. (**b**) Diagram of phagocytosis. Lobes of this amoeba's cytoplasm extend outward and surround its target. The plasma membrane of the extensions fuses, forming a phagocytic vesicle. This type of endocytic vesicle moves deeper into the cytoplasm and fuses with lysosomes. Its contents are digested. The molecular bits, and the vesicle's membrane components, are recycled elsewhere (Figure 5.19).

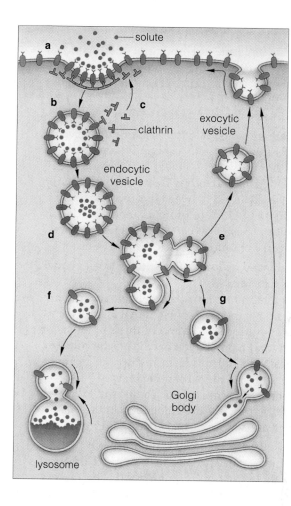

a Molecules get concentrated inside coated pits of plasma membrane.

b Endocytic vesicles form from the pits.

c Vesicles lose molecules of clathrin, which return to plasma membrane.

d Enclosed molecules are sorted and often released from receptors.

e Many sorted molecules are cycled back to the plasma membrane.

f,g Many other sorted molecules are delivered to lysosomes and stay there or are degraded. Still others are routed to spaces in the nuclear envelope and inside ER membranes, and others to Golgi bodies.

Figure 5.19   Cycling of membrane lipids and proteins. This example starts with receptor-mediated endocytosis. The plasma membrane gives up small patches of itself to endocytic vesicles that form from coated pits. It gets membrane back from exocytic vesicles that budded from ER membranes and Golgi bodies. The membrane initially used for endocytic vesicles will cycle receptor proteins and lipids back to the plasma membrane.

cell's membrane. This binding triggers synthesis and the crosslinking of microfilaments into a network just beneath the plasma membrane. Driven by ATP motors, these microfilaments contract, which squeezes some cytoplasm toward the margins of the cell, thus forming the lobes called pseudopods (Figure 5.18). Pseudopods flow over the target and fuse at their tips. The result is one phagocytic vesicle, which sinks into the cytoplasm. There it fuses with lysosomes. Inside these organelles of intracellular digestion, trapped items are digested to fragments and smaller, reusable molecules.

Membrane Cycling

As long as a cell stays alive, exocytosis and endocytosis continually replace and withdraw patches of its plasma membrane. And they apparently do so at rates that can maintain the plasma membrane's total surface area.

As one example, neurons release neurotransmitters in bursts of exocytosis. Each neurotransmitter is a type of signaling molecule released from one cell that acts on neighboring cells. Cell biologists John Heuser and T. S. Reese documented an intense burst of endocytosis

abruptly after a major episode of exocytosis in neurons, and that counterbalanced it. Figure 5.19 provides more examples of ways in which cells cycle their membrane lipids and proteins.

Whereas transport proteins in a cell membrane deal only with ions and small molecules, exocytosis and endocytosis move large packets of materials across a plasma membrane.

By exocytosis, a cytoplasmic vesicle fuses with the plasma membrane, so that its contents are released outside the cell. By endocytosis, a small patch of the plasma membrane sinks inward and seals back on itself, forming a vesicle inside the cytoplasm. Membrane receptors often mediate this process.

SUMMARY *Gold* indicates text section

Membrane Structure and Function

1. The plasma membrane is a structural and functional boundary between the cytoplasm and the surroundings of all cells. In eukaryotic cells, organelle membranes subdivide the fluid portion of the cytoplasm into many functionally diverse compartments. *5.1*

2. A cell membrane consists of two water-impermeable layers of lipids (phospholipids especially) and proteins associated with the layers, as shown in Figure 5.20. *5.1*

   a. Fatty acid tails and other hydrophobic parts of the lipids are sandwiched between hydrophilic heads.

   b. Many different kinds of proteins are embedded in the lipid bilayer or positioned at one of its two surfaces. The proteins carry out most membrane functions.

3. These are the key features of the fluid mosaic model of membrane structure: *5.1*

   a. A cell membrane shows fluid behavior, mainly because its lipid components twist, move laterally, and flex hydrocarbon tails. Also, some of these lipids have ring structures and many have kinked (unsaturated) or short fatty acid tails, all of which disrupt what might otherwise be tight packing within the bilayer.

   b. A membrane is a mosaic, or composite, of lipids and proteins. The proteins are embedded in the bilayer and positioned at its surface. Its two layers differ in the number, kind, and arrangement of lipids and proteins.

4. Each cell membrane incoporates transport proteins and proteins that structurally reinforce it. The plasma membrane also has adhesion proteins, communication proteins, recognition proteins, and diverse receptors. Differences in the number and types of proteins affect responsiveness to substances at the membrane, as well as cell metabolism, pH, and volume. *5.2*

   a. Transport proteins help water-soluble substances cross all membranes by passing through their interior, which opens to both sides of the membrane.

   b. Receptor proteins bind extracellular substances, and binding triggers alterations in metabolic activities.

Recognition proteins are like molecular fingerprints; they identify cells as being of a given type. Adhesion proteins help cells of tissues adhere to one another and to proteins of the extracellular matrix. Communication junctions span the plasma membranes of adjacent cells to transfer substances and signals rapidly from the cytoplasm of one to the other.

Movement of Substances Into and Out of Cells

1. Molecules or ions of a substance tend to move from a region of higher to lower concentration. Diffusion is movement in response to a concentration gradient. *5.4*

   a. Diffusion rates are influenced by the steepness of the concentration gradient, temperature, and molecular size, as well as by gradients in electrical charge and pressure that may occur between two regions.

   b. Cells have built-in mechanisms that work with or against gradients to move solutes across membranes.

   c. Metabolism requires chemical energy inherent in concentration and electric gradients across cell membranes.

2. Oxygen, carbon dioxide, and other small nonpolar molecules diffuse across a membrane's lipid bilayer. Ions and large, polar molecules such as glucose cross it with the passive or active help of transport proteins. Water moves through proteins and across the bilayer. *5.4*

3. A transport protein moves water-soluble substances across a membrane by reversible changes in its own shape. Passive transport doesn't require energy inputs; solutes diffuse through a channel inside the protein's interior. Active transport requires an energy input from ATP. The protein, an ATPase pump, moves the solute against its concentration gradient. *5.6*

4. Osmosis is defined as the diffusion of water across a selectively permeable membrane in response to a water concentration gradient, pressure gradient, or both. *5.7*

5. By exocytosis, a cytoplasmic vesicle moves into the plasma membrane. Its membrane fuses with it, and its contents are automatically released to the outside. 5.8

6. By endocytosis, a patch of plasma membrane forms a vesicle. It does so by receptor recognition of specific solutes, by the indiscriminate uptake of solutes dissolved in the extracellular fluid, or by phagocytosis ("cell eating," an engulfment of cells as well as substances). *5.8*

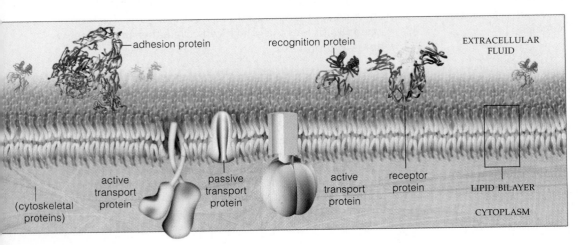

Figure 5.20 Summary of major types of proteins associated with the plasma membrane.

Review Questions

1. Describe the fluid mosaic model of cell membranes. What imparts fluidity to the membrane? What makes it a mosaic? *5.1*

2. State the functions of transport proteins, receptor proteins, recognition proteins, and adhesion proteins. *5.2*

3. Define diffusion. Does diffusion occur in response to a solute concentration gradient, an electric gradient, a pressure gradient, or some combination of these? *5.4*

4. If all transport proteins work by changing shape, then how do passive transporters differ from active transporters? *5.6*

5. Define osmosis. *5.7*

6. Define hypertonic, hypotonic, and isotonic solutions. Does each term refer to a property inherent in a given solution? Or are the terms used only when comparing solutions? *5.7*

7. Is the white blood cell in Figure 5.21 disposing of a worn-out red blood cell by endocytosis, phagocytosis, or both? *5.8*

Self-Quiz   ANSWERS IN APPENDIX III

1. Cell membranes consist mainly of a _____ .
   a. carbohydrate bilayer and proteins
   b. protein bilayer and phospholipids
   c. lipid bilayer and proteins

2. In a lipid bilayer, _____ of lipid molecules are sandwiched between _____ .
   a. hydrophilic tails; hydrophobic heads
   b. hydrophilic heads; hydrophilic tails
   c. hydrophobic tails; hydrophilic heads
   d. hydrophobic heads; hydrophilic tails

3. Most membrane functions are carried out by _____ .
   a. proteins   c. nucleic acids
   b. phospholipids   d. hormones

4. Plasma membranes incorporate _____ .
   a. transport proteins   c. recognition proteins
   b. adhesion proteins   d. all of the above

5. Immerse a living cell in a hypotonic solution, and water will tend to _____ .
   a. move into the cell   c. show no net movement
   b. move out of the cell   d. move in by endocytosis

6. _____ can readily diffuse across a lipid bilayer.
   a. Glucose   c. Carbon dioxide
   b. Oxygen   d. b and c

7. Sodium ions cross a membrane through transport proteins that receive an energy boost. This is an example of _____ .
   a. passive transport   c. facilitated diffusion
   b. active transport   d. a and c

8. Vesicle formation occurs in _____ .
   a. membrane cycling   c. endocytosis and exocytosis
   b. phagocytosis   d. all of the above

Critical Thinking

1. Certain species of bacteria thrive in environments where the temperatures approach the boiling point of water—for example, in the steam-venting fissures of volcanoes and in hot springs of Yellowstone National Park. Assume that the lipid bilayer of the bacterial cell membranes consists mainly of phospholipids. What features might the fatty acid tails of the phospholipids have that help stabilize the membranes at such extreme temperatures?

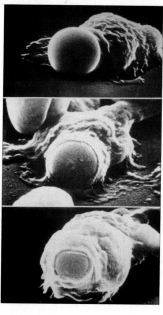

Figure 5.21 Go ahead, name the mystery membrane mechanism.

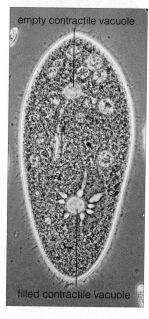

Figure 5.22 *Paramecium* contractile vacuoles.

2. Water moves osmotically into *Paramecium*, a single-celled protistan of aquatic habitats. If unchecked, the influx would bloat the cell and rupture its plasma membrane. An energy-requiring mechanism involving contractile vacuoles expels the excess (Figure 5.22). Water enters tubelike extensions of this organelle and collects in a central space in the vacuole. When full, the vacuole contracts and squirts excess water out of a pore that opens to the outside. Are the fluid surroundings hypotonic, hypertonic, or isotonic relative to *Paramecium*'s cytoplasm?

3. Many cultivated fields in California are heavily irrigated. Over the years, most of the water has evaporated from the soil, leaving behind all of the solutes in the irrigation water. What kinds of problems will the altered soil conditions cause plants?

4. Imagine you're a juvenile shrimp living in an *estuary*, where fresh water draining from the land mixes with saltwater from the sea. Many people own homes around a lake and want boat access to the sea. They ask their city for permission to build a canal to your estuary. If they succeed, what may happen to you?

Selected Key Terms

ABC transporter *CI*	hypotonic solution *5.7*
active transport *5.6*	isotonic solution *5.7*
adhesion protein *5.2*	lipid bilayer *5.1*
biofilm *CI*	osmosis *5.7*
bulk flow *5.7*	osmotic pressure *5.7*
calcium pump *5.6*	passive transport *5.6*
communication protein *5.2*	phagocytosis *5.8*
concentration gradient *5.4*	phospholipid *5.1*
diffusion *5.4*	pressure gradient *5.4*
electric gradient *5.4*	receptor protein *5.2*
endocytosis *5.8*	recognition protein *5.2*
exocytosis *5.8*	selective permeability *5.4*
fluid mosaic model *5.1*	sodium–potassium pump *5.6*
hydrostatic pressure *5.7*	transport protein *5.2*
hypertonic solution *5.7*	

Readings

Nelson, D., and M. Cox. 2000. *Lehninger's Principles of Biochemistry*. Third edition. New York: Worth.

On-Line readings at Student Guide for InfoTrac:
www.brookscole.com/biology

6

GROUND RULES OF METABOLISM

Growing Old With Molecular Mayhem

Somewhere in those slender strands of DNA inside your cells are snippets of instructions for making two kinds of enzymes: superoxide dismutase and catalase (Figure 6.1). Both are *antioxidants*. Like vitamins E and C, they help keep you from growing old before your time by cleaning house, so to speak. Working as a team, they neutralize many potentially toxic wastes through a reaction that uses oxygen (O_2) to strip hydrogen atoms from molecules. Many organisms ranging from bacteria to roundworms to plants also use these two enzymes.

Stripping hydrogen atoms from molecules releases electrons, and O_2 normally picks them up. Sometimes it picks up only one. That one electron is not enough to complete the reaction. But it's enough to give the oxygen a negative charge (O_2^-).

Like other unbound molecular fragments with the wrong number of electrons, O_2^- is a **free radical**. Free radicals escape from a variety of enzyme-catalyzed reactions, including the digestion of fats and amino acids. They slip away from electron transfer chains. And they form when x-rays and other kinds of ionizing radiation bombard water and other molecules.

Free radicals are highly reactive. When they dock with a molecule, they alter its structure and function. They even attack molecules that aren't open to just any reaction. Such molecules include DNA, cell membrane lipids, and membrane receptors that, when activated, tell cells it is time to die (Section 15.6).

Enter superoxide dismutase. Under its biochemical prodding, two rogue oxygen molecules combine with hydrogen ions. Hydrogen peroxide (H_2O_2) and O_2 are the outcome. Hydrogen peroxide, a normal by-product of some reactions, is lethal to cells when it accumulates. Enter catalase. Under its prodding, two molecules of hydrogen peroxide react and split into water and oxygen: $2H_2O_2 \longrightarrow 2H_2O + O_2$. This crucial reaction normally occurs before hydrogen peroxide does major damage. And it occurs often. One catalase molecule can degrade 40 million hydrogen peroxide molecules per second.

As we age, our cells make copies of enzymes in ever diminishing numbers, in crippled form, or both. When this happens to superoxide dismutase and catalase, free radicals and hydrogen peroxide accumulate. Like loose cannons, they career through cells and blast away at

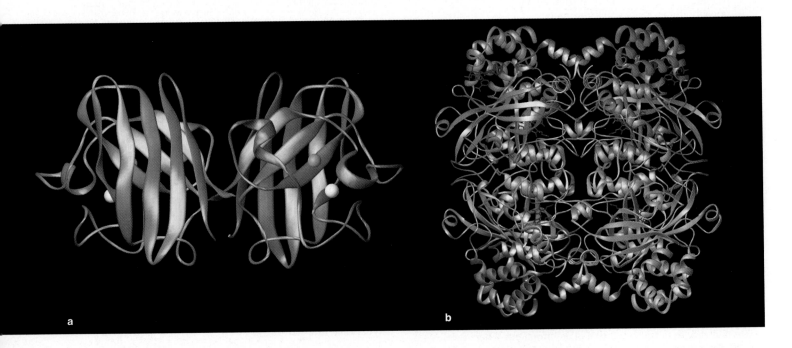

a   b

Figure 6.1 Ribbon models for (**a**) superoxide dismutase and (**b**) catalase. Both enzymes help keep free radicals in check. We find superoxide dismutase molecules in the cytoplasm, nucleus, and peroxisomes, organelles of intracellular digestion in plants and animals. One form is secreted to the extracellular matrix in all mammalian tissues, especially the heart, lungs, pancreas, and placenta. Catalase is concentrated in peroxisomes but is also in the cytoplasm and mitochondrial membranes.

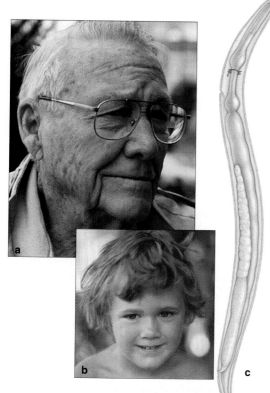

Figure 6.2 (a) Owner of skin that has a spattering of age spots, evidence of free radicals on the loose. At one time he, like the boy in (b), had many smoothly functioning molecules of superoxide dismutase and catalase. (c) The worm of choice for aging studies, *Caenorhabditis elegans*.

the structural integrity of proteins, DNA, lipids, and other molecules. Cells suffer or die.

Those brown "age spots" on an older person's skin are evidence of free radical assaults (Figure 6.2*a*). Each spot is a mass of brownish-black pigments that build up in cells when free radicals take over—all for the want of specific enzymes. Worse yet, free radicals also may usher in heart disease and other disorders.

Knowing this, Simon Melov, Bernard Malfroy, and others experimented with tiny roundworms that don't normally live more than a month (Figure 6.2*c*). The worm's life span increased when synthetic superoxide dismutase and catalase were added to their diet. In some cases it even doubled! The researchers also fed enzymes to worms that were genetically engineered to succumb prematurely to free radicals. Those worms lived just as long as the unaltered ones.

With this example, we turn to **metabolism**, the cell's capacity to acquire energy and use it to build, degrade, store, and release substances in controlled ways. Each extra day in that worm's life reminds us that cells must continually take in energy and use it to drive every one of their activities. At times, the activities may seem remote from your interests. But they help define who you are and who you will become, age spots and all.

Key Concepts

1. Cells engage in metabolism, or chemical work. They use energy to stockpile, build, rearrange, and break apart substances. Cells use energy for mechanical work, as when they move flagella. They also channel energy into electrochemical work, as when they move ions into or out of an organelle compartment.

2. Energy flows in one direction, from usable to less usable forms. Organisms maintain their complex organization by being resupplied with energy lost from someplace else.

3. All organisms secure energy from outside sources. Sunlight energy is the ultimate source for the web of life. Another source is chemical bond energy of molecules in the physical environment or in one species that serves as food for another.

4. ATP, the main carrier of energy in cells, couples reactions that release energy with other reactions that require it. ATP primes molecules to react by transferring a phosphate group to them.

5. Many aspects of metabolism involve electron transfers, or oxidation–reduction reactions. Major transfers occur at electron transfer chains in both photosynthesis and aerobic respiration.

6. On their own, chemical reactions proceed too slowly to sustain life. Enzymes greatly increase the reaction rates in cells. Often they have helpers, such as NAD^+, that assist in the reaction or that transfer electrons and hydrogen released in the reactions to other sites.

7. Enzymes and other kinds of molecules commonly take part in orderly sequences of reactions called metabolic pathways. The coordinated operation of these pathways maintains, increases, and decreases the relative amounts of substances in cells.

8. By controlling a key step of a metabolic pathway, cells can bring about rapid shifts in their activities.

ENERGY AND THE UNDERLYING ORGANIZATION OF LIFE

Defining Energy

If you've ever watched a house cat stalk a mouse, you know it "freezes" its position to avoid detection before springing at its unsuspecting prey. Like everything else in the universe that's stationary, the cat has a store of **potential energy**—a capacity to do work, simply owing to its position in space and the arrangement of its parts. When that cat springs, some of its potential energy is transformed into **kinetic energy**, the energy of motion.

Energy on the move is doing work when it imparts motion to other things. In skeletal muscle cells within the cat, ATP gave up some of its potential energy to molecules of contractile units and set them in motion. The combined motions in many muscle cells resulted in the movement of whole muscles. The transfer of energy from ATP also resulted in the release of another form of energy called **heat**, or *thermal* energy.

The potential energy of molecules has its own name: **chemical energy**. It is measurable, as in kilocalories. A kilocalorie is the same thing as 1,000 calories, which is the amount of energy it takes to heat 1,000 grams of water from 14.5°C to 15.5°C at standardized pressure.

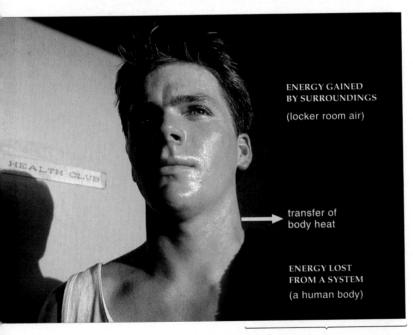

ENERGY GAINED
BY SURROUNDINGS

(locker room air)

transfer of
body heat

ENERGY LOST
FROM A SYSTEM

(a human body)

NET ENERGY CHANGE = 0

Figure 6.3 Example of how the total energy content of any system, *together with its surroundings*, remains constant.

"System" means all matter in a specific region, such as a human body, a plant, a DNA molecule, or a galaxy. "Surroundings" can be a small region in contact with the system or as vast as the whole universe. The system shown (a human male) is giving off heat to the surroundings (a locker room) by evaporative water loss from sweat. What one region loses, the other region gains, so the total energy content of both does not change.

What Can Cells Do With Energy?

All organisms have specific adaptations for securing energy from their environment. Some capture energy from the sun, and others extract energy from inorganic or organic substances in the environment. Regardless of the source, energy inputs are coupled to thousands of energy-requiring processes in cells. Cells use energy for *chemical* work—to stockpile, build, rearrange, and break apart substances. They channel it into *mechanical* work—to move flagella and other cell structures and (in multicelled species) the whole body or portions of it. They channel it into *electrochemical* work—to move charged substances into or out of the cytoplasm or an organelle compartment.

How Much Energy Is Available?

Like single cells, we cannot create energy from scratch; we must get it from someplace else. Why? According to the **first law of thermodynamics**, the total amount of energy in the universe remains constant. More energy cannot be created; existing energy cannot vanish. It can only be converted from one form to some other form.

Think about what the first law means. The universe has only so much energy, distributed in various forms. One form may be converted to another, as when corn plants absorb energy from the sun and convert it to the chemical energy of starch. After you eat and digest corn, your cells extract energy from starch and convert it to other forms, such as mechanical energy for movement.

With each metabolic conversion, a bit of the energy escapes to the surroundings, as heat. Even when you "do nothing," your body gives off about as much heat as a 100-watt lightbulb because of energy conversions in your cells. Released energy is transferred to atoms and molecules making up the air, and the conversion of thermal to kinetic energy "heats up" the surroundings (Figure 6.3). The kinetic energy increases the number of ongoing, random collisions among molecules in the air. And with each collision, a bit more energy is released as heat. However, none of the energy ever vanishes.

The One-Way Flow of Energy

In cells, energy available for conversion resides mainly in the arrangement of atoms and covalent bonds in complex organic compounds, such as glucose, starch, glycogen, and fatty acids. Many of these bonds are said to have a high energy content. When the compounds enter metabolic reactions, the bonds break or become rearranged. During the molecular commotion, some heat energy is lost to the surroundings. In general, cells can't recapture energy lost as heat.

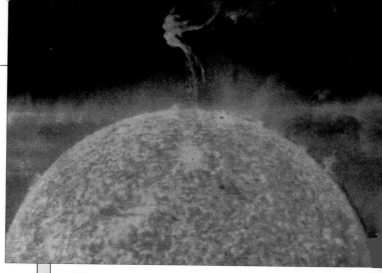

Figure 6.4 Example of the one-way flow of energy into the world of life that compensates for the one-way flow of energy out of it. The sun continually loses energy, much of it in the form of wavelengths of light (Section 7.2). Living cells intercept some of the energy and convert it to useful forms, stored in bonds of organic compounds. Each time a metabolic reaction proceeds in cells, stored energy is released—and some inevitably is lost to the surroundings, mostly as heat.

The lower photograph shows green, water-dwelling, photosynthetic cells (*Volvox*). They live in tiny, spherical colonies. The orange cells serve in reproduction. They form new colonies inside the parent sphere.

ENERGY LOST
one-way flow of energy from sun to Earth's environment

ENERGY GAINED
one-way flow of energy from environment to organisms

Producer organisms harness sun's energy, use it to build organic compounds from simple raw materials available in their environment.

All organisms tap potential energy stored in organic compounds to drive energy conversions that keep them alive. Some energy is lost with each conversion.

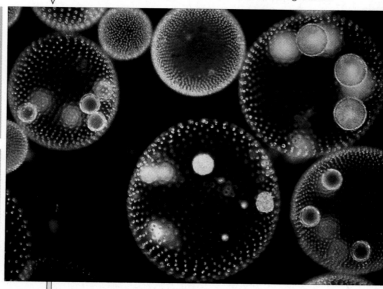

ENERGY LOST
one-way flow of energy from organisms back to the environment

For example, your cells release usable energy from glucose when they break all of its covalent bonds. After many steps, six molecules of carbon dioxide and six of water remain. Compared with glucose, these leftovers are more stable. But chemical energy in the arrangement of their atoms and bonds is much less than the overall chemical energy of glucose. Why? *Some energy was lost at each step leading to their formation.* In other words, glucose is a better source of usable energy.

What about the heat that was transferred from cells to their surroundings when carbon dioxide formed? That heat just isn't useful. Cells can't convert it to other forms, so they can't use it to do work.

Bad news for cells of the remote future: The amount of "low-quality" energy in the universe is increasing. No energy conversion can ever be 100 percent efficient; even highly efficient ones lose heat. From our standpoint, the total amount of energy in the universe is spontaneously flowing from usable forms to unusable forms. Billions of years from now, energy may not be available for conversions; all of it may be dissipated in heat.

Without energy inputs to maintain it, any organized system tends to become more and more disorganized over time. **Entropy** is a measure of the degree of a system's disorder. Think of the Egyptian pyramids—once highly organized, presently crumbling, and many thousands of years from now, dust. It seems the ultimate destination of pyramids and everything else in the universe is a state of maximum entropy. And that, basically, is the point to remember about the **second law of thermodynamics**.

Can life be one glorious pocket of resistance to the depressing flow toward maximum entropy? After all, in every new organism, new bonds form and hold atoms together in precise arrays. So molecules become more organized and have a richer store of energy, not poorer!

Yet a simple example will show that the second law does indeed apply to life on Earth. The primary energy source for life is the sun—which has been releasing energy since it first formed. Plants can capture sunlight energy, convert it to other forms, then lose energy to other organisms that feed, directly or indirectly, on the plants. At each energy transfer from the sun onward, some energy is lost as heat and joins the universal pool. *Overall, energy still flows in one direction.* The world of life maintains its amazing degree of organization only because it is being resupplied with energy that's being lost from someplace else (Figure 6.4).

The amount of energy in the universe remains constant. Energy can undergo conversion from one form to another, but it cannot be created out of nothing or destroyed.

From our perspective, all energy in the universe is flowing spontaneously from usable to unusable forms. A steady, one-way flow of sunlight energy into the interconnected web of life compensates for the steady flow of energy leaving it.

ENERGY INPUTS, OUTPUTS, AND CELLULAR WORK

Cells and Energy Hills

When cells convert one form of energy to another, the amount of potential energy available to them changes. The greater the amount a cell taps into, the larger the energy change, and the more work can be done.

Imagine a Martian watching the NASA Rover scoot around her planet. She decides to push it to the top of a rocky hill, which requires tapping into potential energy stored in her muscle cells (Figure 6.5). Once the Rover is at the top, it has enough potential energy (owing to its position) to roll to the base on its own. The higher it is relative to its final position at the hill's base, the greater the energy change. This same rule applies to the cellular world, in which temperature and pressure remain fairly constant: *Energy changes in living cells tend to proceed spontaneously in the direction that results in a decrease in usable energy.*

Remember glucose ($C_6H_{12}O_6$)? It's made of carbon dioxide ($6CO_2$) and water ($12H_2O$). All three substances have chemical bond energy, but glucose has more than the other two combined. This means making glucose from them isn't something that happens on its own. It is as if carbon dioxide and water are at the base of an "energy hill." On their own, they simply don't have enough energy for an uphill run to make glucose.

In photosynthetic cells, energy inputs from the sun drive reactions that make glucose, the result being a net increase in usable energy (Figure 6.5a). We say the reaction sequence is *endergonic* (meaning energy in).

Cells also run the reactions in reverse, from glucose (at the top of the energy hill) to carbon dioxide and water (at the base). Energetically, the downhill run is favored; it will proceed on its own and end with a net loss in energy. Such reactions are *exergonic*, meaning energy out (Figure 6.5b).

ATP Couples Energy Inputs With Outputs

Cells stay alive by *coupling* energy inputs to energy outputs, primarily with **ATP** (adenosine triphosphate). In this nucleotide, covalent bonds join the five-carbon sugar ribose, the nucleotide base adenine, and three phosphate groups (Figure 6.6). Hundreds of different enzymes can split off the outer phosphate group and join it to another molecule, which thus becomes primed

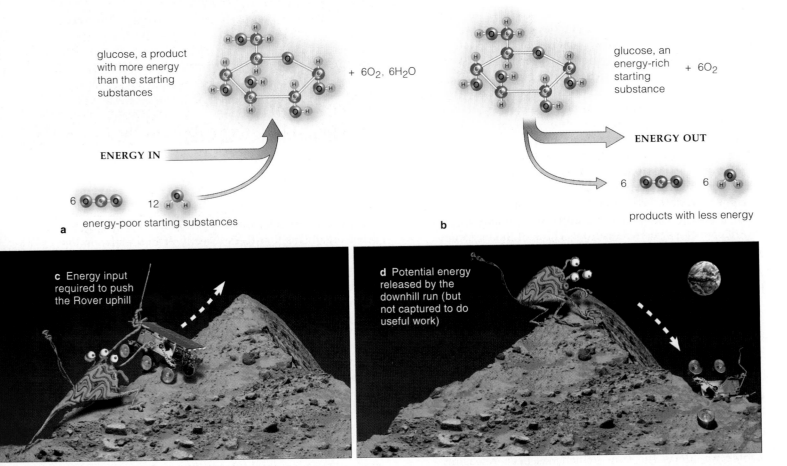

a glucose, a product with more energy than the starting substances + 6O$_2$, 6H$_2$O

ENERGY IN

6   12   energy-poor starting substances

b glucose, an energy-rich starting substance + 6O$_2$

ENERGY OUT

6   6   products with less energy

c Energy input required to push the Rover uphill

d Potential energy released by the downhill run (but not captured to do useful work)

Figure 6.5 Energy changes involved in (**a**,**b**) chemical work and (**c**,**d**) mechanical work.

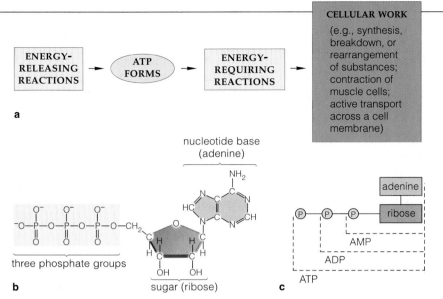

a

nucleotide base
(adenine)

three phosphate groups

b

sugar (ribose)

c

AMP

ADP

ATP

Figure 6.6 (**a**) A key energy relationship in all living cells. ATP, an energy carrier, couples energy-releasing reactions with hundreds of diverse energy-requiring ones. (**b**) Structural formula for ATP. (**c**) Successive phosphate-group transfers turn ATP into ADP (adenosine diphosphate), then AMP (adenosine monophosphate).

Figure 6.7 Example of cellular work initiated when ATP gives up energy to a molecule involved in a specific metabolic reaction. At the plasma membrane, ATP transfers one of its phosphate groups to an active transport protein. Remember Section 5.6? When activated, this transporter changes shape and pumps calcium ions across the membrane and out of the cell, against their concentration gradient.

to enter a reaction. Such a phosphate-group transfer is called a **phosphorylation**.

As you know, covalent bonds are stable interactions between atoms. Why, then, is ATP so ready to give up its phosphates? The phosphates are arranged in a way that puts a number of negative charges close together. The like charges repel each other, which destabilizes the molecule's tail. Getting rid of a phosphate group helps stabilize the molecule and results in more stable products of lower energy. The products are adenosine diphosphate, or **ADP**, and a free inorganic phosphate atom, which we abbreviate as P_i (Figure 6.7). In other words, hydrolysis of ATP is an energetically favorable reaction that releases a large quantity of usable energy.

Think of ATP in cells as being like the currency in an economy. Cells earn it by investing in reactions that release energy. They spend it in reactions that require energy to drive hundreds of activities. That's why we often use a cartoon coin to symbolize ATP.

Because ATP is the main energy carrier for so many reactions, you might speculate that cells should have a way to renew it, and you would be right. When ATP gives up its terminal phosphate group to a molecule, it becomes ADP. This is converted back to ATP when an enzyme attaches inorganic phosphate (or a phosphate group derived from some other molecule) to it.

Regenerating ATP this way is an important aspect of metabolism. It is called the **ATP/ADP cycle**:

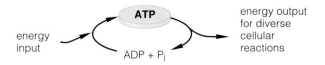

Some Electron Transfers Drive ATP Formation

Energy changes in cells involve transfers of electrons between molecules. These electron transfers are often called **oxidation–reduction reactions**. One molecule is said to be oxidized (it loses one or more electrons) as another molecule is reduced (it gains the electrons).

In photosynthetic organisms—the primary *providers* of electrons—sunlight energy splits water molecules, or H_2O. The oxygen escapes. But electrons and hydrogens are transferred in ways that drive the formation of ATP as well as energy-rich molecules from carbon dioxide. Glucose can be formed this way.

All cells *use* electrons to get at the energy stored in complex molecules. For instance, in aerobic respiration, an energy input jump-starts the release of electrons from glucose. As glucose is degraded to carbon dioxide and water, electron transfers along the way help drive ATP formation. Water forms when the spent electrons are transferred to free oxygen. Section 6.4 takes a closer look at the nature of these electron transfers.

On their own, energy changes in cells spontaneously run in the direction that results in a decrease in usable energy.

ATP, an energy carrier in all living cells, couples energy-releasing reactions with energy-requiring ones. It does so by phosphate-group transfers, which release enough usable energy to activate molecules—to prime them to react.

Cells obtain energy with the help of oxidation–reduction reactions, which are electron transfers from one substance to another. Such reactions are central to the formation of ATP during photosynthesis and aerobic respiration.

CELLS JUGGLE SUBSTANCES AS WELL AS ENERGY

How cells get energy is only one aspect of metabolism. Another is the accumulation, conversion, and disposal of substances by energy-driven reactions.

Participants in Metabolic Reactions

Participants of metabolic reactions go by these names: **Reactants** are the substances that enter a reaction. Any substance that forms during a reaction (or a sequence of reactions) is an **intermediate**. Substances left at the end of a reaction are the **products**. The **energy carriers** activate enzymes and other molecules by phosphate-group transfers. ATP is the main energy carrier. For now, think of enzymes as proteins that speed specific reactions. (We now know a few RNAs also function as enzymes.) The **cofactors** are metal ions and coenzymes, such as NAD^+ and some other organic compounds. They assist enzymes by picking up electrons, atoms, or functional groups from a reaction site and giving them up at different sites. **Transport proteins**, recall, help substances across cell membranes. Controls over these proteins affect metabolism by adjusting concentrations of the substances necessary for specific reactions.

Table 6.1 summarizes these participants. Learn their names; you will be encountering them later on.

What Are Metabolic Pathways?

The concentrations of thousands of substances change continually within living cells. Most of the substances enter or leave by orderly, enzyme-mediated sequences called **metabolic pathways**. Often, small molecules are used to build larger molecules of higher bond energies,

such as complex carbohydrates, complex lipids, and proteins. These are *biosynthetic* (or anabolic) pathways. They cannot proceed without energy inputs. The main biosynthetic pathway is photosynthesis.

Degradative (or catabolic) pathways are exergonic, overall. They break down large molecules to smaller products with lower bond energies. Aerobic respiration is the main degradative pathway. It completely breaks down glucose to carbon dioxide and water, with the release of a considerable amount of usable energy.

The linear pathways advance in a straight line, from substrates to end products. The cyclic pathways occur in a circle; the final step is the conversion of an intermediate back to the reactant. In other words, the cycle regenerates its own point of entry. The branched pathways send a reactant or some intermediate down two or more reaction sequences (Figure 6.8).

Are the Reactions Reversible?

Don't let Figure 6.8 lead you to think that metabolic reactions always proceed in one direction. They might start out in the "forward" direction, from reactants to products. Most can also run in reverse, with products being converted back to reactants. Whenever you see a chemical equation with opposing arrows, this signifies the reaction is reversible. Each arrow means *yields*:

$$A + B \rightleftharpoons C$$

STARTING SUBSTANCES   PRODUCT

Which way such a reaction runs depends partly on the energy content of the participants. It also depends on the reactant-to-product ratio. When the energy level

Table 6.1	*Participants in Metabolic Reactions*
REACTANT	Substance that enters a metabolic reaction or pathway; also called an enzyme's substrate
INTERMEDIATE	Substance formed between reactants and end products of a reaction or pathway
PRODUCT	Substance left at end of reaction or pathway
ENZYME	Usually a protein that enhances reaction rates; a few RNAs also do this
COFACTOR	Coenzyme (such as NAD^+) or metal ion; assists enzymes or taxis electrons, hydrogen, or functional groups between reaction sites
ENERGY CARRIER	Mainly ATP in cells; couples energy-releasing reactions with energy-requiring ones
TRANSPORT PROTEIN	Protein that passively assists substances across a cell membrane or actively pumps them across

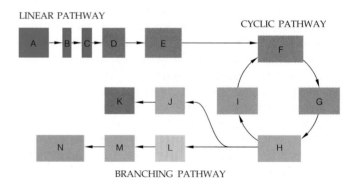

Figure 6.8 Types of reaction sequences—linear, cyclic, and branched—for different metabolic pathways. An arrow signifies an enzyme-mediated step. Here, a linear pathway converts reactant A to product F. In a cyclic pathway, F is a reactant for a series of steps that, as one outcome, regenerate F so the cycle can turn again. One of the cyclic pathway's intermediates, H, becomes a reactant for a branching pathway that splits it and converts it two products, N and K.

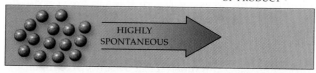

RELATIVE CONCENTRATION
OF REACTANT

RELATIVE CONCENTRATION
OF PRODUCT

HIGHLY
SPONTANEOUS

EQUILIBRIUM

HIGHLY
SPONTANEOUS

Figure 6.9 Chemical equilibrium. When the concentration of reactant molecules is high, a reaction runs most strongly in the forward direction, to products. When the concentration of product molecules is high, it runs most strongly in reverse. At equilibrium, the rates of the forward and reverse reactions are the same.

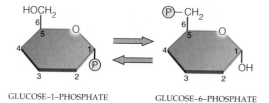

GLUCOSE–1–PHOSPHATE   GLUCOSE–6–PHOSPHATE

Figure 6.10 A reversible reaction. Glucose is primed to enter reactions when a phosphate group becomes attached to it. When there is a high concentration of glucose–1–phosphate, the reaction tends to run in the forward direction. With a high glucose–6–phosphate concentration, it runs in reverse.

The 1 and 6 of these names simply identify which carbon atom of the glucose ring has a phosphate group attached to it.

and concentration of reactant molecules are high, this is an energetically favorable state. That is, the reaction tends to run spontaneously and strongly in the forward direction. When the product concentration gets high enough, however, more molecules or ions of product are available to revert spontaneously to reactants.

Any reversible reaction tends to run spontaneously toward **chemical equilibrium**: the time when it runs at about the same pace in both directions (Figure 6.9). The *amounts* of reactant and product molecules usually differ at that time. It's like a party with people drifting into and out of two rooms. The number in each room stays the same—say, 30 in one and 10 in the other—even as individuals move from one to the other.

Each reversible reaction has a specific equilibrium ratio. For instance, glucose–1–phosphate forms from glucose–6–phosphate. (As Figure 6.10 shows, both are glucose, but they have a phosphate group attached to a different carbon atom.) In this case, the forward and reverse reactions run at the same rate when there are nineteen molecules of glucose–6–phosphate for every glucose–1–phosphate molecule. The equilibrium ratio in this case is 19:1.

Why bother to think about this? *Each cell can bring about big changes in its activities by controlling a few steps of reversible metabolic pathways.*

For instance, when your cells need a very quick fix of energy, they can rapidly break down glucose to two molecules of pyruvate. They can break it down by a sequence of nine enzyme-mediated steps of a pathway called glycolysis (Section 8.2). When glucose supplies

are too low, cells can reverse this pathway and swiftly build glucose from pyruvate and other substances. Six of the reaction steps are reversible; the other three are bypassed. An input of ATP energy drives the bypass reactions in the energetically unfavorable direction.

What would happen if your cells did not have this reverse pathway? During times of starvation, when the blood glucose concentration becomes dangerously low, your cells would not be able to build replacements fast enough to counter the stress.

No Vanishing Atoms at the End of the Run

One other point: When reactions run in the forward or reverse direction, they rearrange atoms, but they never destroy them. By the **law of conservation of mass**, the total mass of all of the substances that enter a reaction equals the total mass of all of the products. When you study any chemical equation, count up the atoms of each reactant and product molecule. There should be as many atoms of each element to the right of the arrow as to the left. When you write equations for metabolic reactions, they must balance this way.

Cells can increase, maintain, and decrease concentrations of substances by coordinating thousands of reactions.

The reactions start with reactants and end with products. Substances formed in between are intermediates. Energy carriers (ATP especially), enzymes, cofactors, and transport proteins are key players in metabolic reactions.

Metabolic pathways are orderly, enzyme-mediated reaction sequences. Biosynthetic (anabolic) pathways build large molecules of higher bond energies from smaller molecules. The degradative (catabolic) pathways break down large molecules to smaller products with lower bond energies.

Cells can rapidly shift their metabolism in major ways by controlling key steps of reversible pathways.

ELECTRON TRANSFER CHAINS IN THE MAIN METABOLIC PATHWAYS

During both photosynthesis and aerobic respiration, certain electron transfers are made at membrane-bound chains of enzymes. Become familiar now with the way these chains work, and you will have a fine head start toward understanding both metabolic processes.

Photosynthetic cells make ATP, the source of energy for making glucose from CO_2 and H_2O. All cells that engage in aerobic respiration break down glucose to make a lot of ATP. It takes a series of steps to build this molecule, and it takes a series of steps to break it down.

Why? Glucose, a reduced, energy-rich molecule, is not as stable as CO_2 and H_2O. (The atmosphere holds so much oxygen that CO_2 is the most stable form of carbon, and H_2O the most stable form of hydrogen.) If you toss a cupful of glucose into a wood-burning fire, its carbon atoms and hydrogen atoms will let go of one another at once and combine with oxygen in the air. However, all of the energy released will be lost as heat (Figure 6.11*a*). By contrast, in cells, energy is released more efficiently by a series of oxidation–reduction reactions. First, glucose is primed to react (Figure 6.10). Then electrons and hydrogen atoms are stripped from it and transferred to coenzymes.

$NADP^+$ is an electron-transferring coenzyme in photosynthesis. NAD^+ and FAD are counterparts in aerobic respiration. We write their reduced forms as NADPH, NADH, and $FADH_2$, respectively. Each functions to transfer hydrogen as well as electrons *to* electron transfer chains. You may hear someone refer to the chains as electron transport systems, but enzymes, not transport proteins, are involved.

Electron transfer chains consist of specific kinds of enzymes and other molecules that are organized to accept and give up electrons, one after the other, at a cell membrane between two compartments. The electrons are at a higher energy level when they enter the chain than when they leave. Think of them as being moved down an energy staircase and losing a bit of energy at certain steps (Figure 6.11*b*).

Your focus on these stepwise transfers should not be the electrons themselves but rather on the attraction that electrons hold for hydrogen. Free hydrogen ions (H^+) associate with electrons entering the transfer chains. At certain transfer steps, components of the chain shunt these ions into the compartment on the other side of the membrane. There, the ions accumulate until H^+ concentration and electric gradients across the membrane have become established. The force of the gradients propels H^+ out of the compartment, through transport proteins that span the membrane. This ion flow drives ATP formation.

In short, the electron transfer chains contribute energy, in the form of H^+ concentration and electric gradients, for making ATP at membrane transport proteins.

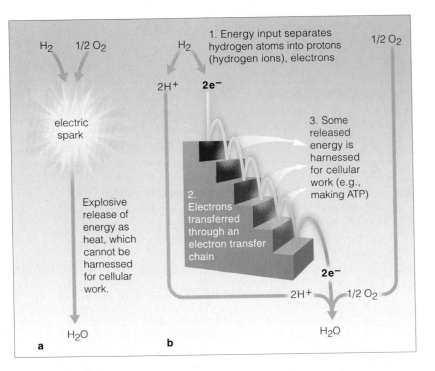

1. Energy input separates hydrogen atoms into protons (hydrogen ions), electrons

H_2 $1/2\ O_2$

H_2 $1/2\ O_2$

$2H^+$ $2e^-$

electric spark

Explosive release of energy as heat, which cannot be harnessed for cellular work.

2. Electrons transferred through an electron transfer chain

3. Some released energy is harnessed for cellular work (e.g., making ATP)

$2e^-$

$2H^+$ $1/2\ O_2$

H_2O

H_2O

a

b

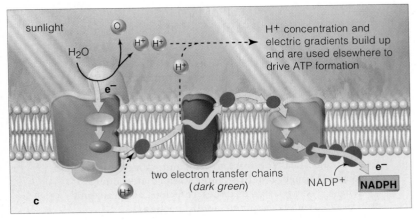

sunlight

H_2O

O

H^+ H^+

H^+

e^-

H^+

two electron transfer chains (*dark green*)

H^+ concentration and electric gradients build up and are used elsewhere to drive ATP formation

$NADP^+$ **NADPH**

e^-

c

Figure 6.11 Uncontrolled versus controlled energy release. (**a**) Oxygen and hydrogen exposed to an electric spark react and release energy all at once. (**b**) Electron transfer chains let the same reaction occur in small, manageable steps that access released energy. (**c**) Electron-donating complexes and electron transfer chains in a chloroplast. The chains accept electrons that were excited to a higher energy level by sunlight and released from water. A coenzyme, $NADP^+$, accepts electrons at the end of the second chain.

Photosynthesis and aerobic respiration rely on electron transfers to secure energy for ATP formation.

Electron transfer chains associated with cell membranes contribute energy, inherent in H^+ concentration and electric gradients, to make ATP at other membrane sites.

ENZYMES HELP WITH ENERGY HILLS

What would happen if you left a cupful of glucose out in the open? Not much. Even though its conversion to CO_2 and H_2O is energetically favored in the presence of oxygen, years would pass before you could expect to see evidence of it. How long does the conversion take in your body? A few seconds. *Enzymes make the difference.*

Without enzymes, you would quickly cease to exist. Reactions would not occur fast enough for you to process food, build new cells and get rid of worn-out ones, keep your brain working, contract muscles, and do everything else you have to do to stay alive.

By definition, **enzymes** are catalytic molecules; they speed the rate at which a specific reaction approaches chemical equilibrium. An enzyme alters only the rate of a reaction, not its outcome.

Except for a few RNAs, enzymes are proteins. Like other proteins, their function depends on their primary structure and final, three-dimensional structure, for the reasons described earlier in Sections 3.6 through 3.8.

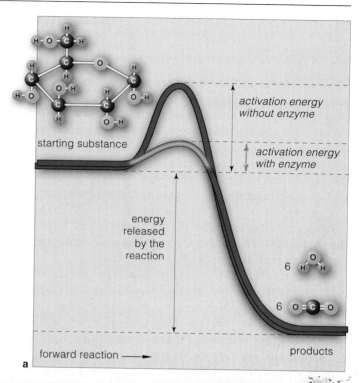

starting substance

activation energy without enzyme

activation energy with enzyme

energy released by the reaction

6 $H \overset{O}{H}$

6 $O = C = O$

forward reaction ⟶

products

a

Figure 6.12 Activation energy. (**a**) A certain amount of energy is required to get a metabolic reaction going. That amount, the activation energy, allows the reaction to spontaneously proceed to the end products. (**b**) An enzyme enhances the reaction rate by lowering the amount of activation energy. Refer back to the analogy in Figure 6.5. Enzymes make the energy hill smaller; they lower the energy barrier.

b

All enzymes share four features. First, enzymes do not make anything happen that could not happen on its own, but they make it happen hundreds to millions of times faster. Second, reactions do not permanently alter or use up enzymes; the same enzyme molecule may work repeatedly. Third, the same type of enzyme usually catalyzes the forward and reverse directions of a reversible reaction. Fourth, an enzyme is picky about its substrates. Its only substrates are specific substances that the enzyme can chemically recognize, bind, and modify in precise ways. For instance, thrombin is one of the enzymes needed to clot blood. It only recognizes and cleaves a side-by-side arrangement of arginine and glycine—two amino acids—in a certain protein. This converts the protein to one of the clotting factors.

To get a sense of how enzymes work, you have to know about activation energy. **Activation energy** is the minimum amount of collision energy required to get a reaction going. The actual amount is not the same for all reactions. And the collision may be spontaneous or enzymes may promote it.

Activation energy is an energy barrier—something like the hill depicted in Figure 6.12. That barrier must be surmounted one way or another before the reaction will proceed. An enzyme lowers the energy barrier, in ways that will be described next.

***

Enzymes enhance reaction rates by lowering the amount of activation energy required to get the reaction going.

***

HOW DO ENZYMES LOWER ENERGY HILLS?

The Active Site

How do enzymes lower the energy hill and so increase rates of reactions? *They present a microenvironment that is energetically more favorable for reaction, compared to the environment at large.* Each contains one or more **active sites**: pockets or crevices where substrates are bound and where specific reactions are catalyzed. Reactants for an enzyme—its **substrates**—have a surface region that is complementary in its size, shape, solubility, and charge to the active site. This complementary fit is why an enzyme can selectively identify its substrate among the thousands of substances in cells. Figure 6.13 depicts one enzyme meeting up with its substrate.

Sometimes functional groups of the enzyme alone carry out a reaction. However, one or more cofactors often help out. Cofactors, again, include metal ions and coenzymes, which are complex organic compounds that may or may not have a vitamin component.

Metal ions, coenzymes, or both are often bound so tightly to an active site that they are *prosthetic* groups, as functionally important as an artificial limb is to an amputee. The "heme" in hemoglobin is an example.

Transition at the Top of the Hill

Think back on the key categories of enzyme-mediated reactions introduced in Chapter 3. During functional group transfers, a molecule gives up a functional group and another molecule accepts it. In electron transfers, one or more electrons are stripped from a molecule and donated to another. In rearrangements, the juggling of internal bonds converts one molecule to another. In condensation, two molecules become covalently bound to form a larger molecule. Finally, in cleavage, a larger molecule splits into two smaller ones.

Thus, when we talk about the amount of activation energy required to convert each kind of substrate to its products, *we really are talking about the energy it takes to align reactive chemical groups, to briefly destabilize electric charges, and to rearrange, create, and break bonds.*

These events can bring a substrate to its **transition state**, the point when a reaction can easily run in either direction, to product or back to a reactant. A substrate is bound most tightly to an enzyme in this state.

How Enzymes Work

During an enzyme-mediated reaction, various bonds form between the substrate and the enzyme, cofactor, or both. Many of these bonds are weak. But energy is released as each forms, and it promotes the formation of an enzyme–substrate complex. Energy released from all of the weak interactions is the **binding energy**. Its release in the transition state is partial payment on the cost of the reaction—the activation energy.

Binding energy can speed a reaction rate by several mechanisms. Depending on the enzyme, the following four mechanisms work alone or in combination with one another to bring about the transition state.

Helping substrates get together. Substrate molecules rarely react if their concentrations are low. Binding at an active site is like a localized boost in concentration. The boost can increase the rate by 10,000 to 10,000,000 times, depending on the particular reaction.

Orienting substrates in positions favoring reaction. On their own, substrates collide from random directions. Weak but extensive bonds at an active site put reactive groups on precise collision courses far more frequently.

Shutting out water. Sometimes substrates are bound so tightly that water molecules are partially or wholly squeezed out of the active site. For some reactions, the resulting nonpolar microenvironment can lower the activation energy by as much as 500,000 times. This is what happens when an enzyme attaches a carboxyl group (—COO⁻) to a molecule.

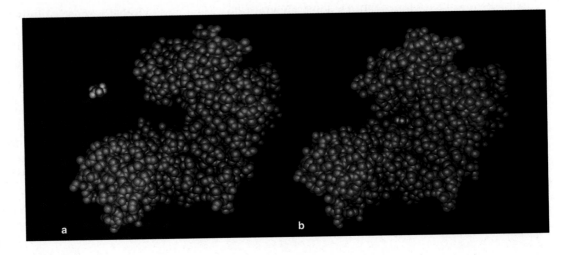

Figure 6.13 Space-filling model for hexokinase, a major enzyme. Hexokinase attaches a phosphate group to glucose. (**a**) The glucose molecule, color-coded *red*, is heading for the enzyme's active site, which is inside a cavity in the enzyme (*green*). (**b**) As glucose enters the cavity, parts of the enzyme briefly close around it.

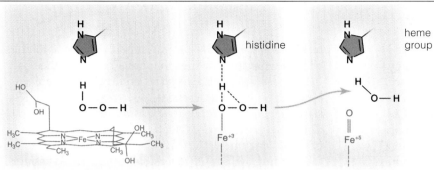

a One molecule of hydrogen peroxide (H_2O_2) enters one of the four cavities of a catalase molecule, where it will become the substrate for the reaction. The heme group enclosed in the cavity is shown in *red.*

b One of the amino acids (histidine) projecting into the cavity attracts one of the substrate's H atoms, leaving an O atom free to bind to the heme group's iron (Fe).

c Binding to the iron causes the substrate's O—O bond to stretch and break. Water (H_2O) forms when the —OH fragment attracts an H atom away from the histidine. (The iron releases the oxygen in a second reaction with another H_2O_2 molecule. Oxygen and another H_2O molecule form.)

Figure 6.14 A closer look at catalase, an enzyme that breaks down hydrogen peroxide to water and oxygen. In cells, four catalase chains associate with each other as a quaternary structure. Each folded chain forms a cavity around one heme group. An iron atom in the heme group assists in catalysis. The electrical properties of the enzyme-bound heme ensure that harmful oxygen radicals are not produced as a by-product of hydrogen peroxide breakdown.

Inducing changes in enzyme shape. In most cases, the weak interactions between an enzyme and its substrate induce change in the enzyme's shape. By the **induced-fit model**, functional groups on the substrate's surface are *not quite c*omplementary to functional groups in the active site. As the enzyme aligns structurally with the substrate, the substrate itself bends, which pushes it toward the transition state. Precise complementarity would not work. If a substrate were to bind too tightly to the active site, the enzyme–substrate complex would become too stable, and reaction would be impeded.

About Those Cofactors

What roles do cofactors play when a substrate has been suitably aligned in an active site? You can get an idea by considering cofactors that transfer electrons.

For instance, unstable, charged intermediates form in many reactions. Left alone, they would swiftly revert to the form of the substrate. Electron (and hydrogen) transfers to and from the substrate or intermediate can tip the reaction toward products rather than substrates.

Cofactors mediate most of these reactions. Example: The metal ions bind so tightly with aspartate, cysteine, histidine, or glutamate that they become built into the structure of many enzymes. They readily give up and accept electrons. When they interact with a substrate or intermediate, they shift electron arrangements in ways that promote product formation. This is what goes on at the heme in catalase. Heme has a complex organic ring structure, with iron (Fe^{++}) at its center. Figure 6.14 shows how this metal ion works. Nearly one-third of all known enzymes use one or more metal ions.

Why Are Enzymes So Big?

Most enzymes, like the one in Figure 6.13, are very big molecules compared to their substrates. Why is this so? Rapid, repeated catalysis demands a stable chemical environment. An enzyme's polypeptide chains must be long enough to fold repeatedly in particular directions, and not just to afford structural stability. Folding also must put specific amino acids and functional groups in locations and orientations that favor interaction with the water that bathes the enzyme's outer surface—and with a substrate molecule that contacts the active site.

The activation energy for a reaction is the energy it takes to align reactive chemical groups, briefly destabilize electric charges, and rearrange, create, and break bonds.

Activation energy drives substrates to the transition state, the point when a reaction can spontaneously run either to product or back to substrate.

Increasing substrate concentrations, orienting substrates, excluding water, and changing the enzyme and substrate shapes are the main mechanisms that lower the activation energy at an active site.

ENZYMES DON'T WORK IN A VACUUM

You probably don't get much done when you feel too hot or cold or out of sorts because you ate too many sour plums or salty potato chips. When the cupboard is bare, you focus on food. Maybe you call a friend to go shopping with you, and when you drive too fast to the grocery store, police tend to slow you down. In such ways, you have a lot in common with enzymes. They, too, respond to shifts in temperature, pH, and salinity, and to the relative abundances of particular substances. Many depend on helpers for specific tasks. And all normal enzymes respond to metabolic police.

How Is Enzyme Activity Controlled?

Each cell controls its enzyme activity. By coordinating control mechanisms, it maintains, lowers, or increases the concentrations of substances. Controls adjust how

fast enzyme molecules are built and become available, or they work on enzymes that are already synthesized.

For example, with *allosteric* control, some molecule binds to an enzyme at a site other than the active site. (*Allo–* means different, *steric* means structure, or state.) The active site changes shape in a way that allows or prevents enzyme action. Figure 6.15 shows models for the binding, which is reversible. Study these models, then picture a bacterium synthesizing tryptophan and other amino acids used to build proteins. Soon, enough proteins are built. But tryptophan synthesis won't stop until the concentration of tryptophan molecules causes **feedback inhibition**. With this feedback mechanism, a change caused by an activity *shuts down the activity*.

In this pathway, a feedback loop starts and ends at an allosteric enzyme. When tryptophan accumulates, unused molecules bind with the allosteric site. Binding shuts down the enzyme, hence the rest of the pathway.

What if tryptophan is scarce when the demand for it steps up? Enzyme molecules are not inactivated, so tryptophan will be built. Such feedback loops quickly adjust concentrations of many substances (Figure 6.16).

In humans and other multicelled organisms, control of enzyme activity is just amazing. Cells not only work to stay alive, they work with other cells in ways that benefit the whole body! For instance, this vast cellular work relies on hormones, a type of signaling molecule. Specialized cells release hormones. Any cell that has receptors for a particular hormone can respond to it. Its program for building a protein or some other activity changes. The hormone trips cell controls into action— and the activity of enzymes increases or slows down.

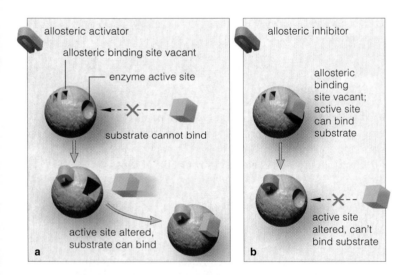

Figure 6.15 Simple sketches of allosteric control of enzyme activity. (**a**) An active site is unblocked when an activator protein binds to a vacant allosteric site. (**b**) An active site is blocked when an inhibitor protein binds to a vacant allosteric site.

Do Temperature and pH Affect Enzymes?

Temperature, recall, is a measure of molecular motion. A rise in temperature boosts reaction rates by making substrates collide more often with active sites. Yet past some temperature (which differs among enzymes), the increased motion disrupts the weak bonds holding the enzyme in its three-dimensional shape. Substrates no longer bind to the active site, so reaction rates decline sharply, as Figure 6.17a shows. When temperatures rise above or fall below the range of tolerance, metabolism is disrupted. That's what happens with extremely high fevers. People usually do not survive when the body's internal temperature reaches 44°C (112°F).

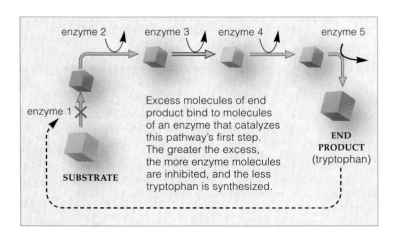

Excess molecules of end product bind to molecules of an enzyme that catalyzes this pathway's first step. The greater the excess, the more enzyme molecules are inhibited, and the less tryptophan is synthesized.

Figure 6.16 Example of feedback inhibition of a metabolic pathway. Five kinds of enzymes act in sequence. They convert a substrate to the product, tryptophan.

a

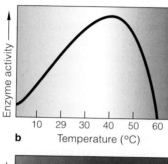

b

Temperature (°C)

10 29 30 40 50 60

Enzyme activity →

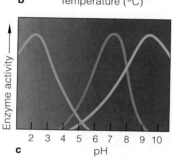

c

pH

2 3 4 5 6 7 8 9 10

Enzyme activity →

Figure 6.17 Enzymes and the environment. (**a**) Siamese cats show observable effects of differences in temperature. Epidermal cells that give rise to their fur produce a brownish-black pigment, melanin. More melanin is produced in fur on ears and paws than elsewhere on the cat body. A heat-sensitive enzyme that helps control melanin production is less active in warmer parts of the body, which end up with lighter fur. (**b**) How one enzyme's activity shifts as the temperature rises. (**c**) How three kinds of enzymes respond to a rise in pH.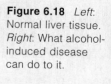

pH values that extend beyond an enzyme's range of tolerance also take a toll (Figure 6.17*b*). Most enzymes work best when pH is between 6 and 8. Trypsin, for instance, is active in the small intestine, where pH is 8 or so. Pepsin, a protein-digesting enzyme, is one of the exceptions. It functions in gastric fluid, a highly acidic liquid (pH of about 1–2) that denatures most enzymes.

In addition, most enzymes do not work well when fluids are saltier than usual. High salinity (extremely high ion concentrations) disrupts the interactions that help hold enzymes in their three-dimensional shapes.

Control mechanisms govern the synthesis of new enzymes and stimulate or inhibit the activity of existing enzymes. By controlling enzymes, cells control the concentrations and kinds of substances available to them.

Enzymes function best when the cellular environment stays within limited ranges of temperature, pH, and salinity. The actual ranges differ from one type of enzyme to the next.

Beer, Enzymes, and Your Liver

That catalase molecule you've been reading about is also a foot soldier against alcoholic attacks on the body. Think of what happens after someone drinks 12 ounces of beer, 6 ounces of wine, or 1.5 ounces of eighty-proof vodka. All contain the same amount of ethanol, or ethyl alcohol (C_2H_6O). This sugar has water-soluble and fat-soluble components. The body absorbs about 20 percent from the stomach and about 80 percent from the small intestine. The bloodstream swiftly transports more than 90 percent of it to the liver, which breaks it down to a nontoxic molecule called acetate (acetic acid). If ethanol were to accumulate to high levels, cells would die and tissues would be damaged. A healthy person's liver detoxifies ethanol before this happens.

The liver can detoxify only so much in a given hour. Its alcohol-metabolizing enzymes set the reaction rate, which is slower than the rate of absorption. Alcohol dehydrogenases in liver cells are key enzymes in the detoxification reactions. They catalyze the conversion of ethyl alcohol to acetaldehyde, which is toxic at high levels. Another enzyme, acetaldehyde dehydrogenase, swiftly breaks down this intermediate to acetate and other nontoxic forms. Also, when the concentration of alcohol in the blood is high, one of the cytochromes assists in the detoxification. So does catalase.

Given the liver's central role in alcohol metabolism, habitually heavy drinkers gamble with alcohol-induced liver diseases. Over time, the capacity to tolerate alcohol diminishes because there are fewer and fewer liver cells —hence fewer enzymes—available for detoxification. In *alcoholic hepatitis*, inflammation and destruction of liver tissue is widespread. Another disease, *alcoholic cirrhosis* permanently scars the liver and eventually destroys its functioning (Figure 6.18).

Such a loss is devastating. The liver is the largest gland in the human body, and its functioning impacts all other organs. As you will see later in the book, it has major roles in digesting and absorbing fats from food. It is central to controlling the synthesis and uptake of carbohydrates, lipids, and proteins by body cells. It is a blood reservoir that is adjusted to help maintain the volume of blood through the body. It helps return excess fluid that collects in tissues to the general circulation. And its enzymes inactivate many toxic compounds besides acetaldehyde. Think about it.

Figure 6.18 *Left*: Normal liver tissue. *Right*: What alcohol-induced disease can do to it.

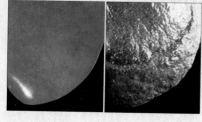

LIGHT UP THE NIGHT—AND THE LAB

BIOLUMINESCENT ORGANISMS In warm waters of the tropical seas, above gardens and fields on summer nights, you may be lucky enough to witness organisms lighting up the night sky with their collective flashings. Many species flash with orange, yellow, yellow-green, or blue light, as in Figure 6.19.

Flashers emit light when enzymes called luciferases convert chemical energy to light energy. Figure 6.19c shows the molecular structure of the firefly luciferase. The reactions start when, in the presence of oxygen, an ATP molecule transfers a phosphate group to a highly fluorescent substance: luciferin. This destabilizes the molecule, which makes it enter reactions that involve electron transfers. At one reaction step, some energy is released as *fluorescent* light. This type of light is emitted when any kind of destabilized molecule reverts to a stable configuration. When organisms flash with such light, this is called **bioluminescence**.

RESEARCH APPLICATIONS It is possible to think about organisms from a "gee-whiz-ain't-nature-grand" point of view. Or you can come up with novel ways to think about what they do and how they do it. The latter way of thinking puts you squarely in the camp of biologists.

And so we have biologists who thought to borrow light from fireflies and use it to make bioluminescent gene transfers. Copies of genes for bioluminescence can now be inserted into a variety of organisms, such as bacteria, plants, and mice (Figure 6.20).

Here is a practical application of their work. Each year, 3 million people die from a lung disease caused by *Mycobacterium tuberculosis*. No single antibiotic is effective against all of the different bacterial strains, so

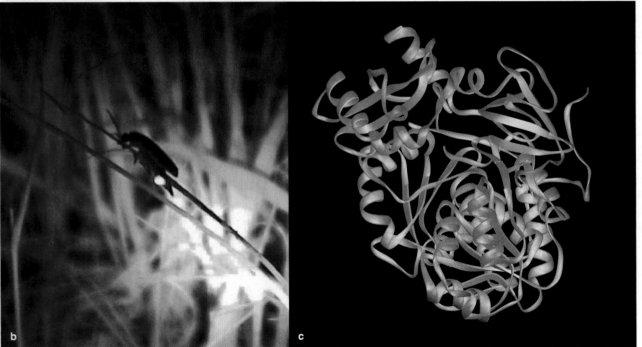

Figure 6.19 (**a**) Stirring up bioluminescent marine organisms near Vieques Island, Puerto Rico. As many as 5,000 free-living cells called dinoflagellates may be in each liter of water, and each flashes with blue light when agitated. (**b**) North American firefly (*Photinus pyralis*) emitting a flash from its light organ. Peroxisomes in this organ are packed with luciferase molecules. Firefly flashes help potential mates find each other in the dark. (**c**) Ribbon model for firefly luciferase. This enzyme catalyzes the reaction that releases light. Its single polypeptide chain is folded into multiple domains.

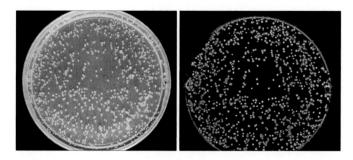

Figure 6.20 Colonies of bioluminescent bacteria in daylight (*left*) and glowing in a culture dish in the dark (*right*).

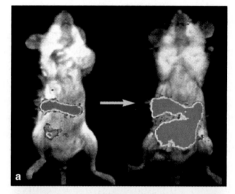

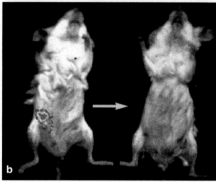

Figure 6.21 Using bioluminescent bacterial cells to chart the location of infectious bacteria inside living laboratory mice and their spread through body tissues. (**a**) False-color images in this pair of photographs show how the infection spread in a control group that had not been given a dose of antibiotics. (**b**) This pair shows how antibiotics had killed most of the infectious bacterial cells.

an infected person can't get effective treatment until the strain that is causing the infection is identified. A fast way to do this is to expose a sample of bacterial cells taken from a patient to luciferase genes. These genes will usually slip into the bacterial DNA of some cells. Clinicians isolate the cells, then expose colonies of the descendants to different antibiotics. When one antibiotic does *not* work, the colonies glow. Their cells have made gene products—including luciferase. An antibiotic works when cells of the colonies don't glow; the antibiotic killed them.

Christopher and Pamela Contag, two postdoctoral students at Stanford University, wanted to light up bacteria that cause *Salmonella* infections in laboratory mice. Why? Researchers of viral or bacterial diseases typically have to infect dozens to hundreds of mice for experiments. The only option has been to kill infected mice and examine tissues to find out whether infection occurred—a costly, tediously painstaking practice that also requires dispatching the experimental animals.

The Contags approached David Benaron, a medical imaging researcher at Stanford, with this hypothesis: If live, infectious bacteria were made bioluminescent, then flashes will shine through the tissues of infected animals. In a preliminary test of this novel idea, the researchers put glowing *Salmonella* cells into a thawed chicken breast from a market. It glowed from inside.

The Contags transferred bioluminescence genes into three strains of *Salmonella*. Then they injected the strains into mice of three experimental groups and used a digital imaging camera to track the infection in each group. The first strain was weak; the mice were able to fight off the infection in less than six days and did not glow. The second strain was not as weak but could not spread through the mouse body; it remained localized. The third strain was dangerous. It spread rapidly through the entire mouse gut—which glowed.

Thus bioluminescent gene transfer, combined with imaging of enzyme activity, can be used to track the course of infection and to evaluate the effectiveness of

drugs in living organisms (Figure 6.21). It might have uses in gene therapy, where copies of functional genes replace defective or cancer-causing ones in patients.

Why use bioluminescent organisms to conclude a chapter? They give us visible signs of metabolism—of the cell's capacity to acquire energy and use it to build, break apart, store, and release substances in controlled ways. Each flash reminds us that living cells are taking in energy-rich solutes, building membranes, storing things, replenishing enzymes, and checking out their DNA. A constant supply of energy drives all of these activities. And the flashes can remind us, too, of how modern-day biologists are busy unlocking the secrets of metabolism and putting them to use in major ways.

Bioluminescence is one outcome of enzyme-mediated reactions that release energy as fluorescent light.

Biologists have transferred genes for bioluminesence, the luciferases, into a variety of organisms. These gene transfers have research applications and practical uses.

SUMMARY   *Gold* indicates text section

1. Cells store, break down, and dispose of substances by acquiring and using energy and raw materials from outside sources. Metabolism, the sum of these energy-driven activities, underlies their survival. *CI*

2. Two laws of thermodynamics affect life. *First*, energy undergoes conversion from one form to another, but its total amount neither increases nor decreases; the total amount in the universe holds constant. *Second*, energy spontaneously flows in one direction, from usable forms to forms that are less and less usable. *6.1*

3. All matter has potential energy, as measured by the capacity to do work, owing to its position in space and the arrangement of its parts. Cells use potential energy to do mechanical, chemical, and electrical work. *6.1*

4. Without energy, a cell (like all organized systems) gets disorganized. It loses chemical potential energy in each metabolic reaction, mainly as heat. It stays alive (organized) by balancing energy outputs and inputs. The sun is life's primary energy source. Plants and other photosynthesizers convert sunlight energy to chemical bond energy of organic compounds. Plants, as well as organisms that feed on plants and each other, use energy in organic compounds to do cellular work. *6.1, 6.2*

5. Cells conserve energy by coupling energy-releasing reactions with energy-requiring ones. ATP is the main coupling agent; it primes molecules to react by giving up a phosphate group to them. *6.2*

6. Cells increase, maintain, and lower concentrations of substances by coordinating thousands of reactions. They rapidly shift rates of metabolism by controlling a few steps of reversible pathways. *6.3, 6.6*

7. Metabolic reactions start with reactants and end with products. Substances that form in between are intermediates. Energy carriers (mainly ATP), enzymes, cofactors, and transport proteins are key players. *6.3*

8. Orderly, enzyme-mediated reaction sequences are called metabolic pathways. The biosynthetic (anabolic) ones build large molecules with higher bond energies from smaller molecules. Photosynthesis is an example. Degradative (catabolic) pathways break down large molecules to small products with lower bond energies. Aerobic respiration is an example. *6.2, 6.3*

9. In photosynthesis and aerobic respiration, electron transfers (oxidation–reduction reactions) are used in forming ATP. At certain membranes, electron transfer chains release energy, inherent in H^+ concentration and electric gradients, to make ATP at other sites. *6.2, 6.4*

10. Enzymes are catalysts; they greatly enhance rates of specific reactions. Pockets or cavities in their surface offer energetically favorable microenvironments for the reaction; these are the active sites. *6.4, 6.5*

a. Nearly all enzymes are protein molecules; some RNAs also are catalytic.

b. The activation energy for a reaction is the energy it takes to align reactive chemical groups, destabilize electric charges, and then rearrange, create, and break bonds. It drives substrates to the transition state, which is the point at which a reaction can spontaneously run either to product or back to a reactant.

c. As bonds briefly form between the enzyme and substrate during the transition state, energy is released. This binding energy is a partial payment on the energy cost of the reaction (the activation energy).

d. Enzymes lower the activation energy by boosting concentrations of substrates in the active site, orienting substrates in positions that favor reaction, shutting out most or all of the water from the active site, and often changing their shape.

11. Sometimes functional groups of the enzyme carry out a reaction. More often, one or more cofactors help out by functional group transfers or electron transfers. Cofactors are metal ions and coenzymes, which are complex organic compounds with or without a vitamin component. When bound tightly to an active site, they are called prosthetic groups. Heme is one example. It is one of nearly a third of all known enzymes that include at least one metal ion. *6.6*

12. Different controls over enzyme action influence the kinds and amounts of substances available at any time in the cell. Enzyme action is best within a limited range of temperature, pH, and salinity. *6.7*

13. Bioluminescence is one result of enzyme-mediated reactions that release energy as fluorescent light. The transfer of genes for bioluminescence into organisms has research and practical applications. *6.9*

Review Questions

1. Define free radical. How does it relate to aging? *CI*

2. State the first and second laws of thermodynamics. Does life violate the second law? *6.1*

3. Define and give a few examples of potential energy. *6.1*

4. Give examples of a change in potential energy involved in (a) mechanical work and (b) chemical work. *6.2*

5. Make a simple diagram of the ATP molecule. Highlight which parts can be transferred to another molecule. *6.2*

6. What is an oxidation–reduction reaction? *6.2*

7. Define the following terms: metabolic pathway, substrate (or reactant), intermediate, and end product. Is photosynthesis a biosynthetic (anabolic) pathway overall? *6.3*

8. What is an electron transfer chain, and how does it work? *6.4*

9. Describe four key features of enzymes. *6.5*

10. Define activation energy, then state four ways in which enzymes may lower it. *6.5, 6.6*

11. Define feedback inhibition as it relates to the activity of an allosteric enzyme. *6.7*

Self-Quiz   ANSWERS IN APPENDIX III

1. _____ is life's primary source of energy.
   a. Food
   c. Sunlight
   b. Water
   d. ATP

2. If we liken chemical equilibrium to the bottom of an energy hill, then an _____ reaction is an uphill run.
   a. endergonic
   c. ATP-assisted
   b. exergonic
   d. both a and c

3. Which statement is *not* correct? A metabolic pathway _____ .
   a. has an orderly sequence of reaction steps
   b. is mediated by one enzyme, which initiates the reactions
   c. may be biosynthetic or degradative, overall
   d. all of the above

4. Which is *not* true of chemical equilibrium?
   a. Product and reactant concentrations are always equal.
   b. The rates of the forward and reverse reactions are the same.
   c. There is no further net change in product and reactant concentrations.

5. Electron transfer chains involve _____ .
   a. enzymes and cofactors
   c. cell membranes
   b. electron transfers
   d. all of the above

6. Enzymes are _____ .
   a. enhancers of reaction rates
   d. not influenced by salinity
   b. influenced by temperature
   e. a through c
   c. influenced by pH
   f. all of the above

7. All enzymes incorporate a(n) _____ .
   a. active site
   c. metal ion
   b. coenzyme
   d. all of the above

8. NAD$^+$, FAD, and NADP$^+$ are _____ .
   a. coenzymes   c. allosteric enzymes
   b. metal ions   d. both a and b

9. Match each substance with the most suitable description.
   ___ coenzyme or metal ion
   ___ adjusts gradients at membrane
   ___ substance entering a reaction
   ___ substance formed while a reaction is proceeding
   ___ substance at end of reaction
   ___ enhances reaction rate
   ___ mainly ATP

   a. reactant or substrate
   b. enzyme
   c. cofactor
   d. intermediate
   e. product
   f. energy carrier
   g. transport protein

Critical Thinking

1. Cyanide, a toxic compound, can bind to an enzyme that is a component of electron transfer chains. The outcome is cyanide *poisoning*. Binding prevents the enzyme from donating electrons to a nearby acceptor molecule in the system. What effect will this have on ATP formation? From what you know of ATP's function, what effect will this have on a person's health?

2. *Pyrococcus furiosus* thrives at 100°C, the boiling point of water. This species of bacterium was discovered growing in a volcanic vent in Italy. Enzymes isolated from *P. furiosus* cells do not function well below 100°C. What is it about the structure of these enzymes that allows them to remain stable and active at such high temperatures? (Hint: Review Section 3.7, which summarizes the interactions that maintain protein structure.)

3. In cells, superoxide dismutase (Figure 6.1 and *above right*) has a quaternary structure; it consists of two polypeptide chains. In each chain, a strandlike domain is arrayed as a barrel around a copper ion and zinc ion (coded *orange* and *white*).

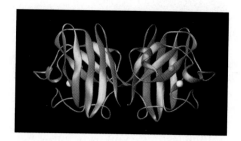

Which part of the barrel is probably hydrophobic? Which part is hydrophilic? Do you suppose substrates bind inside or outside the barrels? Do the metal ions have a role in catalysis?

4. AZT, or azidothymidine, is a drug used to alleviate symptoms of *AIDS*, or acquired immunodeficiency syndrome. AZT is very similar in molecular structure to thymidine, one of the DNA nucleotides. AZT also can fit into the active site of an enzyme produced by the virus that causes AIDS. Infection puts the viral enzyme and viral genetic material (RNA) into a cell. The cell wrongly uses the RNA as a template (structural pattern) for joining nucleotides to form viral DNA. The viral enzyme takes part in the assembly reactions. Propose a model to explain how AZT might inhibit replication of the virus inside cells.

5. The bacterium *Clostridium botulinum* is an obligate anaerobe, meaning it dies quickly in the presence of oxygen. It lives in oxygen-free pockets in soil, and it can enter a metabolically inactive, resting state by forming spores. The spores may end up on the surfaces of garden vegetables. When the picked vegetables are being canned, the spores must be destroyed. If they are not destroyed, *C. botulinum* may grow and produce botulinum, a toxin. In improperly cooked and tainted canned foods, the toxin is active and can cause a type of food poisoning called *botulism*.

   This bacterium is not able to produce superoxide dismutase or catalase. Develop a hypothesis to explain how the absence of these enzymes is related to its anaerobic life-style.

Selected Key Terms

activation energy *6.5*
active site *6.6*
ADP *6.2*
ATP *6.2*
ATP/ADP cycle *6.2*
binding energy *6.6*
bioluminescence *6.9*
chemical energy *6.1*
chemical equilibrium *6.3*
coenzyme *6.6*
cofactor *6.3*
electron transfer chain *6.4*
energy carrier *6.3*
entropy *6.1*
enzyme *6.5*
feedback inhibition *6.7*
first law of thermodynamics *6.1*
free radical *CI*

heat (thermal energy) *6.1*
induced-fit model *6.6*
intermediate *6.3*
kinetic energy *6.1*
law of conservation of mass *6.3*
metabolic pathway *6.3*
metabolism *CI*
oxidation–reduction reaction *6.2*
phosphorylation *6.2*
potential energy *6.1*
product *6.3*
reactant *6.3*
second law of thermodynamics *6.1*
substrate (reactant) *6.6*
transition state *6.6*
transport protein *6.3*

Readings

Adams, S. 22 October 1994. "No Way Back." *New Scientist*. A refreshing look at the second law of thermodynamics.

Branden, C., and J. Tooze, 1999. *Introduction to Protein Structure*. Second edition. New York: Garland.

7

HOW CELLS ACQUIRE ENERGY

Sunlight and Survival

Think about the last time you were hungry and craved a bit of apple, lettuce, chicken, pizza, or any other food. Where did it come from? For the answers, look past the refrigerator, the market or restaurant, or even the farm. Look to individual plants, the starting point for nearly all edibles you put into your mouth. Plants use environmental sources of energy and raw materials to make glucose and other organic compounds. Organic compounds, recall, are built on a framework of carbon atoms. So the questions become these:

1. *Where does the carbon come from in the first place?*

2. *Where does the energy come from to drive the synthesis of carbon-based compounds?*

Answers vary, because they depend on an organism's mode of nutrition.

Plants generally are "self-nourishing" organisms, or **autotrophs**. Their carbon source is carbon dioxide (CO_2), a gaseous compound in the air and dissolved in water. Plants, some bacteria, and many protistans are **photoautotrophs**, meaning they use sunlight for **photosynthesis**. By this process, energy from the sun drives formation of ATP and the coenzyme NADPH. ATP delivers energy to sites where glucose and other carbohydrates are built. NADPH delivers electrons and hydrogen building blocks to the same sites.

Many other organisms are **heterotrophs**, meaning they must feed on autotrophs, one another, and organic wastes. (*Hetero–* means other, as in "being nourished by other organisms.") That is how most bacteria, many protistans, and all fungi and animals stay alive. Unlike plants, heterotrophs can't feed themselves with energy and raw materials from the physical environment.

How do we know such things? We didn't have a clue until the mid-seventeenth century. Before then, most people assumed that plants got raw materials to make food from soil. By 1882 a few chemists had an inkling that plants use sunlight, water, and something in the air to make food. A botanist, T. Engelmann, wondered: What parts of sunlight do plants favor?

As Engelmann knew, free oxygen is released during photosynthesis. He also knew that some bacteria need oxygen for aerobic respiration, as most organisms do. He hypothesized: If the bacteria require oxygen, then we can expect them to gather in places where the most photosynthesis is going on. He put a water droplet containing bacterial cells on a microscope slide with a photosynthesizer, the green alga *Spirogyra*. He used a crystal prism to break up a beam of sunlight and cast a spectrum of colors across the slide.

Bacteria gathered mostly where violet and red light fell on the green alga. Algal cells released more oxygen in the part illuminated by light of those colors—the very best light for photosynthesis (Figure 7.1).

Such observations showed us how photosynthesis works. They also helped reveal one of nature's great patterns, *for photosynthesis is the main pathway by which carbon and energy enter the web of life.* Once a cell makes or takes up organic compounds, it uses or stores them. *All* autotrophs and heterotrophs store energy in organic compounds and release it by other processes, such as aerobic respiration. Figure 7.2 previews the chemical links between photosynthesis and aerobic respiration, the focus of this chapter and the next.

Such processes might seem far removed from your daily interests. But remember, the food that nourishes you and most other organisms can't be produced or used without them. We will return to this point in later chapters. It provides perspective on many issues, such as nutrition, dieting, agriculture, human population growth, genetic engineering, and the impact of pollution on our sources of food—hence on our survival.

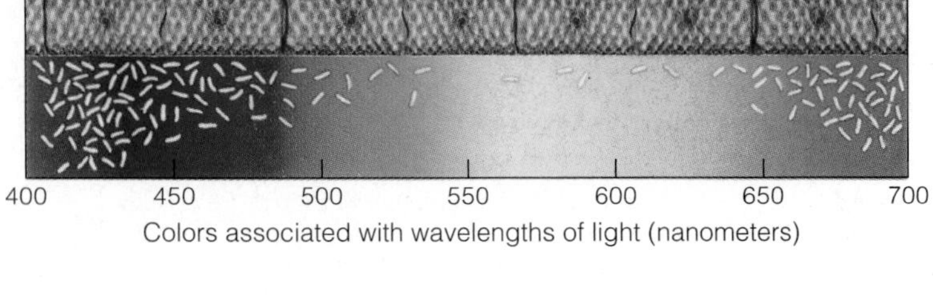

A crystal prism breaks up a beam of light into a spectrum of colors, which are cast across a droplet of water on a microscope slide.

bacteria (*white*)

part of an algal strand stretched out across a microscope slide

| 400 | 450 | 500 | 550 | 600 | 650 | 700 |

Colors associated with wavelengths of light (nanometers)

Figure 7.1 Results from T. Engelmann's observational test that correlated portions of visible light with photosynthesis in *Spirogyra*, a strandlike green alga. Many oxygen-requiring bacterial cells moved to the colors where algal cells released the most oxygen, which is a by-product of their photosynthetic activity.

Plants capture sunlight energy, which drives photosynthesis.

PHOTOSYNTHESIS

1. H_2O is split by light energy. Its oxygen escapes; its electrons, hydrogen are used at transfer chains that release energy for ATP formation. Coenzymes pick up the electrons, hydrogen.

2. ATP energy drives synthesis of glucose from hydrogen and electrons (delivered by the coenzymes), plus carbon and oxygen (from carbon dioxide).

water, carbon dioxide

oxygen

AEROBIC RESPIRATION

1. Glucose is degraded to carbon dioxide, water. Coenzymes pick up its electrons, hydrogen.

2. Coenzymes give up electrons, hydrogen to oxygen-requiring transfer chains that release energy to drive ATP formation.

ATP available to drive many cellular tasks

Figure 7.2 Links between photosynthesis—the main energy-requiring process in the world of life—and aerobic respiration, the main energy-releasing process.

Key Concepts

1. Plants, some bacteria, and many protistans use sunlight energy, carbon dioxide, and water to make glucose and other carbohydrates, which have a backbone of carbon atoms. The metabolic process by which they do this is called photosynthesis.

2. Photosynthesis is the main process by which carbon and energy become available to the web of life.

3. In the first stage of photosynthesis, sunlight energy is trapped and converted to chemical bond energy in ATP molecules. Typically, water molecules are split, their electrons and hydrogen atoms are picked up by the coenzyme $NADP^+$ to form NADPH, and their oxygen atoms are released as a by-product.

4. In the second stage of photosynthesis, ATP delivers energy to the sites of biosynthesis. NADPH delivers electrons and hydrogen, which combine with carbon and oxygen to form glucose. Carbon dioxide provides the carbon and oxygen for the synthesis reactions.

5. In the photosynthetic cells of plants and in many protistans, both stages of the reactions proceed inside organelles called chloroplasts.

6. Often we summarize photosynthesis this way:

$$12H_2O + 6CO_2 \xrightarrow{\text{LIGHT ENERGY}} 6O_2 + C_6H_{12}O_6 + 6H_2O$$

WATER   CARBON DIOXIDE   OXYGEN   GLUCOSE   WATER

PHOTOSYNTHESIS—AN OVERVIEW

Sit outdoors on a warm, cloudless day and you'll get hot but you'll never build your own food. How do plants do it? Think of it! They have to plug into the sun's energy, but not so much that it might cook them. They have to do magic and turn the sun's energy into ATP's chemical bond energy. They have to get carbon, oxygen, and hydrogen from somewhere and combine them into carbohydrates. As you might imagine, these multiple tasks can't all be done in the same place. Different functions, different structures.

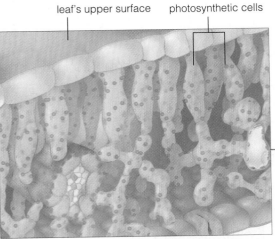

leaf's upper surface   photosynthetic cells

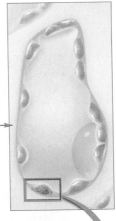

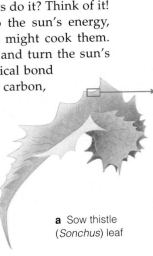

a Sow thistle (*Sonchus*) leaf

b Cutaway of a small section from the leaf. Its upper and lower surfaces enclose many photosynthetic cells.

c One of the photosynthetic cells, with green chloroplasts.

Where the Reactions Take Place

Plants and protistans called algae carry out the tasks in **chloroplasts**: organelles specialized for photosynthesis (Section 4.7). Figure 7.3 shows the structure of one type. All chloroplasts have two outermost membranes that surround a mostly fluid interior, the **stroma**. Another membrane in the stroma is often highly folded into what looks like stacked sacs connected by channels (Figure 7.3*d,e*). We refer to the stacks as thylakoids. But the **thylakoid membrane** makes up only one functional compartment, no matter how it is folded.

The capture and conversion of sunlight energy to chemical energy takes place at thylakoids. With these *light-dependent* reactions, plants get the electrons and hydrogen for making food. Amazingly, sunlight splits water molecules. Water's oxygen diffuses away. But its electrons flow through electron transfer chains at the thylakoid membrane. (Remember Section 6.4?) When electrons flow through the systems, hydrogen moves inside the thylakoid compartment. ATP forms when it flows out. Waste not, want not: The helpful coenzyme NADP$^+$ picks up the electrons and hydrogen.

Making glucose doesn't require light. These *light-independent* reactions proceed elsewhere, in the stroma. This is where ATP gives up energy, the coenzyme gives up electrons and hydrogen, and carbon dioxide (CO_2) is dismantled for its carbon and oxygen atoms.

Overall, you can think about all the reactants and products just described in terms of a simple equation:

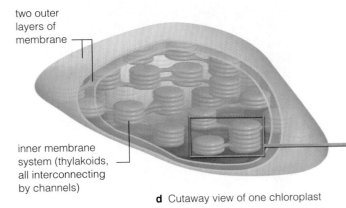

two outer layers of membrane

inner membrane system (thylakoids, all interconnecting by channels)

d Cutaway view of one chloroplast

Figure 7.3 Zooming in on sites of photosynthesis inside a leaf from a typical plant.

In case you're wondering how we know what we know: Remember the description of tracers in Section 2.2? By attaching radioisotopes to atoms of the reactants shown in the summary equation, researchers found out where each atom of the reactants ends up in the products:

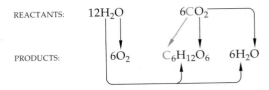

REACTANTS: $12H_2O$ $6CO_2$

PRODUCTS: $6O_2$ $C_6H_{12}O_6$ $6H_2O$

But Things Don't Really End With Glucose

Notice how we show glucose as the end product of the reactions. We do so to keep the chemical bookkeeping simple. But you'll find little glucose in a chloroplast.

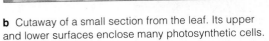

$$12H_2O + 6CO_2 \xrightarrow{\text{LIGHT ENERGY}} 6O_2 + C_6H_{12}O_6 + 6H_2O$$

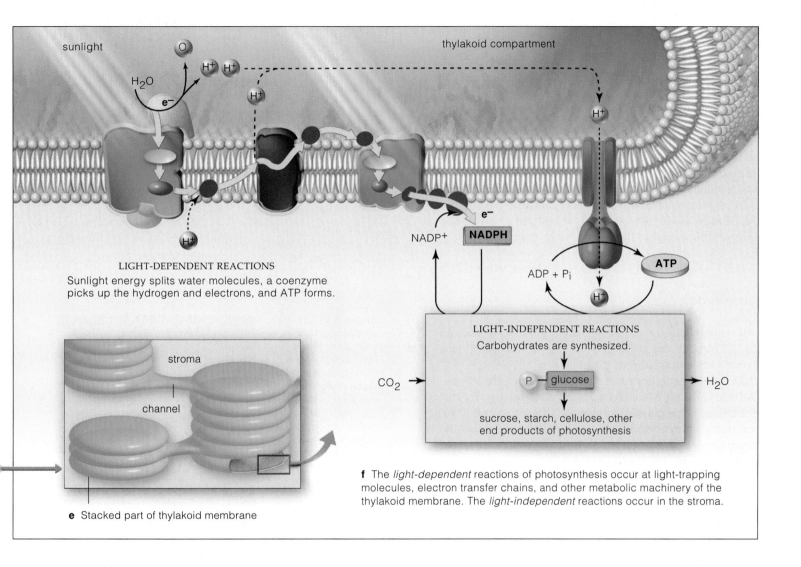

thylakoid compartment

sunlight

O

H^+ H^+

H_2O

H^+

e^-

H^+

H^+

LIGHT-DEPENDENT REACTIONS
Sunlight energy splits water molecules, a coenzyme picks up the hydrogen and electrons, and ATP forms.

e^-

NADP$^+$ **NADPH**

ADP + P$_i$ **ATP**

H^+

LIGHT-INDEPENDENT REACTIONS
Carbohydrates are synthesized.

CO_2

P — glucose

H_2O

sucrose, starch, cellulose, other end products of photosynthesis

stroma

channel

f The *light-dependent* reactions of photosynthesis occur at light-trapping molecules, electron transfer chains, and other metabolic machinery of the thylakoid membrane. The *light-independent* reactions occur in the stroma.

e Stacked part of thylakoid membrane

Each new glucose molecule has an attached phosphate group; it's primed to react with something else. Almost always, it quickly enters reactions that make sucrose, starch, or cellulose. These are the main end products of photosynthesis even though we "stop" with glucose.

The sketch below is a simple version of Figure 7.3. We repeat it in sections that follow to help reinforce your grasp of how the details fit into the big picture:

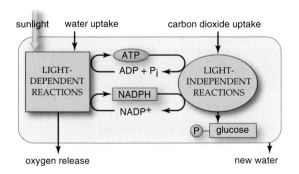

sunlight   water uptake   carbon dioxide uptake

ATP
ADP + P$_i$

LIGHT-DEPENDENT REACTIONS

LIGHT-INDEPENDENT REACTIONS

NADPH
NADP$^+$

P — glucose

oxygen release   new water

We turn next to the details. Before we do, think about this: Two thousand chloroplasts, lined single file, would be no wider than a dime. Imagine all of the chloroplasts in just one corn or rice plant—each a tiny factory for producing sugars and starch—and you may get an idea of the magnitude of the metabolic events required to feed you and every other organism on this planet.

Chloroplasts, the organelles of photosynthesis in plants and many protistans, specialize in food production.

In the first stage of photosynthesis, sunlight energy drives ATP and NADPH formation, and oxygen is released. In chloroplasts, this stage occurs at the thylakoid membrane system that weaves through the chloroplast's stroma.

The second stage occurs in the stroma. Energy from ATP drives glucose formation. For these synthesis reactions, carbon dioxide provides the carbon and oxygen atoms, and NADPH provides the electrons and hydrogen atoms.

SUNLIGHT AS AN ENERGY SOURCE

Properties of Light

The sun is the start of a one-way flow of energy through almost all ecosystems (Figure 7.4). On average, the amount adds up to 7,000 kilocalories per square meter for the whole planet. Photoautotrophs intercept about 1 percent of it.

Radiant energy from the sun undulates across space in a manner analogous to the waves crossing a sea. The horizontal distance between the crests of every two successive waves is defined as a **wavelength**. Figure 7.5a is a diagram of the **electromagnetic spectrum**, the full range of all wavelengths of radiant energy. The shorter the wavelength, the higher its energy:

low-energy wavelength

high-energy wavelength

All wavelengths of visible light combined appear white to us. We and many other organisms see the individual wavelengths as different colors.

Collectively, different photoautotrophs absorb the wavelengths between 380 and 750 nanometers. Shorter wavelengths, including ultraviolet (UV) radiation, are energetic enough to break chemical bonds of organic compounds; they can kill cells. Hundreds of millions of years ago, before the protective layer of ozone (O_3) formed in the atmosphere, the lethal UV bombardment kept photoautotrophs below the surface of the seas.

How do we know these things? Direct a ray of white light through a crystal prism, as Engelmann did, and you see how the prism bends different wavelengths by different degrees. Or look at a rainbow (Figure 7.5b). Rainbows form as the sun's rays pass through water droplets in moisture-rich air. The droplets bend longer wavelengths (yellow to red) more than they bend the shorter (violet to blue) ones, which separates them into nearly circular arcs of color. We usually see only part of the circle, a "bow" (Figure 7.5b). But blue is always on the inside of the arc, and red on the outside.

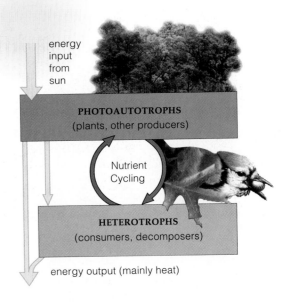

energy input from sun

Figure 7.4 Model for how photoautotrophs are the basis of a one-way flow of energy through nearly all ecosystems, or webs of life, starting with energy inputs from the sun.

PHOTOAUTOTROPHS (plants, other producers)

Nutrient Cycling

HETEROTROPHS (consumers, decomposers)

energy output (mainly heat)

Figure 7.5 **(a)** The electromagnetic spectrum. Visible light is only one of many forms of electromagnetic radiation in the spectrum. **(b)** I can make a rainbow! Observational test of the properties of light. B. Mantoni, a student, created a circular rainbow from a bright flash of light in its center. He pointed his camera at a mirror. The light from the camera's flash unit was reflected off the mirror back to the camera lens, which bent the individual wavelengths to form this circular rainbow.

b

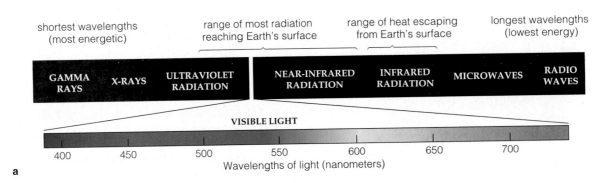

shortest wavelengths (most energetic)

range of most radiation reaching Earth's surface

range of heat escaping from Earth's surface

longest wavelengths (lowest energy)

| GAMMA RAYS | X-RAYS | ULTRAVIOLET RADIATION | NEAR-INFRARED RADIATION | INFRARED RADIATION | MICROWAVES | RADIO WAVES |

VISIBLE LIGHT

400   450   500   550   600   650   700

Wavelengths of light (nanometers)

a

Figure 7.6 (**a**) Absorption spectra that show the efficiency with which two chlorophylls, *a* and *b*, respond to wavelengths. (**b**) Absorption spectra for beta-carotene, which is one of the carotenoids, and for one of the phycobilins.

(**c**) Combined energy absorption efficiency, as indicated by the dashed line, of chlorophylls *a* and *b*, the carotenoids, and the phycobilins.

The background colors indicate the range of wavelengths being absorbed. The graph lines indicate the extent to which each of these is absorbed by a particular pigment. The color of the line itself indicates the main color that the pigment is transmitting. Thus the lines are green for chlorophylls, yellow-orange for the carotenoids, and purple and blue for the phycobilins.

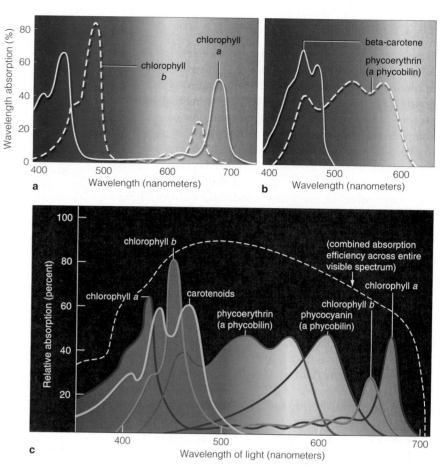

Although light travels in waves, the energy itself has a particle-like quality. When it is absorbed, the energy of visible light can be measured as if it were organized in packets, which we call **photons**. Each type of photon has a fixed amount of energy. The types having the most energy travel as the shortest wavelengths, which correspond to blue-violet light. Photons that have the least amount of energy travel as long wavelengths, and these correspond to red light.

Pigments—Molecular Bridge From Sunlight to Photosynthesis

Engelmann used a prism as part of an experiment to identify which wavelengths drive photosynthesis in a green alga (Figure 7.1). But molecular biology was far in the future, and he was not able to identify the actual bridge between sunlight and photosynthetic activity.

Pigments are the molecular bridge. **Pigments** are a class of molecules that absorb wavelengths of light, and organisms put them to many uses. Most pigments absorb only some wavelengths and transmit the rest. A few absorb so many wavelengths, they appear dark or black. Some phycobilins in algae and the melanins in animals are like this.

An **absorption spectrum** is a diagram that shows how effectively a pigment molecule absorbs different wavelengths in the spectrum of visible light. Consider the chlorophylls, the main pigments of photosynthesis. As Figure 7.6a indicates, the chlorophylls absorb all wavelengths except very little of the yellow-green and green ones, which they mainly transmit. That is why plant parts with an abundance of chlorophylls appear green to us.

Other, *accessory* pigments of different colors also are present in photoautotrophs (Figure 7.6b,c). They differ in kind and amounts among species. The dashed line in Figure 7.6c indicates how their assistance extends the range of wavelengths that can drive photosynthesis. As you can see, the combined absorption efficiency is high.

Radiation from the sun travels in waves, which differ in length and energy content. Shorter wavelengths have the most energy, and long wavelengths have less.

We perceive wavelengths of visible light as different colors and measure their energy content in packets called photons.

Chlorophylls and certain other pigments absorb specific wavelengths of visible light. They are the molecular bridge between the sun's energy and photosynthetic activity.

7.3

THE RAINBOW CATCHERS

The Chemical Basis of Color

How do pigment molecules impart color to plants and other organisms? Each kind has a light-catching array of atoms, often joined by alternating single and double bonds (Figure 7.7). In such arrays, electrons distributed around atomic nuclei can absorb photons of specific energies, which correspond to different colors of light.

Recall, from Section 2.3, that when electrons of an atom absorb energy, they move to a higher energy level. In a pigment molecule, an input of energy destabilizes the distribution of electrons inside the light-catching array. The excited electrons rapidly return to a lower energy level, the electron distribution stabilizes, and some energy is released in the form of light. When any destabilized molecule emits light as it reverts to a more stable configuration, this is called a **fluorescence**.

Excitation occurs only when the amount of energy of a photon matches the amount required to boost an electron to a higher energy level. Suppose a sunbeam strikes a pigment. If photons corresponding to red and orange wavelengths match the amount required for the boost, the pigment will absorb them. In this case, the photons corresponding to blue and violet wavelengths are a mismatch. The pigment molecule cannot absorb them. Because it transmits (reflects) these photons, it appears blue or violet to us.

On the Variety of Photosynthetic Pigments

Most pigments respond to only part of the rainbow of visible light. If acquiring energy is so vital, why doesn't each kind of photosynthetic pigment go after the whole rainbow? In other words, why isn't each one black?

If early photoautotrophs evolved in the seas, then so did their pigments. Ultraviolet and red wavelengths do not penetrate water as deeply as green and blue wavelengths do. Possibly natural selection favored the evolution of different pigments at different depths. Many relatives of the red alga in Figure 7.8a live deep in the sea, and some of them are nearly black. Green

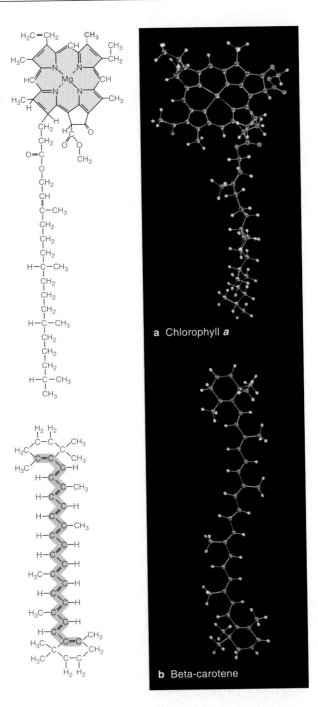

a Chlorophyll *a*

b Beta-carotene

Figure 7.7 Structural formulas and ball-and-stick models for (**a**) chlorophyll *a* and (**b**) beta-carotene. The light-catching array of both pigments is tinted the color of the light it transmits. The backbone, a pure hydrocarbon, dissolves in the lipid bilayer of photosynthetic cell membranes. Chlorophylls *a* and *b* differ only in one functional group at the position shaded *red* ($-CH_3$ for chlorophyll *a* and $-COO^-$ for chlorophyll *b*).

Figure 7.8 (**a**) A red alga from a tropical reef. (**b**) A green alga (*Codium*) from shallow coastal waters.

Figure 7.9 Leaf color. In intense-green leaves, photosynthetic cells continually make chlorophyll, which masks the accessory pigments. In autumn, chlorophyll synthesis lags behind its breakdown in many species. Other pigments are unmasked, and more colors show through. Water-soluble anthocyanins accumulate in leaf cells. They appear red when plant fluids are slightly acidic, blue when they are basic (alkaline), or colors in between in fluids of intermediate pH. Soil conditions contribute to pH values.

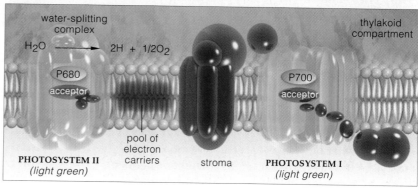

Figure 7.10 Arrangement of the two types of photosystems, II and I, in a chloroplast's thylakoid membranes. Components of neighboring electron transfer chains are coded *dark green*. One of these chains extends from photosystem II to photosystem I. The second chain gives up the electrons to the coenzyme NADP+.

algae, including the one shown in Figure 7.8*b,* live in shallow water. Their chlorophyll molecules absorb red wavelengths, and accessory pigments harvest different ones. Some accessory pigments also function as shields against ultraviolet radiation.

Today, **chlorophylls** are the primary pigments in all but one marginal group of photoautotrophs. Remember, photosynthetic pigments respond best to red and blue-to-violet light. Chlorophyll *a* absorbs both (Figure 7.7*a*). As one of several accessory pigments in chloroplasts, chlorophyll *b* absorbs blue and red-orange wavelengths missed by chlorophyll *a*. Plants, green algae, and a few bacterial photoautotrophs have this pigment.

All photoautotrophs also contain **carotenoids**. These accessory pigments absorb blue-violet and blue-green wavelengths that chlorophylls miss. They reflect red, orange, and yellow ones. Their backbone usually has a carbon ring at each end. Many, including beta-carotene, are fat-soluble, pure hydrocarbons (Figure 7.7*b*). The xanthophylls incorporate oxygen; others have proteins attached and appear purple, violet, blue, green, brown, or black. You can expect an abundance of carotenoids in flowers, fruits, and vegetables in this color range.

Carotenoids are less abundant than chlorophylls in green leaves, but in many plants they become visible in autumn (Figure 7.9). Each year, tourists spend about a billion dollars just to watch the three-week demise of chlorophyll in the deciduous trees of New England.

Anthocyanins and **phycobilins** also are accessory pigments. Many eukaryotic species make an abundance of deep red to purple anthocyanins (think red algae, purple mushroom caps, and blueberries, for example). The blue phycobilins are the signature pigments of red algae and cyanobacteria.

Where Are Photosynthetic Pigments Located?

Some bacteria of ancient lineages have light-trapping pigments embedded in their plasma membrane. The archaebacterium *Halobacterium halobium* has a unique purple pigment (bacteriorhodopsin), not chlorophylls. Cyanobacteria have chloropyll *a,* and other prokaryotes have bacteriochlorophylls that serve the same function but differ slightly in their structure. Various accessory pigments also are embedded in the plasma membrane of photosynthetic eubacteria.

This chapter's focus, remember, is on the thylakoid membrane system inside chloroplasts. Its pigments are organized in clusters known as **photosystems**. In some plant species, the membrane has many thousands of these clusters. Each photosystem consists of proteins and 200 to 300 pigment molecules. As Figure 7.10 shows, it is located next to its functional partner—an electron transfer chain. How these components interact is the topic of the next two sections.

The chlorophylls are the main photosynthetic pigments in photoautotrophs. Carotenoids and other accessory pigments enhance their light-harvesting function.

Photosystems (pigment clusters) are the functional partners of electron transfer chains in the thylakoid membrane.

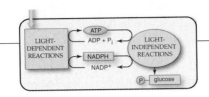

THE LIGHT-DEPENDENT REACTIONS

We turn now to the key events of the **light-dependent reactions**, the first stage of photosynthesis. *First*, many photosystems harvest sunlight energy, which also splits water molecules into oxygen, hydrogen, and electrons. *Second*, the energy input is converted to the chemical bond energy of ATP. *Third*, the coenzyme NADP$^+$ picks up the electrons, and hydrogen accompanies them.

What Happens to the Absorbed Energy?

When a photosynthetic pigment absorbs photons, some of its electrons are boosted to a higher energy level. As electrons return to a lower energy level, recall, they emit the extra energy as fluorescent light.

If nothing else were around to intercept the emitted energy, all of it would escape as light and heat. But a photosystem's pigment clusters prevent this. Most of the pigments harvest energy from photons. Rather than releasing excitation energy as a fluorescent afterglow, a harvester randomly transfers it to a neighbor pigment, which randomly passes it to a neighbor pigment, and so on. Figure 7.11 shows this "random walk" of energy.

Some energy is lost as heat with each transfer. Soon, the energy remaining corresponds to a wavelength that only the photosystem's **reaction center** can trap. That center is a special chlorophyll *a* molecule. It passes the energy, in the form of excited electrons, to an acceptor molecule positioned near an electron transfer chain.

An **electron transfer chain** is an organized array of enzymes, coenzymes, and other proteins embedded in or anchored to a cell membrane (Section 6.4). Electrons are transferred step-by-step through the array, and a bit of energy is released at each step. In chloroplasts, much of the energy drives ATP formation.

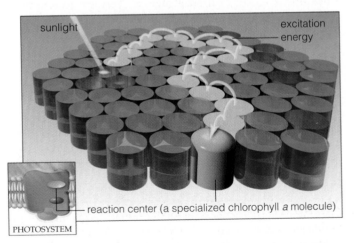

Figure 7.11 A sampling of pigments of a photosystem that harvest photon energy. Notice the random flow of energy among them. The energy of excitation quickly reaches a reaction center which, when suitably activated, releases electrons for photosynthesis.

sunlight — excitation energy — reaction center (a specialized chlorophyll *a* molecule) — PHOTOSYSTEM

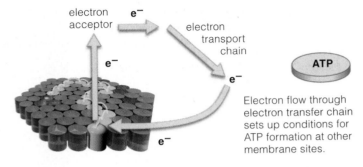

Figure 7.12 Cyclic pathway of ATP formation.

electron acceptor — e$^-$ — electron transport chain — ATP

Electron flow through electron transfer chain sets up conditions for ATP formation at other membrane sites.

Cyclic and Noncyclic Electron Flow

In ancient bacteria, photosynthesis evolved as a way to form ATP, not to make organic compounds. Inorganic compounds provided hydrogen and electrons for the synthesis reactions. Sunlight energy drove electrons in a cycle: from a photosystem called type I, to a transfer chain, and back to the photosystem (Figure 7.12). This *cyclic* pathway has a reaction center designated P700, and it still runs in all photoautotrophs.

Electron flow through one photosystem does not pack enough punch to make NADPH. But more than 2 billion years ago, more energy became available for electron transfers, as if two batteries became hooked together. A different photosystem—type II—evolved. When types II and I operate together, they acquire enough energy to get electrons from water molecules.

The more recent path of electron flow is *noncyclic*. Electrons flow from water to photosystem II, through a transfer chain to photosystem I, then another transfer chain. From there, NADP$^+$ accepts two electrons and a hydrogen ion (H$^+$) to become NADPH (Figure 7.13a).

Let's give the electron flow a closer look. Sunlight striking photosystem II starts **photolysis**. This reaction splits water molecules into oxygen, electrons, and H$^+$ (unbound hydrogen). New electrons from water replace the electrons released from the photosystem's reaction center (P680). ATP still forms. The mechanism is a bit complicated, so we reserve that story for Section 7.5. For now, follow the electrons leaving the first transfer chain. They still have some extra energy. They get more from sunlight at photosystem I. This energy boost puts them at a higher energy level, which lets them enter the second chain. They lose only a small bit of energy before NADP$^+$ picks them up at the end of the second chain.

Today, both the cyclic and noncyclic pathways of electron flow operate in all photosynthetic organisms. Which one dominates at a given time depends on the organism's metabolic demands for ATP and NADPH.

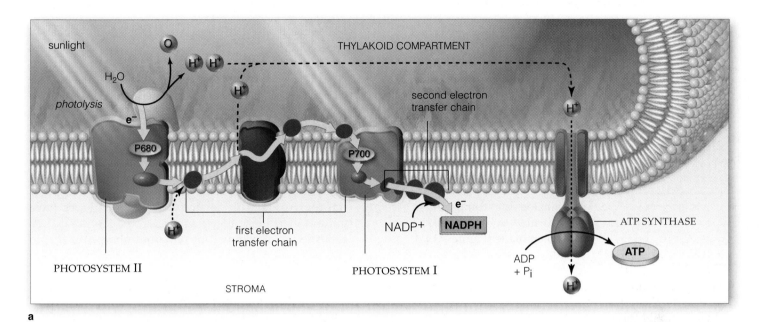

a

Figure 7.13 (**a**) Filling in details for Figure 7.3—how ATP and NADPH form in the noncyclic pathway of photosynthesis. *Yellow* arrows show electron flow. Electrons released from water molecules that were split by photolysis move through two photosystems (*light green*) and two electron transfer chains (*dark green*). The joint operation of two photosystems boosts electrons to an energy level high enough to drive NADPH formation.

As you will read shortly, hydrogen ions also move across the thylakoid membrane in ways that lead to ATP formation at proteins called ATP synthases.

(**b**) Diagram the energy changes in the noncyclic pathway, and you will end up with this "Z scheme."

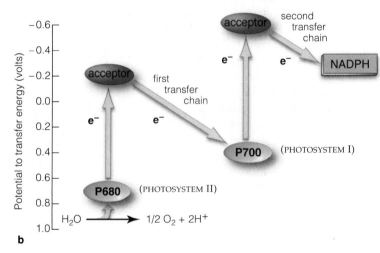

b

The Legacy—A New Atmosphere

On sunny days you can see bubbles of oxygen escaping from aquatic plants (Figure 7.14). When the noncyclic pathway first evolved, the oxygen it released dissolved in seawater, wet mud, and similar prokaryotic habitats. By about 1.5 billion years ago, stupendous quantities of dissolved oxygen were escaping to what had been an oxygen-free atmosphere. The atmosphere changed forever, and so did the world of life. The global rise in free oxygen now favored the chemical evolution of aerobic respiration. This became the most efficient metabolic pathway for releasing usable energy from organic compounds. From an evolutionary perspective, the emergence of the noncyclic pathway allowed you and every other animal to be around today, breathing the oxygen that helps keep your cells alive.

Figure 7.14 Visible evidence of photosynthesis—bubbles of oxygen escaping from the leaves of *Elodea*, a plant that lives in freshwater habitats.

In the light-dependent reactions, sunlight energy drives the release of electrons from photosystems in thylakoid membranes. Electron flow through nearby electron transfer chains results in the formation of ATP, NADPH, or both.

ATP alone forms in a cyclic pathway that uses one type of photosystem. Both ATP and NADPH form by a noncyclic pathway in which electrons flow from water, through two types of photosystems, and finally to NADP+.

All photosynthetic species employ one or both pathways in response to the cell's changing needs for ATP and NADPH.

Oxygen, a by-product of the noncyclic pathway, changed the early atmosphere and made aerobic respiration possible.

CASE STUDY: A CONTROLLED RELEASE OF ENERGY

Spill some sugar into a campfire and you'll release its energy all at once, but not in a form that does any good. Events in the chloroplast control the release of energy initially provided by sunlight, and they direct it into doing useful cellular work.

Inputs of light energy cause enzymes to repeatedly split water molecules. The released oxygen combines to form O_2. In this form it diffuses out of the chloroplast and out of the cell. The released hydrogen does not depart. It gets stockpiled as hydrogen ions (H^+) inside the compartment formed by the thylakoid membrane (Figure 7.15a).

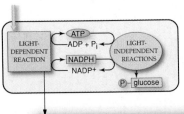

of the chain shunt the ions across the membrane, as in Figure 7.15b.

By a combination of photolysis and electron transport, H^+ is now more concentrated in the thylakoid compartment than in the stroma. Now, this unequal distribution of positively charged ions means there is a difference in charge across the membrane. H^+ concentration and electric gradients have become established.

The combined force of the two gradients propels H^+ into the stroma, through ATP synthases that span the thylakoid membrane. You read about this type of transport protein in Section 5.2. ATP synthases have built-in machinery that responds to the ion flow. They catalyze the attachment of unbound phosphate (P_i) to

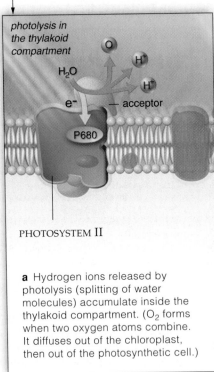

a Hydrogen ions released by photolysis (splitting of water molecules) accumulate inside the thylakoid compartment. (O_2 forms when two oxygen atoms combine. It diffuses out of the chloroplast, then out of the photosynthetic cell.)

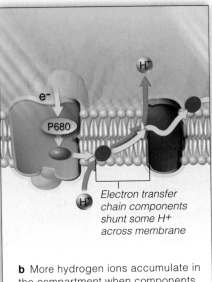

b More hydrogen ions accumulate in the compartment when components of electron transfer chains accept excited electrons (from photolysis). Components of the chain also pick up hydrogen ions in the stroma and shunt them across the membrane.

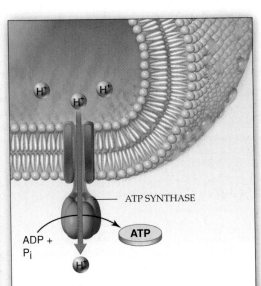

c Ions follow the concentration and electric gradients across the thylakoid membrane. They flow to the stroma, through the interior of ATP synthases that span the membrane. The flow drives the formation of ATP from the cell's pool of ADP and P_i.

Figure 7.15 From photosynthesis, an example of how energy is released and directed to cellular tasks. Sunlight energy at photosystem II sets in motion events that build concentration and electric gradients, which are tapped to make ATP. It also leads to the formation of NADPH.

At the same time, an acceptor molecule picks up the electrons and transfers them to an electron transfer chain positioned next to it in the thylakoid membrane. The membrane has many, many of these chains. While they are all accepting electrons, they also are picking up H^+ from the stroma; these ions are attracted to the negative charge of the electrons. Certain components

a molecule of ADP that is present in the stroma. ATP forms by this phosphate-group transfer (Figure 7.15c).

The sequence of events just described is called the chemiosmotic model for ATP formation in chloroplasts. As you will read in the next chapter, the same model applies to ATP formation in mitochondria.

In chloroplasts, ongoing photolysis and transfers of electrons in transport systems set up H^+ concentration and electric gradients across the thylakoid membrane. Ion flow from the thylakoid compartment to the stroma drives ATP formation.

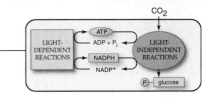

THE LIGHT-INDEPENDENT REACTIONS

The **light-independent reactions** are the "synthesis" part of photosynthesis. They occur as a cyclic pathway called the **Calvin–Benson cycle**. ATP energy drives the reactions. NADPH delivers hydrogen and electrons to build sugars. Carbon dioxide (CO_2), present in the air (or water) around photosynthetic cells, provides the carbon and oxygen. We say these reactions are light-independent because they do not depend directly on sunlight. They proceed just as well in the dark, as long as ATP and NADPH are available.

How Do Plants Capture Carbon?

Suppose a CO_2 molecule diffuses into air spaces inside a leaf, then into a photosynthetic cell, and then into a chloroplast's stroma. An enzyme attaches its carbon atom to **RuBP** (ribulose bisphosphate), a compound having a backbone of five carbon atoms. **Rubisco** (for RuBP carboxylase) is the enzyme's name. The result is an unstable six-carbon intermediate that splits into two molecules of **PGA** (phosphoglycerate). PGA is a stable organic compound with a three-carbon backbone.

Incorporating a carbon atom from CO_2 into a stable organic compound is known as **carbon fixation**. Quite simply, food cannot be produced without this first step of the cyclic light-independent reactions.

How Do Plants Build Glucose?

The cycle yields phosphorylated glucose (Section 6.3). It also regenerates the RuBP. For our purposes, we can simply focus on the carbon atoms of the substrates, intermediates, and end products, as in Figure 7.16.

PGA accepts one phosphate group from ATP, plus hydrogen and electrons from NADPH. The resulting intermediate is **PGAL** (for phosphoglyceraldehyde). To build *one* six-carbon sugar phosphate, carbon atoms from six CO_2 must be fixed, and twelve PGAL must form. Most of the PGAL gets rearranged into RuBP, which is used to fix more carbon. Two PGAL combine, thus forming glucose with a phosphate group attached to its six-carbon backbone.

When phosphorylated this way, glucose is primed to enter reactions. Plants use it as a building block for their main carbohydrates, such as sucrose, cellulose, and starch. *Synthesis of these organic compounds by other pathways marks the end of the light-independent reactions.*

With six turns of the cycle, enough RuBP molecules form to replace the ones used in carbon fixation. The ADP, NADP$^+$, and released phosphate diffuse through the stroma, back to the light-dependent reaction sites. They may be converted again to NADPH and ATP.

During daylight hours, photosynthetic cells convert phosphorylated glucose to sucrose or starch. Sucrose is

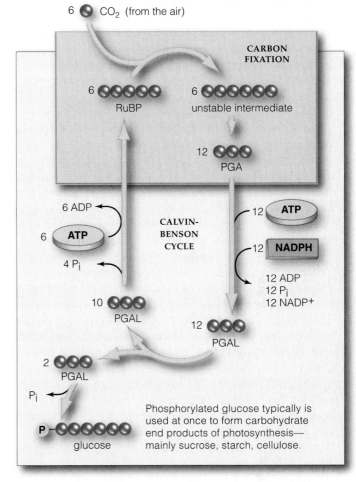

Figure 7.16 Light-independent reactions of photosynthesis. *Brown* circles signify carbon atoms of key molecules. All intermediates have one or two phosphate groups attached; for clarity, we show only the phosphate group on glucose. Many water molecules that formed in the light-dependent reactions enter this pathway; six remain at the end. Appendix VI, Figure C, shows details.

the most readily transportable carbohydrate in plants. Starch is the most common storage form. The cells also convert excess PGAL to starch, which they briefly store as starch grains inside the stroma. After the sun goes down, they convert starch to sucrose for export to other living cells in leaves, stems, and roots.

The products and intermediates of photosynthesis end up as energy sources and as building blocks for all lipids, amino acids, and other organic compounds that plants require for growth, survival, and reproduction.

In light-independent reactions of the Calvin–Benson cycle, glucose is made with hydrogen from NADPH, and carbon and oxygen from CO_2. The reactions regenerate ADP, NADP$^+$, and the RuBP that fixes carbon for this cycle.

7.7

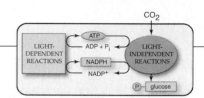

FIXING CARBON—SO NEAR, YET SO FAR

If sunlight intensity, air temperature, rainfall, and soil composition never varied, photosynthesis might be the same in all plants. But environments differ. And details of photosynthesis differ, as you can sense by comparing carbon-fixing adaptations to stressful conditions.

C4 Versus C3 Plants

Plant growth requires carbon, but CO_2 is not always plentiful in leaves. A leaf has a waxy cover that helps plants conserve water. CO_2 diffuses in and O_2 diffuses out at **stomata** (singular, stoma), tiny openings across the leaf surface (Figure 7.17). Stomata close on hot, dry days. Closure of all of the stomata restricts water loss but stops CO_2 from diffusing in. Cells use up the CO_2 inside, and O_2 builds up because it can't diffuse out.

When cells are engaged in photosynthesis, oxygen accumulates in a leaf. A high O_2 concentration triggers *photorespiration*, a process that lowers a plant's sugar-making capacity. Remember rubisco, the enzyme that attaches CO_2 to RuBP in the Calvin–Benson cycle? It also attaches *oxygen* to it instead when CO_2 levels fall and O_2 levels rise. The resulting intermediate breaks down to only one PGA and to one glycolate molecule. Unlike PGA, which helps form sugars, glycolate enters breakdown reactions that release fixed carbon in the form of CO_2—which is back where the plant started but with only half the sugars to show for it.

PGA is the first intermediate of **C3 plants** such as basswood, bluegrass, and beans (Figure 7.18*a*). "C3" refers to its *three* carbons. In **C4 plants**, such as corn, the first to form is *four*-carbon oxaloacetate. The C4 plants also close stomata on hot, dry days. But the CO_2 level doesn't decline as much because these plants fix carbon twice, in two types of photosynthetic cells.

Mesophyll cells use PEP, a three-carbon compound, to fix CO_2. The enzyme catalyzing this step will not use oxygen, regardless of the CO_2/O_2 ratio. Oxaloacetate forms, then malate, which diffuses to adjoining *bundle-sheath cells* (Figure 7.18*b*). There, pyruvate forms. CO_2 is released, then enters the Calvin-Benson cycle. With the help of ATP, the pyruvate regenerates PEP for use in the C4 pathway in the mesophyll cells.

With this pathway, C4 plants can get by with small stomata, lose less water, and make more glucose than C3 plants can when days are hot, bright, and dry.

Experiments show that photorespiration lowers the photosynthetic efficiency of many C3 plants. Example: Grow hothouse tomatoes at CO_2 levels high enough to block photorespiration and growth rates dramatically increase. So why hasn't natural selection eliminated such a wasteful process? Look to rubisco. This enzyme evolved long ago, when the atmosphere had little O_2 and an abundance of CO_2. Maybe some gene coding

a Leaves of basswood (*Tilia*), a typical C3 plant.

upper epidermis

palisade mesophyll

spongy mesophyll

lower epidermis

stoma   vein   air space

b Leaves of corn (*Zea mays*), a typical C4 plant

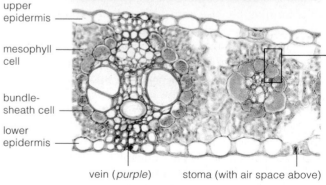

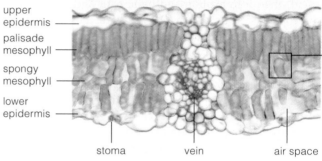

upper epidermis

mesophyll cell

bundle-sheath cell

lower epidermis

vein (*purple*)   stoma (with air space above)

Figure 7.17 Comparison of the internal leaf structure, in cross-section, for a C3 plant and a C4 plant.

Figure 7.19 Prickly pear (*Opuntia*), a CAM plant. These plants open stomata and fix carbon at night. They include orchids, pineapples, and many succulents besides cacti.

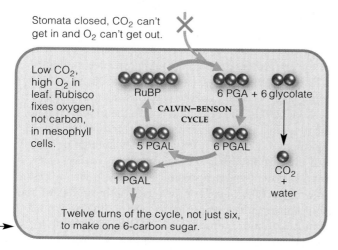

Stomata closed, CO_2 can't get in and O_2 can't get out.

Low CO_2, high O_2 in leaf. Rubisco fixes oxygen, not carbon, in mesophyll cells.

RuBP

6 PGA + 6 glycolate

CALVIN–BENSON CYCLE

5 PGAL

6 PGAL

1 PGAL

CO_2 + water

Twelve turns of the cycle, not just six, to make one 6-carbon sugar.

a C3 carbon fixation with low CO_2 / high O_2 levels in leaf

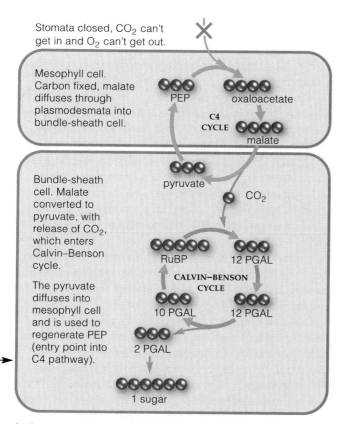

Stomata closed, CO_2 can't get in and O_2 can't get out.

Mesophyll cell. Carbon fixed, malate diffuses through plasmodesmata into bundle-sheath cell.

PEP

oxaloacetate

C4 CYCLE

malate

pyruvate

CO_2

Bundle-sheath cell. Malate converted to pyruvate, with release of CO_2, which enters Calvin–Benson cycle.

The pyruvate diffuses into mesophyll cell and is used to regenerate PEP (entry point into C4 pathway).

RuBP

12 PGAL

CALVIN–BENSON CYCLE

10 PGAL

12 PGAL

2 PGAL

1 sugar

b C4 carbon fixation with low CO_2 / high O_2 levels in leaf

Figure 7.18 Two ways to fix carbon in hot, dry weather, when there is too little CO_2 and too much O_2 in leaves. (**a**) The Calvin–Benson (C3) cycle is common in evergreens and many nonwoody plants of temperate zones, such as basswood and bluegrass. (**b**) The C4 cycle is common in grasses, corn, and other plants that evolved in the tropics and that fix CO_2 twice.

CO_2 uptake at night only

Mesophyll cell stomata open at night only, so CO_2 in and O_2 out but no water loss. C4 cycle operates.

C4 CYCLE

The CO_2 that accumulated overnight in leaf used during day for the C3 cycle in the same cell.

CALVIN–BENSON CYCLE

1 sugar

for its structure can't mutate without bad effects on its primary role—carbon-fixing activity.

Over the past 50 to 60 million years, the C4 cycle evolved independently in many lineages. Before then, atmospheric CO_2 levels were higher, and so C3 plants had the selective advantage in hot climates. Which cycle will be the most adaptive in the future? The CO_2 levels have been rising for decades and may double in the next fifty years. If so, photorespiration will again lose out—and many vital crop plants may benefit.

CAM Plants

We see a carbon-fixing adaptation to desert conditions in cacti. A cactus is one of the *succulents*; it has juicy, water-storing tissues and thick surface layers that limit water loss. It cannot open stomata on hot days without losing water. It opens them and fixes CO_2 *at night*.

Many plants are adapted this way. They are called **CAM plants** (short for Crassulacean Acid Metabolism). At night their mesophyll cells use a C4 cycle much like that shown in Figure 7.19. They store malate and other products until the next day, when their stomata close. Then malate releases CO_2, which enters reactions of the Calvin–Benson (C3) cycle in the same cells. In this way, photosynthesis proceeds without water loss.

Many plants die in prolonged droughts, but some CAM plants survive by closing stomata even at night. They keep fixing the CO_2 from aerobic respiration. Not much forms. It is enough to maintain low metabolic rates, which can only allow slow growth. Try growing cacti in Seattle or other places with a mild climate, and they will compete poorly with C3 and C4 plants.

Compared to C3 plants, the C4 plants and CAM plants have modified ways of fixing carbon for photosynthesis. The modifications counter stresses imposed by hot, dry conditions in their environments.

AUTOTROPHS, HUMANS, AND THE BIOSPHERE

We conclude this chapter with a story to reinforce how photosynthesizers and other autotrophs fit in the world of living things. It is a story of mind-boggling numbers of single-celled and multicelled species on land and in the sunlit waters of the Earth.

In spring, you sense renewed growth of autotrophs on land when leaves unfurl and lawns and fields turn green. You may not be aware that uncountable numbers of single-celled autotrophs also drift through surface waters of the world ocean. You can't see them without a microscope; a row of 7 million cells of one aquatic species would be less than a quarter-inch long. In some regions, a cupful of seawater might hold 24 million cells of one species. And that wouldn't include cells of all the other aquatic species suspended in the cup.

Most of the drifters are photoautotrophic prokaryotic cells and protistans. Together they are the "pastures of the seas," producers that ultimately feed most other marine organisms. The pastures "bloom" in the spring, when nutrient inputs sustain rapid reproduction. At that time seawater becomes warmer and enriched with nutrients that winter currents churn up from the deep.

Until NASA gathered data from space satellites, we had no idea of their numbers and distribution. Figure 7.20a gives visual evidence of their activities one winter in surface waters of the North Atlantic Ocean. Figure 7.20b gives evidence of a springtime bloom stretching from North Carolina all the way past Spain!

Collectively, the cells help shape the global climate, for they deal with staggering numbers of reactant and product molecules. For instance, they sponge up nearly half the carbon dioxide that we humans release each year, as when we burn fossil fuels or burn vast tracts of forests to clear land for farming. Without the aquatic photoautotrophs, atmospheric carbon dioxide would accumulate more rapidly and possibly contribute to global warming. Remember the Chapter 3 introduction? If the atmosphere warms by only a few degrees, then sea levels will rise and all lowlands near the coasts of continents and islands will become submerged.

Although such global change is a real possibility, tons of industrial wastes, sewage, and fertilizers in runoff from croplands still drain into the ocean each day. The pollutants seriously alter the chemical composition of seawater. How long will the marine photoautotrophs be able to function in this chemical brew? The answer will have impact on your own life in more ways than one.

Other autotrophs affect your life in ways you might not expect. On the seafloor, in hot springs, in coal mine wastes are prokaryotic cells called **chemoautotrophs**. Such cells secure carbon dioxide as a carbon source, and they use inorganic compounds as energy sources.

For example, **hydrothermal vents** are fissures in the seafloor where molten rock superheats water. Certain archaebacteria use hydrogen sulfide in the mineral-rich water for hydrogen and electrons. As another example, chemoautotrophs in soil obtain energy from nitrogen-containing wastes and remains of other organisms.

The chemoautotrophs are microscopically small, but they exist in monumental numbers. They affect the cycling of nitrogen, phosphorus, and other elements through the biosphere. We consider their impact on the environment in later chapters. In this unit, we turn next to pathways by which cells release chemical bond energy of glucose and other biological molecules—the chemical legacy of diverse autotrophs everywhere.

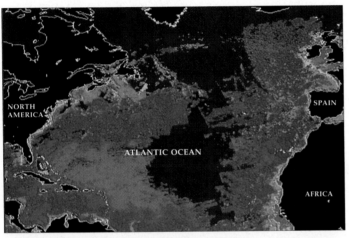

a Photosynthetic activity in winter.

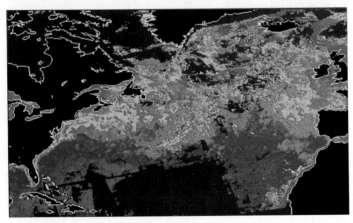

b Photosynthetic activity in spring.

Figure 7.20 Two satellite images that help convey the magnitude of photosynthetic activity during springtime in the North Atlantic portion of the world ocean. In these color-enhanced images, *red-orange* shows where chlorophyll is most concentrated.

Diverse photoautotrophs and chemoautotrophs produce the food that sustains you and all other heterotrophs. Mind-boggling numbers of these producers live in the sunlit waters of the Earth as well as on land.

SUMMARY

Gold indicates text section

1. Cell structure and functions are based on organic compounds, the synthesis of which depends on sources of carbon and energy. Plants and other autotrophs get carbon from carbon dioxide and energy from the sun or inorganic compounds. Animals and other heterotrophs aren't self-nourishing; they get carbon and energy from organic compounds synthesized by autotrophs. *CI*

2. Photosynthesis is the main process by which carbon and energy enter the web of life. Its two stages are the light-dependent and light-independent reactions. This equation and Figure 7.21 summarize the process: *7.1*

$$12H_2O + 6CO_2 \xrightarrow{\text{LIGHT ENERGY}} 6O_2 + C_6H_{12}O_6 + 6H_2O$$

WATER   CARBON DIOXIDE   OXYGEN GLUCOSE WATER

3. Chloroplasts are organelles of photosynthesis in algae and plants. Two outer membranes enclose its semifluid interior (stroma). A third membrane in the stroma forms a single compartment; often it folds back on itself in disk-shaped stacks (thylakoids). The light-dependent reactions take place at this membrane. But the light-independent reactions take place in the stroma. *7.1*

4. The light-dependent reactions start at photosystems that each include 200 to 300 pigments. *7.2–7.5*

   a. Chlorophyll *a*, the main photosynthetic pigment, absorbs all wavelengths of visible light but only a little of the yellow-green and green ones. Accessory pigments (e.g., carotenoids) absorb other wavelengths.

   b. A cyclic pathway forms ATP alone. Photosystem I gives up excited electrons to an electron transfer chain, which sends the "spent" ones back to the photosystem.

   c. The noncyclic pathway forms ATP and NADPH. Water molecules are split into hydrogen, electrons, and oxygen. Excited electrons enter photosystem II, then a transfer chain, photosystem I, then a second transfer chain. NADP+ picks up the electrons and hydrogen to form NADPH. By 1.5 billion years ago, oxygen released by this pathway had accumulated in the atmosphere. It ultimately made possible aerobic respiration.

5. ATP energy drives the light-independent reactions (by phosphate-group transfers). Hydrogen and electrons from NADPH, plus carbon and oxygen (from carbon dioxide), are used as building blocks for glucose. Most glucose is used at once to synthesize starch, cellulose, and other end products of photosynthesis. *7.6*

   a. The enzyme rubisco affixes carbon from CO_2 to the sugar RuBP, the start of the the Calvin–Benson cycle. Two PGA form. These are converted to two PGAL with ATP energy and hydrogen and electrons from NADPH.

   b. For every six carbon atoms that enter the cycle, twelve PGAL form. Two are used to form a six-carbon sugar phosphate. The rest are used to regenerate RuBP.

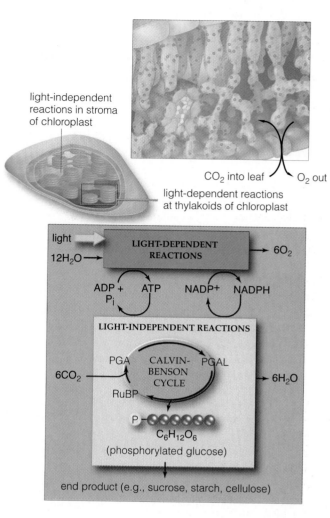

light-independent reactions in stroma of chloroplast

CO_2 into leaf O_2 out

light-dependent reactions at thylakoids of chloroplast

Figure 7.21 Summary of photosynthesis. In the light-dependent reactions, sunlight energy is converted to chemical bond energy of ATP. In a noncyclic pathway, water molecules are split by photolysis. NADP+ conserves the released electrons and hydrogen (as NADPH). Oxygen is released as a by-product.

ATP energy drives the light-independent reactions. Hydrogen and electrons from NADPH, plus carbon and oxygen from carbon dioxide, are used to form glucose. In the Calvin–Benson (C3) cycle, the RuBP on which the cycle turns is regenerated, and new molecules of water form. *Each turn* of the cycle requires one CO_2, three ATP, and two NADPH. Because each glucose molecule has a backbone of six carbon atoms, its formation requires six turns of the cycle.

6. Rubisco, a carbon-fixing enzyme, evolved when air had far more CO_2 and less O_2. Today, when CO_2 is used up and O_2 builds up in leaves, rubisco attaches oxygen (not carbon) to RuBP. This process, photorespiration, is wasteful: one PGA and glycolate form. Glycolate cannot be used to form sugars and is later degraded. *7.6*

7. Photorespiration predominates in C3 plants such as sunflowers under hot, dry conditions. Stomata close, so O_2 from photosynthesis builds up in leaves to levels higher than CO_2 levels. Corn and other C4 plants raise the CO_2 level by fixing carbon twice, in two cell types. CAM plants run the C3 cycle during the day when stomata are closed, using CO_2 fixed by the C4 cycle at night, when stomata are open. *7.7*

Review Questions

1. A cat eats a bird, which earlier ate a caterpillar that chewed on a weed. Which organisms are autotrophs? Which are the heterotrophs? *CI*

2. Summarize the photosynthesis reactions as an equation. Name where each stage takes place inside a chloroplast. *7.1*

3. What is the function of ATP in photosynthesis? What is the function of NADPH? *7.1*

4. Which of the following pigments are most visible in a maple leaf in summer? Which become the most visible in autumn? *7.3*
   a. chlorophylls   c. anthocyanins
   b. phycobilins   d. carotenoids

5. How does chlorophyll *a* differ in function from accessory pigments during the light-dependent reactions? *7.3, 7.4*

6. With respect to the light-dependent reactions, how do the cyclic and noncyclic pathways of electron flow differ? *7.4*

7. What substance does *not* take part in the Calvin–Benson cycle: ATP, NADPH, RuBP, carotenoids, O_2, CO_2, or enzymes? *7.6*

8. Fill in the blanks for the diagram at right. Which substances are the original sources of the carbon atoms and hydrogen atoms used in the synthesis of glucose in the Calvin–Benson cycle? *CI, 7.6*

9. On hot, dry days, oxygen from photosynthesis accumulates in leaves. Explain why, and explain what happens in C3 plants versus C4 plants. *7.7*

10. Are photoautotrophs the only "self feeders"? If not, give examples and where you might find them. *CI, 7.8*

Self-Quiz

1. Photosynthetic autotrophs use _____ from the air as a carbon source and _____ as their energy source.

2. In plants, light-*dependent* reactions proceed at the _____ .
   a. cytoplasm   c. stroma
   b. plasma membrane   d. thylakoid membrane

3. In the light-*dependent* reactions, _____ .
   a. carbon dioxide is fixed   c. CO_2 accepts electrons
   b. ATP and NADPH form   d. sugar phosphates form

4. The light-*independent* reactions proceed in the _____ .
   a. cytoplasm   b. plasma membrane   c. stroma

5. When a photosystem absorbs light, _____ .
   a. sugar phosphates are produced
   b. electrons are transferred to ATP
   c. RuBP accepts electrons
   d. light-dependent reactions begin

6. Identify which of the following substances accumulates inside the thylakoid compartment of chloroplasts during the light-dependent reactions:
   a. glucose   c. chlorophyll   e. hydrogen ions
   b. carotenoids   d. fatty acids

7. The Calvin–Benson cycle starts when _____ .
   a. light is available
   b. light is not available
   c. carbon dioxide is attached to RuBP
   d. electrons leave a photosystem

8. ATP phosphorylates _____ in the light-independent reactions.
   a. RuBP   c. PGA
   b. $NADP^+$ d. PGAL

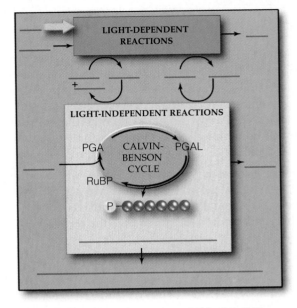

9. Match each event with its most suitable description.
   _____ photon absorption
   _____ NADPH formation
   _____ CO_2 fixation
   _____ PGAL formation
   _____ ATP formation only

   a. rubisco required
   b. ATP, NADPH required
   c. electrons cycled back to photosystem I
   d. energy matches amount required to excite electrons
   e. photolysis required

Critical Thinking

1. Suppose a garden in your neighborhood is filled with red, white, and blue petunias. Explain each of the three floral colors in terms of which wavelengths of light the flower is absorbing and which wavelengths it is transmitting or reflecting.

2. About 200 years ago, Jan Baptista van Helmont performed an experiment on the nature of photosynthesis. He wanted to know where growing plants acquire the raw materials necessary to increase in size. For his experiment, he planted a tree seedling weighing 5 pounds in a barrel filled with 200 pounds of soil. He watered the tree regularly.

   Five years passed. Van Helmont again weighed the tree and the soil. The tree weighed 169 pounds, 3 ounces. The soil weighed 199 pounds, 14 ounces. Because the tree weight had increased so much and soil weight had decreased so little, he concluded the tree had gained weight after absorbing the water he had added to the barrel.

   Given what you know about the composition of biological molecules, why was van Helmont's conclusion misguided? Reflect on the current model of photosynthesis and then give a more plausible explanation of his results.

3. During the late 1980s a graduate student, Cindy Lee Van Dover, was puzzling over an eyeless shrimp (Figure 7.22). Divers discovered it deep in the Atlantic Ocean, near the base of a *hydrothermal vent*. At these fissures in the seafloor, molten rock rises and then mixes with cold seawater. Mineral-rich water, superheated by 650 degrees, spews into the perpetually

Figure 7.22 Preserved specimen of an eyeless shrimp recovered at a hydrothermal vent.

Figure 7.23 Where do the colors of butter, lemon butterfly fishes, morpho butterflies, scarlet macaws, and green pythons come from?

dark surroundings. Chemoautotrophic bacteria are the basis of thriving ecosystems near the vents (Sections 7.9 and 49.11).

When Van Dover saw videotapes of shrimps on the vent, she noticed a pair of bright strips on their back. By studying lab specimens, she found that the strips are connected to a nerve. Could eyeless shrimps have a novel sensory organ?

Steven Chamberlain, a neuroscientist, confirmed that the strips are a sensory organ with light-absorbing pigmented cells (photoreceptors). Later, Ete Szuts, a pigment expert, isolated the pigment. Its absorption spectrum is the same as that for rhodopsin, a visual pigment in eyes like yours.

Van Dover had a hunch: If the strips are light sensitive, then we should expect to find light on the seafloor. On later dives, special cameras confirmed her prediction. Like coils in a toaster, hydrothermal vents release heat (infrared radiation). They also release faint radiation at the low end of the visible spectrum—light up to nineteen times brighter than infrared.

A sunlit tree typically absorbs a billion billion photons per square inch per second. Photoautotrophic bacteria that live 72 meters (240 feet) below the Black Sea's surface encounter a trillion photons per square inch per second. *Just as many photons are available to hydrothermal vent organisms.*

Van Dover casually asked a colleague, "Hey, what if there's enough light for photosynthesis?" The response was, "What a stupid idea."

Euan Nisbet, who studies ancient environments, thought about that. The first cells arose about 3.8 billion years ago. As Nisbet and Van Dover later hypothesized, What if those cells arose at hydrothermal vents? What if they used inorganic compounds such as hydrogen sulfide (for their hydrogen and electrons) and carbon dioxide (for carbon)?

If so, survival probably depended on being able to move away from light at vents and thereby avoid being boiled alive. Millions of years later, some bacteria that may have been their descendants were evolving near the ocean's surface. In time they used their light-detecting machinery to absorb sunlight. They had become adapted to a new energy source—they were photosynthetic.

Did light-sensing machinery of deep-sea bacteria become modified for shallow-water photosynthesis? Some observations support this intriguing hypothesis. A few clues: The absorption spectra for chlorophyll in evolutionarily ancient photosynthetic bacteria correspond to light measured at the vents. Also, photosynthetic machinery incorporates iron, sulfur, manganese, and other minerals—all of which are abundant at hydrothermal vents.

If the earliest photoautotrophs evolved deep in the seas, their pigments evolved in the seas, also. Speculate on how natural selection may have favored the evolution of different pigments at different depths, starting at hydrothermal vents.

4. There are only about eight classes of pigment molecules, but this limited group gets around in the world. For example, animals synthesize the brownish-black melanin and some other pigments, but not carotenoids. These form in photoautotrophs and move up through food webs, as when tiny aquatic snails graze on green algae and then flamingos eat the snails.

Flamingos modify those ingested carotenoids in plenty of ways. For instance, their cells split beta-carotene molecules to form two molecules of vitamin A. This vitamin is the precursor of retinol, a visual pigment that transduces light into electric signals in the flamingo's eyes. Beta-carotene molecules also become dissolved in fat reservoirs under the skin. From there they are taken up by cells that give rise to bright pink feathers.

Pick one of the animals (or the one that produced the butter) shown in Figure 7.23. Do some research into its life cycle and its diet. Use your research to identify some possible sources for the pigments that impart the colors shown.

5. Krishna is using radioactively labeled carbon atoms ($^{14}CO_2$) in the laboratory. The photosynthesizing plants he is studying take them up. Identify the compound in which the labeled carbon will appear first: NADPH, PGAL, pyruvate, or PGA.

Selected Key Terms

absorption spectrum 7.2	light-independent reactions 7.6
anthocyanin 7.3	PGA 7.6
autotroph *CI*	PGAL 7.6
C3 plant 7.7	photoautotroph *CI*
C4 plant 7.7	photolysis 7.4
Calvin–Benson cycle 7.6	photon 7.2
CAM plant 7.7	photosynthesis *CI*
carbon fixation 7.6	photosystem 7.3
carotenoid 7.3	phycobilin 7.3
chemoautotroph 7.8	pigment 7.2
chlorophyll 7.3	reaction center 7.4
chloroplast 7.1	rubisco 7.6
electromagnetic spectrum 7.2	RuBP 7.6
electron transfer chain 7.4	stoma (stomata) 7.7
fluorescence 7.3	stroma 7.1
heterotroph *CI*	thylakoid membrane 7.1
hydrothermal vent 7.8	wavelength 7.2
light-dependent reactions 7.4	

Readings

Bazzaz, F., and E. Jajer. January 1992. "Plant Life in a CO2-Rich World." *Scientific American*, 266:68–74.

Zimmer, C. November 1996. "The Light at the Bottom of the Sea." *Discover*, 63–73.

On-Line readings at Student Guide for InfoTrac:
www.brookscole.com/biology

8

HOW CELLS RELEASE STORED ENERGY

The Killers Are Coming! The Killers Are Coming!

Thanks to selective breeding experiments gone wrong, descendants of "killer" bees that flew out of South America a few decades ago buzzed across the border between Mexico and Texas. By 1995, they had invaded 13,287 square kilometers of southern California and were busily setting up colonies. By 1998, when nectar-rich flowers bloomed profusely in the deserts after heavy El Niño storms, they moved even farther west and north than biologists had predicted.

When provoked, the bees are terrifying. To give an example, when a construction worker started a tractor a few hundred yards from a hive, thousands of agitated bees flew into action. They entered a nearby subway station, where they stung passengers on the platform and in the trains. They killed one person and injured a hundred more.

Where did the bees come from? In the 1950s, some queen bees were shipped over from Africa to Brazil for breeding experiments. Why? Honeybees are serious business. They are the source of nutritious honey. Also, they are rented to commercial orchards, where their collective pollinating activities may significantly enhance fruit production. Imagine yourself enclosing an orchard tree in a cage that keeps out all pollinators. Without pollinators to spread pollen about, less than 1 percent of that tree's flowers will set fruit. Now imagine putting a hive of honeybees inside the cage with the tree. As many as 40 percent of the flowers will set fruit.

Compared to their relatives in Africa, the bees of Brazil are sluggish pollinators and honey producers. By cross-breeding the two strains, researchers thought they might come up with a strain of mild-mannered but zippier bees. So they put some local bees and imported ones together inside netted enclosures, complete with artificial hives. Then they let nature take its course.

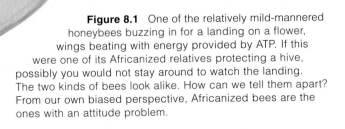

Figure 8.1 One of the relatively mild-mannered honeybees buzzing in for a landing on a flower, wings beating with energy provided by ATP. If this were one of its Africanized relatives protecting a hive, possibly you would not stay around to watch the landing. The two kinds of bees look alike. How can we tell them apart? From our own biased perspective, Africanized bees are the ones with an attitude problem.

Twenty-six African queen bees escaped. That was bad enough. Then beekeepers got wind of preliminary experimental results. After learning that the first few generations of offspring were more energetic but not overly aggressive, they imported hundreds of African queens and encouraged them to mate with the locals. And they set off a genetic time bomb.

Before long, African bees became established in commercial hives—and in wild bee populations. Their traits became dominant. The "Africanized" bees do everything other bees do, but they do more of it faster. Their eggs develop faster into adults. Adults fly faster, outcompete other bees for nectar, and even die sooner.

When something disturbs their hives or swarms, Africanized bees become extremely agitated. They can stay that way for up to eight hours. Whereas a mild-mannered honeybee might chase an intruding animal fifty yards or so, an Africanized bee squadron will chase it a quarter of a mile. If they catch up with it, they can collectively sting it to death.

Doing things faster means having a continuous supply of energy and efficient ways of using it. An Africanized bee's stomach can hold thirty milligrams of sugar-rich nectar—which is enough fuel to fly sixty kilometers. That's more than thirty-five miles! Besides this, compared to other kinds of bees, an Africanized bee's flight muscle cells have larger mitochondria. These organelles specialize in releasing a great deal of energy from sugars and other organic compounds, then converting it to the energy of ATP.

Whenever they tap into the stored energy of organic compounds, Africanized bees reveal their biochemical connection with other organisms. Study a primrose or puppy, a mold growing on stale bread, an amoeba in pondwater, or a bacterium living on your skin, and you will discover that their energy-releasing pathways differ in some details. But all of the pathways require characteristic starting materials. They yield predictable products and by-products. And they yield the universal energy currency of life—ATP.

All through the biosphere, organisms use energy and raw materials in similar ways. *At the biochemical level, we find undeniable unity among all forms of life.* We return to this idea at the chapter's end.

Key Concepts

1. For all organisms, cells release energy stored in glucose and other organic compounds, then use it in ATP production. The energy-releasing metabolic pathways differ from one another. But all of the main ones start with the breakdown of glucose to pyruvate.

2. The initial breakdown reactions, glycolysis, can proceed in the presence of oxygen or in its absence. Said another way, it is the first stage of either aerobic or anaerobic energy-releasing pathways.

3. The types of pathways called fermentation and anaerobic electron transfer do not use oxygen as an electron receptor. They proceed only in the cytoplasm, and none yields more than a small amount of ATP for each glucose molecule metabolized.

4. Another pathway, aerobic respiration, also starts in the cytoplasm. But it runs to completion in organelles called mitochondria. Compared to the other pathways, it releases far more energy from glucose.

5. Aerobic respiration consists of three stages. First, pyruvate forms from glucose (during glycolysis). Second, pyruvate is broken down to carbon dioxide, and coenzymes pick up the released electrons and hydrogen for delivery to electron transfer chains. Third, electron transfer chains help set up conditions that favor ATP formation. Free oxygen accepts the electrons at the end of the line and combines with hydrogen to form water.

6. Over evolutionary time, photosynthesis and aerobic respiration became connected on a global scale. An oxygen-rich atmosphere evolved through long-term photosynthetic activity. It sustains aerobic respiration. In turn, the by-products of this pathway—carbon dioxide and water—serve as raw materials when the photosynthesizers make organic compounds:

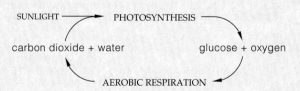

HOW DO CELLS MAKE ATP?

Organisms stay alive by taking in energy. Plants and all other organisms that engage in photosynthesis get energy from the sun. Animals get energy secondhand, thirdhand, and so on, by eating plants and one another. Regardless of its source, energy must be in a form that can drive thousands of life-sustaining reactions. Energy that becomes converted into the chemical bond energy of adenosine triphosphate—ATP—serves that function.

Plants make ATP during photosynthesis and use it to make glucose and other carbohydrates. *But they and every other organism also produce ATP by breaking down glucose and other carbohydrates, lipids, and proteins.* The reactions release electrons as well as hydrogens from intermediates. Both are delivered to electron transfer chains, as described in Section 6.4, and the outcome is central to the energy-releasing pathways.

Comparison of the Main Types of Energy-Releasing Pathways

Energy-releasing pathways originated long before our oxygen-rich atmosphere evolved by 1 billion years ago, when conditions were different. So the pathways must have been *anaerobic*; they did not use oxygen. Many prokaryotes and protistans still live in places where oxygen is absent or not always available. To make ATP, they use anaerobic reactions (fermentation or anaerobic electron transfer). Some of your own cells use anaerobic routes for short periods when they are not receiving enough oxygen. But **aerobic respiration**, an oxygen-dependent pathway of ATP formation, is the dominant pathway. Each breath you take provides your actively respiring cells with a fresh supply of oxygen.

Make note of this point: *The main energy-releasing pathways all start with the same reactions in the cytoplasm.* During this initial stage of reactions, called **glycolysis**, enzymes cleave and rearrange a glucose molecule into two molecules of **pyruvate**, an organic compound with a backbone of three carbon atoms. Once glycolysis is over, the energy-releasing pathways differ. Most importantly, only the aerobic pathway continues in a mitochondrion (Figure 8.2). There, oxygen serves as the final acceptor

of electrons used during the reactions. The anaerobic pathways start and end in the cytoplasm. A substance other than oxygen is the final electron acceptor.

As you examine the energy-releasing pathways in sections to follow, keep in mind that the reaction steps do not proceed by themselves. Enzymes catalyze each step, and intermediates formed at one step function as substrates for the next enzyme in the pathway.

Overview of Aerobic Respiration

Of all energy-releasing pathways, aerobic respiration gets the most ATP for each glucose molecule. Whereas anaerobic routes have a net yield of two ATP, aerobic respiration commonly yields thirty-six or more. If you were a bacterium, you would not require much ATP. Being far larger, more complex, and highly active, you depend on the aerobic pathway's high yield. When a molecule of glucose is used as the starting material, aerobic respiration can be summarized this way:

$$C_6H_{12}O_6 \;+\; 6O_2 \;\longrightarrow\; 6CO_2 \;+\; 6H_2O$$

GLUCOSE   OXYGEN   CARBON DIOXIDE   WATER

However, as you can see, the summary equation only tells us what the substances are at the start and finish of the pathway. In between are three reaction stages.

Figure 8.2 Where the aerobic and anaerobic pathways of ATP formation start and finish.

AEROBIC RESPIRATION	ANAEROBIC ENERGY-RELEASING PATHWAYS
start (glycolysis) in cytoplasm	start (glycolysis) in cytoplasm
completed in mitochondrion	completed in cytoplasm

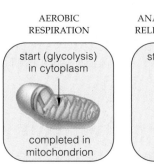
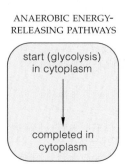

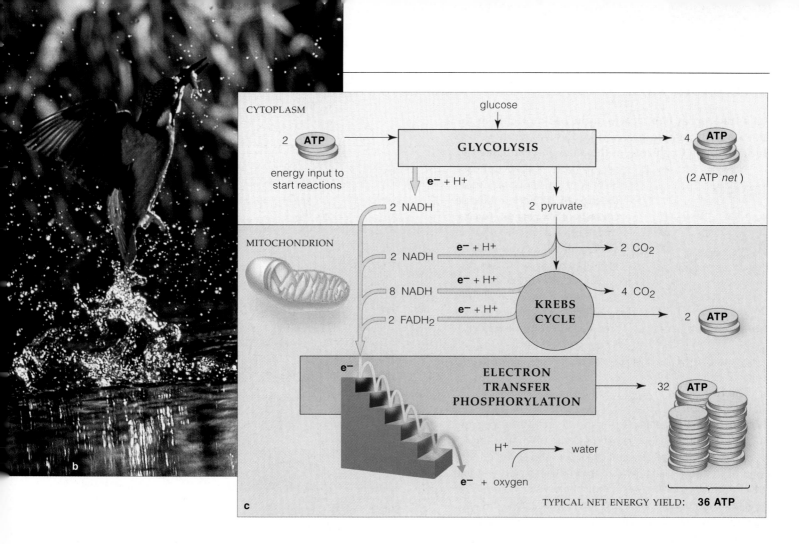

Figure 8.3 Overview of the three stages of aerobic respiration. From start to finish, the typical net energy yield from a glucose molecule is thirty-six ATP. (**a,b**) Only this pathway delivers enough ATP to build and maintain redwoods and all other large, multicelled organisms. It alone delivers enough ATP for highly active animals, including bees, humans, and kingfishers.

(**c**) In the first stage (glycolysis), enzymes partially break down glucose to pyruvate. In the second stage—mainly the Krebs cycle—enzymes degrade pyruvate to carbon dioxide. NAD$^+$ and FAD pick up electrons and hydrogen stripped from intermediates during both stages. In the final stage, electron transfer phosphorylation, the reduced coenzymes (NADH and FADH$_2$) give up electrons and hydrogen to electron transfer chains. Energy released during the flow of electrons through the chains drives ATP formation. Oxygen accepts the electrons at the end of the third stage.

Use Figure 8.3 while we outline the reactions. Again, glycolysis is the first stage. The second stage is mainly a cyclic pathway, the **Krebs cycle**. Enzymes break down pyruvate to carbon dioxide and water, thereby releasing electrons and hydrogen.

NAD$^+$ (for nicotinamide adenine dinucleotide) and **FAD** (flavin adenine dinucleotide) serve in glycolysis and the Krebs cycle. Remember, these coenzymes assist enzymes by accepting electrons and hydrogen derived from the intermediates. Unbound hydrogen atoms are naked protons, or hydrogen ions (H$^+$). When the two coenzymes are carrying electrons and hydrogen, they are abbreviated NADH and FADH$_2$.

Few ATP form during glycolysis or the Krebs cycle. The big energy harvest comes in the third stage, after the coenzymes give up the electrons and hydrogen to electron transfer chains. The chains are the machinery of **electron transfer phosphorylation**. They set up H$^+$ concentration and electric gradients, which drive ATP formation at nearby membrane proteins. It is in this final stage that so many ATP molecules are produced. As it ends, oxygen inside the mitochondrion accepts the "spent" electrons from the last component of each transport system. Oxygen picks up hydrogen at the same time and thereby forms water.

Nearly all metabolic reactions run on energy released from glucose and other organic compounds, then converted to chemical bond energy of ATP. The main energy-releasing pathways start in the cytoplasm with glycolysis, a stage of reactions that break down glucose to pyruvate.

The most common anaerobic pathways, which include the fermentation routes, end in the cytoplasm. Each has a net energy yield of two ATP.

Aerobic respiration, an oxygen-dependent pathway, runs to completion in mitochondria. From start (glycolysis) to finish, it typically has a net energy yield of thirty-six ATP.

GLYCOLYSIS: FIRST STAGE OF ENERGY-RELEASING PATHWAYS

Let's track what happens to a glucose molecule in the first stage of aerobic respiration. Remember, the same things happen to glucose in the anaerobic pathways.

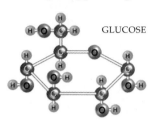

GLUCOSE

As you know, glucose is one of the simple sugars (Section 3.2). Each molecule consists of six carbon, twelve hydrogen, and six oxygen atoms, all covalently bonded together. Its carbons are the backbone. In the cytoplasm, glucose or some other simple sugar is partly broken down during glycolysis to two molecules of pyruvate, a three-carbon compound:

glucose ⟶ glucose–6–phosphate ⟶ 2 pyruvate

The first steps of glycolysis are *energy-requiring*. As Figure 8.4 indicates, they proceed only when two ATP molecules each transfer a phosphate group to glucose. Such phosphate-group transfers, remember, are known as phosphorylations. In this case, they raise the energy content of glucose to a level high enough to allow entry into the *energy-releasing* steps of glycolysis.

The first energy-releasing step cleaves the activated glucose into two molecules. We can call each of them PGAL (phosphoglyceraldehyde). Each PGAL becomes converted to an unstable intermediate that gives up a phosphate group to ADP, and so ATP forms. The next intermediate in the sequence does the same thing.

And so a total of four ATP form by **substrate-level phosphorylation**. We define this metabolic event as the direct transfer of a phosphate group from a substrate of a reaction to some other molecule—in this case, ADP. Remember, though, two ATP were invested to jump-start the reactions. So the *net* energy yield is two ATP.

Meanwhile, the coenzyme NAD^+ picks up electrons and hydrogens liberated from each PGAL molecule, thus becoming NADH. When NADH delivers its cargo to a different reaction site, it reverts to NAD^+. Said another way, like other coenzymes, NAD^+ is reusable.

In sum, glycolysis converts energy stored in glucose to a transportable form of energy, in ATP. NAD^+ picks up electrons and hydrogen stripped from glucose. These have roles in the next stage of reactions. So do the end products of glycolysis—the two molecules of pyruvate.

Glycolysis is a stage of reactions that partially break down glucose or some other carbohydrate to two molecules of pyruvate, thereby releasing energy.

Two NADH and four ATP form. However, when we subtract the two ATP required to start the reactions, the *net* energy yield of glycolysis is two ATP per glucose molecule.

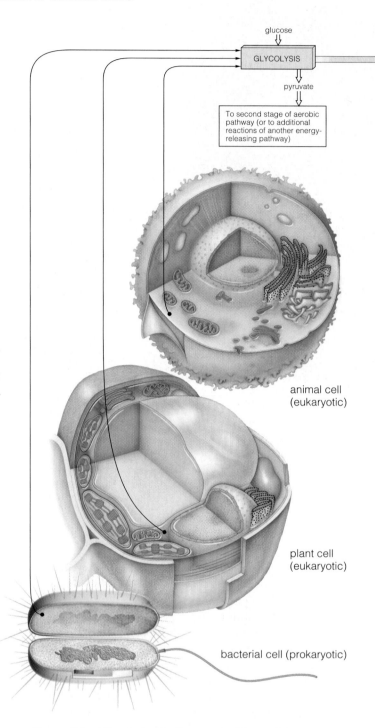

animal cell (eukaryotic)

plant cell (eukaryotic)

bacterial cell (prokaryotic)

Figure 8.4 Glycolysis, the first stage of the main energy-releasing pathways. All prokaryotic and eukaryotic cells use glycolysis, which occurs in the cytoplasm. In this example, glucose is the starting material. Two pyruvate, two NADH, and four ATP form. Because cells invest two ATP to start glycolysis, the *net* energy yield is two ATP. Appendix V (Figure A) gives more details.

Depending on the type of cell and environmental conditions, the pyruvate may enter the second set of reactions of the aerobic pathway, including the Krebs cycle. Or it may be used in other reactions, such as those of fermentation.

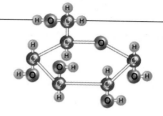

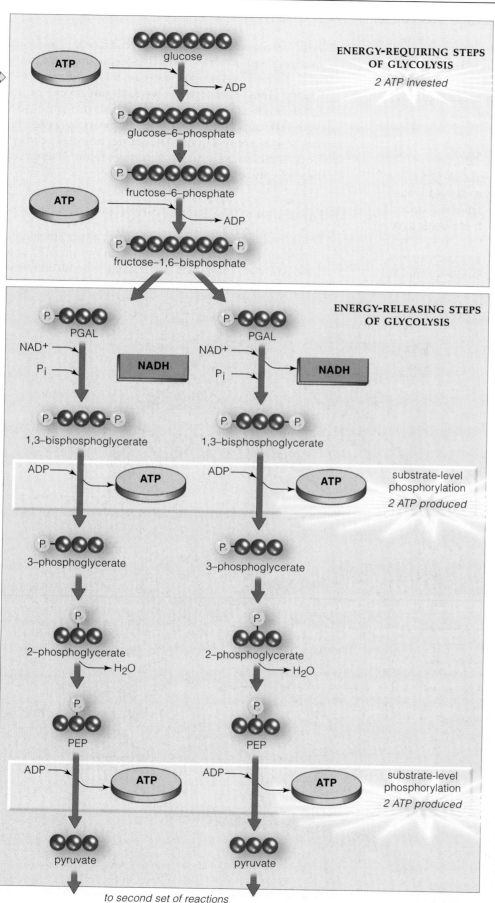

ENERGY-REQUIRING STEPS OF GLYCOLYSIS

2 ATP invested

ATP

glucose

ADP

P-glucose–6–phosphate

P-fructose–6–phosphate

ATP

ADP

P-fructose–1,6–bisphosphate-P

ENERGY-RELEASING STEPS OF GLYCOLYSIS

P-PGAL

NAD⁺

P_i

NADH

P-1,3–bisphosphoglycerate-P

ADP

ATP

P-PGAL

NAD⁺

P_i

NADH

P-1,3–bisphosphoglycerate-P

ADP

ATP

substrate-level phosphorylation

2 ATP produced

P-3–phosphoglycerate

P-2–phosphoglycerate

H_2O

P-PEP

ADP

ATP

P-3–phosphoglycerate

P-2–phosphoglycerate

H_2O

P-PEP

ADP

ATP

substrate-level phosphorylation

2 ATP produced

pyruvate

pyruvate

to second set of reactions

a This diagram tracks the six carbon atoms of the glucose molecule shown above. Glycolysis starts with an energy investment of two ATP.

b One ATP makes a phosphate-group transfer to glucose, and some atoms in glucose are rearranged in response.

c A phosphate-group transfer from the second ATP causes rearrangements that form fructose–1,6–bisphosphate. This intermediate can be split easily.

d It splits at once into two molecules, each with a three-carbon backbone. We can call these two PGAL.

e Two NADH form when each PGAL gives up two electrons and a hydrogen atom to NAD⁺.

f One ATP forms as each PGAL also combines with inorganic phosphate (P_i) and transfers a phosphate group to ADP.

g *Thus, two ATP formed by the direct transfer of a phosphate group from two intermediates of the reactions.* The original energy investment of two ATP has been paid off.

h In the next two reactions, the two intermediates each release a hydrogen atom and an —OH group, which then combine to form water.

i Two 3–phosphoenolpyruvate (PEP) molecules result. Each PEP transfers a phosphate group to ADP.

j *Once again, two ATP have formed by substrate-level phosphorylation.*

k In sum, the net energy yield from glycolysis is two ATP for each glucose molecule entering the reactions. Two molecules of pyruvate, the end product, may enter the next set of reactions in an energy-releasing pathway.

SECOND STAGE OF THE AEROBIC PATHWAY

Suppose two pyruvate molecules formed by glycolysis leave the cytoplasm and enter a **mitochondrion** (plural, mitochondria). In this organelle, the second and third stages of the aerobic pathway are completed. Figure 8.5 shows its structure and functional zones.

Preparatory Steps and the Krebs Cycle

It is during the second stage that the glucose is finally broken down completely to carbon dioxide and water. Two ATP form, but the most striking part of the second stage is the transfer of electrons and hydrogens from intermediates to many coenzymes.

In a few preparatory reactions, an enzyme removes a carbon from each pyruvate molecule. Coenzyme A, an enzyme helper, becomes **acetyl–CoA** by combining

with the two-carbon fragment left after the removal. This fragment is passed to **oxaloacetate**, the entry point for the Krebs cycle. (The name of this cyclic pathway honors Hans Krebs, who worked out many of its details in the 1930s. It also is called the citric acid cycle.) *Six* carbons, three from each pyruvate, enter the second stage of reactions. And *six* depart, in six carbon dioxide molecules, during the preparatory steps and the cycle proper (Figure 8.6).

Functions of the Second Stage

Think of the second stage as having three functions. First, it loads electrons and hydrogen onto NAD^+ and FAD, resulting in NADH and $FADH_2$. It also converts organic carbons into CO_2. Second, it forms two ATP by

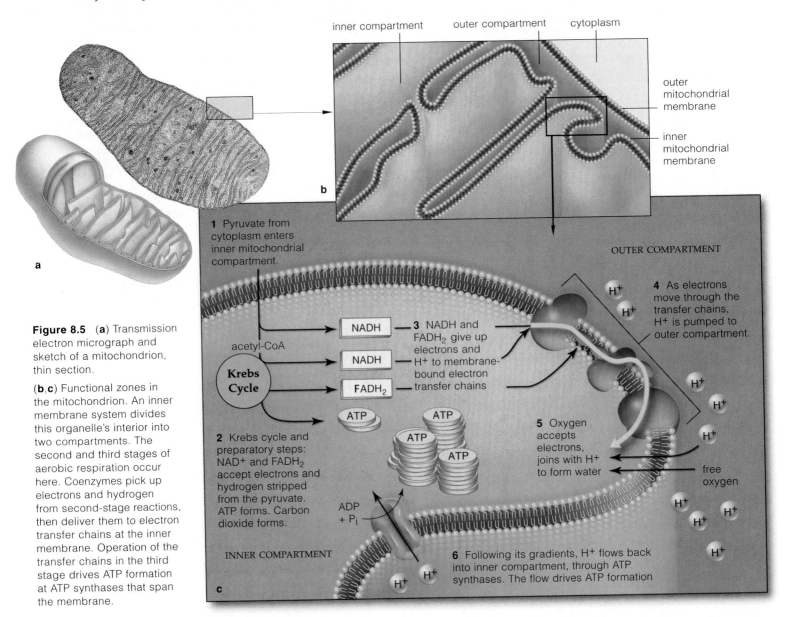

inner compartment   outer compartment   cytoplasm

outer mitochondrial membrane

inner mitochondrial membrane

b

Figure 8.5 (a) Transmission electron micrograph and sketch of a mitochondrion, thin section.

(**b**,**c**) Functional zones in the mitochondrion. An inner membrane system divides this organelle's interior into two compartments. The second and third stages of aerobic respiration occur here. Coenzymes pick up electrons and hydrogen from second-stage reactions, then deliver them to electron transfer chains at the inner membrane. Operation of the transfer chains in the third stage drives ATP formation at ATP synthases that span the membrane.

a

OUTER COMPARTMENT

1 Pyruvate from cytoplasm enters inner mitochondrial compartment.

acetyl-CoA

Krebs Cycle

NADH

NADH

$FADH_2$

ATP

ATP   ATP

ATP   ATP

ADP + P_i

2 Krebs cycle and preparatory steps: NAD^+ and $FADH_2$ accept electrons and hydrogen stripped from the pyruvate. ATP forms. Carbon dioxide forms.

3 NADH and $FADH_2$ give up electrons and H^+ to membrane-bound electron transfer chains

4 As electrons move through the transfer chains, H^+ is pumped to outer compartment.

5 Oxygen accepts electrons, joins with H^+ to form water

free oxygen

H^+

INNER COMPARTMENT

6 Following its gradients, H^+ flows back into inner compartment, through ATP synthases. The flow drives ATP formation

c

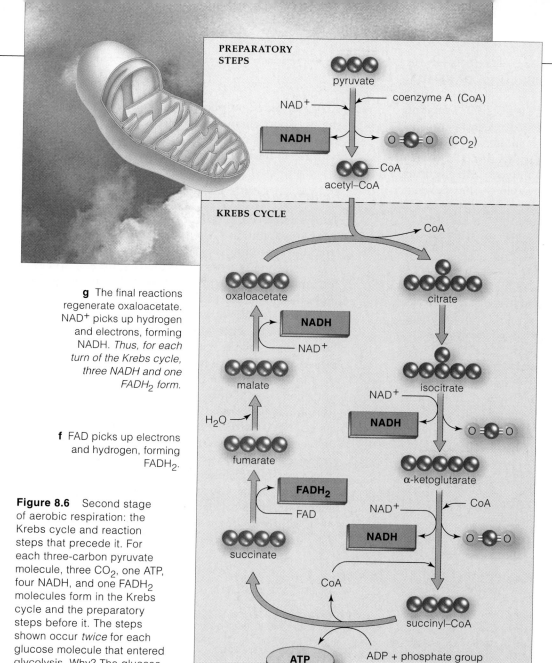

PREPARATORY STEPS

pyruvate

NAD$^+$ → coenzyme A (CoA)

NADH

O = C = O (CO$_2$)

CoA

acetyl–CoA

a Pyruvate from glucose enters a mitochondrion. A carbon atom is released, as CO$_2$. A coenzyme binds with the two-carbon fragment, becoming acetyl–CoA. NADH forms as NAD$^+$ picks up hydrogen and electrons. *Thus, one NADH forms during a few steps preceding the Krebs cycle.*

KREBS CYCLE

CoA

oxaloacetate

citrate

g The final reactions regenerate oxaloacetate. NAD$^+$ picks up hydrogen and electrons, forming NADH. *Thus, for each turn of the Krebs cycle, three NADH and one FADH$_2$ form.*

NADH

NAD$^+$

malate

H$_2$O →

f FAD picks up electrons and hydrogen, forming FADH$_2$.

fumarate

isocitrate

NAD$^+$

NADH

O = C = O

b The four-carbon oxaloacetate is the entry point into the Krebs cycle. Acetyl–CoA transfers two carbons to it, forming citrate, a six-carbon compound. Citrate becomes rearranged into another intermediate.

c Another carbon atom is released, as CO$_2$. NADH forms when NAD$^+$ picks up hydrogen and electrons.

α-ketoglutarate

FADH$_2$

FAD

NAD$^+$ — CoA

NADH

O = C = O

Figure 8.6 Second stage of aerobic respiration: the Krebs cycle and reaction steps that precede it. For each three-carbon pyruvate molecule, three CO$_2$, one ATP, four NADH, and one FADH$_2$ molecules form in the Krebs cycle and the preparatory steps before it. The steps shown occur *twice* for each glucose molecule that entered glycolysis. Why? The glucose was degraded earlier to *two* pyruvate molecules.

succinate

CoA

ATP

ADP + phosphate group

succinyl–CoA

d Another carbon atom is released as CO$_2$, another NADH forms, and a coenzyme A molecule binds to the intermediate. *At this point, for each turn of the cycle, three carbon atoms have been released.* This balances out the three carbons that entered the mitochondrion (in pyruvate).

e A phosphate group replaces the coenzyme A and is attached to ADP. *Thus, for each turn of the Krebs cycle, one ATP forms by substrate-level phosphorylation.*

---

substrate-level phosphorylations. Third, it rearranges the Krebs cycle intermediates into oxaloacetate. This is important; cells have only so much oxaloacetate, which must be regenerated to keep the reactions going.

The two ATP do not add much to the small yield from glycolysis. But the reactions load ten coenzymes with electrons and hydrogen. So far, then, the final stage of the aerobic pathway will get all of these coenzymes:

Glycolysis:	2 NADH
Pyruvate conversion before Krebs cycle:	2 NADH
Krebs cycle:	2 FADH$_2$ + 6 NADH
Coenzymes sent to third stage:	2 FADH$_2$ + 10 NADH

Overall, these are the key points to remember about the second stage of aerobic respiration:

In the second stage of aerobic respiration, two pyruvate molecules from glycolysis enter a mitochondrion.

All of pyruvate's carbon atoms (originally from glucose) are released in the form of carbon dioxide. Two ATP form. The oxaloacetate that is the entry point for the second-stage cyclic reactions is regenerated.

Ten coenzymes are reduced by electrons (and hydrogen) released during pyruvate breakdown. With two coenzymes that formed in glycolysis, they will deliver electrons and hydrogen to sites of the final stage of the aerobic pathway.

THIRD STAGE OF THE AEROBIC PATHWAY

In the aerobic pathway's third stage, ATP formation goes into high gear. This stage uses electron transfer chains and ATP synthases, both located in the inner membrane that divides the mitochondrion into two compartments (Figure 8.7). Both interact with electrons and hydrogen, delivered by coenzymes from the first two stages of the aerobic pathway.

Electron Transfer Phosphorylation

When electrons enter the transfer chains, hydrogen is released and so becomes ionized (H^+). As the electrons pass through the chains, hydrogen ions inside the inner compartment are shuttled to the outer compartment. These repeated shuttlings set up H^+ concentration and electric gradients across the inner membrane.

The only way ions can follow the gradient and flow back to the inner compartment is through the interior of ATP synthases (Figure 8.7). Their flow through these transport proteins drives the formation of ATP from ADP and unbound phosphate. Free oxygen helps keep the transport chains clear for operation. It withdraws spent electrons at the end of the chains and combines with H^+ to form water, a by-product.

Summary of the Energy Harvest

Thirty-two ATP typically form during the third stage of aerobic respiration. When you add these to the yield from the preceding stages, the net harvest is thirty-six ATP from one glucose molecule (Figure 8.8). That's a lot! Anaerobic pathways may use up eighteen glucose molecules to get the same net yield.

Thirty-six ATP is a typical yield only. The amount depends on the type of cell and prevailing conditions. For instance, the yield is lower when an intermediate is pulled away from the reactions and used elsewhere.

Also, NADH from the cytoplasm can't even enter a mitochondrion. It only delivers electrons and hydrogen to it; transport proteins in the membrane shuttle these across. NAD^+ or FAD already inside accepts them and delivers the electrons to transport systems. But FAD drops them off at a *lower* step in a transport system "staircase," so its deliveries produce less ATP.

A final point. Glucose, remember, has more stored energy (in more covalent bonds) compared to products of its full breakdown (carbon dioxide, water). About 686 kilocalories of energy are released as each mole of

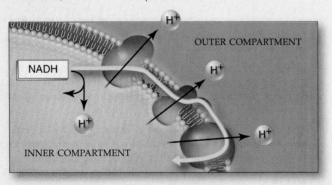

Figure 8.7 Electron transfer phosphorylation, the third and final stage of aerobic respiration. The reactions occur at electron transfer chains and at ATP synthases, a type of transport protein, in the inner mitochondrial membrane. Each electron transfer chain consists of specific enzymes, cytochromes, and other proteins. The membrane divides the mitochondrion's interior into two compartments.

NADH and $FADH_2$ give up electrons and hydrogen to the transfer chains. When electrons are transferred *through* the chains, unbound hydrogen (H^+) is shuttled across the membrane, to the outer compartment:

The H^+ concentration is now greater in the outer compartment. Concentration and electric gradients across the membrane have been set up. H^+ follows these gradients, through the interior of ATP synthases. Energy released by the flow drives the formation of ATP from ADP and unbound phosphate (P_i). Hence the name, electron transfer *phosphorylation:*

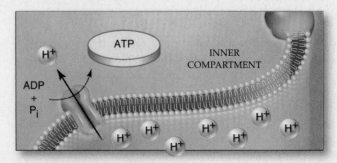

Do these events sound familiar? They should. ATP forms in much the same way in chloroplasts. By the *chemiosmotic* model, H^+ concentration and electric gradients across a cell membrane drive ATP formation. In this case, H^+ flows in the opposite direction compared to the flow in chloroplasts.

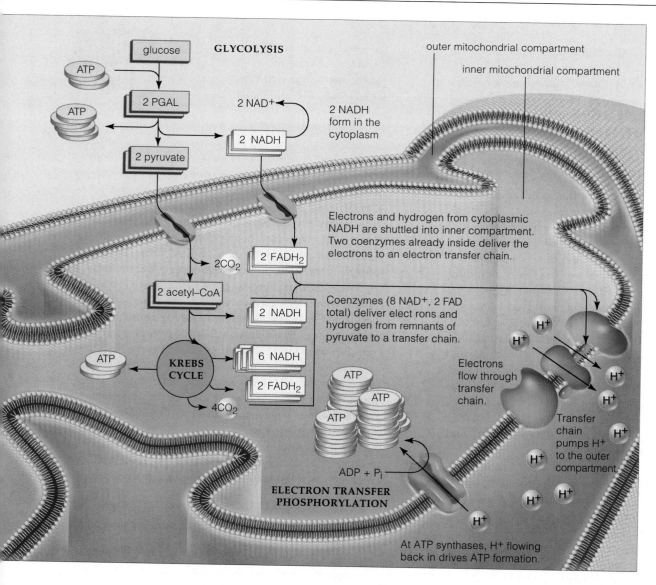

GLYCOLYSIS

glucose

ATP

ATP

2 PGAL

2 pyruvate

2 NAD+

2 NADH

2 NADH form in the cytoplasm

2CO₂

2 acetyl–CoA

KREBS CYCLE

ATP

4CO₂

2 FADH₂

2 NADH

6 NADH

2 FADH₂

outer mitochondrial compartment

inner mitochondrial compartment

Electrons and hydrogen from cytoplasmic NADH are shuttled into inner compartment. Two coenzymes already inside deliver the electrons to an electron transfer chain.

Coenzymes (8 NAD+, 2 FAD total) deliver elect rons and hydrogen from remnants of pyruvate to a transfer chain.

ATP

ATP

ATP

ADP + Pᵢ

ELECTRON TRANSFER PHOSPHORYLATION

Electrons flow through transfer chain.

H⁺
H⁺
H⁺
H⁺
H⁺
H⁺
H⁺
H⁺

Transfer chain pumps H⁺ to the outer compartment.

At ATP synthases, H⁺ flowing back in drives ATP formation.

2 ATP

a In glycolysis, 2 ATP used; 4 ATP form by *substrate-level phosphorylation*. So *net* yield is 2 ATP.

2 ATP

b In Krebs cycle of second stage, 2 ATP form by *substrate-level phosphorylation*.

4 ATP

c In third stage, NADH from glycolysis used to form 4 ATP by *electron transfer phosphorylation*.

28 ATP

d In third stage, NADH and FADH₂ from second stage used to make 28 ATP by *electron transfer phosphorylation*.

TYPICAL NET ENERGY YIELD   **36 ATP**

Figure 8.8 Summary of the net energy harvest from aerobic respiration. Thirty-six ATP per glucose molecule is common. The actual yield varies. Shifting concentrations of reactants, intermediates, and end products affect it. So do membrane crossing mechanisms, which vary among different cell types.

Cells differ in how they use the NADH from glycolysis, which can't enter mitochondria. These NADH give up electrons and hydrogen to transport proteins in the outer mitochondrial membrane, which shuttle them across. NAD+ or FAD already inside accepts them, forming NADH or FADH₂.

Any NADH inside a mitochondrion delivers electrons to the highest entry point into a transfer chain. When it does, enough H⁺ is pumped across the inner membrane to make *three* ATP. FADH₂ delivers them to a lower entry point. Fewer hydrogen ions are pumped, so only *two* ATP can form.

In liver, heart, and kidney cells, for example, electrons and hydrogen from glycolysis enter the highest entry point of transfer chains, so the energy harvest is thirty-eight ATP. More commonly, as in skeletal muscle and brain cells, they are transferred to FAD, so the harvest is thirty-six ATP.

glucose is degraded to these more stable end products. (Section 2.7, *Critical Thinking* question 7, gives a simple definition of molar weight.) Much of the freed energy escapes as heat, but about 39 percent is stored in ATP.

In the final stage of the aerobic pathway, many coenzymes deliver electrons to transfer chains in a mitochondrion's inner membrane. The membrane forms two compartments.

Coenzymes give up hydrogen and electrons to the transfer chains. Electrons passing through these chains cause the unbound hydrogen (H⁺) to be shuttled into the inner compartment. The resulting H⁺ concentration and electric gradients drive ATP formation as H⁺ flows back across the membrane. Oxygen is the final electron acceptor.

Again, from start (glycolysis in the cytoplasm) to finish (in mitochondria), the pathway commonly has a net yield of thirty-six ATP for every glucose molecule metabolized.

8.5

ANAEROBIC ROUTES OF ATP FORMATION

So far, we have tracked the fate of a glucose molecule through the pathway of aerobic respiration. We turn now to its use as a substrate for fermentation pathways. Remember, these are anaerobic pathways; they do *not* use oxygen as the final acceptor of the electrons that ultimately drive the ATP-forming machinery.

Fermentation Pathways

Diverse organisms use fermentation pathways. Many are prokaryotic cells and protistans of marshes, bogs, mud, deep-sea sediments, the animal gut, canned food, sewage treatment ponds, and other oxygen-free places. Some fermenters actually will die if they are exposed to oxygen. Bacteria that cause many diseases, including botulism and tetanus, are like this. Other fermenters, including the acidophilus bacteria employed by yogurt manufacturers, are indifferent to oxygen's presence. Still others use oxygen, but they switch to fermentation when oxygen becomes scarce.

Glycolysis serves as the first stage of fermentation pathways, as it does for aerobic respiration. Glycolysis also requires enzymes that catalyze the breakdown of glucose and rearrangement of the fragments into two pyruvate molecules. Here again, two NADH form, and the net energy yield is two ATP.

However, fermentation reactions do not completely break down glucose to carbon dioxide and water. Also, they produce no more ATP beyond the tiny yield from glycolysis. *Fermentation's final steps simply regenerate NAD⁺, the coenzyme that assists the breakdown reactions.*

Fermentation yields enough energy to sustain many single-celled anaerobic organisms. It even helps some aerobic cells through times of stress. But it isn't enough

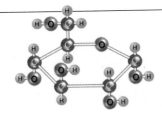

to sustain large, multicelled organisms. And that is one reason why you will never see an anaerobic elephant.

LACTATE FERMENTATION   With these points in mind, take a look at Figure 8.9, which tracks the main steps of **lactate fermentation**. During this anaerobic pathway, the pyruvate molecules from glycolysis—the first stage of its reactions—accept hydrogen and electrons from NADH. This transfer regenerates NAD⁺. At the same time, it converts each pyruvate to lactate, a three-carbon compound. You may hear people call this compound lactic acid, which is the non-ionized form. Lactate, its ionized form, is far more common in cellular fluids.

Some bacteria, such as *Lactobacillus*, rely exclusively on this anaerobic pathway. Their activities often spoil food. Even so, some fermenters have commercial uses, as when they break down glucose in huge vats where cheeses, sauerkraut, and other products are made.

Also, in humans, rabbits, and many other animals, some types of cells use this anaerobic pathway for a quick fix of ATP. When your own demands for energy are intense but brief—say, during a short race—muscle cells use this pathway. They can't do so for long; they would throw away too much of glucose's energy for too little ATP. Muscles fatigue and lose their ability to contract when these cells deplete their glucose supply.

ALCOHOLIC FERMENTATION   In the anaerobic route called **alcoholic fermentation**, enzymes convert each pyruvate molecule from glycolysis to an intermediate form: acetaldehyde. The NADH transfers electrons and hydrogen to acetaldehyde and thereby converts it to an alcoholic end product—ethanol (Figure 8.10).

Some single-celled fungi called yeasts are famous for their use of this pathway. One type, *Saccharomyces cerevisiae*, makes bread dough rise. Bakers mix the yeast with sugar, then blend both into dough. Fermenting yeast cells release carbon dioxide. Bubbles of this gas expand the dough (make it rise). Oven heat forces the gas out of the dough, and a porous product remains.

Beer and wine producers use yeasts on a large scale. Vintners use both wild yeasts and cultivated strains of *S. ellipsoideus*. Both yeasts are active until the alcohol concentration in wine vats exceeds 14 percent. (Wild yeasts die when the concentration exceeds 4 percent.) That is why some birds can get drunk on naturally fermented berries. That is why landscapers don't plant prodigious berry-producing shrubbery near highways;

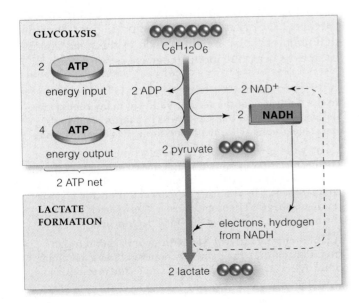

Figure 8.9   Lactate fermentation. In this anaerobic pathway, electrons end up in lactate, the reaction product.

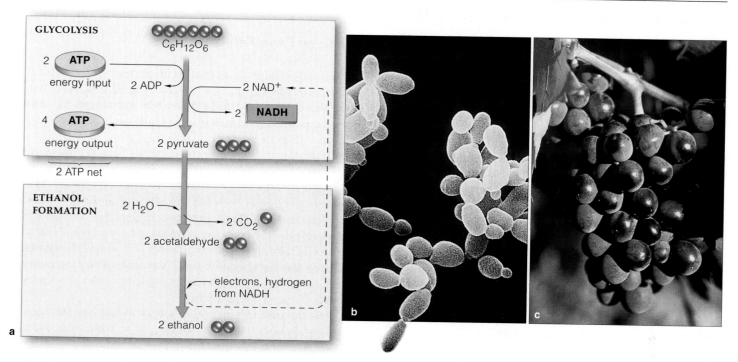

Figure 8.10 (**a**) Alcoholic fermentation, one of the anaerobic pathways. Acetaldehyde, an intermediate of the reactions, is the final acceptor of electrons. Ethanol is the end product. Yeasts, single-celled organisms, use this pathway. (**b**) Activity of one species of *Saccharomyces* makes bread dough rise. (**c**) Another species lives on sugar-rich tissues of ripe grapes.

drunk birds doodle into windshields (Figure 8.11). Wild turkeys similarly have been known to get tipsy when they gobble fermenting apples in untended orchards.

Anaerobic Electron Transfers

Especially among prokaryotic cells, we see less common energy-releasing pathways, some of which are topics of later chapters in this book. For example, many species have key roles in the global cycling of sulfur, nitrogen, and other crucial elements. Collectively, their metabolic

Figure 8.11 Robin feasting on the fermented berries of a pyracantha bush.

activities influence nutrient availability for organisms in ecosystems everywhere.

Some archaebacteria and eubacteria use **anaerobic electron transfers**. Electrons from organic compounds flow through transfer chains in the plasma membrane and H⁺ flows out through ATP synthases to form ATP. Some inorganic compound often is the final electron acceptor. The net energy yield is always small.

Even as you read this, some anaerobic species that live in waterlogged soil are stripping electrons from a variety of compounds. They dump electrons on sulfate. Hydrogen sulfide, a putrid-smelling gas, is the result. Sulfate-reducing species also thrive in aquatic habitats that are enriched with decomposed organic material. They even live on the ocean floor, near hydrothermal vents. As described in Section 49.11, they are part of some unique communities.

In fermentation pathways, an organic substance that forms during the reactions serves as the final acceptor of electrons from glycolysis. The reactions regenerate NAD^+, which is required to keep the pathway operational.

By anaerobic electron transfers, an inorganic substance (but not oxygen) usually serves as the final electron acceptor.

Fermentation pathways typically have a net energy yield of two ATP for each glucose molecule metabolized. The net yield is similarly small for anaerobic electron transfers.

ALTERNATIVE ENERGY SOURCES IN THE HUMAN BODY

So far, you have looked at what happens after a lone glucose molecule enters an energy-releasing pathway. Now you can start thinking about what cells do when they have too many or too few glucose molecules.

Carbohydrate Breakdown in Perspective

THE FATE OF GLUCOSE AT MEALTIME   While you or any other mammal is eating, glucose and other small organic molecules are being absorbed across the gut lining, then blood transports them through the body. A rise in the glucose level in blood prompts the pancreas to release insulin. This hormone stimulates cells to take up glucose faster. Cells convert the incoming glucose to glucose–6–phosphate and trap it inside the cytoplasm. (When phosphorylated, glucose cannot be transported back out, across the plasma membrane.) Look again at Figure 8.4, and you see that glucose–6–phosphate is the first activated intermediate of glycolysis.

If your glucose intake exceeds cellular demands for energy, ATP-producing machinery goes into high gear. Unless a cell is rapidly using its ATP, the cytoplasmic concentration of ATP can rise to high levels. When that happens, the glucose–6–phosphate gets diverted into a biosynthesis pathway. This pathway assembles glucose units into glycogen, a storage polysaccharide (Section 3.4). The pathway is especially favored in muscle and liver cells, which maintain the largest glycogen stores.

THE FATE OF GLUCOSE BETWEEN MEALS   When you are not eating, glucose is not entering the bloodstream, and its level in the blood declines. If the decline were not countered, that would be bad news for the brain, your body's glucose hog. At any time, your brain is taking up more than two-thirds of the freely circulating glucose, because its many hundreds of millions of cells use this sugar alone as their preferred energy source.

The pancreas responds to the decline by secreting glucagon. This hormone prompts liver cells to convert glycogen back to glucose and send it back to the blood. Only liver cells do this; muscle cells won't give it up. The blood glucose level rises, and brain cells keep on functioning. Thus, *hormones control whether your body's cells use free glucose as an energy source or tuck it away.*

A word of caution: Don't let the preceding examples lead you to believe cells squirrel away large amounts of glycogen. In adult humans, glycogen makes up merely 1 percent or so of the body's total energy reserves, the energy equivalent of two cups of cooked pasta. Unless you eat on a regular basis, you will deplete the liver's small glycogen stores in less than twelve hours.

Of the total energy reserves in, say, a typical adult who eats well, 78 percent (about 10,000 kilocalories) is concentrated in body fat and 21 percent in proteins.

Energy From Fats

How does the body access its huge reservoir of fats? A fat molecule, recall, has a glycerol head and one, two, or three fatty acid tails. Most fats get stored in your body as triglycerides, with three tails each. Triglycerides accumulate in fat cells of adipose tissue, which forms at buttocks and other strategic places beneath the skin.

When blood glucose levels decline, triglycerides are tapped as an energy alternative. Enzymes in fat cells cleave the bonds between the glycerol and fatty acids, which enter the blood. Enzymes in the liver convert the glycerol to PGAL, an intermediate of glycolysis. Nearly all cells take up these circulating fatty acids. Enzymes cleave the backbone of the fatty acids. The fragments are converted to acetyl–CoA, which can enter the Krebs cycle (Figures 8.6 and 8.12).

Compared to glucose, a fatty acid tail has far more carbon-bound hydrogen atoms, so it yields far more ATP. Between meals or during sustained exercise, fatty acid conversions supply about one half of the ATP that muscle, liver, and kidney cells require.

What happens if you eat too many carbohydrates? Exceed the glycogen-storing capacity of your liver and muscle cells, and the excess gets converted to fats. *Too much glucose ends up as excess fat.* In the United States, 25 percent of the people have a combination of genes that lets them eat as much as they like without gaining weight. A diet too rich in carbohydrates keeps the other 75 percent fat. Their insulin levels stay elevated, which "tells" the body to store fat, not use it for energy. We return to this topic in Section 36.7 and Chapter 41.

Energy From Proteins

Eat more proteins than your body requires to grow and maintain itself, and its cells won't store them. Enzymes split dietary proteins into amino acid units. Then they remove the amino group ($-NH_3^+$) from each unit, and ammonia (NH_3) forms. What happens to the leftover carbon backbones? The outcome varies, depending on conditions of the moment. Maybe they'll get converted to fats or carbohydrates. Or maybe they'll enter the Krebs cycle (Figure 8.12), where coenzymes pick up the hydrogen and electrons stripped from carbon atoms. The ammonia that forms undergoes conversions and becomes urea. This nitrogen-containing waste product would be toxic if it accumulated to high concentrations. Normally your body excretes urea, in urine.

As this brief discussion makes clear, maintaining and accessing the body's energy reserves is complicated business. Hormonal controls over the disposition of glucose are special only because glucose is the fuel of choice for the all-important brain. However, as you will

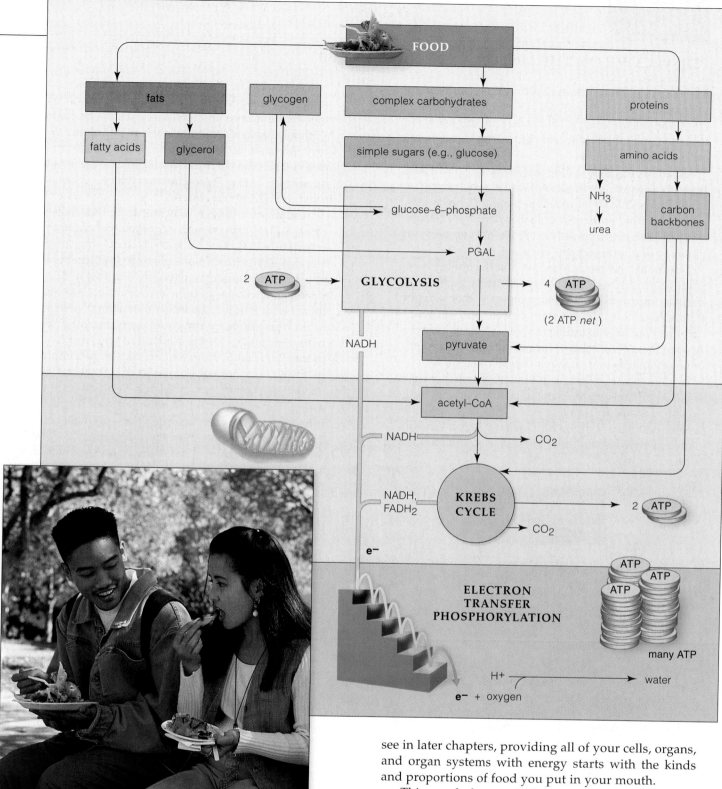

Within the diagram:

FOOD

fats | glycogen | complex carbohydrates | proteins

fatty acids | glycerol | | simple sugars (e.g., glucose) | amino acids

glucose–6–phosphate

NH_3

urea

carbon backbones

PGAL

2 ATP → **GLYCOLYSIS** → 4 ATP

(2 ATP *net*)

NADH

pyruvate

acetyl–CoA

NADH → CO_2

NADH, FADH$_2$ → **KREBS CYCLE** → 2 ATP

CO_2

e⁻

ELECTRON TRANSFER PHOSPHORYLATION

ATP ATP ATP ATP

many ATP

H+ → water

e⁻ + oxygen

Figure 8.12 Alternative energy sources in the human body. The diagram shows reaction sites where a variety of organic compounds can enter the stages of aerobic respiration.

Complex carbohydrates, fats, and proteins cannot enter the aerobic pathway directly. In humans and other mammals, the digestive system, and individual cells, must first break apart these molecules into simpler, degradable subunits.

see in later chapters, providing all of your cells, organs, and organ systems with energy starts with the kinds and proportions of food you put in your mouth.

This concludes our look at aerobic respiration and other energy-releasing pathways. The section to follow may help you get a sense of how they fit into the larger picture of life's evolution and interconnectedness.

In humans and other mammals, the entrance of glucose or other organic compounds into an energy-releasing pathway depends on the kinds and proportions of carbohydrates, fats, and proteins in the diet as well as on the type of cell.

PERSPECTIVE ON THE MOLECULAR UNITY OF LIFE

In this unit you read about photosynthesis and aerobic respiration—the main pathways by which cells trap, store, and release energy. What you may not know is that the two pathways became linked, on a grand scale, over evolutionary time.

When life originated long ago, the atmosphere had little free oxygen. The first single-celled organisms probably made ATP by reactions similar to glycolysis, so fermentation pathways probably dominated. More than a billion years passed before the oxygen-evolving pathway of photosynthesis emerged.

Gradually, oxygen accumulated in the atmosphere. Some cells could use it to accept electrons, perhaps as a chance outcome of mutated proteins in their electron transfer chains. In time, certain descendants of those early aerobic cells abandoned photosynthesis entirely. Among them were the forerunners of animals and all other organisms that engage in aerobic respiration.

With aerobic respiration, a flow of carbon, hydrogen, and oxygen through the metabolic pathways of living organisms came full circle. For the final products of this aerobic pathway—carbon dioxide and water—are the same materials necessary to build organic compounds in photosynthesis:

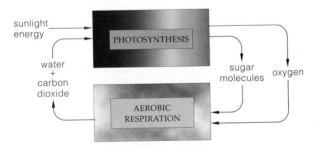

Perhaps you have difficulty seeing the connection between yourself—a highly intelligent being—and such remote-sounding events as energy flow and the cycling of carbon, hydrogen, and oxygen. Is this really the stuff of humanity?

Think back, for a moment, on the structure of a water molecule. Two hydrogen atoms sharing electrons with oxygen may not seem close to your daily life. Yet, through that sharing, water molecules show polarity and hydrogen-bond with one another. Their chemical behavior is a beginning for the organization of lifeless matter that leads to the organization of all living things.

For now you can visualize other diverse molecules interspersed through water. The nonpolar kinds resist interaction with water; polar kinds dissolve in it. On their own, the phospholipids among them assemble into a two-layered film. Such lipid bilayers, recall, are the framework of cell membranes, hence all cells. From the beginning, the cell has been the basic *living* unit.

The essence of life is not some mysterious force. It is molecular organization and metabolic control. With a membrane to contain them, reactions *can* be controlled. With mechanisms built into their membranes, cells respond to energy changes and shifting concentrations of solutes in the environment. Response mechanisms operate by "telling" proteins—enzymes—when and what to build or tear down.

And it is not some mysterious force that creates the proteins. DNA, the double-stranded treasurehouse of inheritance, has the chemical structure—*the chemical message*—that allows molecule to reproduce molecule, one generation after the next. In your own body, DNA strands tell trillions of cells how countless molecules must be built or torn apart for their stored energy.

So yes, carbon, hydrogen, oxygen, and other atoms of organic molecules are the stuff of you, and us, and all of life. But it takes more than molecules to complete the picture. Life continues as long as an unbroken flow of energy sustains its grand organization. Molecules are assembled into cells, cells into organisms, organisms into communities, and so on up through the biosphere. It takes energy inputs from the sun to maintain these levels of organization. And energy flows through time in one direction—from organized to less organized forms. Only as long as energy continues to flow into the web of life can life continue in all its rich expressions.

And so life is no more *and no less* than a marvelously complex system for prolonging order. Sustained by energy transfusions from the sun, life continues by its capacity for self-reproduction. With the hereditary instructions contained in DNA, energy and materials are organized, generation after generation. Even with the death of individuals, life elsewhere is prolonged. With each death, molecules are released and may be cycled once more, as raw materials for new generations.

With this flow of energy and cycling of material through time, each birth is affirmation of our ongoing capacity for organization, each death a renewal.

The diversity of life, and its continuity through time, arises from the unity of life at the molecular level.

SUMMARY

1. Phosphate-group transfers from ATP are central to metabolism. Autotrophic cells alone tap energy from the environment to make ATP for building carbohydrates. *All* cells make ATP by pathways that release chemical energy from organic compounds, such as glucose. *8.1*

2. After glucose enters an energy-releasing pathway, enzymes derive electrons and hydrogen from reaction intermediates formed along the way. Coenzymes pick these up and deliver them to other reaction sites, where the pathway ends. NAD^+ is the main coenzyme; FAD is also used in the aerobic route. Reduced forms of these coenzymes are designated NADH and $FADH_2$. *8.1–8.2*

3. All of the main energy-releasing pathways start with glycolysis, which begins and ends in the cytoplasm and can proceed in the presence or absence of oxygen. *8.2*

   a. In glycolysis, enzymes break down glucose to two pyruvate molecules. Two NADH and four ATP form.

   b. The net energy yield is two ATP (because two ATP had to be invested up front to get the reactions going).

4. Aerobic respiration continues on through two more stages: (1) the Krebs cycle and a few steps preceding it, and (2) electron transfer phosphorylation. These stages occur only in mitochondria of eukaryotic cells. *8.2, 8.3*

5. The second stage of the aerobic pathway starts when an enzyme strips a carbon atom from each pyruvate. Coenzyme A binds the two-carbon fragment, forming acetyl–CoA, and transfers this to oxaloacetate, the entry point of the Krebs cycle. These cyclic reactions and a few steps before them load ten coenzymes with electrons and hydrogen (eight NADH, plus two $FADH_2$). Two ATP form. Three carbon dioxide molecules are released for each pyruvate that entered this second stage. *8.3*

6. The third stage of the aerobic pathway proceeds at a membrane dividing the mitochondrion's interior into two compartments. Electron transfer chains and ATP synthases are embedded in this inner membrane. *8.4*

   a. The chains accept electrons and hydrogen from coenzymes from the first two stages, and establish H^+ concentration and electric gradients.

   b. H^+ follows the gradients and flows from the outer to the inner compartment. It does so through the interior of ATP synthases. Energy released by the ion flow drives formation of ATP from ADP and unbound phosphate.

   c. Oxygen picks up the spent electrons at the end of the transfer chain and combines with hydrogen ions to form water. That is, oxygen is the final acceptor of the electrons that initially resided in glucose.

7. Aerobic respiration has a typical net energy yield of thirty-six ATP for each glucose molecule metabolized. Yields vary by cell type and cellular conditions. *8.4*

8. Fermentation and anaerobic electron transport start with glycolysis but are anaerobic, start to finish. *8.5*

   a. Lactate fermentation has a net energy yield of two ATP, which form in glycolysis. The remaining reaction regenerates NAD^+. The two NADH from glycolysis transfer electrons and hydrogen to two pyruvate from glycolysis. Two lactate molecules are the products.

   b. Alcoholic fermentation has a net energy yield of two ATP from glycolysis, and its remaining reactions serve to regenerate NAD^+. Enzymes convert pyruvate from glycolysis to acetaldehyde, and carbon dioxide is released. NADH from glycolysis transfers electrons and hydrogen to the two acetaldehyde molecules, thus forming two ethanol molecules, the products.

   c. Certain bacteria use anaerobic electron transfer. Electrons stripped from some organic compound flow through transfer chains in the plasma membrane, and H^+ flows out through ATP synthases to form ATP. An inorganic compound in the environment is often the final electron acceptor.

9. In humans and other mammals, simple sugars such as glucose from carbohydrates, glycerol and fatty acids from fats, and carbon backbones of amino acids from proteins can enter ATP-producing pathways. *8.6*

10. Life shows great unity at the molecular level. *8.7*

Review Questions

1. Is this true or false: Aerobic respiration occurs in animals but not plants, which make ATP only by photosynthesis. *8.1*

2. Using the diagram below of the aerobic pathway, fill in the blanks with the number of molecules of pyruvate, coenzymes, and end products. Write in the net ATP formed in each stage, then the net ATP formed from start (glycolysis) to finish. *8.1*

3. Is glycolysis energy-*requiring* or energy-*releasing*? Or do both kinds of reactions occur during glycolysis? *8.2*

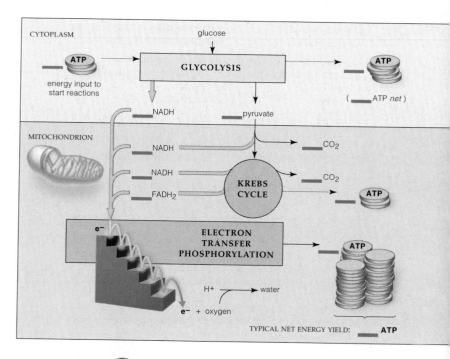

4. In what respect does *electron transfer* phosphorylation differ from *substrate-level* phosphorylation? *8.1, 8.2*

5. Sketch the double-membrane system of the mitochondrion and show where electron transfer chains and ATP synthases are located. *8.3*

6. Name the compound that is the entry point for the Krebs cycle, and state whether it directly accepts the pyruvate from glycolysis. For each glucose molecule, how many carbon atoms enter the Krebs cycle? How many depart from it, and in what molecular form? *8.3*

7. Is this statement true or false: Muscle cells cannot contract at all when deprived of oxygen. If true, explain why. If false, name the alternative(s) available to them. *8.5*

Self-Quiz   ANSWERS IN APPENDIX III

1. Glycolysis starts and ends in the _____ .
   a. nucleus   c. plasma membrane
   b. mitochondrion   d. cytoplasm

2. Which of the following does *not* form during glycolysis?
   a. NADH   c. $FADH_2$
   b. pyruvate   d. ATP

3. Aerobic respiration is completed in the _____ .
   a. nucleus   c. plasma membrane
   b. mitochondrion   d. cytoplasm

4. In the last stage of aerobic respiration, _____ is the final acceptor of electrons that originally resided in glucose.
   a. water   c. oxygen
   b. hydrogen   d. NADH

5. _____ engage in lactate fermentation.
   a. *Lactobacillus* cells   c. Sulfate-reducing bacteria
   b. Muscle cells   d. a and b

6. In alcoholic fermentation, _____ is the final acceptor of the electrons stripped from glucose.
   a. oxygen   c. acetaldehyde
   b. pyruvate   d. sulfate

7. The fermentation pathways produce no more ATP beyond the small yield from glycolysis, but the remaining reactions
   _____ .
   a. regenerate ADP   c. dump electrons on an inorganic
   b. regenerate NAD^+ substance (not oxygen)

8. In certain organisms and under certain conditions, _____ can be used as an energy alternative to glucose.
   a. fatty acids   c. amino acids
   b. glycerol   d. all of the above

9. Match the event with its most suitable metabolic description.
   ___ glycolysis   a. ATP, NADH, $FADH_2$, CO_2,
   ___ fermentation   and water form
   ___ Krebs cycle   b. glucose to two pyruvate
   ___ electron transfer   c. NAD^+ regenerated, two ATP net
       phosphorylation   d. H^+ flows through ATP synthases

Critical Thinking

1. Diana suspects that a visit to her family doctor is in order. After eating carbohydrate-rich food, she always experiences sensations of being intoxicated and becomes nearly incapacitated, as if she had been drinking alcohol. She even wakes up with a hangover the next day. Having completed a course in freshman biology, Diana has an idea that something is affecting the way her body is metabolizing glucose. Explain why.

2. The cells of your body absolutely do not use nucleic acids as alternative energy sources. Suggest why.

3. The human body's energy needs and its programs for growth depend on balancing the levels of amino acids in blood with proteins in cells. Cells of the liver, kidneys, and intestinal lining are especially important in this balancing act. When the levels of amino acids in blood decline, lysozymes in cells can rapidly digest some of their proteins (structural and contractile proteins are spared, except in cases of malnutrition). The amino acids released this way enter the blood and thereby help maintain the required levels.
   Suppose you embark on a body-building program. You already eat plenty of carbohydrates. However, a nutritionist recommends that you follow a protein-rich diet that includes protein supplements. Speculate on how extra dietary proteins will be put to use, and in which tissues.

4. Each year, Canada geese (*above*) lift off in precise formation from their northern breeding grounds. They head south to spend the winter months in warmer climates, then make the return trip in spring. As is the case for other migratory birds, their flight muscle cells are efficient at using fatty acids as an energy source. (Remember, the carbon backbone of fatty acids can be cleaved into small fragments that can be converted to acetyl–CoA for entry into the Krebs cycle.)
   Suppose a lesser Canada goose from Alaska's Point Barrow has been steadily flapping along for three thousand kilometers and is nearing Klamath Falls, Oregon. It looks down and notices a rabbit sprinting like the wind from a coyote with a taste for rabbit. With a stunning burst of speed, the rabbit reaches the safety of its burrow.
   Which energy-releasing pathway predominated in muscle cells in the rabbit's legs? Why was the Canada goose relying on a different pathway for most of its journey? And why wouldn't the pathway of choice in goose flight muscle cells be much good for a rabbit making a mad dash from its enemy?

5. Reflect on this chapter's introduction and question 4 above. Which energy-releasing pathway is predominating in agitated Africanized bees chasing a farmer through a cornfield?

Selected Key Terms

acetyl–CoA *8.3*   Krebs cycle *8.1*
aerobic respiration *8.1*   lactate fermentation *8.5*
alcoholic fermentation *8.5* mitochondrion *8.3*
anaerobic electron transfer *8.5* NAD^+ *8.1*
electron transfer   oxaloacetate *8.3*
   phosphorylation *8.1*   pyruvate *8.1*
FAD *8.1*   substrate-level
glycolysis *8.1*   phosphorylation *8.2*

Readings

DeMauro, S. October 1998. "Hooked on Mitochondria." *Quest.* Volume 5, Number 5. A brief look at mitochondrial disorders.

Wolfe, S. 1995. *An Introduction to Molecular and Cellular Biology.* Belmont, California: Wadsworth. Exceptional reference text.

On-Line readings at Student Guide for InfoTrac:
www.brookscole.com/biology

II Principles of Inheritance

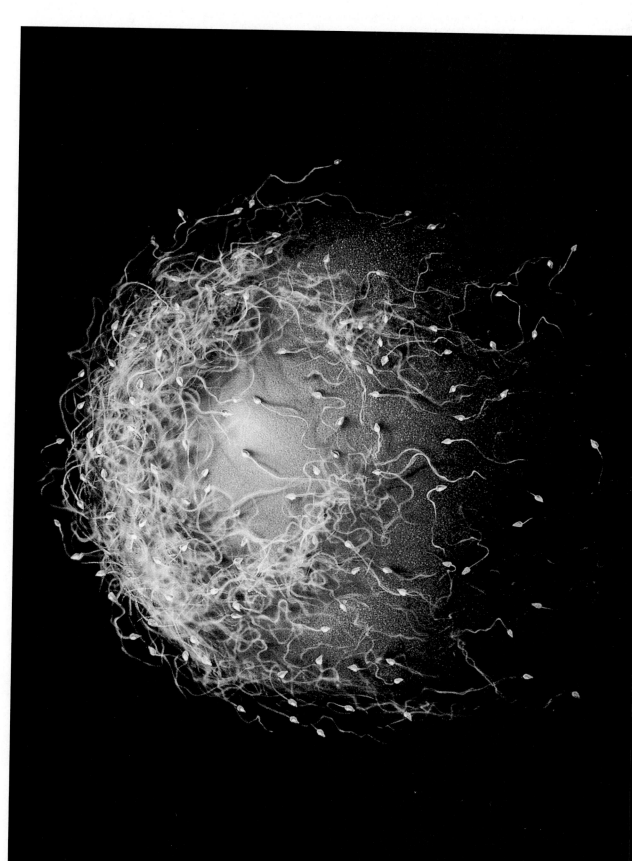

Human sperm, one of which will penetrate this mature egg and so set the stage for the development of a new individual in the image of its parents. This exquisite art is based on a scanning electron micrograph.

CELL DIVISION AND MITOSIS

From Cell to Silver Salmon

Five o'clock, and the first rays from the sun dance over the wild Alagnak River of the Alaskan tundra. It is September, and life is ending and beginning in the cold waters. Thousands of mature silvers—coho salmon—have returned from the open ocean to spawn in their shallow home waters. The females are tinged with red, the color of spawners, and they are dying.

This morning you watch a female salmon releasing translucent eggs into a shallow depression that her fins made in the gravel riverbed (Figure 9.1). A male releases a cloud of sperm, which fertilize the eggs. Trout and other predators eat most of the eggs, but some survive and give rise to a new generation.

Within three years the pea-sized eggs have become sexually mature salmon, fashioned from billions of cells. In time, on some September morning, some of their cells will serve as sperm or eggs and take part in an ongoing story of birth, growth, death, and rebirth.

Your body, too, emerged through *cell divisions*, as it did for salmon and all other multicelled species. Inside your mother, a fertilized egg divided in two, then the

two became four, and so on until billions of cells were growing, developing in specialized ways, and dividing at different times to form different parts. Your body now has more than 65 trillion living cells. Many are still dividing.

Understanding cell division—and, ultimately, how new individuals are put together in the image of their parents—begins with answers to three questions. *First,* what instructions are necessary for inheritance? *Second,* how are those instructions duplicated for distribution into daughter cells? *Third,* by what mechanisms are the duplicated instructions parceled out to daughter cells? We'll need more than one chapter to consider the nature of cell reproduction and other mechanisms of inheritance. Even so, the points made early in this chapter can help you keep the overall picture in focus.

Begin with the word **reproduction**. In biology, this means that parents produce a new generation of cells or multicelled individuals like themselves. The process starts in cells programmed to divide. The ground rule for division is this: *Parent cells must give their daughter*

sexually mature
female salmon

Figure 9.1 The last of one generation and the first of the next in Alaska's Alagnak River—from spawning coho salmon (*Onchorhynchus kisutch*) to eggs, to fingerlings.

cells specific hereditary instructions, encoded in DNA, and enough metabolic machinery to start up their own operation.

DNA, recall, contains instructions for synthesizing proteins. Some proteins are structural materials. Many are enzymes that speed the assembly of specific organic compounds, such as the lipids that cells use as building blocks and sources of energy. Unless a daughter cell receives the necessary instructions for making proteins, it simply will not be able to grow or function properly.

Also, the cytoplasm of a parent cell already contains enzymes, organelles, and other operating machinery. When a daughter cell inherits what looks like a blob of cytoplasm, it really is getting start-up machinery—which will keep that cell operating until it can use its inherited DNA for growing and developing on its own.

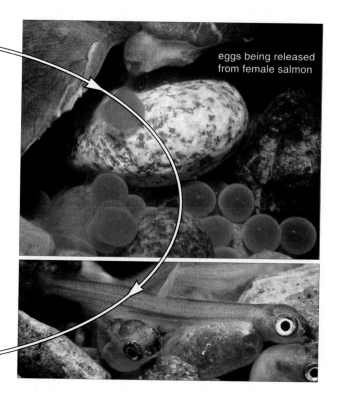

eggs being released from female salmon

Fingerlings—young fishes growing, from fertilized eggs, by way of mitotic cell divisions

Key Concepts

1. The continuity of life depends on reproduction. By this process, parents produce a new generation of cells or multicelled individuals like themselves. Cell division is the bridge between generations.

2. When a cell divides, its two daughter cells must each receive a required number of DNA molecules and some cytoplasm. For eukaryotic cells, a division mechanism called mitosis sorts out the DNA into two new nuclei. A separate mechanism divides the cytoplasm.

3. The cell cycle starts each time a daughter cell forms. It ends when the cell completes its own division. One turn of the cycle proceeds through interphase, mitosis, and cytoplasmic division. A cell spends most of its life in interphase. That is when its mass and the number of its components increase, and that is when the DNA is duplicated.

4. Proteins with structural and functional roles are attached to eukaryotic DNA molecules. Each DNA molecule with its attached proteins is a chromosome.

5. Members of a species have a characteristic number of chromosomes in cells. Their chromosomes differ from one another in length and shape, and they carry different portions of the hereditary instructions.

6. Body cells of many organisms have a diploid chromosome number; they contain two of each type of chromosome characteristic of the species.

7. Mitosis keeps the chromosome number constant from one cell generation to the next. So if a parent cell is diploid, its daughter cells also will be diploid.

8. Mitotic cell division is the basis of growth and tissue repair in multicelled eukaryotes. It also is the means by which single-celled eukaryotes and many multicelled eukaryotes reproduce asexually.

DIVIDING CELLS: THE BRIDGE BETWEEN GENERATIONS

Overview of Division Mechanisms

Hereditary instructions of plants, animals, and all other eukaryotic organisms are distributed among a number of DNA molecules. Before the cells of such organisms reproduce, they must undergo *nuclear* division. **Mitosis** and **meiosis** are two nuclear division mechanisms. Both sort out and then package a parent cell's DNA into new nuclei for their forthcoming daughter cells. A separate mechanism splits the cytoplasm into daughter cells.

Multicelled organisms grow, replace dead or worn-out cells, and repair tissues by way of mitosis and the cytoplasmic division of body cells. We call the cells that make up the body **somatic cells**. Also, many protistans, fungi, plants, and certain animals reproduce asexually by mitotic cell division (Table 9.1).

By contrast, meiosis occurs only in **germ cells**, a cell lineage set aside for the formation of gametes (such as sperm and eggs) and sexual reproduction. As you will read in the next chapter, meiosis has much in common with mitosis, but the end result is different.

What about the prokaryotic cells—the archaebacteria and eubacteria? They reproduce asexually by an entirely different mechanism called prokaryotic fission. We will consider prokaryotic fission later, in Section 21.2.

Table 9.1 *Cell Division Mechanisms*

Mechanisms	Functions
MITOSIS, CYTOPLASMIC DIVISION	In all multicelled eukaryotes, the basis of (1) increases in body size during growth, (2) replacement of dead or worn-out cells, and (3) repair of damaged tissues. Also, the basis of *asexual* reproduction in single-celled and many multicelled eukaryotes.
MEIOSIS, CYTOPLASMIC DIVISION	In single-celled and multicelled eukaryotes, the basis of gamete formation and sexual reproduction.
PROKARYOTIC FISSION	In bacterial cells only, the basis of asexual reproduction.

Some Key Points About Chromosomes

Remember the introduction to chromosomes in Section 4.6? You read that a eukaryotic DNA molecule, together with proteins attached to it, is a **chromosome**. Before a cell enters nuclear division, it duplicates every one of its chromosomes. Each chromosome and its copy stay attached to each other until late in mitosis, as **sister chromatids**. Figure 9.2 is a simple way to think about unduplicated and duplicated chromosomes.

Except in some laboratories, you'll never see naked eukaryotic DNA. Its chromosomal proteins are always bound tightly to it. This is so even when chromosomes are stretched out in threadlike form in a nondividing cell. You can actually see these proteins at extreme magnification; they look like beads on a string (Figure 9.3). The chromosomal proteins are **histones**. Some types of histones are like a spool for DNA. A bit of the DNA molecule winds twice around each one, as Figure 9.3*d* indicates. Each histone–DNA spool is one unit of organization called a **nucleosome**. Nucleosomes can be loosened up in controlled ways to give enzymes access to specific DNA regions (Section 15.1).

Early in mitosis (and in meiosis), the proteins and DNA interact in ways that make the chromosome coil back on itself repeatedly, into a tightly condensed form (Figure 9.3). What's the point of so much condensation? It probably helps keep the chromosomes from getting tangled up when they are being moved and sorted into parcels for daughter cells.

As each type of duplicated chromosome condenses, a pronounced constriction appears in the same region along its length (Figure 9.3*a*). This constricted region is the **centromere**. Disk-shaped structures also show up here; they are docking sites for microtubules with roles in nuclear division. The centromere's location differs among the different types of chromosomes in a cell.

Mitosis and the Chromosome Number

Each species has a characteristic **chromosome number**, the sum total of chromosomes in cells of a given type. Human somatic cells have 46, those of gorillas have 48, and those of pea plants have 14.

a One unduplicated chromosome

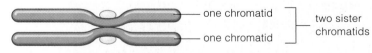

one chromatid

one chromatid

two sister chromatids

b One chromosome (duplicated)

Figure 9.2 A simple way to visualize a chromosome in the unduplicated state and the duplicated state. Chromosomes are duplicated before nuclear division.

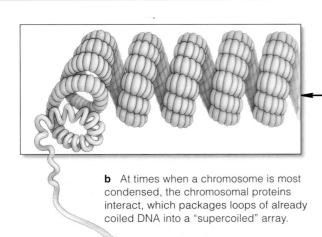

b At times when a chromosome is most condensed, the chromosomal proteins interact, which packages loops of already coiled DNA into a "supercoiled" array.

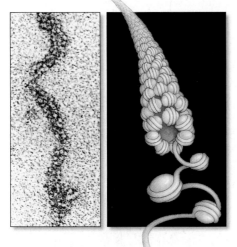

c At a deeper level of structural organization, the chromosomal proteins and DNA are organized as a cylindrical fiber.

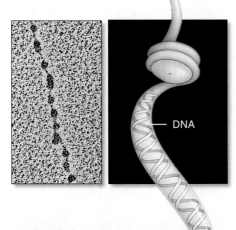

d Immerse a chromosome in saltwater and it loosens up to a beads-on-a-string organization. The "string" is one DNA molecule. Each "bead" is a nucleosome.

DNA

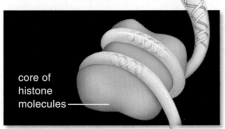

core of histone molecules

e A nucleosome consists of part of a DNA molecule looped twice around a core of histones.

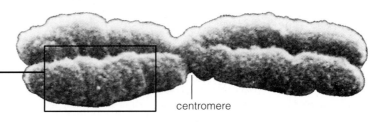

centromere

a A duplicated human chromosome at metaphase, when it is most condensed.

Figure 9.3 One model of levels of organization in a human chromosome at its most condensed form.

Actually, your 46 chromosomes are like volumes of two sets of books. Each set is numbered, from 1 to 23. For example, you have two "volumes" of chromosome 22—that is, *a pair of them*. Except for certain pairings of sex chromosomes, both have the same length and shape. They carry the same portion of hereditary instructions for the same traits. Think of them as two sets of books on how to build a house. Your father gave you one set. Your mother had her own ideas about wiring, storage, plumbing, and so on; she gave you an alternate edition. Her set covers the same topics, but her instructions are slightly different for many of them.

We say the chromosome number is **diploid**, or 2*n*, if a cell has two of each type of chromosome characteristic of the species. The body cells of humans, gorillas, pea plants, and a great many other organisms are like this. By contrast, as Chapter 10 describes, eggs and sperm of such organisms have a *haploid* chromosome number (*n*). This means that they contain only one of each type of chromosome characteristic of the species.

With mitosis, a diploid parent cell can produce two diploid daughter cells. This doesn't mean each merely gets forty-six or forty-eight or fourteen chromosomes. If only the total mattered, one cell might get, say, two pairs of chromosome 22 and no pairs whatsoever of chromosome 9. Neither would be able to function like the parent cell *without two of each type of chromosome*.

A eukaryotic chromosome consists of one DNA molecule and many proteins that structurally organize the DNA.

A chromosome number is the sum total of chromosomes in a body cell or germ cell of a particular organism. Cells with a diploid chromosome number have two of each type of chromosome, usually from two parents.

Mitosis is a nuclear division mechanism that sorts out a required number of chromosomes for each daughter cell. A separate mechanism divides the cytoplasm.

Mitosis keeps the chromosome number constant, division after division, from one cell generation to the next. Thus, if a parent cell is diploid, its daughter cells will be diploid.

THE CELL CYCLE

Let's start thinking about cell reproduction in terms of an orderly sequence of events called a **cell cycle**. It starts every time a new daughter cell forms by mitosis and cytoplasmic division and ends when that cell finishes its own division (Figure 9.4). Thus, *mitosis, cytoplasmic division, and interphase constitute one turn of the cell cycle.*

The Wonder of Interphase

Interphase is the portion of the cell cycle when a cell increases in mass, roughly doubles the number of its cytoplasmic components, and duplicates its DNA. For most types of cells, interphase is the longest portion of the cycle. Biologists divide it into three parts:

G1   Interval ("*Gap*") of cell growth and functioning before the onset of DNA replication

S   Time of "*Synthesis*" (DNA replication)

G2   Second interval (Gap), after DNA replication, when the cell prepares for division

G1, S, and G2 are no more than code names for some amazing events. Consider what cells do with their DNA. If you could coax the DNA molecules from just one of your somatic cells to stretch in a single line, one after

another, that line would extend past the fingertips of your outstretched arms. If you could do the same with salamander DNA, a single line of it would extend ten meters! The wonder is, enzymes and other proteins in cells can selectively access, activate, and silence DNA's instructions. They also make base-by-base copies of the DNA molecules. They do most of this in interphase.

G1, S, and G2 of interphase have distinct patterns of biosynthesis. Most of your cells remain in G1, when they build nearly all of the proteins, carbohydrates, and lipids they use or export. Cells destined to divide enter S, when they copy their DNA as well as the histones and other proteins associated with it. During G2, these cells produce proteins that will drive mitosis to completion.

Once S begins, events normally proceed at about the same rate in all cells of a species and continue through mitosis. Given this observation, you may be wondering whether the cell cycle has built-in molecular brakes. It does. Apply the brakes that are supposed to work in G1, and the cycle stalls in G1. Lift the brakes, and the cell cycle then runs to completion. Said another way, *control mechanisms govern the rate of cell division.*

Imagine a car losing its brakes just as it starts down a steep mountain road. As you will read later on in the book, that is how cancer starts. Controls over division are lost, and the cell cycle cannot stop turning.

The cell cycle lasts about the same length of time for cells of the same type. Its duration differs among cells of different types. Examples: All neurons (nerve cells) in your brain are arrested at G1 of interphase, and usually won't divide again. Each second, 2 million to 3 million precursors of red blood cells form to replace worn-out ones circulating in your body. Early in a sea urchin's development, the number of cells doubles every two hours.

Adverse conditions often disrupt the cell cycle. When deprived of a vital nutrient, for instance, the free-living cells called amoebas do not leave interphase. Even so, if any cell proceeds past a certain point in interphase, the cycle normally continues regardless of outside conditions because of built-in controls over its duration.

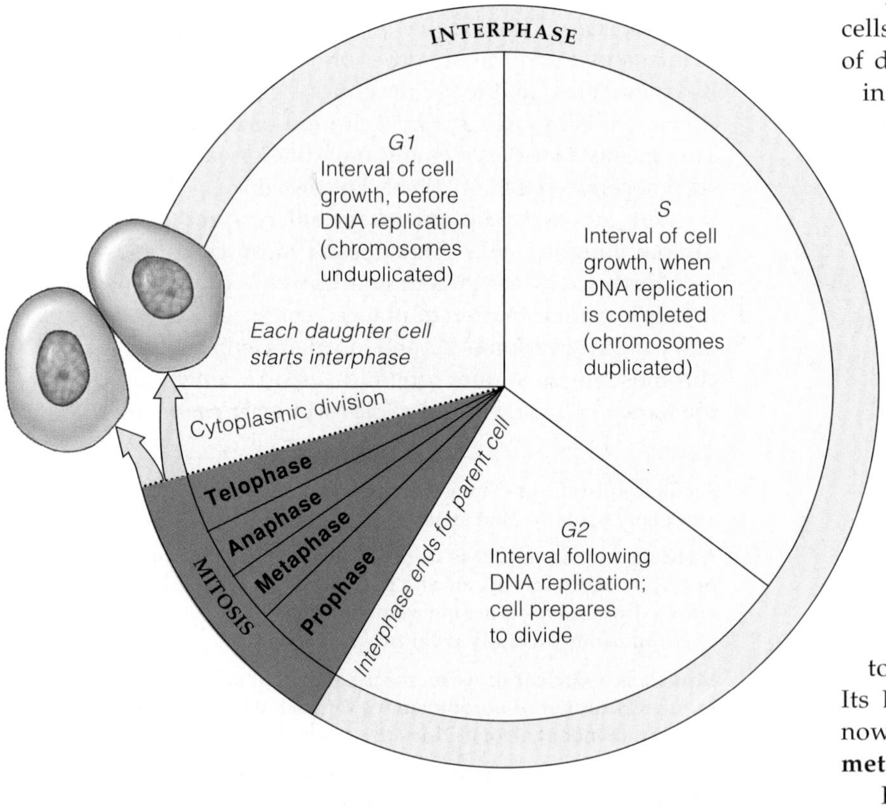

Mitosis Proceeds Through Four Stages

A cell that's making the transition from interphase to mitosis has stopped constructing new cell parts. Its DNA has already been replicated. Major changes now proceed smoothly through four stages: **prophase**, **metaphase**, **anaphase**, and **telophase**.

Figure 9.5 shows these stages in a plant cell. Notice that all the chromosomes are changing positions. They aren't doing so on their own. A **spindle apparatus,** of

Figure 9.4 Eukaryotic cell cycle, generalized. The length of each interval differs among different cell types.

A CELL AT
INTERPHASE:

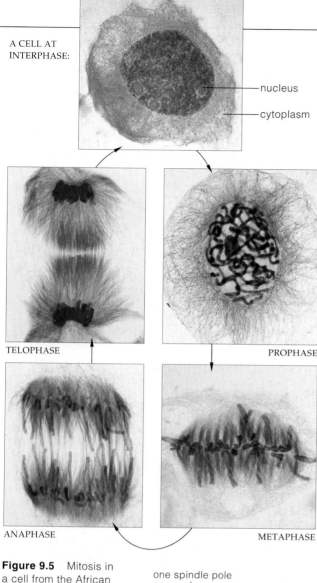

— nucleus

— cytoplasm

TELOPHASE

PROPHASE

ANAPHASE

METAPHASE

Figure 9.5 Mitosis in a cell from the African blood lily, *Haemanthus*. The chromosomes are stained *blue* and the many microtubules, *red*. Before reading further, take a moment to become familiar with the labels on the micrographs.

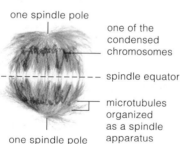

one spindle pole

one of the condensed chromosomes

spindle equator

microtubules organized as a spindle apparatus

one spindle pole

two sets of microtubules, moves them. Each set extends from one of the spindle's two poles (its end points), and each overlaps the other at the spindle equator, midway between the poles. This "bipolar" spindle will establish the final chromosome destinations before mitosis ends and the cell divides in two.

How important are spindle microtubules? One clue comes from plants of the genus *Colchicum*. These plants make colchicine, a poison that evolved against browsing animals. It blocks microtubule assembly and promotes their disassembly. This poison is a favorite of researchers who study cancer and other expressions of cell division.

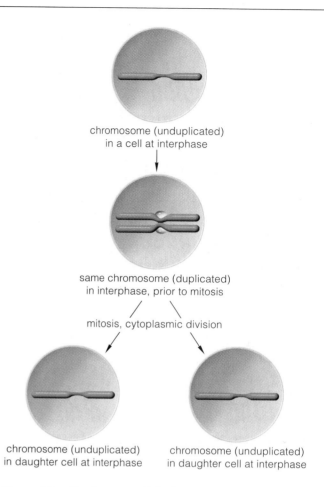

chromosome (unduplicated) in a cell at interphase

same chromosome (duplicated) in interphase, prior to mitosis

mitosis, cytoplasmic division

chromosome (unduplicated) in daughter cell at interphase

chromosome (unduplicated) in daughter cell at interphase

Figure 9.6 Simple way to think about how mitosis maintains the chromosome number, one generation after the next. For clarity, we track only one chromosome.

Another microtubule poison is mentioned in Section 9.6, *Critical Thinking* question 3. Spindles in cells fall apart within seconds or minutes after exposure to it.

Before turning the page to consider the mechanism of mitosis, hold on to these thoughts: Chromosomes are duplicated *before* mitosis. A spindle apparatus moves the sister chromatids of duplicated chromosomes apart *during* mitosis. As Figure 9.6 shows, that is how mitosis maintains the chromosome number through turn after turn of the cell cycle.

Interphase, mitosis, and cytoplasmic division constitute one turn of the cell cycle.

In interphase, a new cell increases its mass, roughly doubles the number of its cytoplasmic components, and duplicates its chromosomes. The cycle ends after the cell undergoes mitosis and then divides. Molecular mechanisms control the rate of cell division.

Mitosis has four consecutive stages: prophase, metaphase, anaphase, and telophase. Its microtubular spindle moves chromosomes that were duplicated earlier, in interphase.

MITOSIS

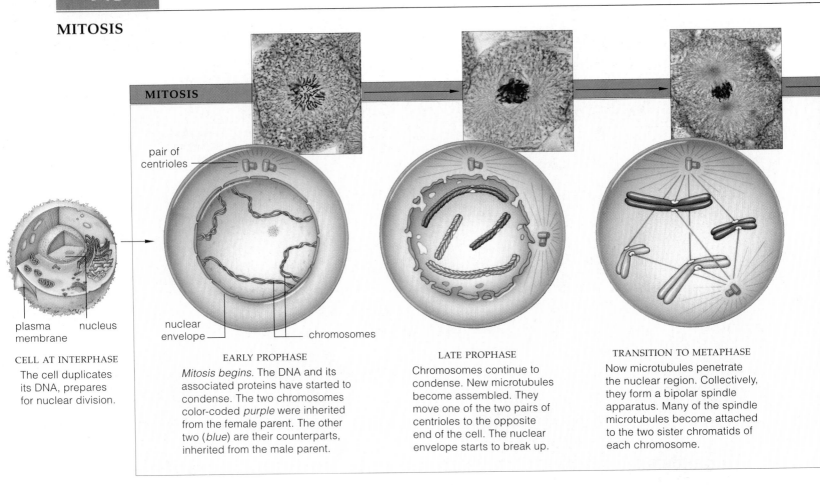

MITOSIS

pair of
centrioles

plasma
membrane   nucleus

CELL AT INTERPHASE
The cell duplicates
its DNA, prepares
for nuclear division.

nuclear
envelope   chromosomes

EARLY PROPHASE
Mitosis begins. The DNA and its
associated proteins have started to
condense. The two chromosomes
color-coded *purple* were inherited
from the female parent. The other
two (*blue*) are their counterparts,
inherited from the male parent.

LATE PROPHASE
Chromosomes continue to
condense. New microtubules
become assembled. They
move one of the two pairs of
centrioles to the opposite
end of the cell. The nuclear
envelope starts to break up.

TRANSITION TO METAPHASE
Now microtubules penetrate
the nuclear region. Collectively,
they form a bipolar spindle
apparatus. Many of the spindle
microtubules become attached
to the two sister chromatids of
each chromosome.

Prophase: Mitosis Begins

We know a cell is in prophase, the first stage of mitosis,
when its chromosomes are visible in light microscopes
as threadlike forms. ("Mitosis" is from the Greek *mitos*,
for thread.) Each chromosome was duplicated earlier, in
interphase; each is two sister chromatids joined at the
centromere. In early prophase, all of these chromatids
twist and fold. By late prophase, they will be condensed
into thicker, compact, rod-shaped forms.

Meanwhile, in the cytoplasm, most microtubules of
the cytoskeleton are breaking down to tubulin subunits
(Section 4.9). The subunits reassemble near the nucleus
as microtubules of the spindle. While new microtubules
are assembling, the nuclear envelope physically prevents
them from interacting with the chromosomes inside the
nucleus. However, the nuclear envelope starts to break
up as prophase draws to a close (Figure 9.7).

Many cells have two barrel-shaped **centrioles**. Each
centriole started duplicating itself during interphase,
so there are two pairs of them when prophase is under
way. Microtubules start moving one pair to the opposite
pole of the newly forming spindle. Centrioles, recall,
give rise to flagella or cilia. If you observe them in cells
of an organism, you can bet that flagellated cells (such
as sperm) or ciliated cells develop during its life cycle.

Figure 9.7 Mitosis in a generalized animal cell. By this nuclear
division mechanism, each daughter cell ends up with the same
chromosome number as the parent cell. For clarity, we track only
two pairs of chromosomes from a diploid (2*n*) cell. The picture
is almost always more complicated, as you may sense from the
above micrographs of mitosis in a whitefish cell.

Transition to Metaphase

So much happens between prophase and metaphase
that we often call this transitional time "prometaphase."
The nuclear envelope breaks up completely into many
small, flattened vesicles, so the chromosomes are now
free to interact with microtubules. Some microtubules
that have been lengthening from the two spindle poles
harness each chromosome and pull on it. The two-way
pulling orients it with respect to the poles. However,
other microtubules extend from both spindle poles as
two sets that overlap at the spindle's midpoint. They
push the poles apart by interacting and ratcheting past
each other. The push–pull forces become balanced when
the chromosomes reach the spindle's midpoint.

When all of the duplicated chromosomes are aligned
midway between the poles of a completed spindle, we
call this metaphase (*meta*– means "midway between").
The alignment is crucial for the next stage of mitosis.

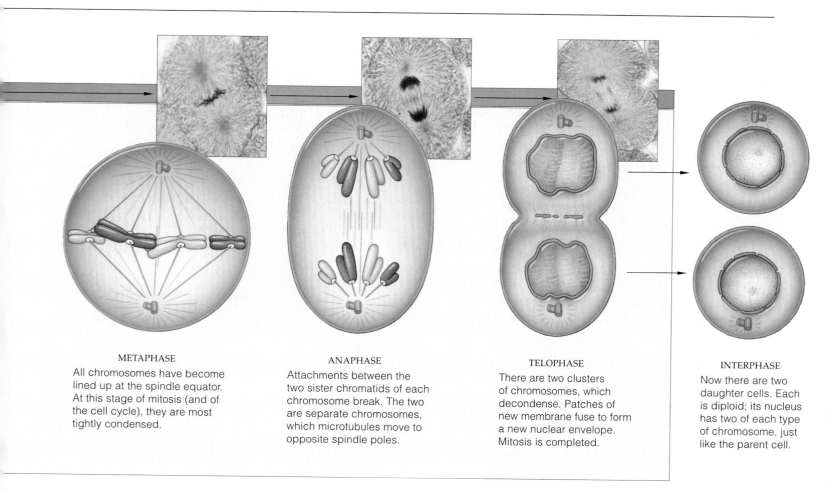

METAPHASE
All chromosomes have become lined up at the spindle equator. At this stage of mitosis (and of the cell cycle), they are most tightly condensed.

ANAPHASE
Attachments between the two sister chromatids of each chromosome break. The two are separate chromosomes, which microtubules move to opposite spindle poles.

TELOPHASE
There are two clusters of chromosomes, which decondense. Patches of new membrane fuse to form a new nuclear envelope. Mitosis is completed.

INTERPHASE
Now there are two daughter cells. Each is diploid; its nucleus has two of each type of chromosome, just like the parent cell.

From Anaphase Through Telophase

At anaphase, the sister chromatids of each chromosome separate from each other and move to opposite spindle poles. Two mechanisms bring this about.

First, microtubules that latched on to the centromere region of each chromosome shorten (they disassemble) and *pull* the chromosome to a spindle pole. Think of the centromere as a train chugging along a railroad track, except the track falls apart after the train passes over it. Activated motor proteins are the engine. They project in orderly arrays from the microtubules and drive the chromosome onward by sliding motions (Section 4.10).

Second, the spindle itself elongates as overlapping microtubules continue to ratchet past one another and *push* the spindle's two poles even farther apart. These microtubules, too, incorporate motor proteins, and they actively slide past one another where they overlap.

Once each chromatid is separated from its sister, we recognize it as a separate chromosome.

Telophase gets under way as soon as each of the two clusters of chromosomes arrives at a spindle pole. The chromosomes, no longer harnessed to the microtubules, return to threadlike form. Vesicles derived from the old nuclear envelope fuse and form patches of membrane around the chromosomes. Patch joins with patch, and soon a new nuclear envelope separates both clusters of chromosomes from the cytoplasm. If the parent cell was diploid, each cluster contains two chromosomes of each type. With mitosis, remember, each new nucleus has the same chromosome number as the parent nucleus. Once two nuclei form, telophase is over—and so is mitosis.

Prior to mitosis, each chromosome in a cell's nucleus is duplicated, so that it consists of two sister chromatids.

In prophase, microtubules assemble outside the nucleus and start to form a bipolar spindle. The nuclear envelope breaks down, so the microtubules are now free to harness and move the duplicated chromosomes.

At metaphase, all chromosomes are aligned and oriented, with respect to the poles, at the spindle equator.

At anaphase, microtubules move sister chromatids of each chromosome apart, to opposite spindle poles. Some pull on them; others push the spindle poles apart, increasing the distance between the former sister chromatids.

At telophase, there are two clusters of chromosomes, and a new nuclear envelope forms around each cluster.

Thus mitosis forms two daughter nuclei. Each has the same chromosome number as the parent cell's nucleus.

9.4

DIVISION OF THE CYTOPLASM

The cytoplasm usually divides at some time between late anaphase and the end of telophase. As you might well conclude by comparing Figure 9.8 with Figure 9.9, the actual mechanism of **cytoplasmic division** (or, as it is often called, cytokinesis) differs among organisms.

Cell Plate Formation in Plants

As described in Section 4.11, most plant cells are walled, which means their cytoplasm can't be pinched in two. Cytoplasmic division of such cells involves **cell plate formation**, as shown in Figure 9.8. By this mechanism, vesicles packed with wall-building materials fuse with one another and with remnants from the microtubular spindle. Together, they form a disklike structure—a cell plate. At this location, deposits of cellulose accumulate. In time, the cellulose deposits are thick enough to form

light micrograph of a cell plate forming in a dividing plant cell

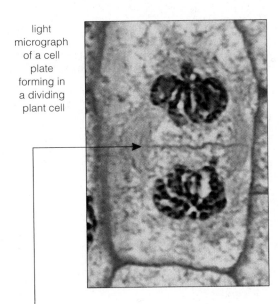

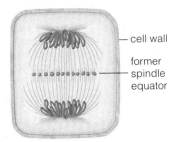 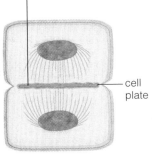 

a As mitosis ends, vesicles converge at the spindle equator. They contain cementing materials and structural materials for a new primary cell wall.

b A cell plate starts forming as membranes of the vesicles fuse. Materials inside the vesicles get sandwiched between two new membranes that elongate along the plane of the cell plate.

c Cellulose is deposited on the inside of the "sandwich." (In time, deposits will form two cell walls. Other deposits will form a middle lamella and cement the walls together; refer to Section 4.11.)

d A cell plate grows at its margins until it fuses with the parent cell's plasma membrane. During growth, new plant cells expand. Their primary wall is still thin, and new material is deposited on it.

Figure 9.8 Cytoplasmic division of a plant cell, as brought about by cell plate formation.

 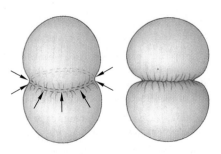   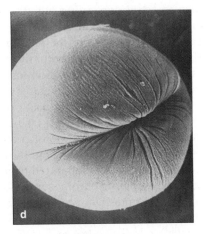

a Mitosis is over, and the spindle is now disassembling.

b At the former spindle equator, a ring of microfilaments attached to the plasma membrane contracts. As its diameter shrinks, it pulls the cell surface inward.

c Contractions continue until the ring cuts the cell in two.

Figure 9.9 (**a–c**) Cytoplasmic division of an animal cell. (**d**) Scanning electron micrograph of the cleavage furrow at the plane of the former spindle's equator. Beneath it, a band of microfilaments attached to the plasma membrane contracts and pulls the surface inward. The furrow deepens until the cell is cut in two.

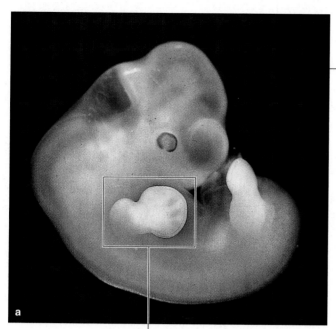

future arm and hand of embryo, five weeks old

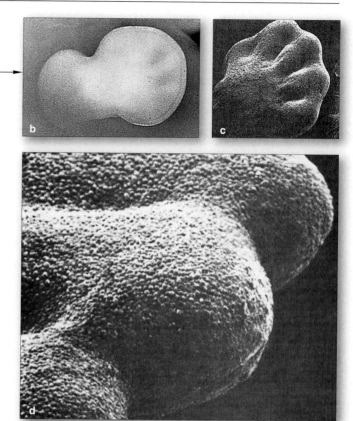

a crosswall. That new crosswall bridges the cytoplasm and divides the parent cell into two daughter cells.

Cytoplasmic Division of Animal Cells

Unlike plant cells, an animal cell isn't confined within a cell wall, and its cytoplasm typically pinches in two. Look at Figure 9.9, a surface view of a newly fertilized animal egg. An indentation is forming about midway between the two poles of the egg. The indentation in its plasma membrane is a **cleavage furrow**. It's the first visible sign that the cytoplasm inside an animal cell is dividing. The furrow will extend all around the cell and continue to deepen along the plane of the former spindle's midpoint until the cell is cut in two.

A band of microfilaments beneath the cell's plasma membrane generates force for a cut. These cytoskeletal elements are organized to interact and slide past one another (Section 4.10). As they do, they pull the plasma membrane inward until there are two daughter cells. Each has a nucleus, cytoplasm, and plasma membrane.

Perspective on Mitotic Cell Division

This concludes our introduction to mitotic cell division. Look now at your hands and try to visualize the cells making up your palms, thumbs, and fingers. Imagine the divisions that produced all the generations of cells that preceded them as you were developing early on, inside your mother (Figure 9.10). And be grateful for the astonishing precision of the mechanisms that led to their formation at certain times, in certain numbers, for the alternatives can be terrible indeed. Why? Good

Figure 9.10 (**a–e**) Transformation from a paddlelike structure into the fingers of a human hand by mitosis, cytoplasmic divisions, and other processes necessary for embryonic development. In (**d**), you can see many individual cells at this high magnification.

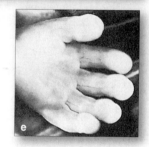

health—and survival itself—depends absolutely on the proper timing and the completion of cell cycle events, including mitosis. Certain genetic disorders arise from mistakes in the duplication or distribution of even one chromosome. Also, when normal controls that prevent cells from dividing are lost, unchecked cell divisions may destroy surrounding tissues and, ultimately, the organism. Section 9.5 describes a landmark case of such losses, which we will explore further in Section 15.6.

After mitosis, a separate mechanism cuts the cytoplasm into two daughter cells, each with a daughter nucleus.

Cytoplasmic division in plants often involves the formation of a cell plate and a crosswall between the adjoining, new plasma membranes of daughter cells.

Cleavage is a form of cytoplasmic division in animals. Rings of microfilaments around a parent cell's midsection slide past one another in a way that pinches the cytoplasm in two.

Henrietta's Immortal Cells

Each human starts out as a single fertilized egg. By the time of birth, mitotic cell divisions and other processes have resulted in a human body of about a trillion cells. Even in adults, billions of cells still divide. For example, cells of the stomach's lining divide every day. Liver cells usually do not divide, but if part of the liver becomes injured or diseased, repeated cell divisions will yield more new cells until the damaged part is finally replaced.

In 1951, George and Margaret Gey of Johns Hopkins University were trying to develop a way to keep human cells dividing *outside* the body. With such isolated cells, these researchers and others could investigate basic life processes. They also could conduct studies of cancer and other diseases without having to experiment directly on patients and risk human lives. The Geys used normal and diseased human cells, which local physicians had sent them. But they couldn't stop the descendants of those precious cells from dying out within a few weeks.

Mary Kubicek, a laboratory assistant, worked with the Geys in their efforts to start a self-perpetuating lineage of cultured human cells. She was about to give up after dozens of failed attempts. Even so, in 1951 she decided to prepare one more sample of cancer cells for culture. She gave the sample the code name **HeLa cells**, for the first two letters of the patient's first and last names.

The HeLa cells began to divide. In four days there were so many cells that Kubicek subdivided them into more culture tubes. Sadly, cancer cells in the patient were dividing as rapidly. Six months after she was diagnosed with cancer, tumor cells had spread to tissues throughout her body. Eight months after the diagnosis, Henrietta Lacks, a young woman from Baltimore, was dead.

Although Henrietta passed away, some of her cells lived on in the Geys' laboratory as the first successful human cell culture. In time, HeLa cells were shipped to research laboratories all over the world. Some even traveled into space for experiments on the *Discoverer XVII* satellite. Each year hundreds of research projects depend on them. Henrietta was only thirty-one when runaway cell divisions killed her. Now, decades later, her legacy is helping humans everywhere, through cellular descendants that are still dividing day after day.

Figure 9.11 Dividing HeLa cells, a legacy of Henrietta Lacks, who was a casualty of cancer. Her cellular contribution to science is still helping others every day.

SUMMARY
Gold indicates text section

1. Through specific division mechanisms, a parent cell provides each of its daughter cells with the hereditary instructions (DNA) and cytoplasmic machinery required to start up its own operation. *CI, 9.1*

   a. In eukaryotic cells, the nucleus divides by mitosis or meiosis. Cytoplasmic division typically follows.

   b. Prokaryotic cells divide by prokaryotic fission.

2. Each eukaryotic chromosome is one DNA molecule with numerous proteins attached. The chromosomes in a cell differ in length, shape, and which portion of the hereditary instructions they carry. *9.1*

   a. We define chromosome number as the sum total of chromosomes in cells of a given type. Cells having a diploid chromosome number ($2n$) contain two of each kind of chromosome.

   b. Mitosis divides the nucleus into two equivalent nuclei, each with the same chromosome number as the parent cell. It maintains the chromosome number from one cell generation to the next.

   c. Mitosis is the basis of growth, tissue repair, and cell replacements among multicelled eukaryotes. It is the basis of asexual reproduction in many single-celled eukaryotes. (Meiosis occurs only in germ cells.)

3. A duplicated chromosome has two DNA molecules attached at the centromere. For as long as the two stay connected to each other, they are sister chromatids. *9.1*

4. A cell cycle starts when a new cell forms. It proceeds through interphase and ends when the cell reproduces by mitosis and cytoplasmic division. In interphase, a cell carries out its functions. When it is to divide again, the cell increases in mass, roughly doubles the number of its cytoplasmic components, then duplicates each of its chromosomes in preparation for division. *9.2*

5. Mitosis has four continuous stages: *9.3*

   a. Prophase. The duplicated, threadlike chromosomes start to condense. A spindle starts to form. The nuclear envelope starts to break up; its remnants form vesicles during the *transition* to metaphase (or prometaphase). Some microtubules from both poles of the developing spindle push the poles apart. Others directly attach to one of two sister chromatids of each chromosome.

   b. Metaphase. *At* metaphase, all chromosomes have become aligned at the spindle equator.

   c. Anaphase. Microtubules pull sister chromatids of each chromosome away from each other, to opposite spindle poles. Now each type of parental chromosome is represented by a daughter chromosome at both poles.

   d. Telophase. Chromosomes decondense to threadlike form. A new nuclear envelope forms around them. Each nucleus has the parental chromosome number.

6. Separate mechanisms divide the cytoplasm near the end of nuclear division or afterward (in plants, by cell plate formation; in animals, by cleavage). *9.4*

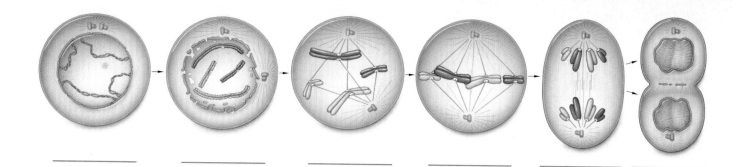

_____   _____   _____   _____   _____   _____

Review Questions

1. Define mitosis and meiosis, two mechanisms that operate in eukaryotic cells. Does either one divide the cytoplasm? *9.1*

2. Define somatic cell and germ cell. *9.1*

3. What are chromosomal proteins? What are histones, and how do they interact with a DNA molecule? *9.1*

4. What is a chromosome called when it is in the unduplicated state? In the duplicated state (with two sister chromatids)? *9.1*

5. Describe the microtubular spindle and its functions. What role do motor proteins play in its operation? *9.2, 9.3*

6. Using the diagram above as a guide, name and describe the key features of the stages of mitosis. *9.3*

7. Briefly explain how cytoplasmic division differs between a typical plant cell and a typical animal cell. *9.4*

Self-Quiz ANSWERS IN APPENDIX III

1. Mitosis and cytoplasmic division function in _____ .
   a. asexual reproduction of single-celled eukaryotes
   b. growth, tissue repair, and sometimes asexual reproduction in many multicelled eukaryotes
   c. gamete formation in prokaryotes
   d. both a and b

2. A duplicated chromosome has _____ chromatid(s).
   a. one   b. two   c. three   d. four

3. In a chromosome, a _____ is a constricted region with attachment sites for microtubules.
   a. chromatid   b. cell plate   c. centromere   d. cleavage

4. A somatic cell having two of each type of chromosome has a(n) _____ chromosome number.
   a. diploid   b. haploid   c. tetraploid   d. abnormal

5. Interphase is the part of the cell cycle when _____ .
   a. a cell ceases to function
   b. a germ cell forms its spindle apparatus
   c. a cell grows and duplicates its DNA
   d. mitosis proceeds

6. After mitosis, the chromosome number of a daughter cell is _____ the parent cell's.
   a. the same as   c. rearranged compared to
   b. one-half   d. doubled compared to

7. Only _____ is not a stage of mitosis.
   a. prophase   b. interphase   c. metaphase   d. anaphase

8. Match each stage with the events listed.
   _____ metaphase   a. sister chromatids move apart
   _____ prophase   b. chromosomes start to condense
   _____ telophase   c. chromosomes decondense and daughter nuclei form
   _____ anaphase   d. all duplicated chromosomes are aligned at the spindle equator

Critical Thinking

1. Suppose you have a way to measure the amount of DNA in a single cell during the cell cycle. You first measure the amount at the G1 phase. At what points during the rest of the cell cycle would you predict changes in the amount of DNA per cell?

2. The cervix is part of the uterus, a chamber in which embryos develop. The *Pap smear* is a screening procedure that can detect *cervical cancer* in its earliest stages. Treatments range from freezing precancerous cells or killing them with a laser beam to removal of the uterus (a hysterectomy). The treatments are 90+ percent effective when this cancer is detected early. Survival chances plummet to less than 9 percent after the cancer spreads.

   Most cervical cancers develop slowly. Unsafe sex increases the risk. A key risk factor is infection by human papillomaviruses that cause genital warts (Section 44.15). In 93 percent of all cases, viral genes coding for the tumor-inducing proteins had become inserted into the DNA of previously normal cervical cells.

   Not all women request Pap smears. Many wrongly believe the procedure is costly. Many don't recognize the importance of abstinence or "safe" sex. Others simply don't want to think about whether they have cancer. Knowing what you've learned so far about the cell cycle and cancer, what would you say to a woman who falls into one or more of these groups?

3. Pacific yews (*Taxus brevifolius*) face extinction. People started stripping its bark and killing trees when they heard that *taxol*, a chemical extract from the bark, may fight breast and ovarian cancer. (Synthesizing taxol in laboratories may save the species.) Taxol is a poison that prevents microtubule disassembly. What does this tell you about its potential as an anticancer drug?

4. X rays and gamma rays emitted from some radioisotopes chemically damage DNA, especially in cells engaged in DNA replication. High-level exposure can result in *radiation poisoning*. Hair loss and damage to the lining of the stomach and intestines are two early symptoms. Speculate why. Also speculate on why highly focused radiation therapy is used against some cancers.

Selected Key Terms

anaphase *9.2*	cytoplasmic	mitosis *9.1*
cell cycle *9.2*	division *9.4*	nucleosome *9.1*
cell plate	diploid (chromosome	prophase *9.2*
formation *9.4*	number) *9.1*	reproduction *CI*
centriole *9.3*	germ cell *9.1*	sister
centromere *9.1*	HeLa cell *9.5*	chromatid *9.1*
chromosome *9.1*	histone *9.1*	somatic cell *9.1*
chromosome	interphase *9.2*	spindle
number *9.1*	meiosis *9.1*	apparatus *9.2*
cleavage furrow *9.4*	metaphase *9.2*	telophase *9.2*

Readings

Murray, A., and M. Kirschner. March 1991. "What Controls the Cell Cycle?" *Scientific American* 264(3): 56–63.

On-Line readings at Student Guide for InfoTrac:
www.brookscole.com/biology

Octopus Sex and Other Stories

The couple clearly are interested in each other. First he caresses her with one tentacle, then with another—and another and another. She reciprocates with a hug here, a squeeze there. This goes on for hours. Finally the male reaches under his mantle, a fold of tissue that drapes around most of his body. He removes a packet of sperm from a reproductive organ and inserts it into an egg chamber beneath the female's mantle. For every sperm that fertilizes an egg, a new octopus may develop.

Unlike the one-to-one coupling between a male and a female octopus, sex for the slipper limpet is a group enterprise. Slipper limpets are marine animals, relatives of the familiar land snails. Before becoming transformed into a sexually mature adult, a slipper limpet must go through a free-living stage of development called a larva. When a limpet larva is about to undergo its programmed transformation, it settles on a rock or pebble or shell. If it settles down all by itself, it will become a female. If another larva settles on the first limpet and proceeds to develop, that second limpet will function right off as a male. However, the second one will switch gears and develop into a female if a third limpet develops into a male on top of *it*. Later, that third limpet will also become a female if still another limpet develops into a male on top of it—and so on amongst ten or more limpets!

Slipper limpets typically live in such piles, with the bottom one always being the oldest female and the uppermost one being the youngest male (Figure 10.1a). Until they switch their sex, male limpets release sperm, which

Figure 10.1 Examples of variations in reproductive modes of eukaryotic organisms. (**a**) Slipper limpets, busily perpetuating the species by group participation in sexual reproduction. The tiny crab in the foreground is merely a passerby. (**b**) Live birth of an aphid, a type of insect that reproduces sexually in autumn but can switch to an asexual mode in summer.

fertilize a female's eggs. Fertilized eggs develop into immature larvae, which further develop into females or males that can later become females—and so it goes, from one sexually flexible generation to the next.

Limpets are not alone in having unusual variations in their mode of reproduction. For example, sexual reproduction is common in many life cycles, but so are asexual episodes based on mitotic cell divisions. Orchids, dandelions, and many other plants reproduce very well with or without sex. Aquatic animals called flatworms can engage in sex or split their small body into two roughly equivalent parts, each of which grows and develops into a new flatworm.

And what about those aphids! In summer, nearly all aphids are females, which produce more females from *unfertilized* egg cells (Figure 10.1b). Only when autumn approaches do male aphids develop and do their part in the sexual phase of the life cycle. Even then, females that manage to survive over the winter can do without the opposite sex. Come summer, they begin another round of producing offspring all by themselves.

These examples only hint at the immense variation in reproductive modes among eukaryotic organisms. And yet, despite the variation, *sexual* reproduction dominates nearly all of the life cycles. And it always involves the same events. Briefly, cells set aside for sexual reproduction duplicate their chromosomes before they divide. **Germ cells**, immature reproductive cells that develop in male and female animals, are a fine example. They undergo meiosis and cytoplasmic division. Cellular descendants of germ cells mature and become **gametes**, or sex cells. When a male and female gamete manage to get together, they form the first cell of a new individual by way of fertilization.

With this chapter, we turn to the kinds of cells that serve as the bridge between generations of organisms. Specialized phases of reproduction and development, including asexual episodes, loop out from the basic life cycle of many eukaryotic species. Regardless of the specialized details, all of the life cycles turn on three events: *meiosis, the formation of gametes, and fertilization.* These three interconnected events are the hallmarks of sexual reproduction. As you will see in many chapters throughout the book, they have contributed to the diversity of life.

Key Concepts

1. Sexual reproduction proceeds through three key events: meiosis, gamete formation, and fertilization. Sperm and eggs are familiar gametes.

2. Meiosis, a nuclear division mechanism, occurs only in cells set aside for sexual reproduction. The immature germ cells of male and female animals are examples. Meiosis sorts out the chromosomes of a germ cell into four new nuclei. After meiosis is completed, gametes form by way of cytoplasmic division and other events.

3. Cells with a diploid chromosome number contain two of each type of chromosome characteristic of the species. The two function as a pair during meiosis. Commonly, one chromosome of the pair is maternal, with hereditary instructions from a female parent. The other is paternal, with the same categories of hereditary instructions from a male parent.

4. Meiosis divides the chromosome number by half for each forthcoming gamete. Thus, if both parents have a diploid chromosome number ($2n$), the gametes that form will be haploid (n). Later, a union of two gametes at fertilization will restore the diploid number in the new individual ($n + n = 2n$).

5. Each pair of chromosomes swaps segments during meiosis and exchanges hereditary information about certain traits. Also, meiosis randomly assigns one of each pair of chromosomes to a forthcoming gamete; *which* one of the pair ends up in a given gamete is a matter of chance. Hereditary instructions are further shuffled at fertilization. All three reproductive events lead to variations in traits among offspring.

6. In most plants, spore formation and other events intervene between meiosis and gamete formation.

COMPARING SEXUAL WITH ASEXUAL REPRODUCTION

When an orchid, flatworm, or aphid reproduces all by itself, what sort of offspring does it get? By the process of **asexual reproduction**, one parent alone produces offspring, and each offspring inherits the same number and kinds of genes as its parent. **Genes** are particular stretches of chromosomes—that is, of DNA molecules. Taken together, the genes for each species contain all the heritable bits of information necessary to make new individuals. Rare mutations aside, this means asexually produced individuals can only be *clones*, or genetically identical copies of the parent.

Inheritance gets much more interesting with **sexual reproduction**. This process involves meiosis, formation of gametes, and fertilization (union of the nuclei of two gametes). In most sexual reproducers, such as humans, the first cell of a new individual contains *pairs of genes* on pairs of homologous chromosomes. Typically, one of each pair is maternal and the other paternal in origin.

If instructions encoded in every pair of genes were identical down to the last detail, sexual reproduction would produce clones, also. Just imagine—you, every single person you know, the entire human population might be a clone, with everybody looking alike. But the two genes of a pair might *not* be identical. Why not? A gene's molecular structure can change; that is what we mean by mutation. Depending on their structure, two genes that happen to be paired in a person's cells may "say" slightly different things about a trait. Each unique molecular form of the same gene is called an **allele**.

Such tiny differences affect thousands of traits. For example, whether your chin has a dimple depends on which pair of alleles you inherited at one chromosome location. One kind of allele at that location says "put a dimple in the chin." Another kind says "no dimple." This leads us to a key reason why members of sexually reproducing species don't all look alike. *Through sexual reproduction, offspring inherit new combinations of alleles, which lead to variations in the details of their traits.*

This chapter gets into the cellular basis of sexual reproduction. More importantly, it starts you thinking about far-reaching effects of gene shufflings at different stages of the process. The process introduces variations in traits among offspring that are typically acted upon by agents of natural selection. Thus, *variation in traits is a foundation for evolutionary change.*

Asexual reproduction produces genetically identical copies of the parent. Sexual reproduction introduces variations in the details of traits among offspring.

Sexual reproduction dominates the life cycles of eukaryotic species. Meiosis, formation of gametes, and fertilization are the basic events of this process.

HOW MEIOSIS HALVES THE CHROMOSOME NUMBER

Think "Homologues"

Think back to the preceding chapter and its focus on mitotic cell division. Unlike mitosis, **meiosis** divides chromosomes into separate parcels not once but *twice* prior to cell division. Unlike mitosis, it is the first step leading to the formation of gametes. Gametes, recall, are sex cells such as sperm or eggs. In most multicelled eukaryotic organisms, gametes develop from cells that arise in specialized reproductive structures or organs. Figure 10.2 shows examples of where gametes form.

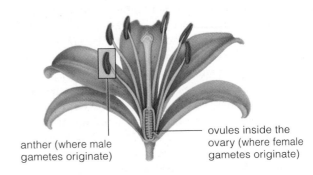

anther (where male gametes originate)

ovules inside the ovary (where female gametes originate)

a Flowering plant

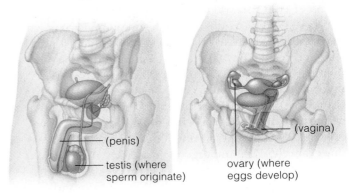

(penis)

testis (where sperm originate)

(vagina)

ovary (where eggs develop)

b Human male   **c** Human female

Figure 10.2   Examples of gamete-producing structures.

As you know, the **chromosome number** is the sum total of chromosomes in cells of a given type (Section 9.1). Germ cells start out with the same chromosome number as somatic cells (the rest of the body's cells). If a cell has a **diploid number** ($2n$), it has a *pair* of each type of chromosome, often from two parents. Except for a pairing of nonidentical sex chromosomes, each pair has the same length, shape, and assortment of genes. And they line up with each other at meiosis. We call them **homologous chromosomes** (*hom–* means alike).

As you can probably deduce from Figure 10.3, your own germ cells have 23 + 23 homologous chromosomes.

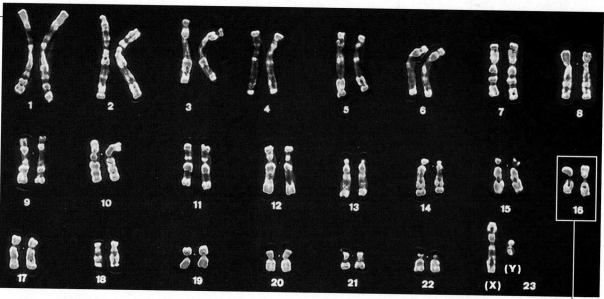

Figure 10.3 From a diploid cell of a human male, twenty-three pairs of homologous chromosomes. The last two are a pair of sex chromosomes (XY or XX). An XY pair differs in length, shape, and which genes they carry. But the two still pair up as homologues during meiosis.

one pair of homologous chromosomes

After meiosis, 23 chromosomes—one of each type—end up in gametes. That is, meiosis halves the chromosome number, so gametes have a **haploid number** (*n*).

Two Divisions, Not One

Meiosis is like mitosis in some ways, but the outcome is different. As in mitosis, a germ cell duplicates its DNA in interphase. The two DNA molecules and their associated proteins remain attached at the centromere, the notably constricted region along their length. For as long as they remain attached, we call them **sister chromatids**:

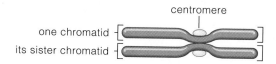

one chromosome in the duplicated state

As in mitosis, the microtubules of a spindle apparatus move the chromosomes in prescribed directions.

With meiosis alone, however, *chromosomes go through two consecutive divisions that end with the formation of four haploid nuclei*. There is no interphase between the nuclear divisions, which we call meiosis I and meiosis II:

interphase (*DNA replication before meiosis I*)	MEIOSIS I	no interphase (*no DNA replication before meiosis II*)	MEIOSIS II
	PROPHASE I		PROPHASE II
	METAPHASE I		METAPHASE II
	ANAPHASE I		ANAPHASE II
	TELOPHASE I		TELOPHASE II

Each duplicated chromosome becomes aligned with its partner—*homologue to homologue*—during meiosis I.

After the two chromosomes of each pair have lined up with each other, they are moved apart:

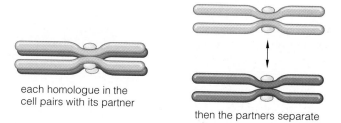

each homologue in the cell pairs with its partner

then the partners separate

The cytoplasm typically starts to divide at some point after each homologue is separated from its partner. The two daughter cells formed this way are haploid; each has *one* of each type of chromosome. But don't forget, the chromosomes are still in the duplicated state.

Next, during meiosis II, *the two sister chromatids of each chromosome are separated from each other*:

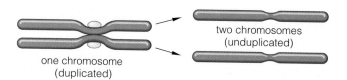

one chromosome (duplicated)

two chromosomes (unduplicated)

Each sister chromatid is now a chromosome in its own right. Four nuclei now form. Often the cytoplasm divides once more. The final outcome is four haploid cells.

Figure 10.4, on the next two pages, gives you a closer look at key events of meiosis and their consequences.

Meiosis is a type of nuclear division mechanism. It reduces the chromosome number of a parental cell by half—to the haploid number (*n*)—in daughter cells.

Meiosis, the first step leading to gamete formation, proceeds only in cells that are set aside for sexual reproduction.

VISUAL TOUR OF THE STAGES OF MEIOSIS

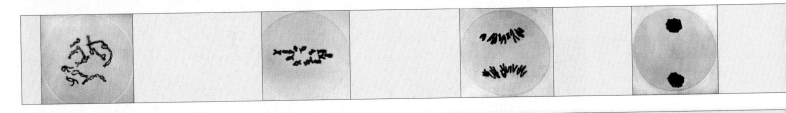

MEIOSIS I

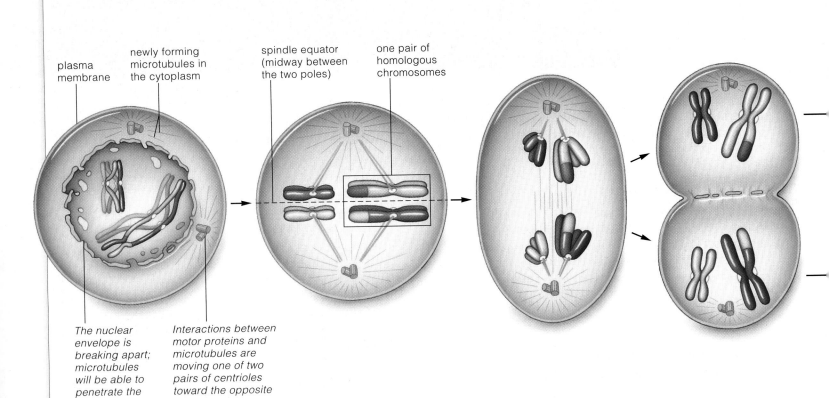

plasma membrane

newly forming microtubules in the cytoplasm

spindle equator (midway between the two poles)

one pair of homologous chromosomes

The nuclear envelope is breaking apart; microtubules will be able to penetrate the nuclear region.

Interactions between motor proteins and microtubules are moving one of two pairs of centrioles toward the opposite spindle pole.

PROPHASE I

Each duplicated chromosome is in threadlike form, but now starts to condense. It pairs with its homologue, and the two typically swap segments. The swapping, called crossing over, is indicated by the break in color on the pair of larger chromosomes. Some of the microtubules of a newly forming spindle become attached to each chromosome's centromere.

METAPHASE I

As in mitosis, motor proteins attached to microtubules move the chromosomes and move the spindle poles apart. They tug the chromosomes into position midway between the spindle poles. Thus the spindle becomes fully formed, owing to dynamic interactions among the motor proteins, microtubules, and the chromosomes themselves.

ANAPHASE I

Microtubules extending from the poles and overlapping at the spindle equator *lengthen* and push the poles apart. Other microtubules extending from the poles to the chromosomes *shorten*, thereby pulling each chromosome away from its homologous partner. These motions move the homologous partners to opposite poles.

TELOPHASE I

The cytoplasm of the germ cell divides at some point. There are now two haploid (*n*) cells. Each cell has one of each type of chromosome that was present in the parent (2*n*) cell. *However, all chromosomes are still in the duplicated state.*

Figure 10.4 Sketches of meiosis in a generalized animal cell. This nuclear division mechanism reduces the chromosome number in immature reproductive cells by half (to the haploid number) for forthcoming gametes. To keep things simple, we track only two pairs of homologous chromosomes. Maternal chromosomes are shaded *purple* and paternal chromosomes *blue*. The light micrographs above show corresponding stages in the formation of pollen grains in a lily (*Lilium regale*).

MEIOSIS II

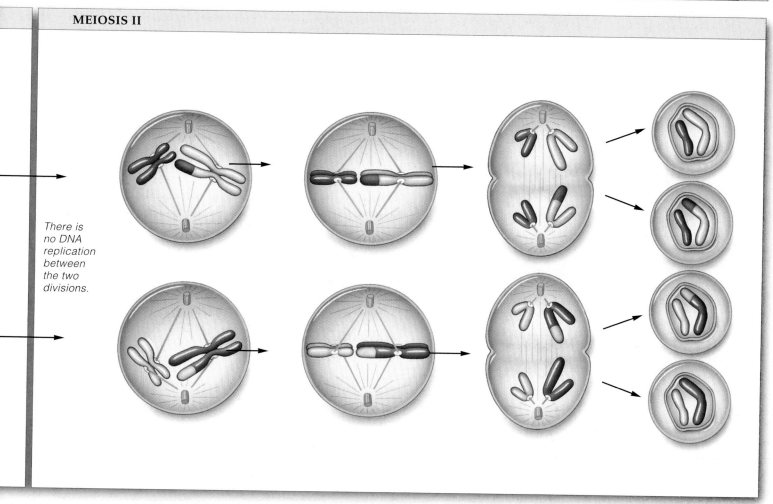

There is no DNA replication between the two divisions.

PROPHASE II
Microtubules have already moved one member of the centriole pair to the opposite pole of the spindle in each of the two daughter cells. Now, during prophase II, microtubules attach to the chromosomes, and motor proteins drive the movement of chromosomes toward the spindle's equator.

METAPHASE II
In each daughter cell, interactions among motor proteins, spindle microtubules, and each duplicated chromosome have moved all of the chromosomes so that they are positioned at the spindle equator, midway between the two poles.

ANAPHASE II
The attachment between the two chromatids of each chromosome breaks. Each of the former "sister chromatids" is now a chromosome in its own right. Motor proteins drive the movement of the newly separated chromosomes to opposite poles of the spindle.

TELOPHASE II
By the time telophase II is over, there will be four daughter nuclei. When cytoplasmic division is completed, each new, daughter cell will have a haploid chromosome number (n). All of the chromosomes will now be in the unduplicated state.

Of the four haploid cells that form by way of meiosis and cytoplasmic divisions, one or all may develop into gametes and function in sexual reproduction. In plants, cells that form after meiosis is over may develop into spores, which take part in a stage of the life cycle that precedes gamete formation. (The telophase II micrograph shows spores that will develop into pollen grains.)

A CLOSER LOOK AT KEY EVENTS OF MEIOSIS I

Study the overview in Sections 10.2 and 10.3, and you can sense the overriding function of meiosis: *a reduction of the chromosome number by half for forthcoming gametes.* However, two other major events occur during meiosis: crossing over at prophase I and the random alignment of homologues at metaphase I. Both contribute greatly to the adaptive advantage of sexual reproduction.

The advantage, recall, is the production of offspring with new combinations of alleles. Those combinations are translated into a new generation of individuals that differ in the details of some number of traits.

Crossing Over in Prophase I

Prophase I of meiosis is a time of major gene shufflings. Reflect on Figure 10.5a, which shows two chromosomes condensed to threadlike form. All chromosomes in a germ cell condense this way. As they do, each is drawn close to its homologue. Molecular interactions stitch homologues together point by point along their length, with little space between. The intimate, parallel array favors **crossing over**, a molecular interaction between two of the *non*sister chromatids of a pair of homologous

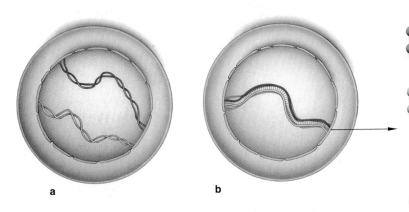

a   b

Figure 10.5 Key events of prophase I, the first stage of meiosis. For clarity, this diagram shows only one pair of homologous chromosomes and one crossover event. (Typically, more than one crossover occurs.) *Blue* signifies the paternal chromosome; *purple* signifies its maternal homologue.

(a) Both chromosomes were duplicated earlier, in interphase. Early in prophase I, the sister chromatids of each duplicated chromosome are in thin, threadlike form. They are positioned so closely together that they look like a single thread.

(b) Each chromosome becomes zippered to its homologue, so all four chromatids are intimately aligned. When the two sex chromosomes have different forms (such as X paired with Y) they still get zippered together, although only in a small region at the ends.

(c,d) We show the pair of chromosomes as if they were already condensed, then teased apart to give you a sense of what goes on. Bear in mind, their double-stranded DNA molecules are still tightly aligned at this stage. The intimate contact allows one crossover (and usually more) to happen at intervals along the length of nonsister chromatids.

(e) Nonsister chromatids exchange segments. As prophase I ends, the chromosomes continue to condense into thicker, rodlike forms. Then they unzipper from each other except at places where they physically cross each other. Such places are called chiasmata (singular, chiasma, meaning "cross"). The chiasmata migrate toward the chromosome ends. They are evidence of crossovers at various places in the chromosomes.

(f) What is the function of crossing over? It breaks up old combinations of alleles and puts new ones together in pairs of homologous chromosomes.

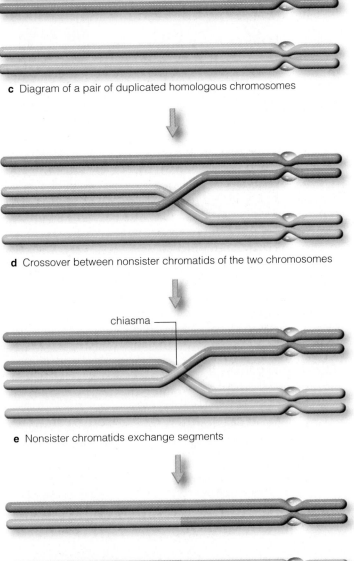

c Diagram of a pair of duplicated homologous chromosomes

d Crossover between nonsister chromatids of the two chromosomes

chiasma

e Nonsister chromatids exchange segments

f Homologues have new combinations of alleles

chromosomes. Nonsister chromatids break at the same places along their length. At these break points, they exchange corresponding segments—that is, genes.

Gene swapping would be pointless if each type of gene never varied. But remember, a gene can come in slightly different forms: alleles. You can bet that some number of the alleles on one chromosome will *not* be identical to their partner alleles on the homologue. Each crossover is a chance to swap slightly different versions of hereditary instructions for particular traits.

We will look at the mechanism of crossing over in later chapters. For now, simply remember this: *Crossing over leads to recombinations among genes of homologous chromosomes, hence to variation in traits among offspring.*

Metaphase I Alignments

Major shufflings of whole chromosomes begin during the transition from prophase I to metaphase I, the second stage of meiosis. Suppose the shufflings are happening right now in one of your germ cells. Crossovers have already made genetic mosaics of the chromosomes, but put this aside in order to simplify tracking. Just call the twenty-three chromosomes you inherited from your mother the *maternal* chromosomes and the twenty-three homologues from your father the *paternal* chromosomes.

Spindle microtubules have already oriented one chromosome of each pair toward one spindle pole and its homologue toward the other (refer to Section 9.3). They are busily moving all of the chromosomes, which soon will become positioned at the spindle's equator.

Will all maternal chromosomes be directed to one spindle pole and all paternal chromosomes directed to the other? Maybe, but probably not. Remember, initial contacts between microtubules and chromosomes are random. Because of the random grabs, the positioning of maternal or paternal chromosomes at the spindle's equator at metaphase I follows no particular pattern. Carry this thought one step further. *Either one* of each pair of homologous chromosomes can end up at either spindle pole after they move apart at anaphase I.

Think about the possibilities when you are tracking merely three pairs of homologues. As you can see from Figure 10.6, by metaphase I, three pairs of homologues may be arranged in any one of four possible positions. Here, eight combinations (2^3) of maternal and paternal chromosomes are possible for forthcoming gametes.

Of course, a human germ cell has twenty-three pairs of homologous chromosomes, not just three. So every time a human germ cell gives rise to sperm or eggs, we can expect a grand total of *8,388,608* (or 2^{23}) possible combinations of maternal and paternal chromosomes!

Moreover, in each sperm or egg, many hundreds of alleles inherited from the mother might not "say" the

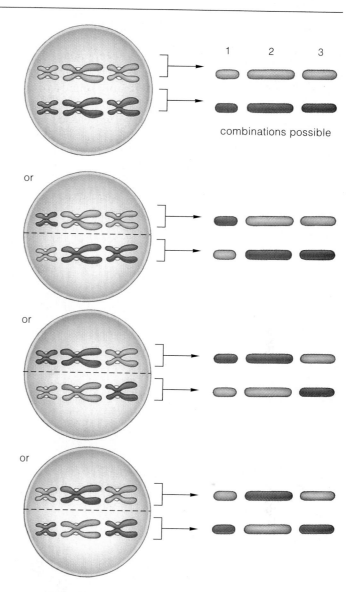

Figure 10.6 Possible outcomes for the random alignment of three pairs of homologous chromosomes at metaphase I. We label three types of chromosomes 1, 2, and 3. The maternal chromosomes are *purple*; paternal ones are *blue*. With merely four possible alignments, eight combinations of maternal and paternal chromosomes are possible in gametes.

exact same thing about hundreds of different traits as the alleles inherited from the father. Are you beginning to get an idea of why such fascinating combinations of traits show up even in the same family?

Crossing over is an interaction between a pair of homologous chromosomes. It breaks up old combinations of alleles and puts new ones together during prophase I of meiosis.

The random attachment and subsequent positioning of each pair of maternal and paternal chromosomes during metaphase I lead to different combinations of maternal and paternal traits in each new generation.

FROM GAMETES TO OFFSPRING

The gametes that form following meiosis are not all the same in their details. For example, human sperm have one tail, opossum sperm have two, and roundworm sperm have none. Crayfish sperm look like pinwheels. Most eggs are microscopic in size, yet an ostrich egg tucked inside its shell is as large as a baseball. From its appearance alone, you might not believe that a plant's gamete is even remotely like an animal's.

Later chapters explain how gametes form in the life cycles of specific organisms, including humans. Figure 10.7 and the rest of this section may help you keep the details in perspective.

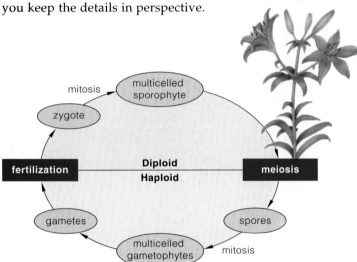

a Generalized life cycle for most kinds of plants

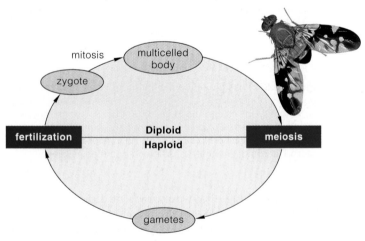

b Generalized life cycle for animals

Figure 10.7 Generalized life cycles for (**a**) most plants and (**b**) animals. The zygote is the first cell that forms when the nuclei of two gametes fuse at fertilization.

For plants, a sporophyte (spore-producing body) develops, by way of mitotic cell divisions, from the zygote. After meiosis, gametophytes (gamete-producing bodies) form. A lily plant is a sporophyte. Gametophytes form in parts of its flowers.

Chapters 21 through 26, 32, 43, and 44 have specific examples of life cycles for representative organisms.

Gamete Formation in Plants

For pine trees, apple trees, roses, dandelions, corn, and nearly all other familiar plants, certain events intervene between meiosis and the time that gametes develop and mature. Among other things, spores form.

Spores are haploid resting cells, often walled, that are good at resisting drought, cold, and other adverse environmental conditions. When favorable conditions return, spores germinate (resume growth) and develop into a haploid body or structure that produces gametes. So *gamete*-producing bodies and *spore*-producing bodies develop during the life cycle of most kinds of plants. Figure 10.7a is a generalized diagram of these events.

Gamete Formation in Animals

In male animals, gametes form by a process known as spermatogenesis. A diploid germ cell grows in size in the male's reproductive system. It becomes a primary spermatocyte. This large immature cell enters meiosis and cytoplasmic divisions. Four haploid cells result and develop into spermatids (Figure 10.8). These immature cells change in form and develop a tail. Each becomes a **sperm**, a common type of mature male gamete.

In female animals, gametes form by a process called oogenesis. In human females, for instance, a diploid germ becomes an **oocyte**, or immature egg. Unlike sperm, an oocyte stockpiles many cytoplasmic components, and its four daughter cells differ in size and function (Figure 10.9). As an oocyte divides after meiosis I, one daughter cell (a secondary oocyte) gets nearly all the cytoplasm. The other cell, the first polar body, is small. Later, both cells undergo meiosis II and cytoplasmic division. One daughter cell of the secondary oocyte develops into a second polar body. The other gets most of the cytoplasm and develops into a gamete. The mature female gamete is called an ovum (plural, ova) or, more often, an **egg**.

And so we have one egg and three polar bodies. The polar bodies don't function as gametes and aren't rich in nutrients or cytoplasm. In time they degenerate. But the very fact that they formed means the egg has a suitable (haploid) chromosome number. Also, by getting most of the cytoplasm, the egg has enough start-up machinery to support the new individual right after fertilization.

More Shufflings at Fertilization

The chromosome number characteristic of the parents is restored at **fertilization**, a time when a female and male gamete unite and their haploid nuclei fuse. Fertilization would double the chromosome number for each new generation if meiosis didn't precede it. Such doublings would disrupt the hereditary instructions, usually for

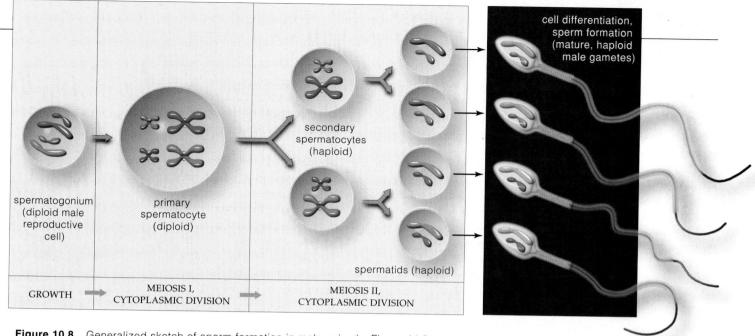

Figure 10.8 Generalized sketch of sperm formation in male animals. Figure 44.2 shows a specific example (how sperm form in human males).

spermatogonium (diploid male reproductive cell)

primary spermatocyte (diploid)

secondary spermatocytes (haploid)

spermatids (haploid)

cell differentiation, sperm formation (mature, haploid male gametes)

GROWTH → MEIOSIS I, CYTOPLASMIC DIVISION → MEIOSIS II, CYTOPLASMIC DIVISION

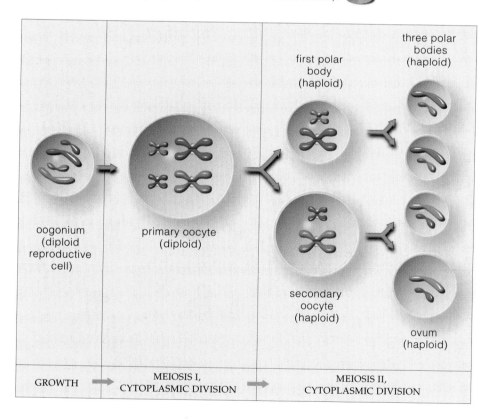

oogonium (diploid reproductive cell)

primary oocyte (diploid)

first polar body (haploid)

three polar bodies (haploid)

secondary oocyte (haploid)

ovum (haploid)

GROWTH → MEIOSIS I, CYTOPLASMIC DIVISION → MEIOSIS II, CYTOPLASMIC DIVISION

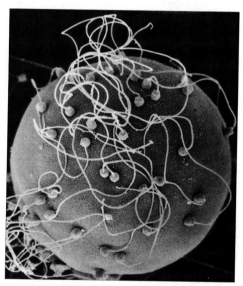

Figure 10.9 Egg formation in female animals. The eggs are far larger than sperm, as the micrograph of sea urchin gametes (*above*) suggests. Also, the three polar bodies are much smaller than the egg. Figure 44.4 gives a specific example (how human eggs form).

the worse. Why? The instructions work as an intricate, fine-tuned package in each individual.

Fertilization also adds to variation among offspring. Reflect upon the possibilities for humans alone. During prophase I, an average of two or three crossovers take place in each human chromosome. Even without these crossovers, the random positioning of pairs of paternal and maternal chromosomes at metaphase I results in one of millions of possible chromosome combinations in each gamete. And of all the male and female gametes that are produced, *which* two actually get together is a matter of chance. The sheer number of combinations that can exist at fertilization is staggering!

Cumulatively, crossing over, the distribution of random mixes of homologous chromosomes into gametes, and fertilization contribute to variation in the traits of offspring.

MEIOSIS AND MITOSIS COMPARED

In this unit our focus has been on two nuclear division mechanisms. Single-celled eukaryotic species reproduce asexually by way of mitosis, followed by cytoplasmic division. Many multicelled eukaryotic species depend on mitosis and cytoplasmic division during episodes of asexual reproduction in their life cycle. All depend on it for growth and tissue repair. By contrast, meiosis occurs only in reproductive cells, such as germ cells that give rise to the gametes used in sexual reproduction. Figure 10.10 summarizes the main similarities and differences between the two nuclear division mechanisms.

The end results of the two mechanisms differ in a crucial way. *Mitotic cell division only produces clones—genetically identical copies of a parent cell. But meiotic cell division, in conjunction with fertilization, promotes variation in traits among offspring.* First, crossing over at prophase I of meiosis puts new combinations of alleles in chromosomes. Second, the random assignment of either member of a pair of homologous chromosomes to either pole of the spindle at metaphase I affects gametes, which end up with mixes of maternal and paternal alleles. And third, different combinations of alleles are brought together simply by chance during fertilization. In later chapters, you will be reading about the ways in which both meiosis and fertilization contribute to the truly stunning diversity and evolution of sexually reproducing organisms.

A *somatic cell* with a diploid chromosome number (2*n*) is at interphase. Before mitotic division begins, its DNA is replicated (all chromosomes are duplicated).

Figure 10.10 Summary of mitosis and meiosis. Both diagrams use a diploid (2*n*) animal cell as the example. They are arranged to help you compare similarities and differences between the division mechanisms. The maternal chromosomes are coded *purple* and the paternal chromosomes *blue*.

MEIOSIS I

A *germ cell* with a diploid chromosome number (2*n*) is at interphase. Before mitotic division begins, its DNA is replicated (all chromosomes are duplicated).

PROPHASE I

Each duplicated chromosome (consisting of two sister chromatids) condenses to threadlike form, then rodlike form. *Crossing over* occurs. Each chromosome unzips from its homologue. Each gets attached to the spindle in transition to metaphase.

METAPHASE I

All chromosomes are now positioned at the spindle's equator.

ANAPHASE I

Each chromosome is separated from its homologue. They are moved to opposite poles of the spindle.

TELOPHASE I

When the cytoplasm divides, there are two cells. Each has a haploid (*n*) number of chromosomes, but these are still in the duplicated state.

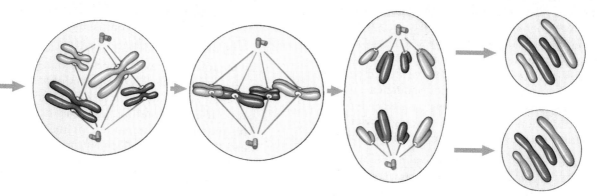

MITOSIS

PROPHASE

Each duplicated chromosome (consisting of two sister chromatids) condenses from threadlike form to rodlike form. Each gets attached to the spindle during the transition to metaphase.

METAPHASE

All chromosomes are now positioned at the spindle's equator.

ANAPHASE

Sister chromatids of each chromosome are separated from each other. These new, daughter chromosomes are moved to opposite poles of the spindle.

TELOPHASE

When the cytoplasm divides, there are two cells. Each is diploid (2n)—*it has the same chromosome number as the parent cell.*

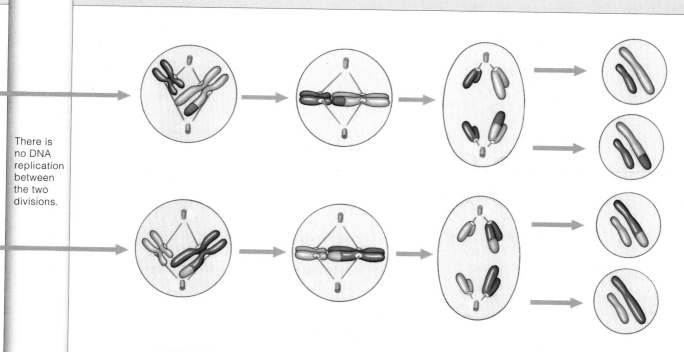

MEIOSIS II

There is no DNA replication between the two divisions.

PROPHASE II

Before prophase II, the two centrioles in each new cell were moved apart and a new spindle formed. Now, each chromosome becomes attached to the spindle and starts moving toward its equator.

METAPHASE II

All chromosomes are now positioned at the spindle's equator.

ANAPHASE II

Sister chromatids of each chromosome are separated from each other. These new, daughter chromosomes are moved to opposite poles of the spindle.

TELOPHASE II

Four daughter nuclei form. When the cytoplasm divides, each new cell is haploid (n). *The original chromosome number has been reduced by half.* One or all of these cells may become gametes.

SUMMARY   *Gold* indicates text section

1. Eukaryotic life cycles often have asexual as well as sexual phases. Asexual reproduction results in clones. Sexual reproduction (by meiosis, gamete formation, and fertilization) leads to variation in traits. *C1, 10.1*

   a. Meiosis, a nuclear division mechanism, reduces the chromosome number of a parent germ cell by half. It precedes the formation of haploid gametes, such as sperm in males and eggs in females.

   b. At fertilization, a sperm and an egg nuclei fuse, which restores the chromosome number (Figure 10.11).

2. A germ cell with a diploid chromosome number (2*n*) has *two* of each type of chromosome characteristic of its species. Commonly, one of each pair of chromosomes is maternal and the other is paternal. *CI, 10.1*

3. Each pair of maternal and paternal chromosomes has homology; the two are alike. Except for a pairing of nonidentical sex chromosomes (e.g., X with Y), the two have the same length, shape, and gene sequence. They interact during meiosis. *10.2*

4. Chromosomes become duplicated during interphase. Each consists of two DNA molecules that stay attached (as sister chromatids) during meiosis. *10.2*

5. Meiosis consists of two consecutive divisions. Both require a microtubular spindle apparatus. *10.2–10.3*

   a. During meiosis I, spindle microtubules attach to the centromere region of each duplicated chromosome. Motor proteins attached to the microtubules separate it from its partner, the homologous chromosome.

   b. In meiosis II, similar interactions move the sister chromatids of each chromosome away from each other.

6. Meiosis I, the first nuclear division, is characterized by the following events and outcomes: *10.2–10.4*

   a. Crossing over occurs in prophase I. Two nonsister chromatids of each pair of homologous chromosomes break at corresponding sites and exchange segments. This puts new combinations of alleles together. Alleles (slightly different molecular forms of the same gene) specify different versions of the same trait.

   b. Different combinations of alleles lead to variation in the details of a given trait among offspring.

   c. Also in prophase I, a microtubular spindle forms outside the nucleus, and the nuclear envelope starts to break up. In cells with duplicated pairs of centrioles, one pair starts moving to the opposite spindle pole.

   d. All of the pairs of homologous chromosomes have become positioned at the spindle equator at metaphase I. Both the maternal chromosome and its homologue have become oriented at random, toward either pole.

   e. In anaphase I, spindle microtubules interact with each duplicated chromosome and move it away from its homologue, to opposite spindle poles.

7. Meiosis II (second nuclear division) is characterized by these events and outcomes: *10.3*

   a. At metaphase II, all the duplicated chromosomes are positioned at the spindle equator.

   b. Sister chromatids are moved apart in anaphase II. Each is now a separate, unduplicated chromosome.

   c. By the end of telophase II, four nuclei—each with a haploid chromosome number (*n*)—have been formed.

8. When the cytoplasm divides, there are four haploid cells. One or all may serve as gametes (or as plant spores that give rise to gamete-producing bodies). *10.3, 10.5*

9. Crossing over, the chance allocation of different mixes of pairs of maternal and paternal chromosomes to different gametes, and the chance of any two gametes meeting at fertilization all contribute to the variation in details of traits among offspring. *10.5, 10.6*

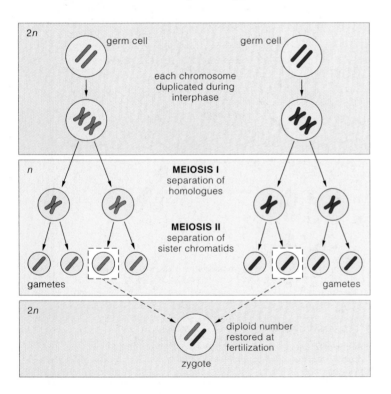

Figure 10.11 Summary of changes in chromosome number at different stages of sexual reproduction, using two diploid (2*n*) germ cells as the example. Meiosis reduces the chromosome number by half (*n*). Union of haploid nuclei of two gametes at fertilization restores the diploid number.

Review Questions

1. The diploid chromosome numbers for the somatic cells of a few organisms are listed at *right*. How many chromosomes will end up in the gametes of each organism? *10.2*

Fruit fly, *Drosophila melanogaster*	8
Garden pea, *Pisum sativum*	14
Corn, *Zea mays*	20
Frog, *Rana pipiens*	26
Earthworm, *Lumbricus terrestris*	36
Human, *Homo sapiens*	46
Chimpanzee, *Pan troglodytes*	48
Amoeba, *Amoeba*	50
Horsetail, *Equisetum*	216

2. A diploid germ cell has four pairs of homologous chromosomes, designated AA, BB, CC, and DD. Which of the chromosomes will be present in gametes? *10.2, 10.3*

3. Look at the chromosomes in the germ cell in the diagram at *right*. Is this cell at anaphase I or anaphase II? *10.3*

The cell is at anaphase ___ rather than anaphase ___ . I know this because:

_____

_____

_____

4. Define meiosis and describe its stages. In what respects is meiosis *not* like mitosis? *10.2, 10.6*

5. Actor Michael Douglas (Figure 10.12*a*) inherited a gene from each parent that influences the chin dimple trait. One form of the gene called for a dimple and the other didn't, but one is all it takes for this particular trait. Figure 10.12*b* shows what the chin of Mr. Douglas might have looked like if he had inherited two ordinary forms of the gene instead. What is the name for the alternative forms of the same gene? *10.1, 10.4*

6. Outline the main steps by which gametes form in plants. Do the same for gamete formation in animals. *10.5*

7. Genetically speaking, what is the key difference between the outcomes of sexual and asexual reproduction? *10.2, 10.6*

Self-Quiz ANSWERS IN APPENDIX III

1. Sexual reproduction requires _____ .
   a. meiosis   c. fertilization
   b. gamete formation d. all of the above

2. An animal cell having two rather than one of each type of chromosome has a _____ chromosome number.
   a. diploid   c. normal gamete
   b. haploid   d. both b and c

3. Generally, a pair of homologous chromosomes _____ .
   a. carries the same genes   c. interacts at meiosis
   b. has the same length, shape d. all of the above

4. Meiosis _____ the parental chromosome number.
   a. doubles   c. maintains
   b. reduces   d. corrupts

5. Meiosis is a division mechanism that produces _____ .
   a. two cells   c. eight cells
   b. two nuclei d. four nuclei

6. Before the onset of meiosis, all chromosomes are _____ .
   a. condensed   c. duplicated
   b. released from protein d. both b and c

7. Duplicated chromosomes move away from their homologue and end up at the opposite spindle pole during _____ .
   a. prophase I   c. anaphase I
   b. prophase II d. anaphase II

8. Sister chromatids of each duplicated chromosome move apart and end up at opposite spindle poles during _____ .
   a. prophase I   c. anaphase I
   b. prophase II d. anaphase II

9. Match each term with its description.
   ____ chromosome number   a. different molecular forms of the same gene
   ____ alleles   b. none between meiosis I and II
   ____ metaphase I   c. pairs of homologous chromosomes are now aligned at the spindle equator
   ____ interphase   d. sum total of chromosomes in all cells of a given type

Figure 10.12 Example of the chin dimple trait (actually a fissure in the chin surface).

Critical Thinking

1. Explain why you can expect meiosis rather than mitosis to give rise to genetic differences between parent cells and their daughter cells.

2. An organism has four pairs of homologous chromosomes (AA, BB, CC, and DD). If it self-fertilizes, what chromosome combinations can you expect to see in its offspring?

3. Assume that you can measure the amount of DNA in the nucleus of a primary oocyte, then in the nucleus of a primary spermatocyte. Each gives you a mass *m*. What mass of DNA would you expect to find inside the nucleus of each mature gamete (egg and sperm) that forms after meiosis? What mass of DNA would you expect to find (1) in the nucleus of the zygote formed at fertilization and (2) in that zygote's nucleus after the first DNA duplication?

4. As you know, aphids can reproduce asexually and sexually. Females reproduce by themselves and give birth to live females in summer. They also reproduce sexually, and lay eggs that survive in summer. They also reproduce sexually, and lay eggs that survive the winter and hatch in spring. Aphids happen to be tasty to various predators. Speculate on how their reproductive flexibility might be an adaptation to agents of predation.

Selected Key Terms

allele *10.1*
asexual reproduction *10.1*
chromosome number *10.2*
crossing over *10.4*
diploid number *10.2*
egg (ovum) *10.5*
fertilization *10.5*
gamete *CI*
gene *10.1*
germ cell *CI*
haploid number *10.2*
homologous chromosome *10.2*
meiosis *10.2*
oocyte *10.5*
sexual reproduction *10.1*
sister chromatid *10.2*
sperm *10.5*
spore *10.5*

Readings

Klug, W., and M. Cummings. 2000. *Concepts of Genetics.* Sixth edition. New York: Macmillan.

Wolfe, S. 1995. *Introduction to Molecular and Cellular Biology.* Belmont, California: Wadsworth.

On-Line readings at Student Guide for InfoTrac:
www.brookscole.com/biology

OBSERVABLE PATTERNS OF INHERITANCE

A Smorgasbord of Ears and Other Traits

Basketball ace Charles Barkley has them. So does actor Tom Cruise. Actress Joan Chen doesn't, and neither did a monk named Gregor Mendel. To see how you fit in with these folks, use a mirror to check out your ears. Is the fleshy lobe at the base of each ear attached to the side of your head? If so, you and Barkley and Cruise have something in common. Or is the fleshy lobe not attached, so that you can flap it back and forth? If so, you are like Chen and Mendel (Figure 11.1).

Tom Cruise

Charles Barkley

Figure 11.1 Attached and detached earlobes of a few representative humans. This sampling provides observable evidence of a trait governed by genes, which occur in different molecular forms in the human population. Which version of the trait do you have? It depends on which molecular forms of certain genes you inherited from your mother and father. As Gregor Mendel perceived, observable traits can be used to identify patterns of inheritance from one generation to the next.

Whether a person is born with attached or detached earlobes depends on which genes he or she inherited. Genes come in slightly different molecular forms—alleles. Some of them have information about detached lobes. The information is put to use as a human body is developing inside the mother. It has to do with a death warrant, a molecular signal sent to all cells positioned between the head and the forming lobes. That signal, as well as the receptors for it on target cells, must be properly synthesized at the proper time. If not, cells don't die and earlobes don't detach.

We all have genes for thousands of traits, including earlobes, cheeks, lashes, and eyeballs. Most traits vary in their details from one person to the next. Remember, we inherit pairs of genes on pairs of chromosomes. In some pairs, one allele has strong effects and overwhelms the other allele's contribution. The outgunned allele is said to be recessive to the dominant one. If you have

detached earlobes, dimpled cheeks, long lashes, or large eyeballs, you have some number of dominant alleles that affect the trait in predictable ways.

When both alleles of a pair are recessive, nothing masks their effect on a trait. You get *attached* earlobes with certain pairs of recessive alleles, *flat* feet with other pairs, a *straight* nose with other pairs, and so on.

How did we discover such remarkable things about our genes? It started with Mendel. By analyzing garden

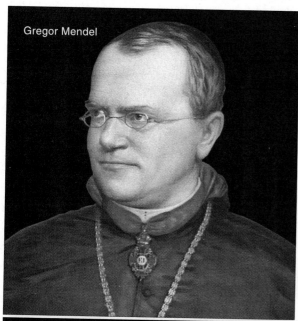

Gregor Mendel

Joan Chen

pea plants generation after generation, Mendel found indirect but *observable* evidence of how parents bestow units of hereditary information—genes—on offspring. This chapter focuses on both the methods and the results of Mendel's experiments. They remain a classic example of how a scientific approach can pry open important secrets about the natural world. And to this day, they serve as the foundation for modern genetics.

Key Concepts

1. Genes are units of information that influence the expression of heritable traits. Alleles, which are slightly different molecular forms of a gene, may affect traits in different ways.

2. Genes have specific locations on the chromosomes of a species. Humans, pea plants, and other organisms with a diploid chromosome number inherit *pairs* of genes, at equivalent locations on pairs of homologous chromosomes.

3. When the two members of each pair of homologous chromosomes are separated from each other during meiosis, their pairs of genes are moved apart, also, and end up in separate gametes. Gregor Mendel found indirect evidence of this gene segregation when he crossbred pea plants showing different versions of the same trait, such as purple versus white flowers.

4. Each pair of homologous chromosomes in a germ cell is sorted out for distribution into one gamete or another independently of how the other pairs of homologous chromosomes are assorted. Mendel discovered indirect evidence of this when he tracked many plants having observable differences in two traits, such as flower color and plant height.

5. Nonidentical alleles affect the forms of traits that Mendel happened to study. One allele is said to be dominant, in that its effect on a trait masks the effect of a recessive allele paired with it.

6. Not all traits have clearly dominant or recessive forms. One allele of a pair may be fully or partially dominant over its partner or codominant with it. Two or more gene pairs often influence the same trait, and some single genes influence many traits.

7. The environment introduces variation in traits.

11.1

MENDEL'S INSIGHT INTO INHERITANCE PATTERNS

More than a century ago, people wondered about the basis of inheritance. It was common knowledge that sperm and eggs both transmit information about traits to offspring. But few suspected that the information is organized in units (genes). Instead, the idea was that a father's blob of information "blended" with the mother's blob, like cream into coffee, at fertilization.

However, carried to its logical conclusion, blending would slowly dilute a population's shared pool of hereditary information until there was only a single intermediate version left of each trait. If that were so, then why did, say, freckles keep showing up among children of nonfreckled parents through the generations? Why weren't all the descendants of a herd of white stallions and black mares gray? The blending theory could scarcely explain the obvious variation in traits that people could observe with their own eyes. Nevertheless, few disputed the theory.

Yet it could not be reconciled with Charles Darwin's theory of natural selection. According to a key premise of Darwin's theory, individuals of a population show variation in heritable traits. Variations that improve the chance of surviving and reproducing show up with greater frequency than those that do not in subsequent generations. Less advantageous variations may persist among fewer individual or may vanish. It is not that separate versions of a given trait are "blended out" of the population. Rather, *each version of a trait may persist in a population, at frequencies that can change over time.*

Just before Darwin presented his theory, someone was gathering evidence that eventually would support his key premise. A monk, Gregor Mendel, had already guessed that sperm and eggs carry distinct "units" of information about heritable traits. By carefully analyzing traits of pea plants generation after generation, Mendel found indirect but *observable* evidence of how parents transmit genes to offspring.

Mendel's Experimental Approach

Mendel spent most of his adult life in a monastery in Brno, a city near Vienna that has since become part of the Czech Republic. However, Mendel was not a man of narrow interests who accidentally stumbled onto principles of great import. The monastery of St. Thomas was close to European capitals that were the centers of scientific inquiry.

Having been raised on a farm, Mendel was aware of agricultural principles and their applications. He kept abreast of the breeding experiments and developments described in the available literature. He was a member of the regional agricultural society. He won awards for developing improved varieties of vegetables and fruits. Shortly after entering the monastery, he took courses in mathematics, physics, and botany at the University of Vienna. Few scholars of his time showed interest in both plant breeding and mathematics.

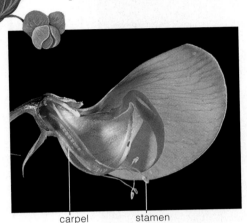

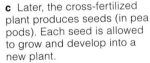
carpel   stamen

Figure 11.2 Garden pea plant (*Pisum sativum*), the organism Mendel chose for experiments on his ideas about inheritance.

a Garden pea flower. The section shows the location of its stamens and carpel. Sperm-producing pollen grains form in stamens. Eggs develop, fertilization takes place, and seeds mature inside carpels.

b Pollen from a garden pea plant that breeds true for purple flowers is brushed onto a floral bud of a plant that breeds true for white flowers. The white flower had its stamens snipped off. This is one way to assure cross-fertilization of plants.

c Later, the cross-fertilized plant produces seeds (in pea pods). Each seed is allowed to grow and develop into a new plant.

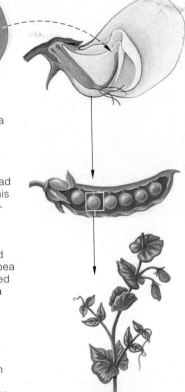

d The flower color of each new plant can be used as visible evidence of patterns in how hereditary material might be transmitted to it from each parent plant.

Shortly after his university training, Mendel began experimenting with the garden pea plant, *Pisum sativum* (Figure 11.2). This plant is self-fertilizing. Its male and female gametes (call them sperm and eggs) develop in different parts of the same flower, where fertilization occurs. Individual pea plants generally breed true for certain traits. In other words, successive generations are just like the parents in one or more traits, as when all offspring grown from seeds of self-fertilized, white-flowered parent plants have white flowers.

Pea plants may be cross-fertilized by transferring pollen from one plant's flower to the flower of another plant. Mendel knew he could open the flower buds of a plant that bred true for a trait, such as white flowers, and snip out its stamens. (Stamens bear pollen grains in which sperm develop.) He could brush those buds with pollen from a plant that bred true for a *different* version of the same trait—say, purple flowers. As he hypothesized, such clearly observable differences would help him track a given trait through many generations. If there were patterns to the trait's inheritance, *then the patterns might tell him something about heredity itself.*

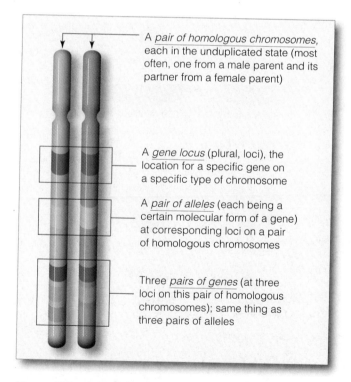

A *pair of homologous chromosomes*, each in the unduplicated state (most often, one from a male parent and its partner from a female parent)

A *gene locus* (plural, loci), the location for a specific gene on a specific type of chromosome

A *pair of alleles* (each being a certain molecular form of a gene) at corresponding loci on a pair of homologous chromosomes

Three *pairs of genes* (at three loci on this pair of homologous chromosomes); same thing as three pairs of alleles

Figure 11.3 A few genetic terms illustrated. Diploid organisms have pairs of genes, on pairs of homologous chromosomes. For example, you inherited one chromosome of each pair from your mother and the other, homologous chromosome from your father.

Most genes come in slightly different molecular forms, called alleles. Different alleles specify different versions of the same trait. An allele at one location on a chromosome may or may not be identical to its partner on the homologous chromosome.

Some Terms Used in Genetics

Having read the chapter on meiosis, you already have insight into the mechanisms of sexual reproduction, which is more than Mendel had. Neither he nor anyone else of his era knew about chromosomes. So he could not have known that a chromosome number is reduced by half in gametes, then restored when gametes meet at fertilization. Even so, Mendel sensed what was going on. As we follow his thinking, let's simplify the story by substituting a few of the modern terms used in studies of inheritance (see also Figure 11.3):

1. **Genes** are units of information about specific traits, and they are passed from parents to offspring. Each gene has a specific location (locus) on a chromosome.

2. Cells with a diploid chromosome number ($2n$) have pairs of genes, on pairs of homologous chromosomes.

3. Mutation can alter a gene's molecular structure. The alteration may change the gene's information about a trait (as when the gene for flower color specifies purple and a mutated version specifies white). All the different molecular forms of the same gene are called **alleles.**

4. When offspring inherit a pair of *identical* alleles for a trait, generation after generation, they represent a **true-breeding lineage.** Offspring of a cross between two different true-breeding individuals are **hybrids,** which have inherited *nonidentical* alleles for a trait.

5. When both alleles of a pair are identical, this is a *homozygous* condition. When the two are not identical, this is a *heterozygous* condition.

6. An allele is *dominant* when its effect on a trait masks that of any *recessive* allele paired with it. Mendel used capital letters for dominant alleles and lowercase letters for recessive ones. *A* and *a* are examples.

7. Putting this all together, a **homozygous dominant** individual has a pair of dominant alleles (*AA*) for the trait being studied. A **homozygous recessive** individual has a pair of recessive alleles (*aa*). And a **heterozygous** individual has a pair of nonidentical alleles (*Aa*).

8. Two terms help keep the distinction clear between genes and the traits they specify. **Genotype** refers to the particular alleles an individual carries. **Phenotype** refers to an individual's observable traits.

9. When tracking the inheritance of traits through generations of offspring, these abbreviations apply:

P	parental generation
F_1	first-generation offspring
F_2	second-generation offspring

MENDEL'S THEORY OF SEGREGATION

Mendel had an idea. In every generation, a plant might inherit two "units" (genes) of information about a trait, one from each parent. He used **monohybrid crosses** to test his idea. For such crosses, two parents that breed true for different forms of a trait produce F_1 offspring that are heterozygous ($AA \times aa \longrightarrow Aa$). The actual experiment is an intercross between two of the identical F_1 heterozygotes—the "monohybrids" ($Aa \times Aa$).

Predicting Outcomes of Monohybrid Crosses

Mendel tracked many traits over two generations. For one set of experiments, he crossed true-breeding purple-flowered plants and true-breeding white-flowered ones. All plants of the next generation had purple flowers. Mendel let those F_1 plants self-fertilize. Later on, some of the F_2 offspring produced white flowers!

If Mendel's hypothesis were correct—if each plant inherited two units of information about flower color—then the "purple" unit had to be dominant. Why? It masked the unit for "white" in the F_1 plants.

Let's rephrase his thinking. Germ cells of pea plants are diploid, with pairs of homologous chromosomes. Assume one parent is homozygous dominant (AA) and the other is homozygous recessive (aa) for flower color. Following meiosis, a sperm or egg carries one allele for flower color (Figure 11.4). So when a sperm fertilizes an egg, only one outcome is possible: $A + a \longrightarrow Aa$.

Mendel knew about sampling error (Section 1.5). He crossed a great many plants and tracked thousands of offspring. He also counted and recorded the number of dominant and recessive plants. On average, three of every four F_2 plants had the dominant phenotype, and one had the recessive phenotype (Figure 11.5).

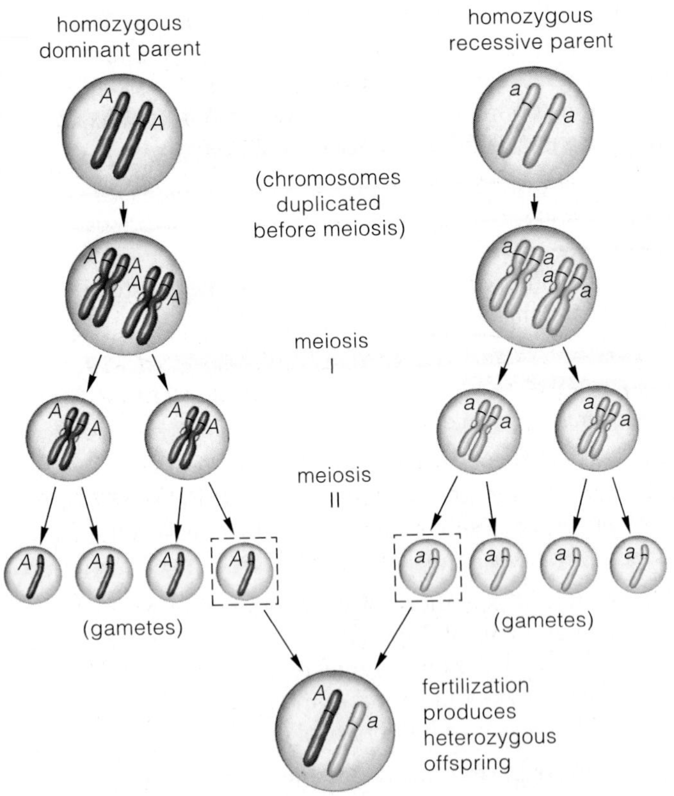

Figure 11.4 Monohybrid cross, showing how one gene of a pair segregates from the other gene. Two parents that breed true for two versions of a trait produce only heterozygous offspring.

Figure 11.5 *Right:* Numerical results from Mendel's monohybrid cross experiments with the garden pea plant (*P. sativum*). The numbers are his counts of the F_2 plants that carried dominant or recessive hereditary "units" (alleles) for the trait. On average, the dominant-to-recessive ratio was 3:1.

Trait Studied	Dominant Form	Recessive Form	F_2 Dominant-to-Recessive Ratio
SEED SHAPE	5,474 round	1,850 wrinkled	2.96:1
SEED COLOR	6,022 yellow	2,001 green	3.01:1
POD SHAPE	882 inflated	299 wrinkled	2.95:1
POD COLOR	428 green	152 yellow	2.82:1
FLOWER COLOR	705 purple	224 white	3.15:1
FLOWER POSITION	651 along stem	207 at tip	3.14:1
STEM LENGTH	787 tall	277 dwarf	2.84:1
		Average ratio for all traits studied:	**3:1**

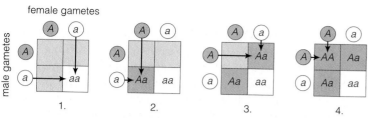

female gametes

male gametes

1.   2.   3.   4.

Figure 11.6 Punnett-square method of predicting the probable outcome of a genetic cross. Circles represent gametes. *Italic* letters on gametes represent dominant or recessive alleles. In the squares are the different genotypes possible among offspring. In this case, gametes are from a self-fertilizing heterozygous (*Aa*) plant.

Figure 11.7 *Right*: Results from one of Mendel's monohybrid crosses. On average, the dominant-to-recessive ratio among the second-generation (F₂) plants was 3:1.

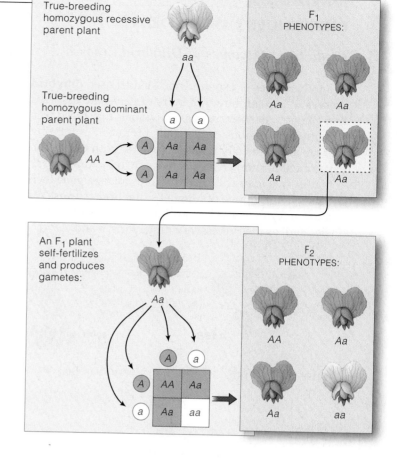

To Mendel, the ratio suggested that fertilization is a chance event with a number of possible outcomes. And he had an understanding of probability, which applies to chance events *and therefore could help him predict the possible outcomes of genetic crosses*. **Probability** simply means this: The chance that each outcome of a given event will occur is proportional to the number of ways in which the event can be reached.

The **Punnett-square method**, explained in Figure 11.6 and applied in Figure 11.7, may help you visualize the possibilities. As you can see, if half of a plant's sperm (or eggs) were *a* and half were *A*, then four outcomes would be possible each time a sperm fertilized an egg:

POSSIBLE EVENT:	PROBABLE OUTCOME:
sperm *A* meets egg *A*	1/4 *AA* offspring
sperm *A* meets egg *a*	1/4 *Aa*
sperm *a* meets egg *A*	1/4 *Aa*
sperm *a* meets egg *a*	1/4 *aa*

By this prediction, an F₂ plant has three chances in four of getting at least one dominant allele (purple flowers). It has one chance in four of getting two recessive alleles (white flowers). That is a probable phenotypic ratio of three purple to one white, or 3:1.

Mendel's observed ratios were not *exactly* 3:1. You can see this for yourself by looking at the numerical results listed in Figure 11.5. Why did Mendel put aside the deviations? To understand why, flip a coin several times. As we all know, a coin is just as likely to end up heads as tails. But often a coin ends up heads, or tails, several times in a row. So if you flip the coin only a few times, the observed ratio might differ greatly from the predicted ratio of 1:1. Flip the coin many, many times, and you are more likely to come close to the predicted ratio. Mendel understood the rules of probability—and observed a large number of offspring. Almost certainly, this kept him from being confused by minor deviations from the predicted results of the experimental crosses.

Testcrosses

By running **testcrosses**, Mendel gained support for his prediction. In this type of experimental test, an organism shows dominance for a specified trait but its genotype is unknown, so it is crossed to a known homozygous recessive individual. Test results may reveal whether the organism is homozygous dominant or heterozygous.

Mendel tested one prediction that purple-flowered F₁ offspring were heterozygous by crossing them with true-breeding, white-flowered plants. If they were all homozygous dominant, then all the testcross offspring would show the dominant form of the trait. If they were heterozygous, then there would be about as many dominant as recessive plants. Sure enough, when old enough to flower, about half of the testcross offspring had purple flowers (*Aa*) and half had white (*aa*). Can you construct two Punnett squares to show the possible outcomes of this testcross?

The results from Mendel's monohybrid crosses and testcrosses became the basis of a theory of **segregation**, which we state here in modern terms:

MENDEL'S THEORY OF SEGREGATION **Diploid cells have pairs of genes, on pairs of homologous chromosomes. The two genes of each pair are separated from each other during meiosis, so they end up in different gametes.**

INDEPENDENT ASSORTMENT

Predicting Outcomes of Dihybrid Crosses

In another series of experiments, Mendel used **dihybrid crosses** to explain how *two* pairs of genes are assorted into gametes. In such crosses, individuals that breed true for different versions of *two* traits produce F_1 offspring that are all identically heterozygous for both traits. The experiment is an intercross between two F_1 "dihybrids" —that is, two identical heterozygotes for two gene loci.

Let's diagram one of Mendel's dihybrid crosses. We can use A for flower color and B for height as dominant alleles and, as their recessive counterparts, a and b:

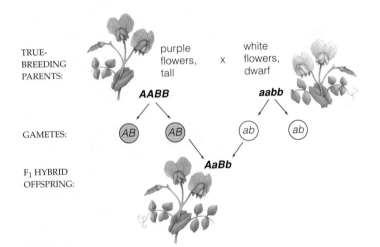

TRUE-BREEDING PARENTS: purple flowers, tall *AABB* × white flowers, dwarf *aabb*

GAMETES: AB AB ab ab

F_1 HYBRID OFFSPRING: *AaBb*

As Mendel would have predicted, the F_1 offspring from this cross are all purple-flowered and tall ($AaBb$).

When those F_1 plants reproduce, how will the two gene pairs be assorted into gametes? The answer partly depends on which chromosomes carry the pairs. Assume that one pair of homologous chromosomes carry the Aa alleles and a different pair carry the Bb alleles. Now think of how all chromosomes become positioned at the spindle equator during metaphase I of meiosis (Figures 10.6 and 11.8). The chromosome with the A allele might be positioned to move to either spindle pole (then into one of four gametes). Its homologue with the a allele might also move to either spindle pole. And the same can happen to the homologous chromosomes that carry the B and b alleles. After meiosis and gamete formation, then, four combinations of alleles are possible in sperm or eggs: $1/4\,AB$, $1/4\,Ab$, $1/4\,aB$, and $1/4\,ab$.

Given the alternative alignments of chromosomes at metaphase I, several allelic combinations are possible at fertilization. Simple multiplication (four kinds of sperm times four kinds of eggs) tells us sixteen combinations of gametes are possible in the F_2 offspring of a dihybrid cross. Use the Punnett-square method to diagram the probabilities (Figure 11.9). Now add up all the possible phenotypes and you get 9/16 tall purple-flowered, 3/16 dwarf purple-flowered, 3/16 tall white-flowered, and 1/16 dwarf white-flowered plants. That is a probable phenotypic ratio of 9:3:3:1. Results from one dihybrid cross that Mendel described were close to this ratio.

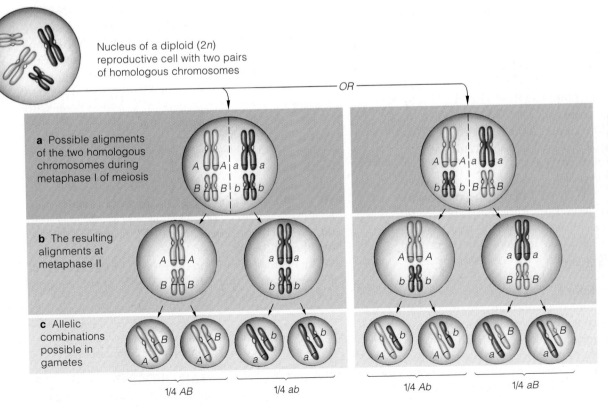

Nucleus of a diploid (2n) reproductive cell with two pairs of homologous chromosomes

— OR —

a Possible alignments of the two homologous chromosomes during metaphase I of meiosis

b The resulting alignments at metaphase II

c Allelic combinations possible in gametes

1/4 AB   1/4 ab   1/4 Ab   1/4 aB

Figure 11.8 One case of independent assortment. This example tracks two pairs of homologous chromosomes. An allele at a given locus on a chromosome may or may not be identical with its partner allele on the homologous chromosome. Either chromosome of a pair may become attached to either pole of the spindle during meiosis. As you can see, when just two pairs are being tracked, two different metaphase I lineups are possible.

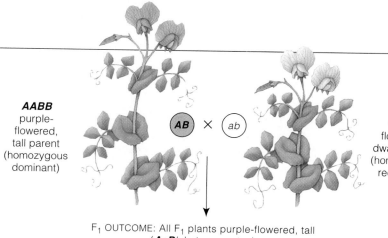

AABB
purple-
flowered,
tall parent
(homozygous
dominant)

AB × ab

aabb
white-
flowered,
dwarf parent
(homozygous
recessive)

Figure 11.9 Results from Mendel's dihybrid cross between parent plants that bred true for different versions of two traits: flower color and plant height. *A* and *a* represent the dominant and recessive alleles for flower color. *B* and *b* signify the dominant and recessive alleles for plant height. As the Punnett square indicates, the probabilities of certain combinations of phenotypes among the F_2 offspring occur in a 9:3:3:1 ratio, on average.

F_1 OUTCOME: All F_1 plants purple-flowered, tall
(**AaBb** heterozygotes)

AaBb

AaBb

meiosis,
gamete formation

meiosis,
gamete formation

	1/4 AB	1/4 Ab	1/4 aB	1/4 ab
1/4 **AB**	1/16 **AABB**	1/16 **AABb**	1/16 **AaBB**	1/16 **AaBb**
1/4 **Ab**	1/16 **AABb**	1/16 **AAbb**	1/16 **AaBb**	1/16 **Aabb**
1/4 **aB**	1/16 **AaBB**	1/16 **AaBb**	1/16 **aaBB**	1/16 **aaBb**
1/4 **ab**	1/16 **AaBb**	1/16 **Aabb**	1/16 **aaBb**	1/16 **aabb**

Possible outcomes of cross-fertilization

ADDING UP THE F_2 COMBINATIONS POSSIBLE:

- 9/16 or 9 purple-flowered, tall
- 3/16 or 3 purple-flowered, dwarf
- 3/16 or 3 white-flowered, tall
- 1/16 or 1 white-flowered, dwarf

The Theory in Modern Form

Mendel could do no more than analyze the numerical results from his dihybrid crosses, because he didn't know that seven pairs of homologous chromosomes carry the pea plant's "units" of inheritance. It just seemed to him that the two units for the first trait he was tracking had been assorted into gametes independently of the two units for the other trait. In time his interpretation became known as the theory of **independent assortment**, which we state here in modern terms: By the end of meiosis, each pair of homologous chromosomes—and the genes they carry—have been sorted for shipment into gametes independently of how the other pairs were sorted out.

Independent assortment and hybrid intercrosses lead to great genetic variation. In a monohybrid cross that involves only one gene pair, three genotypes are possible: *AA*, *Aa*, and *aa*. We can represent this as 3^n, where *n* is the number of gene pairs. With more pairs of genes, the number of possible combinations increases dramatically. When the parents differ in ten gene pairs, almost 60,000 genotypes are possible. When they differ in twenty gene pairs, the number approaches 3.5 billion!

In 1865, Mendel presented his ideas to the Brünn Natural History Society. His ideas had little impact. The next year he published a paper, and apparently it was read by few and understood by no one. In 1871 he became abbot of the monastery, and his experiments gradually gave way to administrative tasks. He died in 1884, never to know his experiments would become the starting point for the development of modern genetics.

Mendel's segregation theory still stands. Hereditary material is indeed organized in units (genes) that retain their identity and are segregated for distribution into gametes. But the theory of independent assortment does not apply to *all* gene combinations, as you will see in the next chapter.

MENDEL'S THEORY OF INDEPENDENT ASSORTMENT **By the end of meiosis, genes on pairs of homologous chromosomes have been sorted out for distribution into one gamete or another independently of gene pairs of other chromosomes.**

DOMINANCE RELATIONS

For the most part, Mendel studied traits having clearly dominant or recessive forms. As the remaining sections of this chapter will make clear, however, the expression of other traits is not as straightforward.

Incomplete Dominance

With **incomplete dominance**, one allele of a pair is not fully dominant over its partner, so the phenotype of the heterozygote is *somewhere in between* the phenotypes of the two homozygotes. Example: Cross a true-breeding red snapdragon and a true-breeding white one. All F_1 offspring will have pink flowers. Cross two F_1 plants and you can expect red, pink, and white snapdragons in a predictable ratio (Figure 11.10). What causes this inheritance pattern? Red snapdragons have two alleles that allow them to make an abundance of red pigment molecules. White snapdragons have two mutant alleles that make them pigment-free. Pink snapdragons are heterozygous; the red allele they carry specifies enough pigment to make flowers pink, but not red.

homozygous parent   ×   homozygous parent

All F_1 offspring
are heterozygous
for flower color:

Cross two of the
F_1 plants, and
the F_2 offspring
will show three
phenotypes in a
1:2:1 ratio:

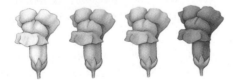

Figure 11.10 Visible evidence of incomplete dominance in heterozygous (pink) snapdragons, in which an allele for red pigment is paired with a "white" allele.

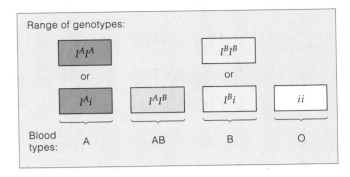

Range of genotypes:

$I^A I^A$ $I^B I^B$

or   or

$I^A i$ $I^A I^B$ $I^B i$ ii

Blood
types: A   AB   B   O

Figure 11.11 Possible allelic combinations for ABO blood typing.

ABO Blood Types: A Case of Codominance

In **codominance**, a pair of nonidentical alleles specify two phenotypes, which are both expressed at the same time in heterozygotes. Example: One of the glycolipids at the plasma membrane of your red blood cells helps give these cells a unique identity. However, it comes in slightly different molecular forms. An analytical method, *ABO blood typing*, reveals which form a person has.

An enzyme dictates the glycolipid's final structure. In humans, the gene for that enzyme has three alleles. Two alleles, I^A and I^B, are codominant when paired. The third, i, is recessive; a pairing with either I^A or I^B masks its effect. Together, they are a **multiple allele system**, which we define as the presence of three or more alleles of a gene among individuals of a population.

Before each glycolipid became positioned at the cell surface, it was modified in the endomembrane system (Section 4.5). An oligosaccharide chain was attached to a lipid, then a sugar was attached to the chain. Alleles I^A and I^B specify different versions of the enzyme that attaches the sugar. The enzymes attach *different* sugars, which gives the glycolipid different identities: A or B.

Which alleles do you have? With either $I^A I^A$ or $I^A i$, you have type A blood. With $I^B I^B$ or $I^B i$, your blood is type B. With codominant alleles $I^A I^B$, it's AB—meaning you have both versions of the sugar-attaching enzyme. But if you are homozygous recessive (ii), the molecules never did get the final sugar attached to them. Your blood type isn't A or B; that is what type "O" means. Figure 11.11 summarizes the possibilities.

With *transfusions*, the blood of two people mixes. The recipient's immune system perceives any incompatible red blood cells as "nonself." In such cases, it acts against them and may cause death (Section 38.4).

One allele may be fully dominant, incompletely dominant, or codominant with its partner on the homologous chromosome.

MULTIPLE EFFECTS OF SINGLE GENES

Expression of the alleles at just a single location on a chromosome may have positive or negative effects on two or more traits. This phenotypic outcome of a single gene's activity is known as **pleiotropy** (after the Greek *pleio–*, meaning more, and *–tropic*, meaning to change).

Consider the gene for fibrillin 1 (Figure 11.12). This protein occurs in the extracellular matrix of connective tissues—the most abundant and widely distributed of all tissues in the vertebrate body. By itself or together with another protein (elastin), it becomes assembled into long, thin strands that are especially abundant in the heart, skin, blood vessels, tendons, and around the eyes. The effects of fibrillin may come into play when embryos are developing. They may be important in the formation of the elastic fibers of connective tissues.

Sometimes the gene for fibrillin mutates. When that happens, many connective tissues weaken throughout the body. The result is the genetic disorder known as *Marfan syndrome*. Marfan syndrome affects 1 in 10,000 people throughout the world. It affects both women and men of any ethnicity.

Table 11.1 lists some of the key symptoms of Marfan syndrome. That single gene mutation has often severe repercussions upon the skeleton, cardiovascular system (the heart and blood vessels), lungs, eyes, and skin. For example, the large blood vessel that transports blood away from the heart, the aorta, may have a weakened wall region, which has been known to rupture during strenuous exercise (Figure 11.12). Until recent medical advances, most patients died of such cardiovascular complications before they were fifty years old.

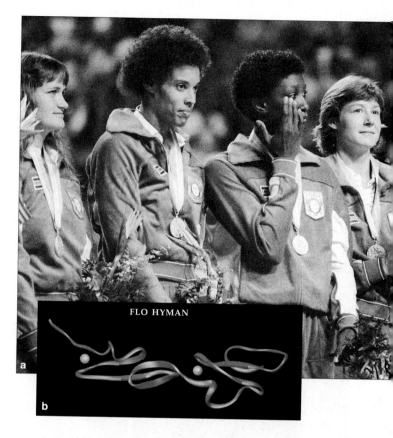

FLO HYMAN

a

b

Figure 11.12 (**a**) Flo Hyman, captain of the U.S. volleyball team that won an Olympic silver medal in 1984. Two years later, during a game in Japan, she slid silently to the floor and died. A dime-size weak spot in the wall of her aorta, the main artery transporting blood away from the heart, had abruptly burst. (**b**) Ribbon model for two calcium-binding domains that are part of the protein fibrillin. A mutation in this region is thought to give rise to Marfan syndrome. Besides Hyman, at least two college basketball stars have died abruptly as a result of this syndrome.

Table 11.1	*Pleiotropic Effects Often Associated With Marfan Syndrome*

SKELETON Affected people typically lanky, loose jointed, with limbs, fingers, toes often disproportionately long; teeth often crowded; protruding or sunken breastbone; abnormally curved spine; joint pain; flat feet.

HEART AND BLOOD VESSELS Abnormal, often leaky valve between heart chambers; wall of aorta weakened, stretched, extremely vulnerable to tearing or rupturing.

LUNGS Tiny airs sacs where gases are exchanged sometimes become stretched or swollen, so risk of lung collapse.

NERVOUS SYSTEM Protective connective tissue around brain, spinal cord weakens, stretches, presses on lower backbone, causing discomfort or pain in abdominal region.

EYES Lens displacement in more than 50 percent of those affected (higher or lower, shifted to one side because fibers that hold it in place are damaged); nearsightedness common; glaucoma (high fluid pressure inside eyeball) or cataracts (cloudy lens) may develop early.

SKIN Stretch marks common.

More than 220 fibrillin 1 gene mutations have been identified. One adversely affects the biosynthesis of the protein, its secretion from cells, and its deposition into the extracellular matrix.

Think about what happens to the aorta. First, the structure and activity of all smooth muscle cells in the blood vessel wall change. Then the cells infiltrate and multiply in the epithelial lining of the wall. Calcium deposits build up in the lining, and the wall becomes enflamed. Elastic fibers are split into fragments, which leads to a thin, weakened wall.

The alleles at a single gene location may have positive or negative effects on two or more traits.

The effects may not be simultaneous. Rather, they may have repercussions over time. The gene may lead to an alteration in one trait. That change may alter another trait, and so on.

INTERACTIONS BETWEEN GENE PAIRS

Often a trait results from interactions among products of two or more gene pairs. For example, two alleles of a gene may mask expression of another gene's alleles, so some expected phenotypes may not appear at all. Such interactions between the product of pairs of genes are called **epistasis** (meaning the act of stopping).

Hair Color in Mammals

Epistasis is common among the gene pairs responsible for skin or fur color in mammals. Consider the black, brown, or yellow fur of Labrador retrievers (Figure 11.13). The different colors arise from variations in the amount and distribution of melanin, a brownish-black pigment. A variety of enzymes and other products of many gene pairs affect different steps in the production of melanin and its deposition in certain body regions.

The alleles of one gene specify an enzyme required to produce melanin. Expression of allele B (black) has a more pronounced effect and is dominant to b (brown). Alleles of a different gene control the extent to which molecules of melanin will be deposited in a retriever's hairs. Allele E permits full deposition. Two recessive alleles (ee) reduce deposition, and fur will be yellow.

In some individuals, those two gene pairs are not able to interact, owing to a certain allelic combination at still another gene locus. There, a gene (C) calls for tyrosinase, the first of several enzymes in a melanin-producing pathway. An individual bearing one or two dominant alleles (CC or Cc) can make the functional enzyme. An individual bearing two recessive alleles (cc) cannot. When the biosynthetic pathway for melanin production gets blocked, then *albinism*—the absence of melanin—is the resulting phenotype (Figure 11.14).

a BLACK LABRADOR

b YELLOW LABRADOR

c CHOCOLATE LABRADOR

Figure 11.13 The heritable basis of coat color among Labrador retrievers. The trait arises through interactions among the alleles of two pairs of genes.

One kind of gene is involved in melanin production. Allele B (black) of this gene is dominant to allele b (brown). A different kind of gene influences the deposition of melanin pigment in individual hairs. Allele E of this gene promotes melanin deposition, but a pairing of recessive alleles (ee) of the gene blocks deposition, and a yellow coat results.

F$_1$ offspring of a dihybrid cross produce F$_2$ offspring in a 9:3:4 ratio, as the Punnett-square diagram at *right* indicates.

The yellow Labrador in (**b**) probably has genotype BBee, because it can produce melanin but cannot deposit pigment in hairs. After thinking about this Punnett square, can you guess why?

HOMOZYGOUS PARENTS: BBEE × bbee
↓
F$_1$ PUPPIES: BbEe
↓

ALLELIC COMBINATIONS POSSIBLE AMONG F$_2$ PUPPIES:

	BE	Be	bE	be
BE	BBEE	BBEe	BbEE	BbEe
Be	BBEe	BBee	BbEe	Bbee
bE	BbEE	BbEe	bbEE	bbEe
be	BbEe	Bbee	bbEe	bbee

RESULTING PHENOTYPES:

☐ 9/16 or 9 black
☐ 3/16 or 3 brown
☐ 4/16 or 4 yellow

Figure 11.14 A rare albino rattlesnake. Like other animals that can't produce melanin, its body surface is white, overall, and its eyes are pink. In birds and mammals, surface coloration arises largely from pigments in feathers, fur, or skin. In fishes, amphibians, and reptiles, it depends on color-bearing cells. Some of these cells contain melanin or yellow-to-red pigments. Others contain crystals that reflect light and thus alter the surface coloration.

The mutation affecting melanin production in the snake shown here had no effect on the production of its yellow-to-red pigments and light-reflecting crystals. That's why the snake's skin appears to be iridescent yellow as well as white. Its eyes look pink because melanin is absent from a tissue layer in each eyeball. Without melanin to absorb it, red light is reflected from blood vessels in the eyes.

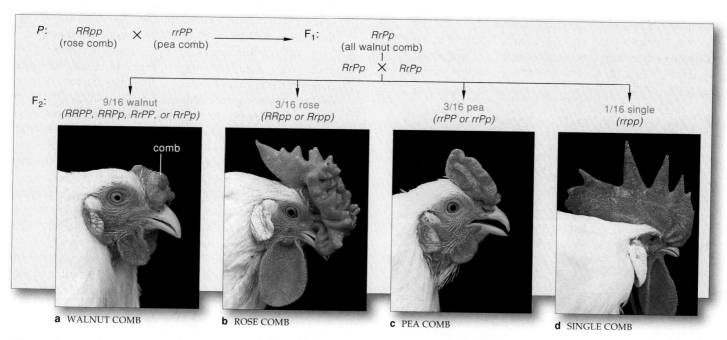

Figure 11.15 Interaction between two genes that affect the same trait in domestic chicken breeds. The initial cross is between a Wyandotte (with a rose comb, **b**, on the crest of its head) and brahma (pea comb, **c**). With complete dominance at the locus for pea comb and at the locus for rose comb, products of the two gene pairs interact and give rise to a walnut comb (**a**). With full recessiveness at both gene loci, products of the genes interact and give rise to a single comb (**d**).

Comb Shape in Poultry

In some cases, interaction between two gene pairs results in a phenotype that neither pair can produce by itself. Geneticists W. Bateson and R. Punnett identified two interacting gene pairs (*R* and *P*) that affect comb shape in chickens. Allelic combinations of *rr* at one gene locus and *pp* at the other locus result in the least common phenotype, the single comb. The presence of dominant alleles (*R*, *P*, or both) results in varied phenotypes.

Take a look at Figure 11.15. The diagram shows the combinations of alleles that interact to specify the rose, pea, and walnut combs shown in the photographs.

Gene interactions affect phenotype, as when alleles of one gene mask the expression of another gene, and when some expected phenotypes may not appear at all.

HOW CAN WE EXPLAIN LESS PREDICTABLE VARIATIONS?

Regarding the Unexpected Phenotype

As Mendel demonstrated, the phenotypic effects of one or two pairs of certain genes show up in predictable ratios when you track them from one generation to the next. Besides this, interactions among two or more gene pairs also can produce phenotypes in predictable ratios, as the example of Labrador coat color demonstrated in Section 11.6.

However, even if you were to track a single gene over the generations, you might find that the resulting phenotypes were not quite what you had expected.

Consider *camptodactyly*, a rare genetic abnormality that affects the shape and the movement of fingers. People who carry the mutant allele for this heritable trait have immobile, bent fingers on both hands. Other people have immobile, bent fingers on the left or right hand only. The fingers of others who carry the mutant allele aren't affected in any obvious way.

What causes such odd variation? Remember, most organic compounds are synthesized by a sequence of metabolic steps. *And different enzymes, each the product of a gene, regulate different steps.* Maybe one gene mutated in one of a number of possible ways. Maybe the gene product blocks the pathway or causes it to run nonstop or not long enough. Maybe poor nutrition or another factor that is variable in the individual's environment affects a crucial enzyme in the pathway. These are the kinds of variable factors that often introduce far less predictable variations in the phenotypes resulting from gene expression.

Continuous Variation in Populations

Generally, individuals of a population display a range of small differences in most traits. This characteristic of populations is called **continuous variation**. It's mainly an outcome of the number of genes affecting a trait and the number of environmental factors influencing their expression. Usually, the greater the number of genes and environmental factors, the more continuous will be the expected distribution of all the versions of that trait.

Look in a mirror at your eye color. The colored part is the iris, a doughnut-shaped, pigmented structure just beneath the cornea. Its color is the cumulative outcome of a number of gene products. Some products take part in the stepwise production and distribution of melanin, the same light-absorbing pigment that influences coat color in mammals. Dark eyes that seem to be almost black have dense deposits of melanin molecules inside the iris. These molecules absorb most of the incoming light. Melanin deposits aren't as extensive in brown eyes, and some unabsorbed light is being reflected out. Light brown or hazel eyes have even less (Figure 11.16).

Green, gray, or blue eyes don't have green, gray, or blue pigments. Their iris incorporates some amount of melanin, but not much. As a result, many or most of the blue wavelengths of light that do enter the eyeball are reflected out.

How might you describe the continuous variation of some trait within a group, such as the students in Figure 11.17a? The students range from very short to very tall, with average heights much more common than either extreme. You might start out by dividing the full range of different phenotypes into measurable categories. Next, you can count all of the individual students in each category. Doing so will give you the relative frequencies of all the phenotypes, distributed across the range of measurable values.

The bar chart in Figure 11.17c plots the proportion of students in each category against the range of the measured phenotypes. Here, the shortest vertical bars represent categories with the least number of students. The tallest bar represents the category with the greatest number of students. Finally, draw a graph line around all the bars and you end up with a bell-shaped curve.

Figure 11.16 A small sampling from the range of continuous variation in the color of human eyes. The products of different pairs of genes interact in producing and then distributing the pigment melanin. Among other things, melanin helps color the eye's iris. Different combinations of alleles result in small differences in eye color. Thus, the frequency distribution for the eye-color trait is continuous over a range from black to light blue.

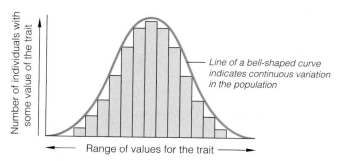

a Two examples of continuous variation: Biology students (males, *above*; females, *right*) organized according to height.

Line of a bell-shaped curve indicates continuous variation in the population

Number of individuals with some value of the trait

← Range of values for the trait →

b Idealized bell-shaped curve for a population that displays continuous variation in a trait.

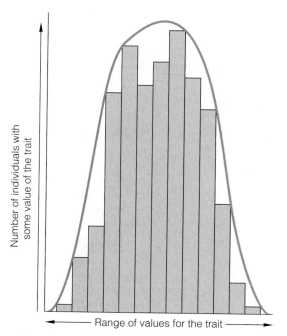

Number of individuals with some value of the trait

← Range of values for the trait →

c Specific bell curve corresponding to the distribution of a trait (height) among the females in the far-right photograph in (**a**).

Figure 11.17 Continuous variation in body height, a trait that is one of the characteristics of the human population.

(**a**) Jon Reiskind and Greg Pryor decided to show the frequency distribution for height among biology students at the University of Florida. They divided students into two groups: male and female. For each group, they divided the range of possible heights, measured the students, and assigned each to the appropriate category.

(**b**) A bar graph is commonly used to depict continuous variation in a population. In such graphs, the proportion of individuals in each category is plotted against the range of measured phenotypes. The curved line above this particular set of bars is an idealized example of the kind of "bell-shaped" curve that emerges for populations showing continuous variation in a trait. The bell-shaped curve in (**c**) is a specific example of this type of diagram.

Such "bell curves" are typical of populations that show continuous variation in a trait.

Enzymes and other products of genes regulate each step of most metabolic pathways. Mutations, gene interactions, and environmental conditions may affect one or more of the steps. The outcome is variation in phenotypes.

For most traits, individuals of a population show continuous variation—that is, a range of small differences.

The greater the number of genes and environmental factors that can influence a trait, the more continuous will be the expected distribution of all versions of that trait.

ENVIRONMENTAL EFFECTS ON PHENOTYPE

We have mentioned, in passing, that the environment often contributes to variable gene expression among a population's individuals. Before leaving this chapter, consider a few cases of phenotypic variations that arise from environmental effects on genotype.

Possibly you have thought about the fur color of a Himalayan rabbit or Siamese cat. These mammals both have dark fur in some body regions and lighter fur in others. Let's just focus on Himalayan rabbits. They are homozygous for the c^h allele of the gene for tyrosinase. Tyrosinase is one of the enzymes necessary to make melanin. The c^h allele specifies a heat-sensitive form of the enzyme. And that enzyme is active only when the air temperature around the body is below about 33°C.

When cells that give rise to this rabbit's hairs grow under warmer conditions, they cannot make melanin, so the hairs appear light. This happens in body regions that are massive enough to conserve a fair amount of metabolic heat. Ears and other slender extremities are cooler because they tend to lose metabolic heat faster.

Figure 11.18 gives an experiment that demonstrated an environmental effect on this allele.

We also can identify environmental effects on genes that govern plant phenotypes. Consider an experiment with yarrow plants. Yarrow plants are able to grow from cuttings. This makes them a good experimental organism. Why? Cuttings from the same plant will all have the same genotype, so experimenters can discount genes as a basis for differences that appear among them.

In this case, three yarrow cuttings were grown at three elevations. The two plants grown at the lowest and highest elevation fared best; the one at the medium elevation grew poorly, as Figure 11.19 shows.

But remember the sampling error trap (Chapter 1)? The researchers did the same growth experiments for *many* yarrow plants. They saw no consistent pattern in phenotypic variation; it was too great. For instance, a cutting from one plant developed best at the medium elevation. The conclusion? Different yarrow genotypes react differently across a range of environments.

Icepack is strapped onto a hair-free patch.

New hair growing in patch exposed to cold is black.

Figure 11.18 Observable effect of differences in environmental conditions on the expression of genes in animals. A Himalayan rabbit normally has black hair only on its long ears, nose, tail, and lower leg limbs. For one experiment, a patch of a rabbit's white fur was plucked clean, then an icepack was secured over the hairless patch. Where the colder temperature had been maintained, the hairs that grew back were black.

Himalayan rabbits are homozygous for an allele of the gene for tyrosinase, an enzyme required to make melanin. As described in the text, this allele specifies a heat-sensitive form of the enzyme, which functions only when air temperature is below about 33°C.

a Mature cutting at high elevation (3,050 meters)

b Mature cutting at medium elevation (1,400 meters)

c Mature cutting at low elevation (30 meters above sea level)

Figure 11.19 Experiment demonstrating the effect of environmental conditions on gene expression in a yarrow plant (*Achillea millefolium*). Cuttings from the same parent plant were grown in the same soil batch but at three different elevations.

Figure 11.20 Environmental effect on gene expression in *Hydrangea macrophylla*, a common garden plant. Different plants that carry the same alleles may have floral colors ranging from pink to blue. This case of color variation arises because of differences in the acidity of soil in which a plant happens to be growing.

Figure 11.21 A San Diego State University student exhibiting the tongue-rolling trait to a tongue-roll-challenged student. Once thought to be genetically determined, this trait is mostly learned in the individual's environment.

As another example, plant a hydrangea in a garden and it may make pink blossoms instead of the expected blue ones, depending on its environment (Figure 11.20). Genes for its floral color are affected by soil acidity.

As one more example, for years, tongue-rolling was viewed as a genetically determined trait (Figure 11.21). It now appears to be mostly learned by individuals.

And so we conclude this chapter, which has dealt with heritable and environmental factors that give rise to variations in phenotype. What is the take-home lesson? Simply this: An individual's phenotype is an outcome of complex interactions among its genes, enzymes and other gene products, and environmental factors.

Individuals of most populations or species show complex variation for many traits. The variation arises not only from gene mutations and cumulative gene interactions. It arises also in response to variations in environmental conditions.

$\mathcal{SUMMARY}$ *Gold* indicates text section

1. A gene is a unit of information about a heritable trait. Alleles of a gene are different molecular versions of that information. Through experimental crosses with pea plants, Mendel gathered indirect evidence that diploid organisms have pairs of genes, and that genes retain their identity when transmitted to offspring. *CI, 11.2*

2. For a particular trait being studied, an individual who has inherited two dominant alleles (AA) is said to be homozygous dominant. A homozygous recessive individual has inherited two recessive alleles (aa), and a heterozygote has two nonidentical alleles (Aa). *11.1*

3. An individual's specific combination of alleles is its genotype. Its observable traits are its phenotype. *11.1*

4. A hybrid is an individual offspring from any cross between parents of different genotypes. In monohybrid crosses, two individuals that breed true for different versions of the same trait produce F_1 offspring that are identically heterozygous at one gene pair. *11.1, 11.2*

5. Mendel's monohybrid crosses between garden pea plants gave indirect evidence that some forms of a gene may be dominant over other, recessive forms. *11.2*

6. All F_1 offspring of the parental cross $AA \times aa$ were Aa. Crosses between F_1 monohybrids resulted in these combinations of alleles in F_2 offspring: *11.2*

	A	a
A	AA	Aa
a	Aa	aa

AA (dominant)
Aa (dominant)   ⎫ the expected
Aa (dominant)   ⎬ phenotypic
aa (recessive)   ⎭ ratio of 3:1

7. Results from Mendel's monohybrid crosses led to the formulation of a theory of segregation. In modern terms, diploid organisms have pairs of genes, on pairs of homologous chromosomes. The genes of each pair segregate from each other at meiosis, so each gamete formed ends up with one or the other gene. *11.2*

8. For dihybrid crosses, individuals that breed true for different versions of *two* traits produce F_1 offspring that are all identical heterozygotes for both genes. The experiment is an intercross between two F_1 dihybrids. Phenotypes of the F_2 offspring from Mendel's dihybrid crosses were close to a 9:3:3:1 ratio: *11.3*

    9 dominant for both traits
    3 dominant for A, recessive for b
    3 dominant for B, recessive for a
    1 recessive for both traits

9. Mendel's dihybrid crosses led to his formulation of the theory of independent assortment. In modern terms, by the end of meiosis, the gene pairs of two homologous chromosomes have been sorted out for distribution into one gamete or another, independently of how the gene pairs of other chromosomes were sorted out. *11.3*

10. Certain factors influence gene expression. *11.4–11.8*

    a. In some cases, one allele of a pair is incompletely dominant or codominant. *11.4*

    b. Products of pairs of genes often interact in ways that influence the same trait. *11.5–11.8*

    c. One gene may have positive or negative effects on two or more traits, a condition called pleiotropy. *11.5*

    d. Environmental conditions to which an individual is subjected may affect gene expression. *11.8*

Review Questions

1. Distinguish between these terms: *11.1*
    a. gene and allele
    b. dominant allele and recessive allele
    c. homozygote and heterozygote
    d. genotype and phenotype

2. What is a true-breeding lineage? A hybrid? *11.1*

3. Distinguish between monohybrid cross, dihybrid cross, and testcross. *11.2, 11.3*

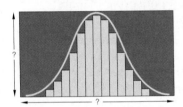

4. Do segregation and independent assortment occur during mitosis, meiosis, or both? *11.2, 11.3*

5. In the bell-shaped curve at left, a diagram of continuous variation in a population, what do the bars and the curved line represent? *11.7*

Self-Quiz ANSWERS IN APPENDIX III

1. Alleles are _____ .
    a. different molecular forms of a gene
    b. different phenotypes
    c. self-fertilizing, true-breeding homozygotes

2. A heterozygote has a _____ for the trait being studied.
    a. pair of identical alleles
    b. pair of nonidentical alleles
    c. haploid condition, in genetic terms
    d. a and c

3. The observable traits of an organism are its _____ .
    a. phenotype   c. genotype
    b. sociobiology   d. pedigree

4. Second-generation offspring from a cross are the _____ .
    a. F_1 generation   c. hybrid generation
    b. F_2 generation   d. none of the above

5. F_1 offspring of the monohybrid cross $AA \times aa$ are _____ .
    a. all AA c. all Aa
    b. all aa d. 1/2 AA and 1/2 aa

6. Refer to question 5. Assuming complete dominance, the F_2 generation will show a phenotypic ratio of _____ .
    a. 3:1   b. 9:1   c. 1:2:1   d. 9:3:3:1

7. Crosses between F_1 pea plants resulting from the cross $AABB \times aabb$ lead to F_2 phenotypic ratios close to _____ .
    a. 1:2:1   b. 3:1   c. 1:1:1:1   d. 9:3:3:1

8. Match each example with the most suitable description.
    ____ dihybrid cross   a. *bb*
    ____ monohybrid cross   b. *AABB* × *aabb*
    ____ homozygous condition   c. *Aa*
    ____ heterozygous condition   d. *Aa* × *Aa*

Critical Thinking—Genetics Problems
ANSWERS IN APPENDIX IV

1. One gene has alleles *A* and *a*. Another has alleles *B* and *b*. For each genotype listed, what type(s) of gametes will be produced? Assume independent assortment occurs before gametes form.
    a. *AABB*   c. *Aabb*
    b. *AaBB*   d. *AaBb*

2. Refer to Problem 1. What will be the genotypes of offspring from the following matings? Indicate the frequencies of each genotype among them.
    a. *AABB* × *aaBB*   c. *AaBb* × *aabb*
    b. *AaBB* × *AABb*   d. *AaBb* × *AaBb*

3. In one experiment, Mendel crossed a pea plant that bred true for green pods with one that bred true for yellow pods. All the F_1 plants had green pods. Which form of the trait (green or yellow pods) is recessive? Explain how you arrived at your conclusion.

4. Return to Problem 1. Assume you now study a third gene having alleles *C* and *c*. For each genotype listed, what type(s) of gametes will be produced?

    a. *AABB CC*   c. *Aa BB Cc*
    b. *Aa BB cc*   d. *Aa Bb Cc*

5. Mendel crossed a true-breeding tall, purple-flowered pea plant with a true-breeding dwarf, white-flowered plant. All F_1 plants were tall and had purple flowers. If an F_1 plant self-fertilizes, then what is the probability that a randomly selected F_2 offspring will be heterozygous for the genes specifying height and flower color?

6. *DNA fingerprinting* is a method of identifying individuals by locating unique base sequences in their DNA molecules (Section 16.3). Before researchers refined the method, attorneys often relied on the ABO blood-typing system to settle disputes over paternity. Suppose that you, as a geneticist, are asked to testify during a paternity case in which the mother has type A blood, the child has type O blood, and the alleged father has type B blood. How would you respond to the following statements?
    a. Attorney of the alleged father: "The mother's blood is type A, so the child's type O blood must have come from the father. My client has type B blood; he could not be the father."
    b. Mother's attorney: "Because further tests prove this man is heterozygous, he must be the father."

7. Suppose you identify a new gene in mice. One of its alleles specifies white fur. A second allele specifies brown fur. You want to determine whether the relationship between the two alleles is one of simple dominance or incomplete dominance. What sorts of genetic crosses would give you the answer? On what types of observations would you base your conclusions?

8. Your sister moves away and gives you her purebred Labrador retriever, a female named Dandelion. Suppose you decide to breed Dandelion and sell puppies to help pay for your college tuition. Then you discover that two of her four brothers and sisters show *hip dysplasia*, a heritable disorder arising from a number of gene interactions. If Dandelion mates with a male Labrador known to be free of the harmful alleles, can you guarantee to a buyer that puppies will not develop the disorder? Explain your answer.

9. A dominant allele *W* confers black fur on guinea pigs. A guinea pig that is homozygous recessive (*ww*) has white fur. Fred would like to know whether his pet black-furred guinea pig is homozygous dominant (*WW*) or heterozygous (*Ww*). How might he determine his pet's genotype?

10. Red-flowering snapdragons are homozygous for allele R^1. White-flowering snapdragons are homozygous for a different allele (R^2). Heterozygous plants (R^1R^2) bear pink flowers. What

phenotypes should appear among first-generation offspring of the crosses listed? What are the expected proportions for each phenotype?

a. $R^1R^1 \times R^1R^2$ c. $R^1R^2 \times R^1R^2$
b. $R^1R^1 \times R^2R^2$ d. $R^1R^2 \times R^2R^2$

Notice, in Problem 10, that in cases of incomplete dominance it is inappropriate to refer to either allele of a pair as dominant or recessive. When the phenotype of a heterozygous individual is halfway between those of the two homozygotes, then there is no dominance. Such alleles are usually designated by superscript numerals, as shown here, rather than by uppercase letters for dominance and lowercase letters for recessiveness.

11. Two pairs of genes affect comb type in chickens (Figure 11.15). When both genes are recessive, a chicken has a single comb. A dominant allele of one gene, P, gives rise to a pea comb. Yet a dominant allele of the other (R) gives rise to a rose comb. An epistatic interaction occurs when a chicken has at least one of both dominants, $P\_ R \_$, which gives rise to a walnut comb. Predict the ratios resulting from a cross between two walnut-combed chickens that are heterozygous for both genes ($PpRr$).

12. As described in Section 3.8, a single mutant allele gives rise to an abnormal form of hemoglobin (Hb^S instead of Hb^A). Homozygotes (Hb^SHb^S) develop sickle-cell anemia. But the heterozygotes (Hb^AHb^S) show few outward symptoms.

Suppose a woman's mother is homozygous for the Hb^A allele. She marries a male who is heterozygous for the allele, and they plan to have children. For *each* of her pregnancies, state the probability that this couple will have a child who is:

a. homozygous for the Hb^S allele
b. homozygous for the Hb^A allele
c. heterozygous Hb^AHb^S

13. Certain dominant alleles are so vital for normal development that an individual who is homozygous recessive for a mutant recessive form of the allele is unable to survive. Such recessive, *lethal alleles* can be perpetuated by heterozygotes.

Consider the Manx allele (M^L) in cats. Homozygous cats (M^LM^L) die when they are still embryos inside the mother cat. In heterozygotes (M^LM), the spine develops abnormally, and the cats end up with no tail whatsoever (Figure 11.22).

Two M^LM cats mate. Among their *surviving* progeny, what is the probability that any one kitten will be heterozygous?

14. A recessive allele c is responsible for *albinism*, an inability to produce or deposit melanin in tissues. Humans and some other organisms can have this phenotype (Figure 11.23). In each of the following cases, what are the possible genotypes of the father, of the mother, and of their children?

Figure 11.22 Manx cat, which has no tail.

Figure 11.23 An albino male in India.

a. Both parents have normal phenotypes; some of their children are albino and others are unaffected.

b. Both parents are albino and have only albino children.

c. The woman is unaffected, the man is albino, and they have one albino child and three unaffected children.

15. Kernel color in wheat plants is determined by two pairs of genes. Alleles of one pair show incomplete dominance over alleles of the other pair.

For the gene pair at one locus on the chromosome, allele A^1 imparts one dose of red color to the kernel, whereas allele A^2 does not. At the second locus, allele B^1 gives one dose of red color to the kernel, whereas allele B^2 does not. One kernel with genotype $A^1A^1B^1B^1$ is dark red. A different kernel with genotype $A^2A^2B^2B^2$ is white. All other genotypes have kernel colors in between the two extremes.

a. Suppose you cross a plant grown from a dark red kernel with a plant grown from a white kernel. What genotypes and what phenotypes would you expect among the offspring?

b. If a plant with genotype $A^1A^1B^1B^2$ self-fertilizes, what genotypes and what phenotypes would be expected among the offspring? In what proportions?

Selected Key Terms

Readings

Fairbanks, D. J., and W. R. Andersen. 1999. *Genetics: The Continuity of Life.* Monterey, California: Brooks-Cole.

Orel, V. 1996. *Gregor Mendel: The First Geneticist.* New York: Oxford University Press.

12

HUMAN GENETICS

The Philadelphia Story

Positioned at strategic locations in chromosomes are genes concerned with the cell cycle—that is, with cell growth and division. Some specify enzymes and other proteins that perform these tasks. Other genes control whether, when, and how fast the tasks are completed. When something disrupts the controls, cell growth and division can spiral out of control and lead to cancer.

The first abnormal chromosome tied to cancer was named the *Philadelphia chromosome* after the city where someone discovered it. This chromosome shows up in cells of some people affected by a type of leukemia.

The disorder starts with stem cells in bone marrow. All stem cells are unspecialized and retain the capacity for mitotic cell division. A portion of their descendants divide and become specialized. With leukemias, there are too many descendants—far too many of the white blood cells in charge of housekeeping and defense.

Leukemic cells may crowd out the stem cells that give rise to red blood cells and platelets. Anemia and internal bleeding follow. Leukemic cells also infiltrate blood and the liver, lymph nodes, spleen, and other organs, where they interfere with basic functions. Left to itself, the cancerous transformation kills the patient.

No one knew about the Philadelphia chromosome until microscopists learned how to identify its physical appearance. Chromosomes, recall, are most condensed at metaphase of mitosis. Then, their size, length, and centromere location are easiest to identify. A **karyotype** is a preparation of metaphase chromosomes based on their defining features. In this chapter, you will learn how to make a karyotype diagram from a photograph of metaphase chromosomes, just as microscopists do. Such diagrams yield useful information when they are compared with a standard karyotype for a species.

The Philadelphia chromosome shows up clearly with *spectral* karyotyping. This newer research and diagnostic tool artificially colors chromosomes, and the colors are clues to their structure (Figure 12.1).

As it turns out, the Philadelphia chromosome is physically longer than its normal counterpart, human chromosome 9. The extra length is actually a piece of chromosome 22! What happened? By chance, both chromosomes broke inside a stem cell. Each broken piece was reattached, but on the wrong chromosome. At the broken end of chromosome 9, a gene with a role in cell division fused with the control region of a gene at the broken end of chromosome 22. In its mutant fusion form, that gene is expressed far more than it should be. Uncontrolled divisions of white blood cells are the outcome (Figure 12.2).

The Philadelphia story is a glimpse into the world of modern genetics research. Reading it invites you to think about how far you have come in this unit of the book. You first looked at cell division, the starting point of inheritance. You looked at how chromosomes and the genes they carry are shuffled during meiosis, then at fertilization. You also mulled over Mendel's insights into inheritance and some exceptions to his conclusions. Here, you have a current example of what we know about the chromosomal basis of inheritance.

How did we get to that understanding? The answer requires a bit of history, which picks up from where we left Mendel.

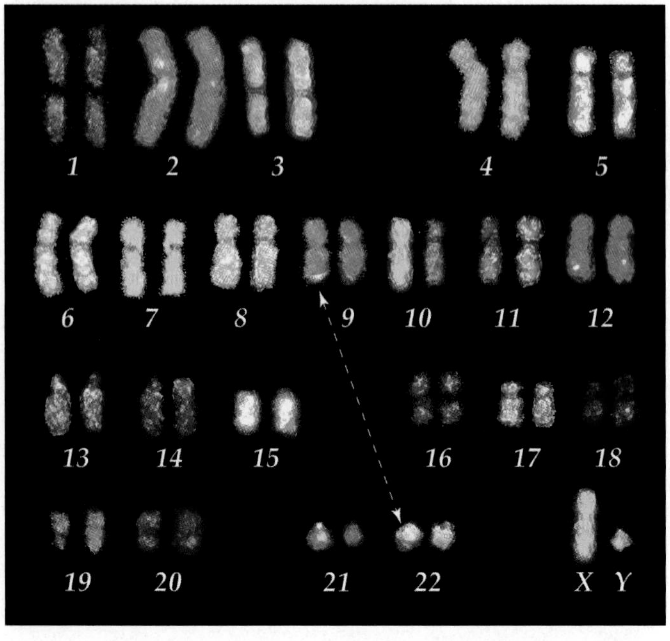

Figure 12.1 Revealing the origin of a killer—how reciprocal translocation between human chromosome 9 and chromosome 22 might appear with the help of spectral karyotyping, a new imaging technique. This particular translocation, the Philadelphia chromosome, results in a gene abnormality that gives rise to chronic myelogenous leukemia (CML), a type of cancer. Gleevec, a new oral drug, inactivates the abnormal protein product of the gene, thus arresting the dangerous cell proliferation. In preliminary tests, the drug put fifty of fifty-three CML patients into remission. So far, its side effects have been mild, compared to severe side effects of chemotherapy.

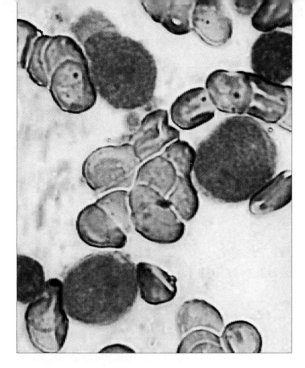

Figure 12.2 Blood typical of chronic myelogenous leukemia. Abnormal, immature white blood cells are starting to crowd out the normal cell types.

Key Concepts

1. The cells of humans and many other sexually reproducing species contain pairs of homologous chromosomes that interact during meiosis. Typically, one chromosome of each pair is maternal in origin, and its homologue is paternal in origin.

2. Each gene has its own specific location, or locus, in a particular chromosome.

3. The molecular form of a gene occupying a given locus may be slightly different from one chromosome to the next. All of the different molecular forms of a gene are called alleles.

4. The combination of alleles along the length of a chromosome does not necessarily remain intact through meiosis and gamete formation. By the event called crossing over, some alleles along its length swap places with their partner on the homologous chromosome. Alleles that swap places may or may not be identical.

5. Allelic recombinations contribute to variations in the phenotypes of offspring.

6. A chromosome may change structurally, as when a segment of it is deleted, duplicated, inverted, or moved to a different location. Also, the chromosome number of an individual's cells may change as a result of an improper separation of duplicated chromosomes during meiosis or mitosis.

7. Chromosome structure and chromosome number rarely change. When changes do occur, they may result in genetic abnormalities or genetic disorders.

In 1884 Mendel had just passed away, and his paper on pea plants had been gathering dust in a hundred libraries for nearly two decades. Then the resolving power of microscopes improved and rekindled interest in the hereditary material. Walther Flemming had seen threadlike bodies—chromosomes—in dividing cells. Could chromosomes be the hereditary material?

Microscopists soon realized that each gamete has half the number of chromosomes of a fertilized egg. In 1887, August Weismann hypothesized that a special division process halves the chromosome number before gametes form. Sure enough, meiosis was discovered that same year. Weismann promoted another hypothesis: If the halved chromosome number is restored at fertilization, then half of the chromosomes in our cells must come from the father and half from the mother. His view was hotly debated. It prompted a flurry of experimental crosses—just like the ones Mendel had carried out.

Finally, in 1900, researchers came across Mendel's paper while checking literature related to their own genetic crosses. To their surprise, their experimental results confirmed what Mendel's results had already suggested: Diploid cells generally have two copies of each gene, and the two copies segregate from each other before gametes form.

In the decades to follow, researchers learned more about chromosomes. You'll come across a few high points of their work as we continue with our look at inheritance. As the Philadelphia story tells you, the methods of analysis are not remote from your interests. An inherited collection of information in chromosomal DNA gives rise to traits that, for better or worse, define each organism, young and old alike.

CHROMOSOMES AND INHERITANCE

Genes and Their Chromosome Locations

Earlier chapters described the structure of chromosomes and what happens to them during meiosis. To refresh your memory and get a general sense of where you are going from here, take a moment to read this list:

1. **Genes** are units of information about heritable traits. The genes of eukaryotic cells are distributed among chromosomes. Each gene has its own location —a gene locus—in one type of chromosome.

2. Any cell with a diploid chromosome number ($2n$) has inherited pairs of **homologous chromosomes**. All but one pair are identical in length, shape, and gene sequence. The single exception is a pairing of nonidentical sex chromosomes, such as X with Y. The two members of a pair of homologous chromosomes interact and segregate from each other during meiosis.

3. A gene at one locus may have the same form or a slightly different one compared to its partner gene on the homologous chromosome. When considering a population as a whole, *which* forms are inherited usually varies from one individual to the next.

4. All the different molecular forms of a gene that are possible at a given locus are called **alleles**. New alleles arise through mutation.

5. A *wild-type* allele is the most common form of a gene, either in a natural population or in a standard, laboratory-bred strain of a species. Any one of the less common forms of a gene is a *mutant* allele.

6. Genes on the same chromosome are physically tied together. The farther apart two of the genes are, the more vulnerable they are to **crossing over**. By this event, homologous chromosomes exchange corresponding segments (Figure 12.3a,b).

7. Crossing over results in **genetic recombination**— combinations of alleles in chromosomes that were not present in the parental cell (Figure 12.3c).

8. **Independent assortment** refers to the random alignment of each pair of homologous chromosomes at metaphase I of meiosis. It results in nonparental combinations of alleles in gametes and offspring.

9. On rare occasions, the structure of chromosomes changes abnormally during mitosis or meiosis. So does the parental chromosome number.

Autosomes and Sex Chromosomes

In all but one case, a pair of homologous chromosomes normally are alike in length, shape, and gene sequence. Microscopists discovered the exception in the late 1800s with analytical methods such as karyotyping (Section 12.2). A distinctive chromosome is present in females *or* males of many species, but not both. As an example,

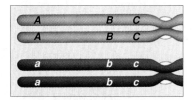

a A pair of duplicated homologous chromosomes (two sister chromatids each). This example has nonidentical alleles at three gene loci (*A* with *a*, *B* with *b*, and *C* with *c*).

b In prophase I of meiosis, a crossover event occurs: two nonsister chromatids exchange corresponding segments.

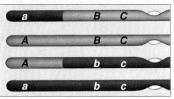

c This is the outcome of the crossover: genetic recombination between nonsister chromatids (which are shown after meiosis as unduplicated, separate chromosomes).

Figure 12.3 Review of crossing over. As shown earlier in Section 10.4, this event occurs in prophase I of meiosis.

a human male diploid cell has one **X chromosome** and one **Y chromosome** (XY). A human female diploid cell has two X chromosomes (XX). This inheritance pattern is common among mammals, fruit flies, and many other species. We see a different pattern in butterflies, moths, some birds, and certain fishes. For them, inheriting two identical sex chromosomes results in a male. Inheriting two nonidentical sex chromosomes results in a female.

The human X and Y chromosomes differ physically; one is shorter than the other (Figure 12.4f). They also differ in which genes they carry. Even so, they are able to synapse (become zippered together briefly) in a small region along their length. That bit of zippering allows them to function as homologues during meiosis.

The human X and Y chromosomes fall into the more general category of **sex chromosomes**. The term refers to types of chromosomes that, in certain combinations, determine a new individual's sex—whether a male or a female will develop. All other chromosomes in cells are the same in both sexes; they are **autosomes**.

Diploid cells have pairs of genes, on pairs of homologous chromosomes. At each gene locus, the alleles (alternative forms of a gene) may be either identical or nonidentical.

Crossing over and other events during meiosis give offspring new combinations of alleles and parental chromosomes.

Abnormal events during meiosis or mitosis can change the structure and number of chromosomes.

Autosomes are the pairs of chromosomes that are the same in males and females of a species. One other pair, the sex chromosomes, govern the sex of a new individual.

Karyotyping Made Easy

Karyotype diagrams help answer questions about an individual's chromosomes. Chromosomes are the most condensed and easiest to identify in cells going through metaphase of mitosis. Technicians don't count on finding a cell that happens to be dividing in the body when they go looking for it. They culture cells in vitro (literally, "in glass"). They put a sample of cells, usually from blood, into a glass container. The container holds a solution that stimulates cell growth and mitotic cell divisions.

Adding colchicine to a culture medium arrests cell division at metaphase. This extract of *Colchicum* plants also blocks spindle formation (Section 4.9). If a spindle can't form, sister chromatids of duplicated chromosomes won't be able to separate at anaphase. By using suitable colchicine concentrations and exposure times, technicians can stockpile metaphase cells. This improves the odds of finding candidates for karyotype diagrams.

Following colchicine treatment, the culture medium is transferred to tubes of a centrifuge (Figure 12.4a). Cells have greater mass and density than the solution bathing them. The spinning force moves them farthest from the center of rotation, to the bottom of the attached tubes.

The cells are transferred to a saline solution. When immersed in this hypotonic fluid, they swell (by osmosis) and move apart. The metaphase chromosomes also move apart. The cells are ready to be mounted on a microscope slide, fixed as by air-drying, and stained.

Chromosomes take up some stains uniformly along their length, which allows identification of chromosome size and shape. By other staining procedures, horizontal bands show up along the length of the chromosomes of certain species. If researchers direct a ray of ultraviolet light at the chromosomes treated with special fluorescent stains, the bands will fluoresce, as in Figure 10.3.

Next, the chromosomes are photographed through the microscope, and the image is enlarged. The photographed chromosomes are cut apart and arranged by size, shape, and length of their arms. All the pairs of homologous chromosomes are horizontally aligned by centromeres, as in Figure 12.4f. Today, such images of chromosomes are being cut and pasted electronically, on computers.

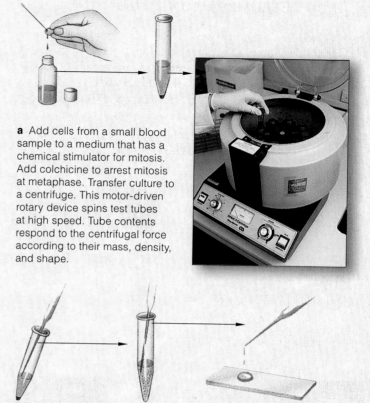

a Add cells from a small blood sample to a medium that has a chemical stimulator for mitosis. Add colchicine to arrest mitosis at metaphase. Transfer culture to a centrifuge. This motor-driven rotary device spins test tubes at high speed. Tube contents respond to the centrifugal force according to their mass, density, and shape.

b Centrifugation forces cells to bottom of tube. Draw off culture medium. Add a dilute saline solution to tube. Add a fixative.

c Prepare and stain cells for microscopy.

d Put cells on a microscope slide. Observe.

e Photograph one cell through microscope. Enlarge image of its chromosomes. Cut the image apart. Arrange chromosomes as a set.

Figure 12.4 (a–e) How to prepare karyotypes. (f) Human karyotype. Human somatic cells hold 22 pairs of autosomes and 1 pair of sex chromosomes (XX or XY). That is a diploid number of 46. These are metaphase chromosomes; each is in the duplicated state, consisting of two sister chromatids joined at the centromere.

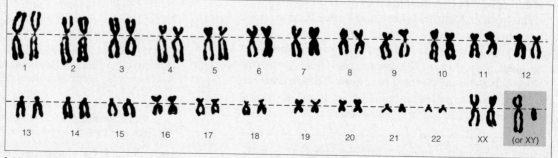

f Human karyotype

SEX DETERMINATION IN HUMANS

Genetic analysis of human cells has yielded evidence that every normal egg produced by a female has one X chromosome. Half the sperm cells produced by a male carry an X chromosome and half carry a Y.

If an X-bearing sperm fertilizes an X-bearing egg, the new individual will develop into a female. If the sperm happens to carry a Y chromosome, the individual will develop into a male (Figure 12.5).

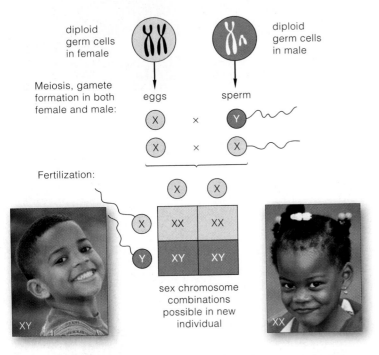

Figure 12.5 Pattern of sex determination in humans.

The human Y chromosome carries only 330 genes. However, one of them is the master gene for male sex determination. Its expression leads to the formation of testes, the primary male reproductive organs (Figure 12.6). In that gene's absence, ovaries form. Ovaries are the primary female reproductive organs. Testes and ovaries both produce important sex hormones, which influence the development of particular sexual traits.

The human X chromosome carries 2,062 genes. Like other chromosomes, it carries some genes associated with sexual traits, such as the distribution of body fat and hair. But most of its genes deal with *nonsexual* traits, such as blood-clotting functions. These genes can be expressed in males as well as in females. Males, remember, also carry one X chromosome.

A certain gene on the human Y chromosome dictates that a new individual will develop into a male. In the absence of the Y chromosome (and the gene), a female develops.

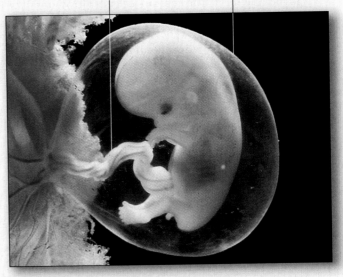

umbilical cord (lifeline between the embryo and the mother's tissues)   amnion (a protective, fluid-filled sac surrounding and cushioning the embryo)

a A human embryo, eight weeks old and about an inch long. The mass at left is part of the placenta, an organ that forms from maternal and embryonic tissues.

Figure 12.6 Boys, girls, and the Y chromosome.

For about the first four weeks of its existence, a human embryo has neither male nor female traits, regardless of whether it is XY or XX. Then ducts and other internal structures that can develop either way start forming.

(a–c) In an XX embryo, ovaries (the primary female reproductive organs) start to form *in the absence of a Y chromosome*. By contrast, in an XY embryo, testes (the primary male reproductive organs) start to form during the next four to six weeks. Apparently, a gene region on the Y chromosome governs a fork in the developmental road that can lead to maleness.

The newly forming testes start to produce testosterone and other sex hormones. These hormones are crucial for the development of a male reproductive system. By contrast, in an XX embryo, newly forming ovaries start to produce different kinds of sex hormones, particularly estrogens. Estrogens are crucial for the development of a female reproductive system.

The master gene for male sex determination is named *SRY* (short for the *S*ex-determining *R*egion of the *Y* chromosome). The same gene has been identified in DNA from male humans, chimpanzees, mice, rabbits, pigs, horses, cattle, and tigers, among others. None of the females tested had the gene. Tests with mice indicate that the gene region becomes active about the time that testes are starting to develop.

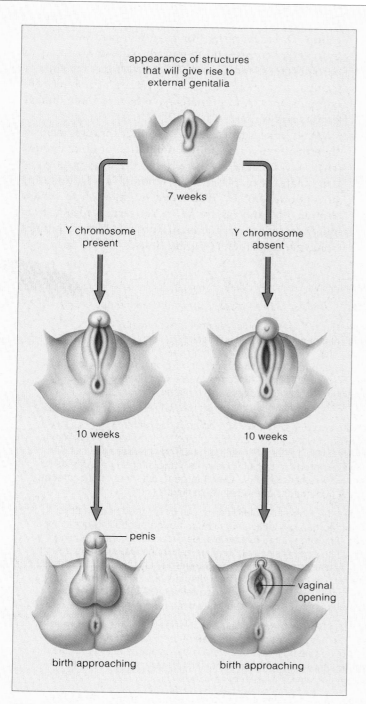

appearance of structures that will give rise to external genitalia

7 weeks

Y chromosome present

Y chromosome absent

10 weeks

10 weeks

penis

vaginal opening

birth approaching

birth approaching

b External appearance of reproductive organs forming in human embryos.

The *SRY* gene encodes certain regulatory proteins. As you will read in Chapter 15, regulatory proteins influence the activity of genes and their products. The *SRY* gene product regulates a cascade of reactions necessary for male sex determination.

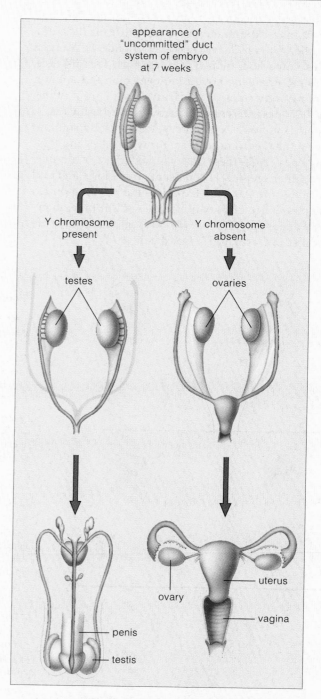

appearance of "uncommitted" duct system of embryo at 7 weeks

Y chromosome present

Y chromosome absent

testes

ovaries

ovary

penis

testis

uterus

vagina

c A duct system that forms early in the human embryo. It may develop into male *or* female primary reproductive organs, depending on whether the *SRY* gene is present or absent.

WHAT MENDEL DIDN'T KNOW: CROSSOVERS AND RECOMBINATIONS

By the early 1900s, researchers had an inkling that each gene has a specific location on a chromosome. Through hybridization experiments with mutant fruit flies, of a type called *Drosophila melanogaster*, Thomas H. Morgan and his coworkers helped confirm this. For instance, they found one gene governing eye color and another governing body color on this fruit fly's X chromosome. Figure 12.7 describes one series of their experiments.

Some *Drosophila* experiments dealt with two mutant genes on the X chromosome: *w* for white eyes and *y* for yellow body. Were these genes linked? That is, do they stay together on an X chromosome in meiosis and end up in the same gamete? Experimental results pointed to the possibility of "sex-linked genes." In time, many genes were assigned to sex chromosomes, and they are now called *X-linked* and *Y-linked* genes.

Researchers eventually identified a large number of genes on each type of chromosome—that is, a **linkage group**. *D. melanogaster*, for example, has four linkage groups that correspond to its four pairs of homologous chromosomes. Indian corn (*Zea mays*) has ten linkage groups, corresponding to its ten pairs of homologous chromosomes. And humans have twenty-three linkage groups and so on.

If all linked genes stayed together through meiosis, then there would be no recombination of linked genes. That is, you could always predict the ratio of possible combinations of phenotypes among, say, F₂ offspring of dihybrid crosses. (Here you might wish to review Section 10.4 and Figure 11.3.) Yet certain results from the *Drosophila* experiments did not match expectations. Some genes were not "tightly linked."

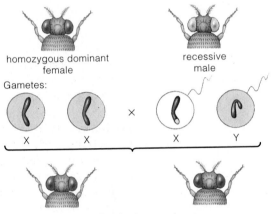

homozygous dominant female   recessive male

Gametes:

X   X   ×   X   Y

All F₁ offspring have red eyes.

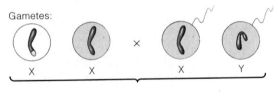

Gametes:

X   X   ×   X   Y

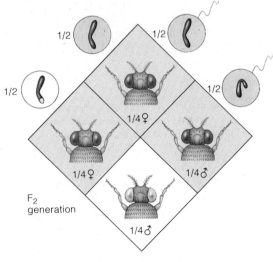

1/2   1/2

1/2   1/2

1/4♀

1/4♀   1/4♂

F₂ generation

1/4♂

Figure 12.7 X-linked genes—clues to inheritance patterns.

Thomas Morgan, an embryologist, discovered a genetic basis for a relationship between sex determination and some of the *nonsexual* traits. For instance, human males and females have blood-clotting factors, but hemophilia, a rare blood-clotting disorder, occurs most often in males. This sex-linked outcome was not like anything Mendel saw in his hybrid crosses of pea plants. For pea plants, it made no difference which parent carried a recessive allele.

Morgan was studying eye color and other nonsexual traits of *Drosophila melanogaster*. This fruit fly is a great choice for laboratory experiments. It can live in small bottles on agar, cornmeal, molasses, and yeast. A female lays hundreds of eggs in a few days and offspring reproduce in less than two weeks. In a single year, Morgan could track traits through nearly thirty generations of thousands of flies.

At first all of the flies were wild type for eye color; they had brick-red eyes. Then Morgan got lucky; mutation in a gene controlling eye color presented him with a white-eyed male. He established true-breeding strains of white-eyed males and females for **reciprocal crosses**. (In the first of such paired crosses, one parent displays the trait of interest. In the second cross, the other parent displays it.) He let white-eyed males mate with homozygous red-eyed females. All F₁ offspring had red eyes, and some F₂ males had white eyes (*see diagram*).

Then Morgan let true-breeding red-eyed males mate with white-eyed females. The results surprised him. Half of the F₁ offspring turned out to be red-eyed females, and half were white-eyed males. And of the F₂ offspring, 1/4 were red-eyed females, 1/4 white-eyed females, 1/4 red-eyed males, and 1/4 white-eyed males.

The results implied a relationship between an eye-color gene and sex determination. Probably the gene locus was on a sex chromosome. Which one? Because females (XX) could be white-eyed, the recessive allele had to be on one of their X chromosomes. What if white-eyed males (XY) had the recessive allele on their X chromosome and their Y chromosome had no corresponding eye-color allele? In that case, they would have white eyes. They would have no dominant allele to mask the effect of the recessive one, as the diagram shows.

And so Morgan's idea of an X-linked gene dovetailed with Mendel's concept of segregation. By proposing that a specific gene is located on an X chromosome but not on the Y, Morgan explained his reciprocal crosses. His experimental results matched predicted outcomes.

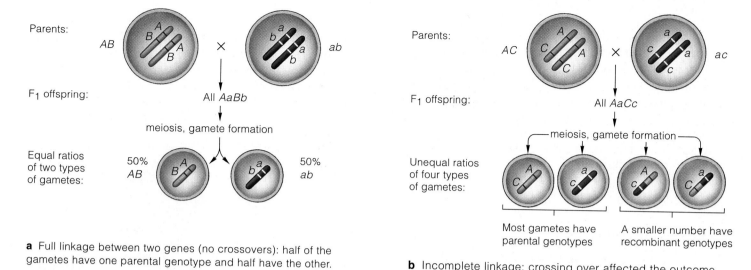

a Full linkage between two genes (no crossovers): half of the gametes have one parental genotype and half have the other.

b Incomplete linkage; crossing over affected the outcome.

Figure 12.8 Examples of gene linkages that affect a dihybrid cross. In (**a**), linkage between two gene loci is complete. In (**b**), it is incomplete.

Example: In one set of experiments, true-breeding mutant females (white eyes, yellow body) were crossed with wild-type males (red eyes, gray body). As the researchers expected, 50 percent of the F₁ offspring had one or the other parental phenotype. However, of 2,205 F₂ offspring, 129 were recombinants. That's 1.3 percent.

As it turns out, some alleles tend to remain together more often than others through meiosis. They are closer together on the same chromosome and less vulnerable to crossovers! Imagine any two genes at two different locations on the same chromosome. *The probability that a crossover will disrupt their linkage is proportional to the distance that separates the two loci.* Suppose genes *A* and *B* are twice as far apart as two other genes, *C* and *D*:

$$A \qquad B \qquad\qquad\qquad C \quad D$$

We would expect crossing over to disrupt the linkage between *A* and *B* much more often.

Two genes are very closely linked when the distance between them is small. Their allelic combinations nearly always end up inside the same gamete. Linkage is more vulnerable to crossover when the distance between them is greater (Figure 12.8). When two gene loci are very far apart, crossing over is so frequent that the genes assort independently of each other into gametes.

Unlike fruit flies, humans do not lend themselves to experimental crosses. Even so, all gene linkages have been identified by tracking the phenotypes in certain families over the generations.

For example, recessive alleles at two gene loci on the human X chromosome give rise to *color blindness* and *hemophilia*, described shortly. One female carried

both alleles but was symptom-free. Her father was, too, so she must have inherited a normal X chromosome from him (remember, males have only one X and one Y). The X chromosome inherited from her mother must have carried both mutant alleles. She gave birth to six sons. Three developed color blindness and hemophilia; two were unaffected. Here was phenotypic evidence of no recombination. However, her sixth son started life as a fertilized, recombinant egg; he was color-blind only.

Tracking many families affected by these disorders revealed that recombination frequencies of these two genes are low because they are very closely linked at one end of the X chromosome. More generally, studies showed that crossovers are not rare. For humans and most other eukaryotic species, meiosis cannot even be completed properly unless every pair of homologous chromosomes takes part in at least one crossover.

The human linkage map was completed during the 1990s. Today, linkage studies are being enhanced by gene sequencing, which reveals the physical distances between genes. (Take a look at www.ncbi.nlm.nih.gov.) These genetic and physical maps are both part of the Human Genome Initiative, a topic of Chapter 16.

Genes at different loci on the same chromosome belong to the same linkage group and do not assort independently during meiosis.

Crossing over between homologous chromosomes disrupts gene linkages and results in nonparental combinations of genes in chromosomes.

The farther apart two genes are on the same chromosome, the greater will be the frequency of crossing over and genetic recombination between them.

HUMAN GENETIC ANALYSIS

Some organisms, including pea plants and fruit flies, are ideal for genetic analysis. They grow and reproduce rapidly in small spaces, under controlled conditions. It does not take very long to track a trait through many generations. Humans are another story. We live under variable conditions in diverse environments. We select our own mates and reproduce if and when we want to. Humans live as long as the geneticists who study them, so tracking traits through generations is tedious. Most human families are not large, so there are not enough offspring for easy inferences about inheritance.

Constructing Pedigrees

To get around the problems associated with analyzing human inheritance, geneticists put together pedigrees. A **pedigree** is a chart of the genetic connections among individuals. Standardized methods, definitions, and symbols for representing individuals are utilized when constructing it (Figure 12.9*a*).

When geneticists analyze pedigrees, they rely on their knowledge of probability and Mendelian inheritance patterns, which may yield clues to the genetic basis for a trait. For example, clues might suggest that an allele responsible for a certain disorder is dominant or recessive or that it is located on a certain autosome or sex chromosome.

Gathering a great many family pedigrees increases the numerical base for analysis. When any trait follows a simple Mendelian inheritance pattern, a geneticist has greater confidence for predicting the probability of its occurrence among children of prospective parents. We will return to this topic later.

Figure 12.9 (**a**) Some standardized symbols used in constructing pedigree diagrams. (**b**) Example of a pedigree for *polydactyly*. With this condition, an individual has extra fingers, extra toes, or both. Expression of the gene for this trait varies among individuals. *Black* numerals signify the known number of fingers on each hand; *blue* numerals signify the number of toes on each foot.

(**c**) From human genetic researcher Nancy Wexler, a pedigree for *Huntington disease*, by which the human nervous system progressively degenerates. Wexler and her team pieced together an extended family tree for nearly 10,000 Venezuelans. Analysis of affected and unaffected individuals revealed that a dominant allele on human chromosome 4 is the genetic culprit. Wexler has a special interest in Huntington disease; she herself has a 50 percent chance of developing it.

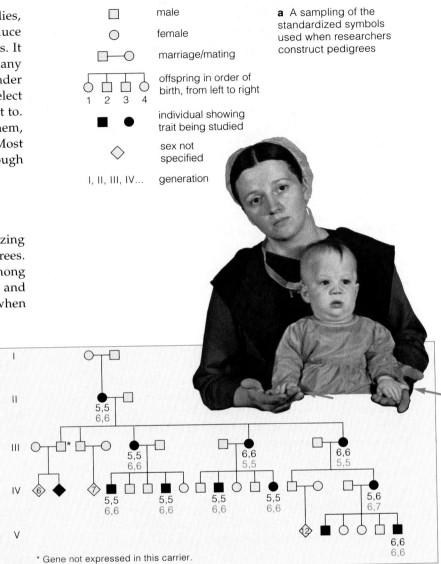

a A sampling of the standardized symbols used when researchers construct pedigrees

b Pedigree for a family in which polydactyly recurs as one symptom of Ellis–van Creveld syndrome (Section 17.11)

* Gene not expressed in this carrier.

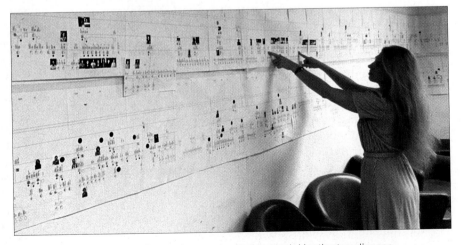

c Nancy Wexler and a pedigree she constructed to track Huntington disease

Table 12.1 Examples of Human Genetic Disorders and Genetic Abnormalities

Disorder or Abnormality*	Main Consequences	Disorder or Abnormality*	Main Consequences
AUTOSOMAL RECESSIVE INHERITANCE		**X-LINKED RECESSIVE INHERITANCE**	
Albinism *11.6; 11.9 CT*	Absence of pigmentation	Androgen insensitivity syndrome *36.2*	XY individual but having some female traits; sterility
Blue offspring *17.12 CT*	Bright blue skin coloration	Color blindness *12.4, 12.6, 35.9*	Inability to distinguish among some or all colors
Cystic fibrosis *5 CI*	Excessive glandular secretions leading to tissue, organ damage	Fragile X syndrome *12.6*	Mental impairment
Ellis–van Creveld syndrome *12.5, 17.11*	Extra fingers, toes, short limbs	Hemophilia *12.4, 12.6, 12.12 CT*	Impaired blood-clotting ability
Fanconi anemia *12.11*	Physical abnormalities, bone marrow failure	Muscular dystrophies *12.12 CT, 15.7 CT, 37.10*	Progressive loss of muscle function
Galactosemia *12.6*	Brain, liver, eye damage	X-linked anhidrotic dysplasia *15.4*	Mosaic skin (patches with or without sweat glands); other effects
Phenylketonuria (PKU) *12.10*	Mental impairment		
Sickle-cell anemia *3.8. 12.11, 14.4, 17.9, 38.3*	Adverse pleiotropic effects on organs throughout body	**CHANGES IN CHROMOSOME NUMBER**	
		Down syndrome *12.9, 44.4*	Mental impairment; heart defects
AUTOSOMAL DOMINANT INHERITANCE		Turner syndrome *12.10*	Sterility; abnormal ovaries, abnormal sexual traits
Achondroplasia *12.6*	One form of dwarfism		
Camptodactyly *11.7*	Rigid, bent fingers	Klinefelter syndrome *12.10*	Sterility; mild mental impairment
Familial hypercholesterolemia *16 CI*	High cholesterol levels in blood; eventually clogged arteries	XXX syndrome *12.10*	Minimal abnormalities
Huntington disease *12.5, 12.6*	Nervous system degenerates progressively, irreversibly	XYY condition *12.10*	Mild mental impairment or no effect
Marfan syndrome *11.5, 12.12 CT*	Abnormal or no connective tissue	**CHANGES IN CHROMOSOME STRUCTURE**	
Polydactyly *12.5*	Extra fingers, toes, or both	Chronic myelogenous leukemia *12 CI*	Overproduction of white blood cells in bone marrow; organ malfunctions
Progeria *12.7*	Drastic premature aging	Cri-du-chat syndrome *12.8*	Mental impairment; abnormally shaped larynx
Neurofibromatosis *14.5*	Tumors of nervous system, skin		

*Italic numbers indicate sections in which a disorder is described. *CI* signifies Chapter *I*ntroduction. *CT* signifies an end-of-chapter *C*ritical *T*hinking question.

Regarding Human Genetic Disorders

Table 12.1 is a list of some heritable traits that have been studied in detail. A few are abnormalities, or deviations from the average condition. Said another way, a **genetic abnormality** is nothing more than a rare or uncommon version of a trait, as when a person is born with six toes on each foot instead of five. Whether an individual or society at large views an abnormal trait as disfiguring or merely interesting is subjective. As the classic novel *The Hunchback of Notre Dame* suggests, there is nothing inherently life-threatening or even ugly about it.

By comparison, a **genetic disorder** is an inherited condition that sooner or later will cause mild to severe medical problems. A **syndrome** is a recognized set of symptoms that characterize a given disorder.

Because alleles underlying severe genetic disorders put people at great risk, they are rare in populations. Why, then, don't they disappear entirely? First, we can expect rare mutations to introduce new copies of the alleles into the population. Second, in heterozygotes, a harmful allele is paired with a normal one that may cover its functions, so it still can be passed to offspring.

You may hear someone refer to a genetic disorder as a disease, but the terms aren't always interchangeable. A disease also is an abnormal alteration in the way the body functions, and it, too, is characterized by a set of symptoms. But **disease** is illness caused by infectious, dietary, or environmental factors, not by inheritance of mutant genes. That said, it is still appropriate to call an illness a *genetic* disease if factors alter previously workable genes in a way that disrupts body functions.

With these qualifications in mind, we turn next to examples of inheritance in the human population. As you'll see, some show simple Mendelian patterns. Many traits have been traced to a dominant or recessive allele on an autosome or X chromosome. Others arise from changes in the structure or number of chromosomes.

For many genes, pedigree analysis might reveal simple Mendelian inheritance patterns that will allow inferences about the probability of their transmission to children.

A genetic abnormality is a rare or less common version of an inherited trait. A genetic disorder is an inherited condition that results in mild to severe medical problems.

EXAMPLES OF INHERITANCE PATTERNS

Autosomal Recessive Inheritance

For some traits, inheritance patterns reveal two clues that point to a recessive allele on an autosome. First, if both parents are heterozygous, any child of theirs will have a 50 percent chance of being heterozygous and a 25 percent chance of being homozygous recessive, as Figure 12.10*a* indicates. Second, if the parents are both homozygous recessive, any child of theirs will be, also.

On average, 1 in 100,000 newborns is homozygous for a recessive allele that causes *galactosemia*. It cannot make functional molecules of an enzyme that prevents a product of lactose breakdown from accumulating to toxic levels. Lactose normally is converted to glucose and galactose, then to glucose–1–phosphate (which is broken down by glycolysis or converted to glycogen). The full conversion is blocked in galactosemics:

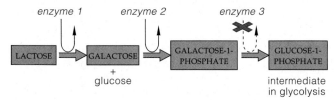

Galactose in blood rises to levels that can be detected in urine. Excessive amounts damage the eyes, liver, and brain, and cause malnutrition, diarrhea, and vomiting. Untreated galactosemics often die in childhood. When they are quickly put on a restricted diet that excludes dairy products, they can grow up symptom-free.

Autosomal Dominant Inheritance

Two clues point to an autosomal dominant allele for a trait. First, the trait typically appears each generation; usually the allele is expressed, even in heterozygotes. Second, if one parent is heterozygous and the other is homozygous recessive, any child of theirs will have a 50 percent chance of being heterozygous (Figure 12.10*b*).

A few dominant alleles persist in populations even though they cause severe genetic disorders. Some persist by spontaneous mutations. For others, expression of the dominant allele may not interfere with reproduction, or affected people reproduce before symptoms are severe.

For example, *Huntington disease* is characterized by progressive involuntary movements and deterioration of the nervous system; death is inevitable (Figure 12.9). Symptoms may not start to show up until an affected individual is past age thirty. Most people have already reproduced by then. Affected individuals usually die in their forties or fifties, sometimes before they realize they transmitted the mutant allele to their children.

As another example, *achondroplasia* affects about 1 in 10,000 people. Commonly, the homozygous dominant

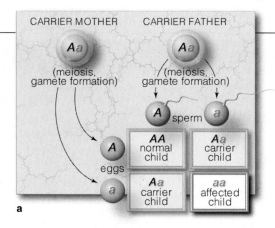

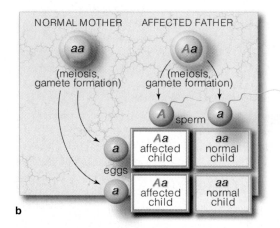

Figure 12.10 (**a**) One pattern for autosomal recessive inheritance. In this case, both parents are heterozygous carriers of the recessive allele (coded *red*). (**b**) A pattern for autosomal dominant inheritance. This dominant allele (*red*) is fully expressed in the carriers.

Figure 12.11 Infanta Margarita Teresa of the Spanish court and maids, including the achondroplasic woman (*right*).

condition leads to stillbirth, yet heterozygotes are able to reproduce. While heterozygotes are young, the cartilage of the skeleton forms improperly. At maturity, affected people have abnormally short arms and legs relative to other body parts (Figure 12.11). Adult achondroplasics are less than 4 feet, 4 inches tall. Often the dominant allele has no other phenotypic effects.

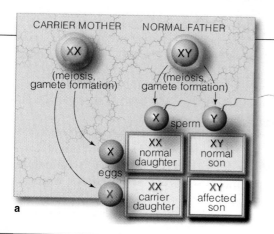

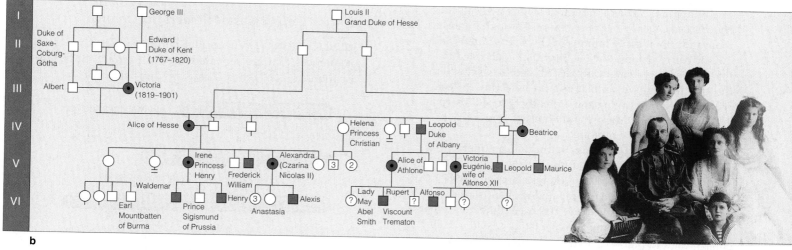

Figure 12.12 (a) One pattern for X-linked inheritance. In this example, assume the mother carries the recessive allele on one of her X chromosomes (*red*).

(b) Partial pedigree for Queen Victoria's descendants, showing carriers and affected males who carried the X allele for hemophilia A. Of the Russian royal family members in the photograph, the mother was a carrier; Crown Prince Alexis was hemophilic. ●

Figure 12.13 Fragile X chromosome from a cultured cell. The arrow points to the fragile site.

X-Linked Recessive Inheritance

Distinctive clues often show up when a recessive allele on an X chromosome causes a genetic disorder. First, males show the recessive phenotype more often than females. A dominant allele on the other X chromosome can mask the allele in females. The allele is not masked in males, who have one X chromosome (Figure 12.12a). Second, a son can't inherit the recessive allele from his father. A daughter can. If she is heterozygous, there is a 50 percent chance each son of hers will inherit the allele.

Color blindness, an inability to distinguish between some or all colors, is a common X-linked recessive trait. For instance, in red–green color blindness, an individual lacks some or all of the sensory receptors that normally respond to visible light of red and green wavelengths.

Hemophilia A, a blood-clotting disorder, is a case of X-linked recessive inheritance. Normally a blood-clotting mechanism quickly stops bleeding from minor injuries. Clotting requires several gene products, some on the X chromosome. If one of the X-linked genes is mutated in a human male, bleeding will be prolonged. About 1 in 7,000 males is affected by hemophilia A. Clotting time is close to normal in heterozygous females.

The frequency of hemophilia A was high in royal families of nineteenth-century Europe, in which close relatives often married. Queen Victoria of England was a notable carrier (Figure 12.12b). At one time eighteen of her sixty-nine descendants housed the recessive allele.

Fragile X syndrome, an X-linked recessive disorder that causes mental retardation, affects 1 in 1,500 males in the United States. In cultured cells, an X chromosome carrying the mutant allele is constricted near the end of its long arm (Figure 12.13). This constriction is called a fragile site because the end of the chromosome tends to break away. The term is misleading, for the end breaks away only in cultured cells, not in cells in the body.

A mutant gene, not breakage, causes the syndrome. It specifies a protein required for normal development of brain cells. Within that gene, a segment of DNA is repeated several times. Certain mutations result in the addition of many more repeats in the DNA; hence their name, expansion mutations. The outcome is a mutant allele that cannot function properly. Because that allele is recessive, males who inherit it and females who are homozygous for it have fragile X syndrome. Expansion mutations are now known to be the cause of Huntington disease and some other genetic disorders.

Genetic analyses of family pedigrees have revealed simple Mendelian inheritance patterns for certain traits, as well as for many genetic disorders that arise from expression of specific alleles on an autosome or X chromosome.

Progeria—Too Young to Be Old

Imagine being ten years old with a mind trapped inside a body that is rapidly getting a bit more shriveled, more frail—old—with each passing day. You are just barely tall enough to peer over the top of the kitchen counter, and you weigh less than thirty-five pounds. Already you are bald and have a crinkled nose. Possibly you have a few more years to live. Would you, like Mickey Hayes and Fransie Geringer, still laugh with your friends?

Of every 8 million newborn humans, one is destined to grow old far too soon. On one of its autosomes, that rare individual carries a mutant gene that gives rise to *Hutchinson–Gilford progeria syndrome*. Through hundreds, thousands, then many billions of DNA replications and mitotic cell divisions, terrible information encoded in that gene was systematically distributed to every cell in the growing embryo, and later in the newborn. Its legacy will be accelerated aging and a greatly reduced life span. The photograph of Mickey and Fransie in Figure 12.14 shows some of the symptoms.

The mutation causes gross disruptions in interactions among genes that bring about the body's growth and development. Observable symptoms start before age two. Skin that should be plump and resilient starts to thin. Skeletal muscles weaken. Tissues in limb bones that should lengthen and grow stronger start to soften. Hair loss is pronounced; extremely premature baldness is inevitable. There are no documented cases of progeria running in families, so a gene must undergo random, spontaneous mutation. Probably the gene is dominant over a normal partner on the homologous chromosome.

Most progeriacs can expect to die in their early teens as a result of strokes or heart attacks. These final insults are brought on by a hardening of the walls of arteries, a condition typical of advanced age. When Mickey turned eighteen, he was the oldest living progeriac. Fransie was seventeen when he died.

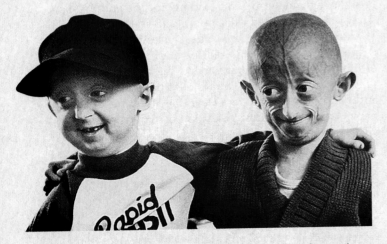

Figure 12.14 Two boys who met at a gathering of progeriacs at Disneyland, California, when they were not yet ten years old.

CHANGES IN CHROMOSOME STRUCTURE

On rare occasions, the physical structure of one or more chromosomes changes. The result is a genetic disorder or abnormality. Such changes occur spontaneously in nature. They also are induced in research laboratories by exposure to chemicals or by irradiation. Either way, changes may be detected by microscopic examination and karyotype analysis of cells at mitosis or meiosis. Let's now review four kinds of structural change. As you will see, some have severe or lethal consequences.

Major Categories of Structural Change

DUPLICATION   Even normal chromosomes have gene sequences that are repeated several to many hundreds or thousands of times. These are **duplications**:

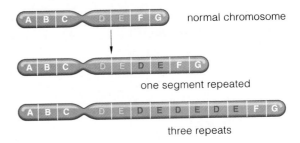

normal chromosome

one segment repeated

three repeats

INVERSION   With an **inversion**, a linear stretch of DNA within the chromosome becomes oriented in the reverse direction, with no molecular loss:

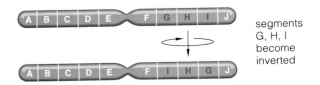

segments G, H, I become inverted

TRANSLOCATION   As the chapter introduction showed, the Philadelphia chromosome that has been linked to a form of leukemia results from **translocation**, whereby a broken part of a chromosome becomes attached to a *non*homologous chromosome. Most translocations are reciprocal (both chromosomes exchange broken parts):

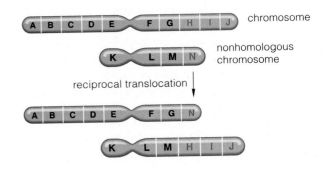

chromosome

nonhomologous chromosome

reciprocal translocation

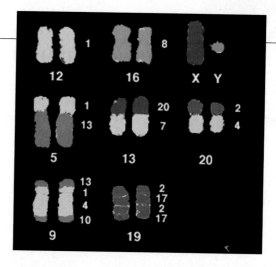

Figure 12.16 Spectral karyotype of duplicated chromosomes of the gibbon, one of the apes. The colors identify regions of gibbon chromosomes that are structurally identical with human chromosomes (compare Figure 12.1).

Top row: Chromosomes 12, 16, X, and Y are the same in both primates. *Second row:* Translocations are present in gibbon chromosomes 5, 13, and 20, and they correspond to regions of human chromosomes 1, 13, 20, 7, 2, and 4.

Third row: Gibbon chromosome 9 corresponds to regions of several human chromosomes. Duplications in gibbon chromosome 19 occur in human chromosomes 2 and 17.

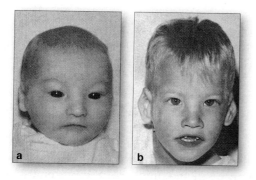

Figure 12.15 (**a**) Male infant who later developed cri-du-chat syndrome. Ears are low on the side of the head relative to the eyes. (**b**) The same boy, four years later. By this age, affected humans stop making mewing sounds typical of the syndrome.

DELETION Viral attacks, irradiation (ionizing radiation especially), chemical assaults, and other environmental agents may cause a **deletion**, the loss of some segment of a chromosome:

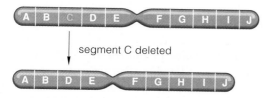

segment C deleted

Most deletions are lethal or cause serious disorders in mammals, for they disrupt normal gene interactions that underlie a program of growth, development, and maintenance activities. For example, one deletion from human chromosome 5 results in mental impairment and the development of an abnormally shaped larynx. When affected infants cry, they produce sounds rather like a cat's meow. Hence *cri-du-chat* (cat-cry), the name of this disorder. Figure 12.15 shows an affected child.

Does Chromosome Structure Evolve?

Some changes in chromosome structure tend not to be conserved; they are selected against over evolutionary time. Even so, among many species, we see interesting signs of past changes. For example, duplications are common. Many have not harmed their bearers. Over billions of years, neutral ones have been accumulating and are now built into the DNA of all species.

Duplicates of gene sequences with neutral effects might have an adaptive advantage. Having two of the same gene could free up one of them for potentially beneficial mutations. The other gene would still issue the required product. The duplicated gene region could become slightly modified, then the products of the two genes could function in slightly different or novel ways.

Apparently, several kinds of duplicated, modified genes were pivotal in evolution. Think back on those gene regions for the polypeptide chains of hemoglobin (Section 3.8). In humans and other primates, the regions have multiple nucleotide sequences that are strikingly similar. The sequences specify whole families of chains, each with a slight structural difference. The structural differences translate into slightly different efficiencies in hemoglobin's capacity to bind and transport oxygen under a range of cellular conditions.

It also appears that certain duplications, inversions, and translocations helped put the primate ancestors of humans on a unique evolutionary road. They seem to have contributed to divergences that led to the modern apes and humans. Of the twenty-three pairs of human chromosomes, eighteen are nearly identical with their counterparts in chimpanzees and gorillas. The other five pairs differ at inverted and translocated regions. You can observe the dramatic similarities for yourself by comparing chromosomes of a human with those of a gibbon, one of the apes (Figure 12.16).

On rare occasions, a segment of a chromosome may become duplicated, inverted, moved to a new location, or deleted.

Most chromosome changes are harmful or lethal when they alter gene interactions that underlie growth, development, and maintenance activities.

Over evolutionary time, many changes have been conserved; they confer adaptive advantages or have had neutral effects.

ANGES IN CHROMOSOME NUMBER

Occasionally, abnormal events occur before or during cell division, then gametes and new individuals end up with the wrong chromosome number. The consequences range from minor to lethal physical changes.

Categories and Mechanisms of Change

With **aneuploidy**, individuals usually have one extra or one less chromosome. This condition is a major cause of human reproductive failure. Possibly it affects about one-half of all fertilized eggs. As autopsies reveal, most *miscarriages* (spontaneous aborting of embryos before pregnancy reaches full term) are aneuploids.

With **polyploidy**, individuals have three or more of each type of chromosome. About one-half of all species of flowering plants are polyploid (Section 18.3). Often, researchers can induce polyploidy in undifferentiated plant cells by exposing them to colchicine (Section 12.2). Some species of insects, fishes, and other animals are polyploids. However, polyploidy is lethal for humans. All but about 1 percent of human polyploids die before birth, and the rare newborns die soon after birth.

Chromosome numbers can change during mitotic or meiotic cell divisions. Suppose a cell cycle proceeds through DNA duplication and mitosis but is arrested before the cytoplasm divides. The cell is now *tetra*ploid, with four of each type of chromosome. Suppose one or more pairs of chromosomes fail to separate in mitosis or meiosis, an event called **nondisjunction**. Some or all of the forthcoming cells will have too many or too few chromosomes, as in the Figure 12.17 example.

The chromosome number also may change during fertilization. Visualize a normal gamete that unites by chance with an $n + 1$ gamete (one extra chromosome).

The new individual will be "trisomic" ($2n + 1$); it will have three of one type of chromosome and two of every other type. And what if an $n - 1$ gamete unites with a normal n gamete? Then, the new individual will turn out to be "monosomic" ($2n - 1$).

Nearly all inherited changes in chromosome number arise by nondisjunction at meiosis or when gametes are forming. Mitotic cell divisions perpetuate the mistake as an embryo is developing. Let's start with one of the changes in the number of autosomes. The next section looks mainly at altered numbers of sex chromosomes.

Case Study: Down Syndrome

A trisomic 21 newborn, with three chromosomes 21, can be expected to develop *Down syndrome*, the most frequent change in human chromosome number. This autosomal disorder occurs once in every 800 to 1,000 births. It affects more than 350,000 people in the United States alone. Figure 12.18 shows a typical trisomic 21 karyotype and a few affected individuals.

Nondisjunction during meiosis accounts for about 95 percent of all Down syndrome cases. Translocation and mosaicism account for the remainder. With **mosaicism**, nondisjunction occurs in one of the embryonic cells that form after fertilization. But only the descendants of that altered cell inherit three chromosomes 21; other cells in the body have the normal chromosome number. Translocation may occur while gametes are forming or after fertilization. Part of chromosome 21 breaks away and then becomes attached to a different chromosome in the gamete or embryonic cell.

At this time we do not know why nondisjunction occurs. We know its incidence rises with advancing age

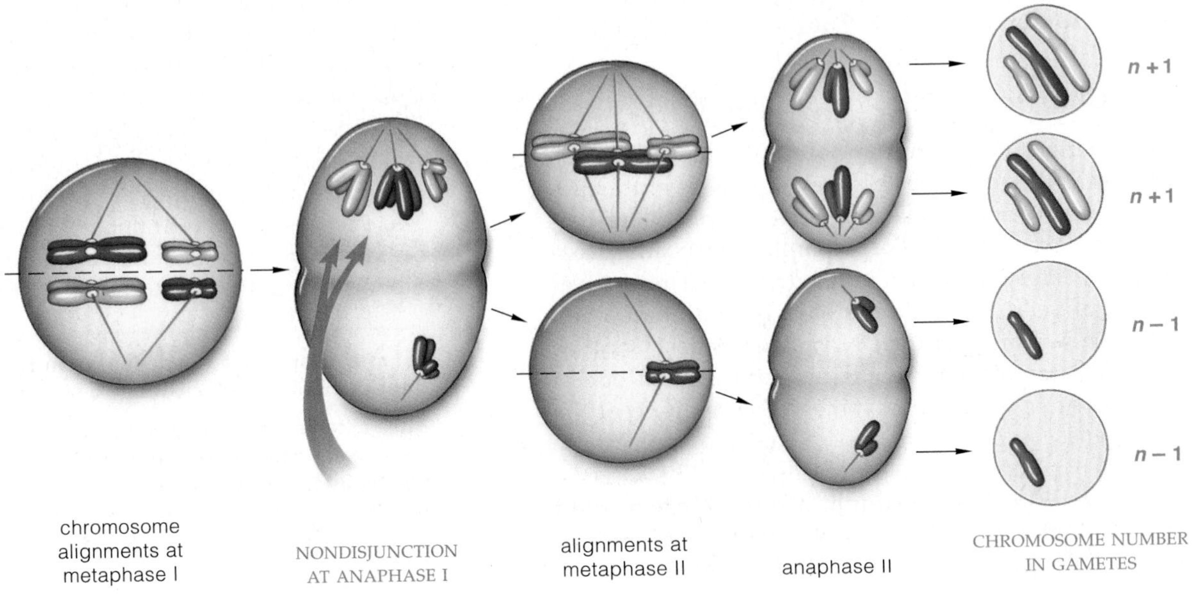

Figure 12.17 Example of nondisjunction. Of two pairs of homologous chromosomes shown, one pair fails to separate at anaphase I of meiosis. The chromosome number is changed in gametes. (Make a simple sketch of nondisjunction during anaphase II. What will the chromosome numbers be in gametes?)

chromosome alignments at metaphase I

NONDISJUNCTION AT ANAPHASE I

alignments at metaphase II

anaphase II

CHROMOSOME NUMBER IN GAMETES

$n + 1$

$n + 1$

$n - 1$

$n - 1$

Figure 12.18 Karyotype revealing the trisomic 21 condition of a young male. Both the young girl and the teenager shown here were lively participants in the Special Olympics held annually in San Mateo, California.

of the mother (Figure 12.19). Nondisjunction also may originate with the father, but not nearly as frequently. Whatever causes it, we see the same outcome. While a new individual was developing, mitotic cell divisions put an extra chromosome 21 (or an extra portion of it) in some or all of its new cells.

Characteristics associated with the disorder include upwardly slanted eyes, a skin fold that starts at the inner corner of the eye, a deep crease across each palm and foot sole, one (not two) horizontal furrows across the fifth finger, a tongue that is large relative to the oral cavity, somewhat flattened facial features, and poor muscle tone. Affected children often have heart defects and respiratory and digestive problems, most of which are now treatable. Adults often develop Alzheimer's disease. With good medical care, trisomics 21 now have a life expectancy of fifty-five years, on average.

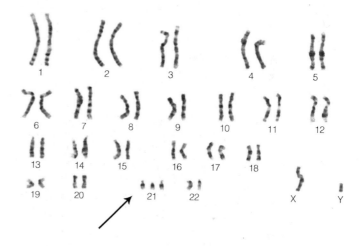

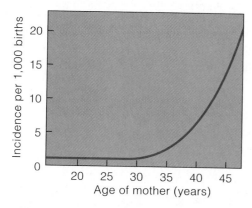

Figure 12.19 Relationship between the frequency of Down syndrome and mother's age at the time of childbirth. Results shown here are from a study of 1,119 affected children who were born in Victoria, Australia, between 1942 and 1957. The risk of giving birth to a trisomic 21 individual increases with the mother's age. This may seem counterintuitive, for about 80 percent of trisomics 21 are born to women not yet 35 years old. However, these women are in age categories with the highest fertility rates; they simply have more babies.

Not all trisomic 21 individuals develop all of the symptoms of Down syndrome. Some newborns show only a few. Besides this, some of the so-called defining features also appear in the population at large.

That said, most trisomics 21 display moderate to severe mental impairment. Heart defects are common. The skeleton develops abnormally, so older children have shortened body parts, loose joints, and poorly aligned bones of the hips, fingers, and toes. Muscles and muscle reflexes are weaker than normal. Speech and other motor skills develop slowly.

Trisomy 21 is one of many conditions that may be detected through prenatal diagnosis, as described in Section 12.11. With medical intervention and special training that begins early on, trisomic 21 individuals can take part in normal activities. As a group, they tend to be cheerful and affectionate people, and they derive great pleasure from socializing (Figure 12.18).

Nondisjunction prior to cell division in reproductive cells or early embryonic cells results in an altered number of autosomes or sex chromosomes. The change affects the course of development and the resulting phenotypes.

CASE STUDIES: CHANGES IN THE NUMBER OF SEX CHROMOSOMES

Besides causing most of the alterations in the number of autosomes, nondisjunction also causes most alterations in the number of X and Y chromosomes. The frequency of such changes is 1 in 400 live births. Most often they lead to difficulties in learning and motor functioning, including speech, although problems may be so subtle that the underlying cause is not even diagnosed.

Changes in sex chromosome number are only a bit less common than they are for the autosomal changes, partly because the phenotypes of affected individuals usually are not severely altered. Most of these changes can be diagnosed before birth by procedures described in the next section.

Female Sex Chromosome Abnormalities

TURNER SYNDROME   Inheriting an X chromosome and no corresponding X or Y chromosome gives rise to *Turner syndrome*, which affects 1 in 2,500 to 10,000 or so newborn girls. Figure 12.20 shows one XO karyotype. Nondisjunction that originates with the father accounts for 75 percent of all the cases. We see fewer people with Turner syndrome, compared to other sex chromosome abnormalities. The likely reason is at least 98 percent of all XO zygotes spontaneously abort early in pregnancy. From one study, we know that about 20 percent of all spontaneously aborted embryos in which chromosome abnormalities were detected were XO.

Despite the near lethality, XO survivors are not as disadvantaged as other aneuploids. They grow up well proportioned, albeit short—4 feet, 8 inches tall, on the average. Their childhood behavior is typically normal.

Most Turner females don't have functional ovaries, can't produce normal amounts of sex hormones, and are infertile. The reduction of sex hormones adversely affects development of secondary sexual traits, such as breast enlargement. These females do produce a few immature eggs, but by the time they are two years old,

the eggs are destroyed. Possibly as a consequence of their arrested sexual development and small stature, XO teenagers often are passive and easily intimidated by peers. Hormone replacement therapy or corrective surgery, or both, may mitigate some of the symptoms that cause these individuals distress.

XXX SYNDROME   A few females inherit three, four, or five X chromosomes. The XXX condition occurs at a frequency of about 1 in 1,000 live births. Most adults are an inch or so taller and more slender than average. They are fertile. Except for slight learning difficulties, these females have a normal appearance and tend to fall within the normal range of social behavior.

Male Sex Chromosome Abnormalities

KLINEFELTER SYNDROME   One in 500 to 2,000 males has inherited one Y and two or more X chromosomes, mainly as an outcome of nondisjunction. They have an XXY or, more rarely, XXXY, XXXXY, or XY/XXY mosaic genotype. About 67 percent of the affected individuals inherited the extra chromosome from their mother.

The symptoms of the resulting *Klinefelter syndrome* develop after the onset of puberty. For example, XXY males tend to be taller than average and overweight. Most have a normal outward appearance. However, the testes and prostate gland usually are much smaller than average; the penis and scrotum are not.

In affected individuals, testosterone production is lower than normal and estrogen production higher. This has feminizing effects. For example, sperm counts are low. Hair on the face and elsewhere is sparse, the voice is pitched high, and the breasts are somewhat enlarged. Testosterone injections from puberty onward can reverse the feminized traits, although not the low fertility. Even without such reversal, XXY phenotypes are not that different. Many affected individuals were

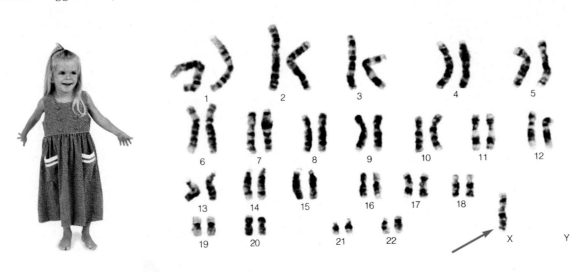

Figure 12.20
Example of an XO karyotype from a female affected by Turner syndrome.

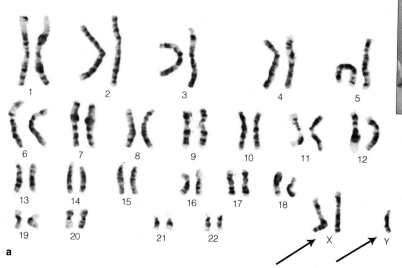

a

not even aware of their chromosomal abnormality until they were tested for infertility.

Although many XXY males fall within the normal range of intelligence, some have problems with speech and short-term memory. Those who are not diagnosed and given guidance early on often have difficulties of the sort highlighted in Figure 12.21.

XYY CONDITION About 1 in 500 to 1,000 males has one X and two Y chromosomes. Those with this *XYY condition* tend to be taller than average. Some show mild mental impairment, but most are phenotypically normal. XYY males were once thought to be genetically predisposed to become criminals. The conclusion was erroneous. It was based on small numbers of cases in narrowly selected groups, including prison inmates. The investigators often knew who the XYY males were, which may have biased their evaluations.

Also, there were no **double-blind studies**, by which different investigators gather data independently of one another and match them up only after both sets of data are completed. In this case, the same investigators gathered karyotypes *and* personal histories. Fanning the stereotype was a sensationalized report in 1968 that a mass-murderer of young nurses was XYY. He wasn't.

In 1976 a Danish geneticist reported on a large-scale study. It was based on the records of 4,139 tall males, twenty-six years old, who had reported to their draft board. Besides giving results of physical examinations and intelligence testing, the records provided clues to social and economic status, educational background, and any criminal convictions. Only twelve of the males were XYY, which meant there were more than 4,000 males in the control group. The only significant finding was that tall, mentally impaired males who engage in criminal activity are just more likely to get caught—irrespective of karyotype.

Figure 12.21 (a) Karyotype of a male with Klinefelter syndrome. (b) One person's story: Stefan as a baby, and later in life. Until his teenage years, he was shy and reserved, yet would become enraged for no apparent reason. His parents sensed something was wrong. Psychologists and doctors could not pinpoint the problem. They assumed he had learning disabilities that affected comprehension, auditory processing, memory, and abstract thinking. A psychologist told him he was stupid, lazy, and would be lucky to graduate from high school.

Stefan did graduate from high school and went on to college, where he obtained a BS degree in business administration and also in sports management. He never discussed his learning disabilities. Rather, he took pride in doing the work on his own and not being treated differently.

It was not until Stefan was 25 years old that he was diagnosed with Klinefelter syndrome. During a routine physical examination, a physician noticed that Stefan's testes were smaller than normal. Subsequently, laboratory tests and karyotyping revealed his 46XY/47XXY mosaic condition.

That same year, Stefan began a job as a software engineer. It was tough, but he felt he would not have been nearly as successful if he hadn't learned to work on his own. Having a full-time position helped him open doors to volunteer work with the Klinefelter syndrome network. During his volunteer work, he met his future fiancée, whose son also has the syndrome.

Recent research suggests that as many as 64 percent to 85 percent of all XXY, XXX, and XYY children have not even been properly diagnosed. Some dismiss them unfairly as being underachievers without having a clue to the genetic basis for their learning disabilities.

Sex chromosome abnormalities are most often caused by nondisjunction during meiosis. They typically cause subtle difficulties with learning, speech, and other motor skills.

These abnormalities are only slightly less common than autosomal abnormalities, partly because they are rarely lethal and usually have much less severe effects.

Prospects in Human Genetics

With the arrival of their newborn, parents typically ask, "Is our baby normal?" Quite naturally, they want their baby to be free of genetic disorders, and most babies are. But what are the options when they are not?

We do not approach heritable disorders and diseases the same way. We attack diseases with antibiotics, surgery, and other weapons. But how do we attack a heritable "enemy" that can be transmitted to offspring? Should we institute regional, national, or global programs to identify people who might be carrying harmful alleles? Do we tell them they are "defective" and run a risk of bestowing a disorder on their children? Who decides which alleles are harmful? Should society bear the cost of treating genetic disorders before and after birth? If so, should society also have a say in whether an affected embryo will be born at all, or whether it should be aborted? An **abortion** is the expulsion of a pre-term embryo or fetus from the uterus.

Such questions are only the tip of an ethical iceberg. We don't have answers that are universally acceptable.

PHENOTYPIC TREATMENTS Often, symptoms of genetic disorders can be minimized or suppressed by dietary controls, adjustments to environmental conditions, and surgical intervention or hormone replacement therapy.

For example, dietary control works in *phenylketonuria*, or PKU. A certain gene specifies an enzyme that converts one amino acid to another—phenylalanine to tyrosine. If an individual is homozygous recessive for a mutated form of the gene, the first of these amino acids accumulates inside the body. If excess amounts are diverted into other pathways, then phenylpyruvate and other compounds may form. High levels of phenylpyruvate in the blood can impair the functioning of the brain. When affected people restrict their intake of phenylalanine, they are not required to dispose of excess amounts, so they can lead normal lives. Among other things, they can avoid soft drinks and other food products that are sweetened with aspartame, a compound that contains phenylalanine.

Environmental adjustments can counter or minimize symptoms of some disorders, as when albinos avoid direct sunlight. Surgery can repair conditions such as a *cleft lip*, an opening in the upper lip that did not seal, as it should have, during embryonic development. This condition usually arises by interactions among multiple genes from both parents and environmental factors.

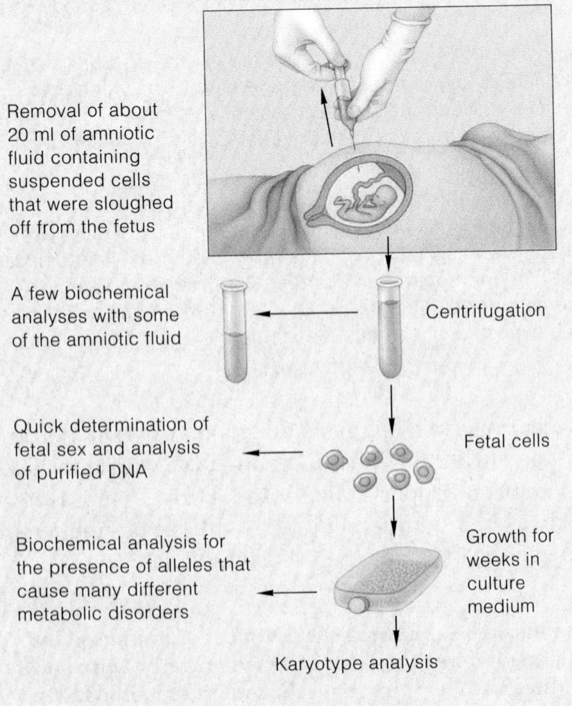

GENETIC SCREENING Through large-scale screening programs in the general population, affected persons or carriers of a harmful allele often can be detected early enough to start preventive measures before symptoms develop. For example, most hospitals in the United States routinely screen newborns for PKU, so today it is less common to see people with symptoms of this disorder.

GENETIC COUNSELING If a first child or close relative has a severe heritable problem, prospective parents may worry about their next child. They may request help in evaluating their options from a qualified professional counselor. *Genetic counseling* often includes diagnosis of parental genotypes, detailed pedigrees, and genetic testing for hundreds of known metabolic disorders. Geneticists, too, help predict risks for genetic disorders. Counselors must remind prospective parents that the same risk usually applies to each pregnancy.

PRENATAL DIAGNOSIS Methods of *prenatal diagnosis* are used to determine the sex of embryos or fetuses and more than a hundred genetic conditions. (*Prenatal* means before birth. The term *embryo* applies until eight weeks after fertilization, after which *fetus* is appropriate.)

Suppose a woman who is forty-five years old becomes pregnant and worries about Down syndrome. Usually between 8 and 12 weeks after pregnancy, she might ask for prenatal diagnosis by *amniocentesis* (Figure 12.22). With this procedure, a clinician withdraws a tiny sample of fluid inside the amnion, a membranous sac enclosing the fetus. Some cells that the fetus shed are suspended in the sample. The cells are cultured and analyzed.

Chorionic villi sampling (CVS) is a different diagnostic procedure. A clinician withdraws cells from the chorion, a fluid-filled, membranous sac that surrounds the amnion.

Removal of about 20 ml of amniotic fluid containing suspended cells that were sloughed off from the fetus

A few biochemical analyses with some of the amniotic fluid

Centrifugation

Quick determination of fetal sex and analysis of purified DNA

Fetal cells

Biochemical analysis for the presence of alleles that cause many different metabolic disorders

Growth for weeks in culture medium

Karyotype analysis

Figure 12.22 Amniocentesis, a prenatal diagnostic tool.

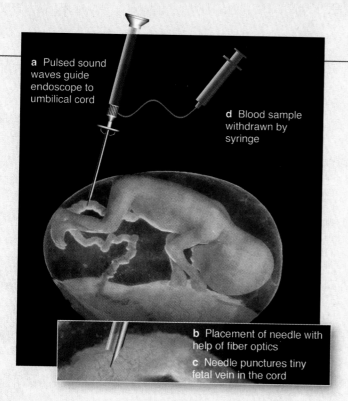

a Pulsed sound waves guide endoscope to umbilical cord

d Blood sample withdrawn by syringe

b Placement of needle with help of fiber optics

c Needle punctures tiny fetal vein in the cord

Figure 12.23 Fetoscopy for prenatal diagnosis.

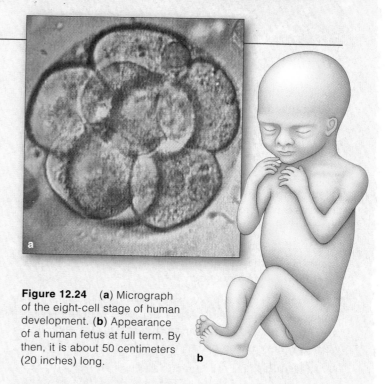

Figure 12.24 (a) Micrograph of the eight-cell stage of human development. (b) Appearance of a human fetus at full term. By then, it is about 50 centimeters (20 inches) long.

CVS can be performed weeks before amniocentesis. It yields results as early as the eighth week of pregnancy.

Direct visualization of a developing fetus is possible with *fetoscopy*. An endoscope, a fiberoptic device, uses pulsed sound waves to scan the uterus and visually locate particular parts of the fetus, umbilical cord, or placenta (Figure 12.23). Fetoscopy has been used to diagnose blood cell disorders, such as sickle-cell anemia and hemophilia.

All three procedures may cause infection or puncture the fetus. If a punctured amnion doesn't reseal itself fast, too much amniotic fluid leaks out, which harms the fetus. Amniocentesis increases the risk of miscarriage by 1 to 2 percent. CVS has a 0.3 percent risk that a future child will have missing or underdeveloped fingers and toes. Fetoscopy raises risk of a miscarriage by 2 to 10 percent.

Parents-to-be probably should seek counseling from a doctor to help them weigh the risks and benefits of such procedures in terms of their own circumstances. They may wish to ask about the small overall risk of 3 percent that any child will have some kind of birth defect. They may ask about the severity of a genetic disorder that a child might be at risk of developing. And they might consider how old the woman is at the time of pregnancy.

REGARDING ABORTION What happens when prenatal diagnosis does reveal a serious problem? Do prospective parents opt for induced abortion? We can only say here that they must weigh their awareness of the severity of the genetic disorder against ethical and religious beliefs. Worse, they must play out their personal tragedy on a larger stage, dominated by a nationwide battle between fiercely vocal "pro-life" and "pro-choice" factions. We return to this volatile issue in Sections 44.14 and 44.16.

PREIMPLANTATION DIAGNOSIS Another procedure, *preimplantation diagnosis*, relies on **in-vitro fertilization**. In vitro, recall, means "in glass." Sperm and eggs from prospective parents are placed in an enriched medium in a glass or plastic petri dish. One or more eggs may get fertilized. In two days, mitotic cell divisions may convert one of these into a ball of eight cells (Figure 12.24a).

According to one view, the tiny, free-floating ball is a *pre*-pregnancy stage. Like the unfertilized eggs discarded monthly from a woman, it isn't attached to the uterus. Its cells all have the same genes and are not yet committed to giving rise to specialized cells of a heart, lungs, and other organs. Doctors take one of the undifferentiated cells and analyze its genes. If that cell has no detectable genetic defects, the ball is inserted into the uterus.

Some couples who are at risk of passing on muscular dystrophy, cystic fibrosis, and other disorders have opted for the procedure. Many *"test-tube" babies* have been born in good health and are free of the mutant alleles.

In 2000, genetic screening of a dozen fertilized eggs revealed an embryo free of a certain mutated gene. Both prospective parents carried the gene, and their first-born child, Molly, had developed *Fanconi anemia*. She had no thumbs, deformed arms, an incomplete brain, and the prospect of dying soon from leukemia. When her new brother was born, blood stem cells from the umbilical cord were transfused into her. The cells took hold. At this time Molly's bone marrow is producing red blood cells and platelets that are countering the anemia. The first embryo genetically selected to save a life is doing just that.

SUMMARY
Gold indicates text section

1. Genes, the units of instruction for heritable traits, are arranged one after the other along chromosomes. Each gene has its own specific location, or locus, on one type of chromosome. Different molecular forms of that gene, called alleles, may occupy the locus. *12.1*

2. Human somatic cells are diploid (2*n*). They contain twenty-three pairs of homologous chromosomes that interact during meiosis. Each pair has the same length, shape, and genes (except for an XY pairing). *12.1*

3. An allele on one chromosome may or may not be identical to the allele at the equivalent locus on the homologous chromosome. *12.1*

4. Human females have two X chromosomes. Males have one X paired with one Y. All other chromosomes are autosomes (the same in both females and males). A gene on the Y chromosome determines sex. *12.1, 12.3*

5. Karyotypes, used in genetic analysis, are constructed to compare an individual's chromosomes (based upon their defining structural features) against standardized preparations of metaphase chromosomes. *CI, 12.2*

6. Genes on the same chromosome represent a linkage group. However, crossing over (breakage and exchange of segments between homologues) disrupts linkages. The farther apart two gene loci are along the length of a single chromosome, the greater will be the frequency of crossovers between them. *12.4*

7. Pedigrees are charts of genetic connections through lines of descent. Pedigrees provide clues to inheritance of a trait. *12.5*

8. Mendelian patterns of inheritance are characteristic of certain dominant or recessive alleles on autosomes or on the X chromosome. *12.5, 12.6*

9. A chromosome's structure may be altered on rare occasions. A segment may be deleted, inverted, moved to a new location (translocated), or duplicated. *12.8*

10. The chromosome number can change. Gametes and offspring may get one more or one less chromosome than the parents (aneuploidy). They may get three or more of each type of chromosome (polyploidy). Nondisjunction during meiosis accounts for most of these changes in chromosome number. *12.9*

11. Changes in chromosome structure or number often result in genetic abnormalities or disorders. Phenotypic treatments, genetic screening, genetic counseling, and prenatal diagnosis are social responses. *12.8–12.10*

12. Crossing over adds to potentially adaptive variation in traits in a population. Most but not all changes in chromosome number or structure are harmful or lethal. Over evolutionary time, some have become established in chromosomes of all species. *12.8*

Review Questions

1. What is a gene? What are alleles? *12.1*

2. Distinguish between: *12.1, 12.2*
   a. homologous and nonhomologous chromosomes
   b. sex chromosomes and autosomes
   c. karyotype and karyotype diagram

3. Define genetic recombination, and describe how crossing over can bring it about. *12.4*

4. Define pedigree. Explain the difference between genetic abnormality and genetic disorder, using examples. *12.5*

5. Contrast a typical pattern of autosomal recessive inheritance with that of autosomal dominant inheritance. *12.6*

6. Describe two clues that often show up when a recessive allele on an X chromosome causes a genetic disorder. *12.6*

7. Distinguish among a chromosomal deletion, duplication, inversion, and translocation. *12.8*

8. Define aneuploidy and polyploidy. Make a simple sketch of an example of nondisjunction. *12.9*

Self-Quiz   ANSWERS IN APPENDIX III

1. _____ align and segregate during _____ .
   a. Homologues; mitosis
   b. Genes on nonhomologous chromosomes; meiosis
   c. Homologues; meiosis
   d. Genes on one chromosome; mitosis

2. The probability of a crossover occurring between two genes on the same chromosome is _____ .
   a. unrelated to the distance between them
   b. increased if they are closer together on the chromosome
   c. increased if they are farther apart on the chromosome

3. Genetic disorders are caused by _____ .
   a. altered chromosome number   c. mutation
   b. altered chromosome structure   d. all of the above

4. A recognized set of symptoms that characterize a specific disorder is a _____ .
   a. syndrome   b. disease   c. pedigree

5. Chromosome structure can be altered by a _____ .
   a. deletion   c. inversion   e. all of the above
   b. duplication   d. translocation

6. Nondisjunction is the name for _____ .
   a. crossing over in mitosis
   b. segregation in meiosis
   c. failure of chromosomes to separate during meiosis
   d. multiple independent assortments

7. A gamete affected by nondisjunction would have _____ .
   a. a change from the normal chromosome number
   b. one extra or one missing chromosome
   c. the potential for a genetic disorder
   d. all of the above

8. Match the chromosome terms appropriately.
   ____ crossing over   a. number and defining features of an
   ____ deletion   individual's metaphase chromosomes
   ____ nondisjunction   b. chromosome segment moves to a
   ____ translocation   nonhomologous chromosome
   ____ karyotype   c. disrupts gene linkages at meiosis
   ____ linkage group   d. causes gametes to have abnormal
                              chromosome numbers
                           e. loss of a chromosome segment
                           f. all genes on a given chromosome

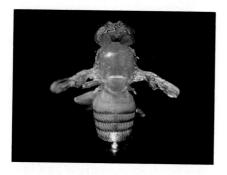

Figure 12.25 Mutant fruit fly with vestigial wings.

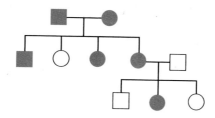

Figure 12.26 Go ahead, identify the mystery pedigree.

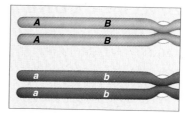

Figure 12.27 Two of your linked genes, on a pair of homologous chromosomes.

Critical Thinking—Genetics Problems

ANSWERS IN APPENDIX IV

1. Human females are XX and males are XY.
    a. Does a male inherit the X from his mother or his father?
    b. With respect to X-linked alleles, how many different types of gametes can a male produce?
    c. If a female is homozygous for an X-linked allele, how many types of gametes can she produce with respect to that allele?
    d. If a female is heterozygous for an X-linked allele, how many types of gametes can she produce with respect to that allele?

2. The wild-type allele of a gene locus governing wing length in *D. melanogaster* results in normal long wings. Figure 12.25 shows how homozygosity for a mutant recessive allele results in formation of vestigial (short) wings. Suppose you expose a homozygous dominant, long-winged fly to x-rays, then cross it with a homozygous recessive, vestigial-winged fly. The eggs develop into adults. Most are heterozygous, with long wings. A few have vestigial wings. What might explain these results?

3. *Marfan syndrome* is a genetic disorder with pleiotropic effects (Section 11.15). Affected people are often very tall and thin, with double-jointed fingers. The spine curves abnormally. Arms, fingers, and lower limbs are disproportionately long. Eye lenses dislocate easily. Thin tissue flaps in the heart that act like valves to direct blood flow may billow the wrong way when the heart contracts. The heart beats irregularly. The main artery from the heart is fragile. Its diameter widens; its wall may tear.
    Genetic analysis shows the disorder follows a pattern of autosomal dominant inheritance. What is the chance any child will inherit the allele if one parent is heterozygous for it?

4. In the Figure 12.26 pedigree, does the phenotype indicated by *red* circles and squares follow a Mendelian inheritance pattern that is autosomal dominant, autosomal recessive, or X-linked?

5. One kind of *muscular dystrophy*, a genetic disorder, is due to a recessive X-linked allele. Usually, symptoms start in childhood. Over time, a slowly progressing loss of muscle function leads to death, usually by age twenty or so. Unlike color blindness, this disorder is nearly always restricted to males. Suggest why.

6. Suppose you carry two linked genes with alleles Aa and Bb, respectively, as in Figure 12.27. If the crossover frequency between these two genes is zero, what genotypes would be expected among the gametes you produce, and with what frequencies?

7. Individuals showing *Down syndrome* usually have an extra chromosome 21, so their body cells contain 47 chromosomes.
    a. At which stages of meiosis I and II could a mistake occur that could result in the altered chromosome number?
    b. In a few cases, 46 chromosomes are present, including two chromosomes 21 with a normal appearance and a longer-than-normal chromosome 14. Explain how this chromosome abnormality may arise.

8. In the human population, mutation of two different genes on the X chromosome causes two types of X-linked *hemophilia* (types A and B). In a few known cases, a woman is heterozygous for both mutant alleles (one on each of her two X chromosomes). All her sons should have either hemophilia A or B. Yet, on very rare occasions, such a woman gives birth to a son who does not have hemophilia, and his one X chromosome does not have either mutant allele. Explain how such an X chromosome could arise.

9. Think back on the chapter definitions of genetic disorder and genetic abnormality. Then think about how subjective we are in terms of our acceptance of a condition that is out of the ordinary. Example: Would you find a cleft lip as acceptable, even attractive, on someone not as gorgeous as Joaquin Phoenix, shown at right? You will not find an answer to this question in Appendix IV. We present it simply as a way to invite critical thinking about what we as a society consider to be "ideal" phenotypes.

Selected Key Terms

abortion *12.11*	genetic abnormality *12.5*	linkage group *12.4*
allele *12.1*	genetic disorder *12.5*	mosaicism *12.9*
aneuploidy *12.9*	genetic	nondisjunction *12.9*
autosome *12.1*	recombination *12.1*	pedigree *12.5*
crossing over *12.1*	homologous	polyploidy *12.9*
deletion *12.8*	chromosome *12.1*	reciprocal cross *12.4*
disease *12.5*	independent	sex chromosome *12.1*
double-blind	assortment *12.1*	syndrome *12.5*
study *12.10*	inversion *12.8*	translocation *12.8*
duplication *12.8*	in-vitro fertilization *12.11*	X chromosome *12.1*
gene *12.1*	karyotype *CI*	Y chromosome *12.1*

Readings

Fairbanks, D., and W. R. Andersen. 1999. *Genetics: The Continuity of Life.* Monterey, California: Brooks-Cole.

DNA STRUCTURE AND FUNCTI

Cardboard Atoms and Bent Wire Bonds

One might have wondered, in the spring of 1868, why Johann Friedrich Miescher was collecting cells from the pus of open wounds and, later, from the sperm of a fish. Miescher, a physician, wanted to identify the chemical composition of the nucleus. These particular cells have very little cytoplasm, which makes it easier to isolate the nuclear material for analysis.

Miescher finally succeeded in isolating an organic compound having the properties of an acid. Unlike most substances in cells, it contained a notable amount of phosphorus. Miescher called the substance nuclein. He had discovered what came to be known many years later as **deoxyribonucleic acid**, or **DNA**.

The discovery did not cause even a ripple through the scientific community. At the time, no one really knew much about the physical basis of inheritance— that is, *which chemical substance encodes the instructions for reproducing parental traits in offspring.* Few even suspected that the cell nucleus might hold the answer. For a time, researchers generally believed hereditary instructions had to be encoded in the structure of some unknown class of proteins. After all, heritable traits are spectacularly diverse. Surely the molecules encoding information about those traits were structurally diverse also. Proteins are put together from potentially limitless combinations of twenty different amino acids, so the thinking was that they could function as the sentences (genes) in each cell's book of inheritance.

By the early 1950s, however, the results of a few ingenious experiments clearly indicated that DNA is the substance of inheritance. Moreover, in 1951, Linus Pauling did something no one had done before. Through his training in biochemistry, a talent for model building, and a few great educated guesses, Pauling deduced the three-dimensional structure of the protein collagen. His discovery was truly electrifying. If someone could pry open the secrets of proteins, then why not assume the same might be done for DNA? And if the structural details of the DNA molecule were worked out, then wouldn't they provide clues to its biological functions? *Who would go down in history as having discovered the very secrets of inheritance?*

Scientists around the world started scrambling after that ultimate prize. Among them were James Watson, a young postdoctoral student from Indiana University, and Francis Crick, an unflappably exuberant researcher at Cambridge University. Exactly how could DNA, a molecule that consists of only four kinds of subunits, hold genetic information? Watson and Crick spent long hours arguing over everything they had read about the size, shape, and bonding requirements of the subunits of DNA. They fiddled with cardboard cutouts of the subunits. They badgered chemists to help them identify any potential bonds they might have overlooked. Then they assembled models from bits of metal connected by wire "bonds" bent at seemingly suitable angles.

In 1953, Watson and Crick put together a model that fit all the pertinent biochemical rules and all the facts about DNA they had gleaned from other sources. They had discovered the structure of DNA (Figures 13.1 and 13.2). And the breathtaking simplicity of that structure enabled them to solve another long-standing riddle— *how the world of life can show such unity at the molecular level and yet show such spectacular diversity at the level of whole organisms.*

Figure 13.1 James Watson and Francis Crick posing in 1953 by their newly unveiled structural model of DNA.

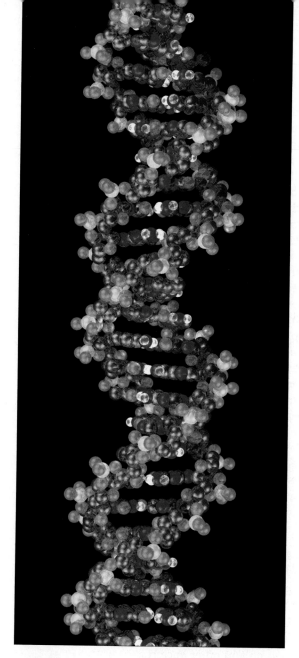

Figure 13.2 A recent space-filling model for DNA, corresponding to the prototype that Watson and Crick put together decades ago.

Key Concepts

1. In all living cells, DNA molecules are storehouses of information about heritable traits.

2. In a DNA molecule, two strands of nucleotides twist together, like a spiral stairway. Each strand of the molecule consists of four kinds of nucleotides that are the same except for one component—a nitrogen-containing base. The four bases are adenine, guanine, thymine, and cytosine.

3. Many nucleotides are arranged one after another in each strand of the DNA molecule. In at least some regions, the order in which one kind of nucleotide follows another is unique for each species. Hereditary information is encoded in the particular sequence of nucleotide bases.

4. Hydrogen bonds connect bases of one strand of the DNA molecule to bases of the other strand. As a rule, adenine pairs (hydrogen-bonds) with thymine, and guanine with cytosine.

5. Before any cell divides, its DNA is replicated with the assistance of enzymes and other proteins. Each double-stranded DNA molecule starts unwinding. As it does so, a new, complementary strand is assembled bit by bit on the exposed bases of each parent strand according to the base-pairing rule stated above.

With this chapter, we turn to investigations that led to our current understanding of DNA. The story is more than a march through details of its structure and function. *It also is revealing of how ideas are generated in science.* On the one hand, having a shot at fame and fortune quickens the pulse of men and women in any profession, and scientists are no exception. On the other hand, science proceeds as a community effort, with individuals sharing not only what they can explain but also what they do not understand. Even when an experiment fails to produce the anticipated results, it may turn up information that others can use or lead to questions that others can answer. Unexpected results, too, may be clues to something important about the natural world.

DISCOVERY OF DNA FUNCTION

Early and Puzzling Clues

The year was 1928. Frederick Griffith, an army medical officer, was attempting to develop a vaccine against *Streptococcus pneumoniae*, a bacterium that is one cause of the lung disease pneumonia. (When introduced into the body, vaccines mobilize internal defenses against a real attack. Many vaccines are preparations of killed or weakened bacterial cells.) Griffith never did develop a vaccine. But his work unexpectedly opened a door to the molecular world of heredity.

Griffith isolated and cultured two different strains of the bacterium. He noticed that colonies of one strain had a rough surface appearance, but those of the other strain appeared smooth. He designated the two strains R and S and used them in a series of four experiments:

1. Laboratory mice were injected with live R cells. The mice did not develop pneumonia, as Figure 13.3 shows. *The R strain was harmless.*

2. Other mice were injected with live S cells. The mice died. Blood samples taken from them teemed with live S cells. *The S strain was pathogenic* (disease-causing).

3. S cells were killed by exposure to high temperature. Mice injected with these cells did not die.

4. Live R cells were mixed with heat-killed S cells and injected into mice. The mice died—and blood samples from them teemed with *live* S cells!

What was going on in the fourth experiment? Maybe heat-killed S cells in the mixture weren't really dead. But if that were true, then mice injected with heat-killed S cells alone (experiment 3) would have died. Maybe harmless R cells in the mixture had mutated into a killer form. But if that were true, then mice injected with the R cells alone (experiment 1) would have died.

The simplest explanation was as follows: *Heat killed the S cells but did not destroy their hereditary material— including the part that specified "how to cause infection."* Somehow, that material had been transferred from the dead S cells to living R cells, which put it to use.

Further experiments showed the harmless cells had indeed picked up information on causing infections and were permanently transformed into pathogens. After a few hundreds of generations, descendants of those transformed bacterial cells were still infectious!

The unexpected results of Griffith's experiments intrigued Oswald Avery and his fellow biochemists. Later, they also transformed harmless bacterial cells with *extracts* of killed pathogenic cells. Finally in 1944, after rigorous chemical analyses, they felt confident in reporting that the hereditary substance in their extracts probably was DNA—not proteins, as was then widely believed. To give experimental evidence for their conclusion, they reported that they had added certain protein-digesting enzymes to some extracts, but cells exposed to those extracts were transformed anyway. To other extracts, they had added an enzyme that digests DNA but not proteins. Doing so blocked hereditary transformation.

Despite these impressive experimental results, many biochemists refused to give up on the proteins. Avery's findings, many said, probably applied only to bacteria.

Confirmation of DNA Function

By the early 1950s molecular detectives, including Max Delbrück, Alfred Hershey, Martha Chase, and Salvador Luria, were using viruses as experimental subjects. The viruses they had selected, called **bacteriophages**, infect *Escherichia coli* and other bacteria.

Viruses are biochemically simple infectious agents. They aren't alive, but they hold hereditary information about building more new virus particles. At some point after a virus infects a host cell, viral enzymes take over the cell's metabolic machinery, which starts churning out substances necessary to make new virus particles.

genetic material

viral coat

sheath
base plate

tail fiber

a

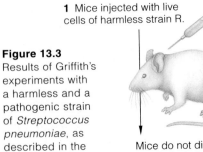 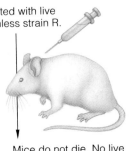 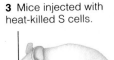 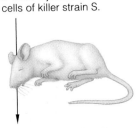 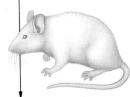

1 Mice injected with live cells of harmless strain R.

2 Mice injected with live cells of killer strain S.

3 Mice injected with heat-killed S cells.

4 Mice injected with live R cells *plus* heat-killed S cells.

Figure 13.3
Results of Griffith's experiments with a harmless and a pathogenic strain of *Streptococcus pneumoniae,* as described in the text above.

Mice do not die. No live R cells in their blood.

Mice die. Live S cells in their blood.

Mice do not die. No live S cells in their blood.

Mice die. Live S cells in their blood.

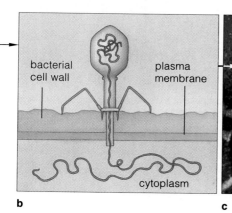

b

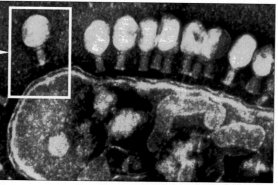

c

Figure 13.4 (**a,b**) *Far left*: Structural organization of a T4 bacteriophage. The diagram shows the genetic material of this type of virus being injected into the cytoplasm of a host cell. The genetic material of this bacteriophage is DNA (the *blue*, threadlike strand). (**c**) Micrograph of T4 virus particles that infected the bacterium *Escherichia coli*, which thereby became an unwilling host.

Figure 13.5 Examples of the landmark experiments that pointed to DNA as the hereditary substance. In the 1940s, Alfred Hershey and Martha Chase were studying the biochemical basis of inheritance. They knew that certain bacteriophages consist of proteins and DNA. *Did the proteins, DNA, or both contain viral genetic information?*

To find a possible answer, Hershey and Chase designed two experiments. They started with two known biochemical facts: First, bacteriophage proteins incorporate sulfur (S) but not phosphorus (P). Second, by contrast, bacteriophage DNA incorporates phosphorus but not sulfur.

(**a**) In one experiment, some bacterial cells were grown on a culture medium that included a radioisotope of sulfur, ^{35}S. When bacterial cells synthesized proteins, they had to take up the radioisotope—which the researchers used as a tracer. (Here you may wish to review Section 2.2.)

After the cells were labeled with the tracer, bacteriophages were allowed to infect them. As the infection ran its course, the host cells synthesized viral proteins. These proteins also became labeled with ^{35}S. So did the new generation of virus particles.

Labeled bacteriophages were allowed to infect a new batch of unlabeled bacteria that were suspended in a fluid culture medium. Afterward, Hershey and Chase whirred the fluid in a kitchen blender. Whirring dislodged the viral protein coats from the cells, so the particles became suspended in the fluid medium. Chemical analysis revealed the presence of labeled protein in the fluid. There was very little labeled protein in the preparation of bacterial cells.

(**b**) For the second experiment, Hershey and Chase cultured more bacterial cells. The phosphorus available to the cells for synthesizing DNA included the radioisotope ^{32}P. Later, bacteriophages were allowed to infect the cells.

As predicted, viral DNA synthesized inside infected cells became labeled. So did a new generation of virus particles. The labeled particles were allowed to infect bacteria that were suspended in a fluid medium. They were dislodged from the host cells.

Analysis showed the vast majority of labeled viral DNA stayed inside host cells, where its hereditary instructions had to be used to make more virus particles. Here was evidence that DNA is the genetic material of this type of virus.

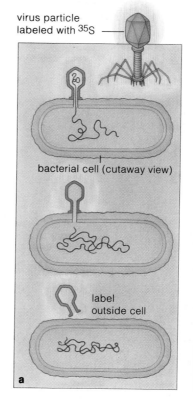

virus particle labeled with ^{35}S

bacterial cell (cutaway view)

label outside cell

a

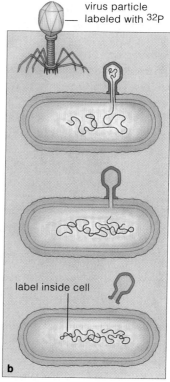

virus particle labeled with ^{32}P

label inside cell

b

By 1952, researchers knew that some bacteriophages consist only of DNA and a protein coat. Also, electron micrographs revealed that the main part of the viruses remains *outside* the cells they are infecting (Figure 13.4). Possibly, such viruses were injecting genetic material alone *into* host cells. If that were true, was the material DNA, protein, or both? Through many experiments, researchers accumulated strong evidence that DNA, not proteins, serves as the molecule of inheritance. Figure 13.5 describes two of these landmark experiments.

Information for producing the heritable traits of single-celled and multicelled organisms is encoded in DNA.

DNA STRUCTURE

What Are the Components of DNA?

Long before the bacteriophage studies were under way, biochemists knew that DNA contains only four types of nucleotides—the building blocks of nucleic acids. Each **nucleotide** consists of a five-carbon sugar (which, in DNA, is deoxyribose), a phosphate group, and one of the following nitrogen-containing bases:

adenine	**guanine**	**thymine**	**cytosine**
A	G	T	C

All four types of nucleotides have their component parts organized the same way (Figure 13.6). But **T** and **C** are pyrimidines, which are single-ring structures. **A** and **G** are purines, which are larger, bulkier molecules; they have double-ring structures.

By 1949, Erwin Chargaff, a biochemist, had shared with the scientific community two crucial insights into the composition of DNA. First, the amount of adenine relative to guanine differs from one species to the next. However, the amount of thymine is equal to that of adenine, and amount of cytosine is equal to the amount of guanine. We may show this as:

$$A = T \quad \text{and} \quad G = C$$

The proportions of those four kinds of nucleotides relative to one another were tantalizing clues. In some way, the proportions almost certainly were related to the arrangement of the nucleotides in a DNA molecule.

The first convincing evidence of that arrangement emerged from Maurice Wilkins's research laboratory in England. Rosalind Franklin, one of Wilkins's colleagues, had obtained especially good **x-ray diffraction images** of DNA fibers. (Maybe for the reasons sketched out in Section 13.3, Franklin's contribution has only recently been acknowledged.) X-ray diffraction images can be made by directing a beam of x-rays at a molecule. The molecule scatters the beam in patterns that are captured on film. The pattern consists only of dots and streaks; it alone does not reveal molecular structure. However, researchers can use photographic images of the patterns to calculate the positions of the molecule's atoms.

DNA does not readily lend itself to x-ray diffraction. But researchers can rapidly spin a suspension of DNA

Figure 13.6 Four kinds of nucleotides used as building blocks for DNA. Small numerals on the structural formulas identify carbon atoms to which other parts of the molecule are attached.

Each nucleotide in a DNA molecule has a five-carbon sugar (coded *red*) and a phosphate group attached to the fifth carbon atom of its carbon ring structure. Each nucleotide also has one of four kinds of nitrogen-containing bases (*blue*) attached to the first carbon atom. The four kinds of nucleotides in DNA differ only in which base they have: adenine, guanine, thymine, or cytosine.

All chromosomes in a cell contain DNA. What does DNA contain? Four kinds of nucleotides, A, G, T, and C. Here are the structural formulas for those nucleotides:

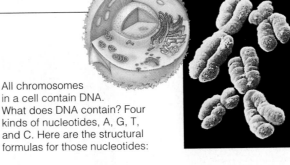

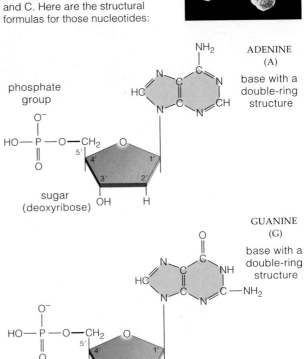

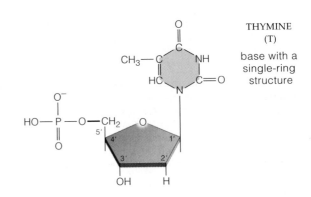

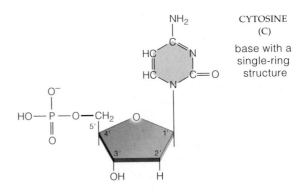

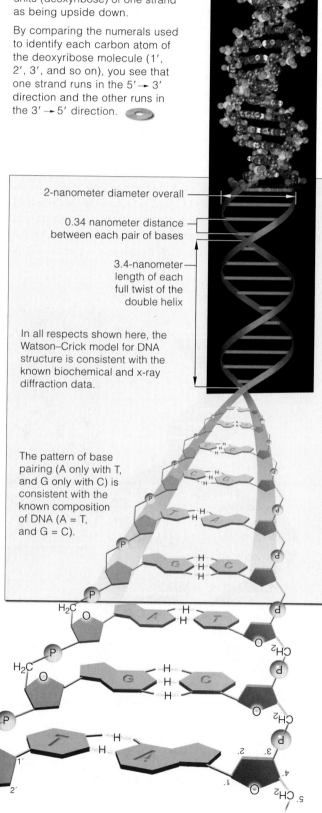

Figure 13.7 Composite of different models for a DNA double helix. The two sugar–phosphate backbones run in *opposing* directions. Think of the sugar units (deoxyribose) of one strand as being upside down.

By comparing the numerals used to identify each carbon atom of the deoxyribose molecule (1', 2', 3', and so on), you see that one strand runs in the 5' → 3' direction and the other runs in the 3' → 5' direction.

2-nanometer diameter overall

0.34 nanometer distance between each pair of bases

3.4-nanometer length of each full twist of the double helix

In all respects shown here, the Watson–Crick model for DNA structure is consistent with the known biochemical and x-ray diffraction data.

The pattern of base pairing (A only with T, and G only with C) is consistent with the known composition of DNA (A = T, and G = C).

molecules, spool them onto a rod, and gently pull them into gossamer fibers, like cotton candy. If the atoms in DNA were arranged in a regular order, x-rays directed at a fiber should scatter in a regular pattern that could be captured on film. As calculations based on Franklin's images strongly indicated, a DNA molecule is long and thin, with a 2-nanometer diameter. Some molecular configuration repeats every 0.34 nanometer along its length, and another every 3.4 nanometers.

Could the sequence of nucleotide bases be twisting, like a circular stairway? Certainly Pauling thought so. After all, he had discovered a helical shape in collagen. He and everyone else—including Wilkins, Watson, and Crick—were thinking "helix." Watson later wrote, "We thought, why not try it on DNA? We were worried that *Pauling* would say, why not try it on DNA? Certainly he was a very clever man. He was a hero of mine. But we beat him at his own game. I still can't figure out why."

Pauling, it turned out, made a big chemical mistake. His model had hydrogen bonds at phosphate groups holding DNA's structure together. That does happen in highly acidic solutions. It doesn't happen in cells.

Patterns of Base Pairing

As Watson and Crick perceived, DNA consists of *two* strands of nucleotides, held together at their bases by hydrogen bonds (Figure 13.7). These bonds form when the two strands run in opposing directions and twist to form a double helix. Two kinds of base pairings form along the length of the molecule: **A—T** and **G—C**. This bonding pattern allows variation in the order of bases. For example, even a tiny stretch of DNA from a rose, a gorilla, a human, or any other organism might be:

 or 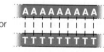 or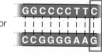

one base pair

All DNA molecules show the same bonding pattern, but each species has unique base sequences in its DNA. *This molecular constancy and variation among species is the foundation for the unity and diversity of life.*

Intriguingly, recent computer simulations show that if you want to pack a string into the least space, coil it into a helix. Could this space-saving advantage also be a factor in nature? Did natural selection favor it in the molecular evolution of the DNA double helix? Maybe.

The pattern of base pairing between the two strands in DNA is constant for all species—A with T, and G with C. Each species also has some unique differences in *which* base pair follows the next along the length of its DNA molecules.

Rosalind's Story

In 1951, Rosalind Franklin arrived at King's Laboratory of Cambridge University with impressive credentials. Earlier, in Paris, she had refined existing procedures for x-ray diffraction while studying the structure of coal. She also had devised a new mathematical approach to interpreting x-ray diffraction images and had built three-dimensional models of molecules, as Pauling had done. Now she had been asked to run an x-ray crystallography laboratory, which she would create with state-of-the-art equipment. Her assignment? Investigate the structure of DNA.

No one bothered to tell her that, down the hall, Maurice Wilkins was already working on the puzzle. Even the graduate student assigned to assist her failed to mention it. And no one bothered to tell Wilkins about Franklin's assignment, so he assumed she was a technician hired to do his x-ray crystallography work because he did not know how to do it himself. And so began a poisonous clash. To Franklin, Wilkins seemed inexplicably prickly. To Wilkins, Franklin displayed an appalling lack of the deference that technicians usually show to researchers.

Wilkins had a prized cache of crystalline DNA fibers —each having parallel arrays of hundreds of millions of DNA molecules—which he gave to his "technician."

Five months later, Franklin gave a talk on what she had learned so far. DNA, she said, may have two, three, or four parallel chains twisted in a helix, with phosphate groups projecting outward. She had measured DNA's density and assigned DNA fibers to 1 of 230 categories of crystals, based on the symmetry of their parallel chains.

With his background in crystallography, Crick would have recognized the significance of that symmetry *if* he had been present. (To wit, *paired* chains running in opposite directions would look the same even if flipped 180°. Two paired chains? No. DNA's density ruled that out. But *one pair* of chains? Yes!) Watson was in the audience, but he didn't comprehend what Franklin was talking about.

Later, Franklin created an outstanding x-ray diffraction image of wet DNA fibers that fairly screamed *Helix!* She also worked out DNA's length and diameter. But she had been working with dry fibers for so long she didn't dwell on her new data. Wilkins did. In 1953, he allowed Watson to see Franklin's exceptional x-ray diffraction image and reminded him of what she had reported fourteen months earlier. And when Watson and Crick finally did focus on her data, they had the final bits of information necessary to start building a DNA model—one that had two helically twisted chains running in opposing directions.

Figure 13.8 Researcher Rosalind Franklin.

DNA REPLICATION AND REPAIR

How Is a DNA Molecule Duplicated?

The discovery of DNA structure was a turning point in studies of inheritance. Until then, no one could explain **DNA replication**, or how the molecule of inheritance is duplicated before the cell divides. Once Watson and Crick had assembled their model, Crick understood at once how this might be done.

As he knew, enzymes can easily break the hydrogen bonds between the two nucleotide strands of a DNA molecule. When these enzymes and other proteins act on the molecule, one strand can unwind from the other, thereby exposing stretches of nucleotide bases. Cells have stockpiles of free nucleotides, and these can pair with the exposed bases.

Each parent strand remains intact, and a companion strand is assembled on each one according to this base-pairing rule: **A** to **T**, and **G** to **C**. As soon as a stretch of a new, partner strand forms on a stretch of the parent strand, the two twist together into a double helix, in the manner shown in Figure 13.9. Because the parent DNA

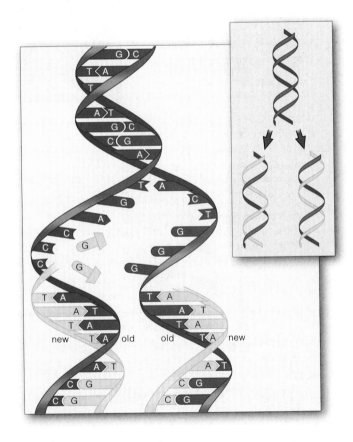

Figure 13.9 Overview of the semiconservative nature of DNA replication. The original two-stranded DNA molecule is shown in *blue*. Each parent strand remains intact. A new strand (*yellow*) is assembled on each one.

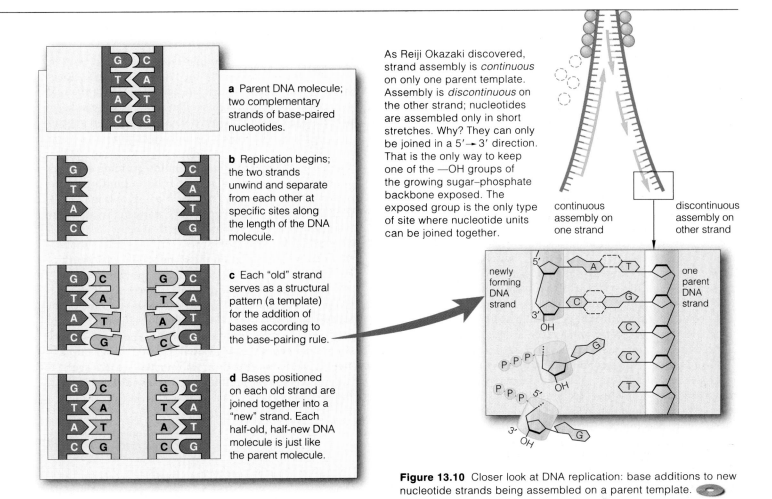

a Parent DNA molecule; two complementary strands of base-paired nucleotides.

b Replication begins; the two strands unwind and separate from each other at specific sites along the length of the DNA molecule.

c Each "old" strand serves as a structural pattern (a template) for the addition of bases according to the base-pairing rule.

d Bases positioned on each old strand are joined together into a "new" strand. Each half-old, half-new DNA molecule is just like the parent molecule.

As Reiji Okazaki discovered, strand assembly is *continuous* on only one parent template. Assembly is *discontinuous* on the other strand; nucleotides are assembled only in short stretches. Why? They can only be joined in a 5'→3' direction. That is the only way to keep one of the —OH groups of the growing sugar–phosphate backbone exposed. The exposed group is the only type of site where nucleotide units can be joined together.

continuous assembly on one strand

discontinuous assembly on other strand

newly forming DNA strand

one parent DNA strand

Figure 13.10 Closer look at DNA replication: base additions to new nucleotide strands being assembled on a parent template.

strand is conserved during the replication process, half of every double-stranded DNA molecule is "old" and half is "new" (Figure 13.9). That is why biologists refer to the process as *semiconservative* replication.

DNA replication uses a team of molecular workers. In response to cellular signals, the replication enzymes become active along the length of the DNA molecule. Together with other proteins, some enzymes unwind the strands in both directions and prevent them from rewinding. Enzyme action jump-starts the unwinding but isn't necessary to unzip hydrogen bonds between the strands; hydrogen bonds are individually weak.

Now enzymes called **DNA polymerases** attach short stretches of free nucleotides to the unwound portions of a parent template (Figure 13.10). The free nucleotides themselves actually drive the strand assembly. Each has three phosphate groups. DNA polymerase splits off two of them, releasing energy that drives the attachments.

DNA ligases fill in tiny gaps between the new short stretches to form a continuous strand. Then enzymes wind the template strand and complementary strand together to form a DNA double helix.

As you will read in Section 15.1, some replication enzymes have uses in recombinant DNA technology.

Monitoring and Fixing the DNA

Cells have trouble replicating DNA with structurally broken or altered strands. **DNA repair** processes have evolved that minimize damage. DNA ligases fix some breaks in strands. Specialized DNA polymerases can fix mismatched base pairs or replace mutated bases with undamaged ones, as described in Section 14.4. During replication, some even extend a growing strand past a lesion. This confers a survival advantage on the cell, which commits suicide (by issuing signals for its own death) when damage arrests replication. But the special polymerases do their bypass trick on *undamaged* DNA as well. Over time, such bypasses allow spontaneous mutations to accumulate. They can give rise to genetic disorders when other repair systems aren't operating.

DNA is replicated prior to cell division. Enzymes unwind its two strands. Each strand remains intact throughout the process—it is conserved—and enzymes assemble a new, complementary strand on each one.

Enzymes involved in replication also repair the DNA where base-pairing errors have crept into the nucleotide sequence.

Cloning Mammals—A Question of Reprogramming DNA

Imagine the possibility of **cloning**—making a genetically identical copy—of yourself. Is the image that far-fetched? Consider this: Researchers have been cloning complex animals for more than a decade. For example, some use in vitro fertilization methods to grow cattle embryos in petri dishes. After a fertilized egg starts dividing, they split the early cluster of cells. The two clusters develop as two identical-twin cattle embryos, get implanted in surrogate mothers, and are born as cloned calves (Figure 13.11a).

A researcher who clones farm animals derived from embryonic cells has to wait for the clones to grow up to see if they display a desired trait. Using a differentiated cell from an adult would be faster, for a prized genotype would already be known. At one time, though, tricking a differentiated cell into reprogramming its DNA to direct the development of a whole embryo seemed impossible.

What does "differentiated" mean? When an embryo first grows from a fertilized egg, all of its cells have the same DNA and are pretty much alike. Then different embryonic cells start using different parts of their DNA. Their unique selections commit them to being liver cells, heart cells, brain cells, and other specialists in structure, composition, and function (Sections 15.3 and 43.4).

In 1997 in Scotland, Ian Wilmut coaxed a differentiated sheep cell to become the "uncommitted" first cell of an embryo. His group had slipped nuclei from differentiated cells into unfertilized eggs from which the nucleus had been removed (compare Figure 13.12). Of hundreds of modifed eggs, one developed into a whole animal. The cloned lamb, named Dolly, grew into a healthy adult and gave birth to a lamb of her own (Figure 13.11b).

In science, extraordinary claims call for extraordinary proof—in this case, successful repeats of the experiment at a time when no one thought mammals could ever be cloned. Researchers around the world have since cloned sheep and mice, cows, pigs, and goats. Some of the mice have been cloned through six generations. Researchers have now successfully cloned some endangered species.

But cloning processes may introduce random errors in gene expression. As you will read in Chapter 43, the cytoplasm of a mammalian egg holds proteins, mRNAs, and other components that have roles in guiding gene expression, starting with the early embryo. It may take months or years for an egg that is maturing in an ovary to stockpile required components in specific locations in the cytoplasm. Yet cloning processes use differentiated eggs that must reprogram the donor cell's DNA within minutes or hours after it is inserted into them.

That may be why fewer than 3 percent of all cloning efforts result in healthy animals. Even then, some of the "successful" clones develop medical problems, including heart and lung abnormalities, sudden and gross obesity, and a damaged immune system. In 2002, for example, Dolly developed arthritis at an unusually early age.

Being mammals, are humans also candidates for cloning? We return to this question in Section 16.10.

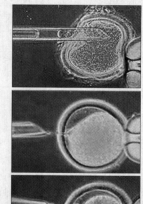

a A pipette holds an egg in place while suction is used to pull its nucleus into a fine, hollow needle.

b Only the cytoplasm remains inside the plasma membrane of the egg.

c A skin cell from an animal to be cloned is transferred into the egg.

d Electric shock triggers fusion of the skin cell with the egg cytoplasm.

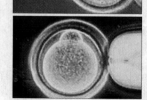

e The recipient egg starts to divide within a few hours. Seven days later, the cloned first cell of the embryo gets transferred into a host animal.

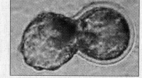

Figure 13.11 (a) Student with two genetically identical Holsteins, prized milk producers, obtained by embryos split during in vitro fertilization. (b) Dolly, a cloned sheep. She started life as a differentiated cell that was extracted from an adult ewe, then induced to start mitotic cell divisions. She is shown with her first lamb. She is able to breed normally and reproduce the old-fashioned way.

Figure 13.12 One type of nuclear transfer process.

SUMMARY

1. For all living cells, hereditary information is encoded in DNA (deoxyribonucleic acid). *CI*

2. DNA consists of nucleotide subunits. Each of these has a five-carbon sugar (deoxyribose), one phosphate group, and one of four kinds of nitrogen-containing bases (adenine, thymine, guanine, or cytosine). *13.2*

3. A DNA molecule consists of two nucleotide strands twisted together as a double helix. Bases of one strand pair (hydrogen-bond) with bases of the other. *13.2*

4. The bases of the two strands in a DNA double helix pair in constant fashion. Adenine pairs with thymine (**A** to **T**), and guanine with cytosine (**G** to **C**). *Which* base pair follows the next (**A–T**, **T–A**, **G–C**, or **C–G**) varies along the length of the strands. *13.2*

5. Overall, the DNA of one species includes a number of unique stretches of base pairs that set it apart from the DNA of all other species. *13.2*

6. During DNA replication, enzymes unwind the two strands of a double helix and assemble a new strand of complementary sequence on each parent strand. Two double-stranded molecules result. One strand of each molecule is old (is conserved); the other is new. *13.4*

7. Repair systems fix damaged DNA strands during replication. Special DNA polymerases bypass lesions, which allows mutations to accumulate in DNA. *13.4*

Review Questions

1. Name the three molecular parts of a nucleotide in DNA. Also name the four different bases in these nucleotides. *13.2*

2. What kind of bond joins two DNA strands in a double helix? Which nucleotide base-pairs with adenine? With guanine? *13.2*

3. Explain how DNA molecules can show both constancy and variation from one species to the next. *13.2*

Self-Quiz   ANSWERS IN APPENDIX III

1. Which is *not* a nucleotide base in DNA?
   a. adenine   c. uracil   e. cytosine
   b. guanine   d. thymine   f. All are in DNA.

2. What are the base-pairing rules for DNA?
   a. A–G, T–C   c. A–U, C–G
   b. A–C, T–G   d. A–T, G–C

3. One species' DNA differs from others in its _____ .
   a. sugars   c. base sequence
   b. phosphates   d. all of the above

4. When DNA replication begins, _____ .
   a. the two DNA strands unwind from each other
   b. the two DNA strands condense for base transfers
   c. two DNA molecules bond
   d. old strands move to find new strands

5. DNA replication requires _____ .
   a. free nucleotides   c. many enzymes
   b. new hydrogen bonds   d. all of the above

6. Cell differentiation involves _____ .
   a. cloning   c. selective gene expression
   b. nuclear transfers   d. both b and c

7. Match the DNA terms appropriately.
   _____ DNA polymerase   a. two nucleotide strands that
   _____ constancy in   are twisted together
          base pairing   b. A with T, G with C
   _____ replication   c. hereditary material duplicated
   _____ DNA double helix   d. replication enzyme

Critical Thinking

1. Chargaff's data suggested that adenine pairs with thymine, and guanine pairs with cytosine. What other data available to Watson and Crick suggested that adenine–guanine and cytosine–thymine pairs normally do not form?

2. One of Matthew Meselson and Frank Stahl's experiments supported the semiconservative model of DNA replication. The researchers made "heavy" DNA by growing *Escherichia coli* in a medium enriched with ^{15}N, a heavy isotope of nitrogen. They prepared "light" DNA by growing *E. coli* in the presence of ^{14}N, the more common isotope. An available technique helped them identify which replicated molecules were heavy, light, or hybrid (one heavy strand, one light). Use two pencils of different colors, one for heavy strands and one for light. Starting with a DNA molecule having two heavy strands, sketch daughter molecules that would form after replication in a ^{14}N-containing medium. Sketch the four DNA molecules that would form if the daughter molecules were replicated a second time in the ^{14}N medium.

3. Mutations, permanent changes in base sequences of genes, are the original source of genetic variation—the raw material of evolution. Yet how can mutations accumulate, given that cells have repair systems that can rapidly fix structurally altered or discontinuous DNA strands during replication?

4. As Section 4.12 indicates, a pathogenic strain of *E. coli* has acquired an ability to produce a dangerous toxin that causes medical problems and fatalities. This is especially the case for young children who have ingested undercooked, contaminated beef. Develop hypotheses to explain how a normally harmless bacterium such as *E. coli* can become a pathogen.

5. In 1999, scientists discovered a woolly mammoth that had been frozen in glacial ice for the past 20,000 years. They thawed it very carefully so they could use its DNA to clone a woolly mammoth. It turns out there wasn't enough material to work with. But they plan to try again the next time a frozen woolly mammoth comes along. Consider Section 13.5, then speculate on the pros and cons of cloning an extinct animal.

Selected Key Terms

adenine (A) *13.2*   DNA repair *13.4*
bacteriophage *13.1*   DNA replication *13.4*
cloning *13.5*   guanine (G) *13.2*
cytosine (C) *13.2*   nucleotide *13.2*
deoxyribonucleic acid (DNA) *CI*   thymine (T) *13.2*
DNA ligase *13.4*   x-ray diffraction
DNA polymerase *13.4*   image *13.2*

Readings

Watson, J. 1978. *The Double Helix.* New York: Atheneum. Highly personal view of scientists and their methods, interwoven into an account of how DNA structure was discovered.

FROM DNA TO PROTEINS

Beyond Byssus

Picture a mussel, of the sort shown in Figure 14.1. Hard-shelled but soft of body, it is using its muscular foot to probe a wave-scoured rock. At any moment, pounding waves can whack the mussel into the water, hurl it repeatedly against the rock with shell-shattering force, and so offer up a gooey lunch for gulls.

By chance, the mussel's foot comes across a crevice in the rock. The foot moves, broomlike, and sweeps the crevice clean. It presses down, forcing air out from underneath it, then arches up. The result is a vacuum-sealed chamber, much like the one that forms when a plumber's rubber plunger is being squished down and up to unclog a drain. Into this vacuum chamber the mussel spews a fluid that's made of keratin and other proteins. The fluid bubbles into a sticky foam. Now, by curling its foot into a tubelike shape and pumping the foam through it, the mussel produces sticky threads about as wide as a human whisker. It varnishes these threads with another type of protein and ends up with an adhesive. With this adhesive, which we call byssus, the mussel anchors itself to the rock.

Byssus is the world's premier underwater adhesive. Nothing humans have manufactured comes close to it; water degrades or deforms synthetic adhesives. Byssus fascinates biochemists, dentists, and surgeons looking for better ways to do tissue grafts and rejoin severed nerves. Genetic engineers insert mussel DNA into yeast cells. These cells, which reproduce in huge numbers, are

Figure 14.1 Mussels (*Mytilus californianus*) busily demonstrating the importance of proteins for survival. When mussels come across a suitable anchoring site, they use their muscular foot like a plumber's plunger and create a vacuum chamber. In this chamber they manufacture the world's best underwater adhesive from a mix of proteins. The adhesive anchors the mussels to rocks in their wave-swept habitat.

"factories" for translating mussel genes into useful quantities of proteins. This exciting work, like the mussel's own byssus-building efforts, starts with one of life's universal precepts: *Every protein is synthesized in accordance with instructions in DNA*.

You are about to trace the steps leading from DNA to proteins. Many enzymes are players in this pathway. So is another kind of nucleic acid besides DNA. The same steps produce *all* proteins, from mussel-inspired adhesives to the keratin in your hair and fingernails to the insect-digesting enzymes of a Venus flytrap.

Start out by thinking of each cell's DNA as a book of protein-building instructions. The alphabet used to create the book is simple enough: A, T, G, and C (for the nucleotide bases adenine, thymine, guanine, and cytosine). How do you get from that alphabet to a protein? The answer starts with DNA's structure.

DNA, recall, is a double-stranded molecule. Which kind of nucleotide base follows the next along the length of a strand—the **base sequence**—differs from one kind of organism to the next. The two strands unwind entirely from each other when DNA is being replicated. However, at other times in a cell's life, the two strands unwind only in certain regions to expose certain base sequences—genes. Most genes contain instructions for building proteins.

It takes two steps, **transcription** and **translation**, to carry out a gene's protein-building instructions. In eukaryotic cells, transcription proceeds in the nucleus. A newly exposed base sequence in DNA serves as a structural pattern—a template—for assembling a strand of **ribonucleic acid** (RNA) from the cell's pool of free nucleotides. Sooner or later the RNA moves into the cytoplasm, where translation proceeds. At this second step, RNA directs the assembly of amino acids into polypeptide chains. The newly formed chains become folded into the three-dimensional shapes of proteins.

In short, DNA guides the synthesis of RNA, then RNA guides the synthesis of proteins:

$$\text{DNA} \xrightarrow{\textit{transcription}} \text{RNA} \xrightarrow{\textit{translation}} \text{PROTEIN}$$

The newly synthesized proteins will play structural and functional roles in cells. Some even will have roles in synthesizing more DNA, RNA, and proteins.

Key Concepts

1. Organisms cannot stay alive without enzymes and other proteins. Proteins consist of polypeptide chains, which consist of amino acids. The sequence of amino acids corresponds to a gene, which is a sequence of nucleotide bases in a DNA molecule.

2. The path leading from genes to proteins consists of two steps, called transcription and translation.

3. During transcription, the double-stranded DNA molecule is unwound at a gene region, and then an RNA molecule is assembled on the exposed bases of one of the strands.

4. Translation uses three classes of RNA molecules: messenger RNA, transfer RNA, and ribosomal RNA.

5. During translation, amino acids are joined together sequentially into a polypeptide chain, in a sequence specified by messenger RNA. Transfer RNA delivers the amino acids one at a time to the construction site. Ribosomal RNA catalyzes the chain-building reaction.

6. With few exceptions, the genetic "code words" by which DNA's instructions are translated into proteins are the same in all species.

7. A mutation is a permanent alteration in a gene's base sequence. Mutations are the original source of genetic variation in populations.

8. Mutations introduce changes in protein structure, protein function, or both. The changes may lead to small or large differences in traits among individuals of a population.

HOW IS RNA TRANSCRIBED FROM DNA?

The Three Classes of RNA

Before turning to the details of protein synthesis, be clear on one point. The chapter introduction may have left you with the impression that synthesis of proteins requires only one class of RNA molecules. Actually, it requires three. Transcription of most genes produces **messenger RNA**, or **mRNA**—the only class of RNA that carries *protein-building* instructions. Transcription of some other genes produces **ribosomal RNA**, or **rRNA**, a major component of ribosomes. Ribosomes, recall, are the structural units upon which polypeptide chains are assembled. Transcription of still other genes produces **transfer RNA**, or **tRNA**, which delivers amino acids one by one to a ribosome in the order specified by mRNA.

The Nature of Transcription

An RNA molecule is almost but not quite like a single strand of DNA. RNA, too, consists of only four types of nucleotides. Each nucleotide has a five-carbon sugar, ribose (not DNA's deoxyribose), a phosphate group, and a base. Three types of bases—adenine, cytosine, and guanine—are the same in RNA and DNA. But in RNA, the fourth type of base is **uracil**, not thymine (Figure 14.2). Like thymine, uracil can pair with adenine. This means a new RNA strand can be put together on a DNA region according to base-pairing rules (Figure 14.3).

Transcription resembles DNA replication in another way. Enzymes add nucleotides to a growing RNA strand one at a time, in the 5′ → 3′ direction. Section 13.4 has a simple explanation of strand assembly.

But transcription *differs* from DNA replication in three key respects. First, only a selected stretch of one DNA strand rather than the whole molecule is used as the template. Second, instead of DNA polymerases, the type of enzyme known as **RNA polymerase** catalyzes nucleotide additions to a growing RNA strand. Third, at the end of transcription there is a single, free strand of RNA nucleotides, not a double helix.

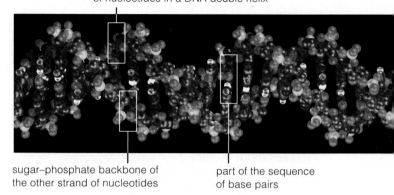

sugar–phosphate backbone of one strand of nucleotides in a DNA double helix

sugar–phosphate backbone of the other strand of nucleotides

part of the sequence of base pairs

a Location of nucleotide bases in DNA

Transcription is initiated at a **promoter**. This base sequence in DNA signals the start of a gene. Proteins position an RNA polymerase on DNA and help start transcription at the promoter. The enzyme moves along the DNA strand, joining one nucleotide after another (Figure 14.4). When it reaches a certain point in the gene, the new RNA molecule is released as a free transcript.

Finishing Touches on mRNA Transcripts

In eukaryotic cells, each new molecule of mRNA is not in its final form. This "pre-mRNA" must be modified before its protein-building instructions can be put to use. Just as a dressmaker might snip off some threads or add bows on a dress before it leaves the shop, so do eukaryotic cells tailor their pre-mRNA.

For example, enzymes attach a cap to the 5′ end of pre-mRNA. The cap, a nucleotide, incorporates a methyl group and phosphate groups. Enzymes also attach a tail of about 100 to 300 nucleotides to the 3′ end of the pre-mRNA transcript. The new tail becomes wound up with proteins. Later on, in the cytoplasm, the cap will help bind the mRNA to a ribosome. Enzymes will also slowly destroy the wound-up tail from the tip on back.

Figure 14.2 Structural formula for one of the four types of RNA nucleotides. The three others have a different base (adenine, guanine, or cytosine instead of the uracil shown here). Compare Section 13.2, which shows DNA's four nucleotides. Notice how the sugars of DNA and RNA differ at one group only (*yellow*).

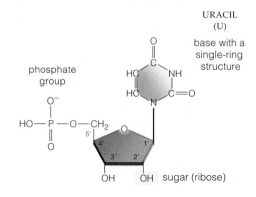

URACIL
(U)

phosphate group

base with a single-ring structure

sugar (ribose)

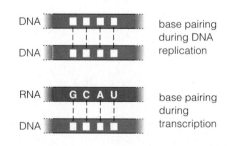

base pairing during DNA replication

base pairing during transcription

Figure 14.3 An example of base pairing of RNA with DNA during transcription, compared to base pairing during DNA replication.

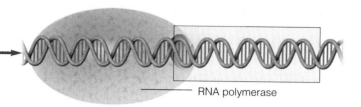

b RNA polymerase initiates transcription at a promoter region in the DNA. It will recognize the base sequence located downstream from that site as a template for linking together the nucleotides adenine, cytosine, guanine, and uracil into a strand of RNA.

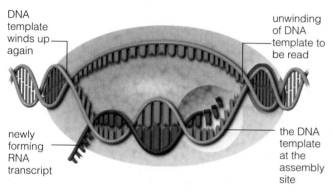

DNA template winds up again

unwinding of DNA template to be read

newly forming RNA transcript

the DNA template at the assembly site

c All through transcription, the DNA double helix becomes unwound in front of the RNA polymerase. Short lengths of the newly forming RNA strand briefly wind up with its DNA template strand. New stretches of RNA unwind from the template (and the two DNA strands wind up again).

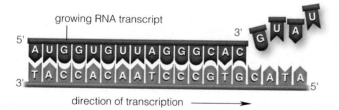

growing RNA transcript

5′ A U G G U G U U A G G G C A C 3′ G U A U

3′ T A C C A C A A T C C C G T G C A T A 5′

direction of transcription ——→

d What happened at the assembly site? RNA polymerase catalyzed the base-pairing of RNA nucleotides, one after another, with exposed bases on the DNA template strand.

5′ A U G G U G U U A G G G C A C G U A U 3′

e At the end of the gene region, the last stretch of the new mRNA transcript is unwound and released from the DNA.

Figure 14.4 The process of gene transcription, by which an RNA molecule is assembled on a DNA template. The diagram in (**a**) shows a gene region in part of a DNA double helix. In this region, the base sequence of one of the two nucleotide strands (not both) is about to be used as a template for transcription of a molecule of RNA, in the manner shown in (**b**) through (**e**).

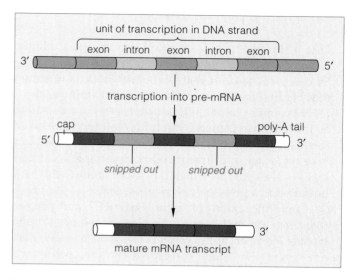

unit of transcription in DNA strand

exon | intron | exon | intron | exon

3′ ——————————————————————— 5′

transcription into pre-mRNA

cap

5′ —————————————————————— 3′ poly-A tail

snipped out   *snipped out*

—————————————————— 3′

mature mRNA transcript

Figure 14.5 Transcription and modification of new mRNA in the nucleus of eukaryotic cells. Its cap is a nucleotide with functional groups attached. Its tail is a sequence of adenine nucleotides (hence the name, poly-A tail).

Such tails "pace" the access of enzymes to the mRNA. That controlled access determines how long an mRNA molecule will last. It helps keeps the protein-building messages intact for as long as the cell requires them.

Besides these alterations, the pre-mRNA itself gets modified. Most eukaryotic genes contain one or more **introns**, base sequences that must be removed before a pre-mRNA molecule can be translated. These introns intervene between **exons**, the parts that are still in the mRNA when it's translated into protein. As Figure 14.5 shows, introns are transcribed along with the exons but are snipped out before the mRNA leaves the nucleus in mature form. (Think of it this way: *Ex*ons are *ex*ported from the nucleus and *in*trons stay *in* the nucleus, where they are degraded.)

Many introns are actually sites where instructions for building a protein can be snipped apart and spliced together in more than one way. With this alternative splicing, different cells in the body use the same gene for making different versions of a pre-mRNA transcript, and so the resulting proteins differ slightly in form and function. We return to this topic in Section 15.3.

During gene transcription, a sequence of exposed bases in one of the two strands of a DNA molecule serves as the template upon which RNA polymerase assembles a single strand of RNA. In this case, adenine base-pairs with uracil, and cytosine with guanine.

Before leaving the nucleus, each new mRNA transcript, or pre-mRNA, undergoes modification into final form.

DECIPHERING THE mRNA TRANSCRIPTS

What Is the Genetic Code?

Like a strand of DNA, an mRNA molecule is a linear sequence of nucleotides. What are the protein-building "words" encoded in that sequence? Each is a certain number of nucleotides that codes for an amino acid.

Ribosomes "read" nucleotide bases *three at a time*, as triplets. Base triplets in an mRNA strand were given this name: **codons**. Figure 14.6 will give you an idea of how the order of different codons in an mRNA strand dictates the order in which particular amino acids will be added to a growing polypeptide chain.

Count the codons listed in Figure 14.7, and you see that there are sixty-four kinds. Notice how most of the twenty kinds of amino acids correspond to more than one codon. Glutamate corresponds to the code words GAA *and* GAG, for example. AUG codes for the amino acid methionine and is the start point for translation of the mRNA transcripts. Said another way, the "three-bases-at-a-time" selections start at a particular AUG in the transcript's nucleotide sequence. Methionine is the first amino acid in new polypeptide chains. Codons UAA, UAG, and UGA do not correspond to an amino acid. They serve as STOP signals that prevent further additions of amino acids to a new polypeptide chain.

The set of sixty-four different codons is the **genetic code**. It is the basis of protein synthesis in all organisms.

Structure and Function of tRNA and rRNA

In a cell's cytoplasm are pools of free amino acids and free tRNA molecules. The tRNAs each have a molecular "hook," an attachment site for amino acids. They also have an **anticodon**, a nucleotide triplet that can base-pair with a codon (Figure 14.8). When tRNAs bind to the codons, they automatically position their attached amino acids in the order specified by mRNA.

A cell has a cytoplasmic pool of sixty-four kinds of codons, but it is able to utilize fewer kinds of tRNAs.

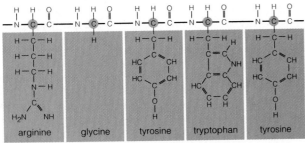

a Part of the amino acid sequence that was put together when mRNA was translated into a polypeptide chain

b Part of the mRNA strand that was transcribed from DNA

c The base sequence of a gene region in DNA

Figure 14.6 The correspondence between genes and proteins, as deduced by Marshall Nirenberg, Philip Leder, Severo Ochoa, and Gobind Korana. (**a**) A sequence of amino acids, part of a protein's polypeptide chain. (**b**) An mRNA transcript. Every three nucleotide bases, equaling one codon, calls for one of the chain's amino acids. (**c**) Exposed bases on one strand of a DNA double helix that is unwound during transcription. It is the template for assembling an mRNA strand. Referring to Figure 14.7, can you fill in the blank codon for tryptophan in the mRNA strand in (**b**)?

first base	second base				third base
	U	C	A	G	
U	phenylalanine	serine	tyrosine	cysteine	U
	phenylalanine	serine	tyrosine	cysteine	C
	leucine	serine	STOP	STOP	A
	leucine	serine	STOP	tryptophan	G
C	leucine	proline	histidine	arginine	U
	leucine	proline	histidine	arginine	C
	leucine	proline	glutamine	arginine	A
	leucine	proline	glutamine	arginine	G
A	isoleucine	threonine	asparagine	serine	U
	isoleucine	threonine	asparagine	serine	C
	isoleucine	threonine	lysine	arginine	A
	methionine (or START)	threonine	lysine	arginine	G
G	valine	alanine	aspartate	glycine	U
	valine	alanine	aspartate	glycine	C
	valine	alanine	glutamate	glycine	A
	valine	alanine	glutamate	glycine	G

Figure 14.7 The genetic code. Codons in mRNA are nucleotide bases "read" in blocks of three. Sixty-one of the base triplets correspond to specific amino acids. Three others are signals to stop translation. The *left* vertical column lists choices for the first of three nucleotides in an mRNA codon. The top horizontal row lists choices for the second codon. The *right* vertical column lists choices for the third. Example: reading from left to right, the triplet UGG corresponds to tryptophan. Both UUU and UUC correspond to phenylalanine.

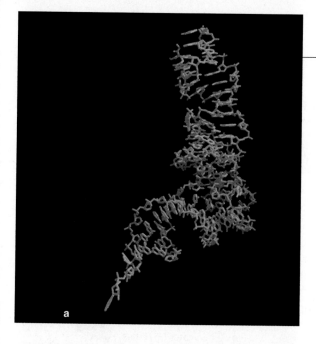

a

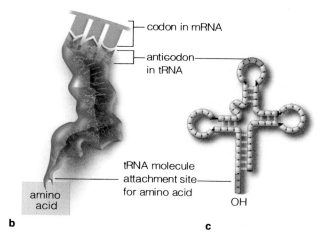

b

- codon in mRNA
- anticodon in tRNA
- tRNA molecule attachment site for amino acid

amino acid

c

OH

Figure 14.8 (**a**) Stick model for one type of tRNA molecule. (**b**) An icon for tRNA that you will come across in illustrations to follow. The "hook" at the lower end of this icon represents a binding site for a specific amino acid. (**c**) Structural features common to all tRNAs.

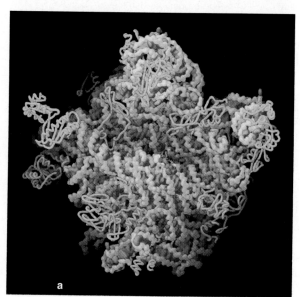

a

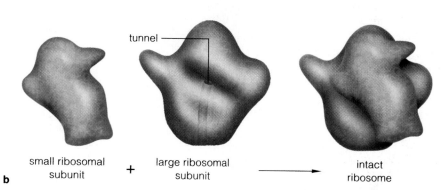

tunnel

small ribosomal subunit + large ribosomal subunit → intact ribosome

b

Figure 14.9 (**a**) Ribbon model for the large subunit of a bacterial ribosome. It consists of two rRNA molecules (*gray*) and thirty-one structural proteins (*gold*), which stabilize the structure. At one end of a tunnel inside this subunit, rRNA catalyzes polypeptide chain assembly. This is an ancient, highly conserved molecular structure. Its role is so vital that the corresponding subunit of eukaryotic ribosomes, which is bigger, is probably similar in structure and function. (**b**) Model for the small and large subunits of a ribosome.

How do tRNAs match up with more than one type of codon? According to base-pairing rules, adenine must pair with uracil, and cytosine with guanine. For codon–anticodon interactions, however, the rules loosen up for the third base of the codon. To give one example, AUU, AUC, and AUA specify isoleucine. All three of these codons can pair with a single type of tRNA that carries isoleucine. Such freedom in codon–anticodon pairing at a base is known as the "wobble effect."

Before anticodons interact with codons of an mRNA strand, that strand must bind to a ribosome. As shown in Figure 14.9, each ribosome has two subunits. These are assembled in the nucleus from rRNA and structural proteins, which stabilize the ribosome's structure. The enzyme action of rRNA drives protein synthesis.

At some point the subunits are shipped separately to the cytoplasm. There, intact, functional ribosomes are put together, each from two subunits, when messages encoded in mRNA are to be translated.

The nucleotide sequence of both DNA and mRNA encodes protein-building instructions. The genetic code is a set of sixty-four base triplets, which are nucleotide bases read in blocks of three. A codon is a base triplet in mRNA.

Different combinations of codons specify the amino acid sequence of different polypeptide chains, start to finish.

mRNAs are the only molecules that carry protein-building instructions from DNA into the cytoplasm.

tRNAs deliver amino acids to ribosomes, where they base-pair with codons in the order specified by mRNA.

Ribosomes, composed of rRNA and proteins, are structures upon which amino acids are assembled into polypeptide chains. An rRNA catalyzes chain assembly.

14.3

HOW IS mRNA TRANSLATED?

Stages of Translation

The protein-building code built into mRNA transcripts of DNA becomes translated at intact ribosomes in the cytoplasm. Translation proceeds through three stages: initiation, elongation, and termination.

During the stage called *initiation*, an initiator tRNA (the only one that can start transcription) and an mRNA transcript are both loaded onto a ribosome. First, the initiator tRNA binds with the small ribosomal subunit.

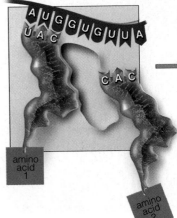

binding site for mRNA

ELONGATION

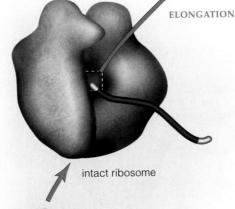

P (first binding site for tRNA)   *A* (second binding site for tRNA)

d Simplified model for binding sites at one end of the tunnel through the large ribosomal subunit, as shown in Figure 14.9. One site is for an mRNA transcript. Two others are for tRNAs that deliver amino acids to the intact ribosome.

e The initiator tRNA has become positioned in the first tRNA binding site (called *P*) on the ribosomal platform. Its anticodon matches up with the START codon (AUG) of the mRNA, which also has become positioned in *its* binding site. Another tRNA is about to move into the platform's second tRNA binding site (called *A*). It is one that can bind with the codon following the START codon.

c As the final step of the initiation stage, a large ribosomal subunit joins with the small one. Once this initiation complex has formed, chain *elongation*— the second stage of translation—can get under way.

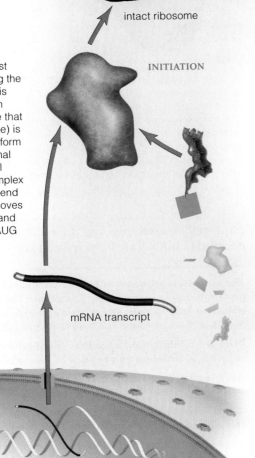

intact ribosome

INITIATION

b *Initiation*, the first stage of translating the mRNA transcript, is about to begin. An initiator tRNA (one that can start this stage) is loaded onto a platform of a small ribosomal subunit. The small subunit/tRNA complex attaches to the 5' end of the mRNA. It moves along the mRNA and "scans" it for an AUG START codon.

a A mature mRNA transcript leaves the nucleus by passing through pores across the nuclear envelope. Thus it enters the cytoplasm, which contains pools of many free amino acids, tRNAs, and ribosomal subunits.

mRNA transcript

Figure 14.10 Translation, the second step of protein synthesis.

The START codon for the transcript, AUG, matches up with that tRNA's anticodon. Second, a large ribosomal subunit binds with the small subunit. When joined this way, the ribosome, mRNA, and tRNA are an initiation complex (Figure 14.10*b*). The next stage can begin.

In the *elongation* stage of translation, a polypeptide chain is assembled as the mRNA passes between the two ribosomal subunits, a bit like a thread being moved through the eye of a needle. Part of the rRNA molecule located at the center of the large ribosomal subunit has unusual acidity. This region functions as an enzyme. It catalyzes the joining of individual amino acids, and does so in the sequence that is dictated by the codon sequence in the mRNA molecule.

Figure 14.10*f–i* shows how a peptide bond forms between the most recently attached amino acid and the next one delivered to the intact ribosome while the polypeptide chain is growing. Here, you might wish to look once more at Section 3.6 (Figure 3.18), which has a sketch and description of peptide bond formation.

During the last stage of translation, *termination*, a STOP codon in the mRNA moves onto the platform. No tRNA has a corresponding anticodon. Proteins known as release factors bind to the ribosome. They trigger enzyme activity that detaches the mRNA *and* the chain from the ribosome (Figure 14.10*j–l*).

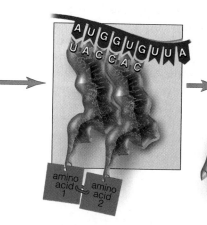

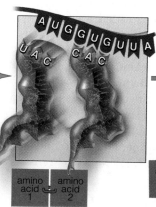

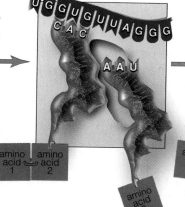

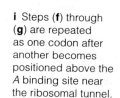

f Enzyme action breaks the bond between the initiator tRNA and the amino acid hooked to it. At the same time, enzyme action also catalyzes the formation of a peptide bond between that amino acid and the one hooked to the second tRNA. Then the initiator tRNA is released from the ribosome.

g Now the first amino acid is attached only to the second one—which is still hooked to the second tRNA. The ribosome is about to move this tRNA into the *P* site and slide the mRNA along with it by one codon. This will align the third codon in the *A* site.

h A third tRNA is about to move into the vacated *A* site. Its anticodon can base-pair with the third codon of the mRNA transcript. Now the ribosome will catalyze the formation of a peptide bond between amino acids 2 and 3.

i Steps (**f**) through (**g**) are repeated as one codon after another becomes positioned above the *A* binding site near the ribosomal tunnel.

What Happens to the New Polypeptides?

Unfertilized eggs and other cells that will be called upon to rapidly synthesize many copies of different proteins usually stockpile mRNA transcripts in their cytoplasm. In cells that are rapidly using or secreting proteins, you often observe polysomes. Each polysome is a cluster of many ribosomes translating one mRNA transcript at the same time. The transcript threads through all of them, one after another, like the thread of a pearl necklace.

After new polypeptide chains are synthesized, many join the cytoplasmic pool of free proteins. Many others enter the ribosome-studded, flattened sacs of rough ER, part of the endomembrane system (Section 4.5). There they take on final form before they are shipped to their ultimate destinations inside or outside the cell.

Translation is initiated when a small ribosomal subunit and an initiator tRNA arrive at an mRNA's START codon and then a large ribosomal subunit binds to them.

tRNAs deliver amino acids to the ribosome in the order dictated by the sequence of mRNA codons, to which the tRNA anticodons base-pair. A polypeptide chain lengthens as peptide bonds form between the amino acids.

Translation ends when a STOP codon triggers events that cause the chain and mRNA to detach from the ribosome.

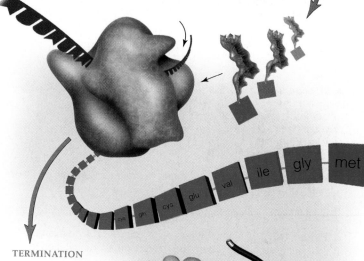

TERMINATION

j A STOP codon moves into the area where the chain is being built. It is the signal to release the mRNA transcript from the ribosome.

k The newly formed polypeptide chain also is released from the ribosome. It is free to join the pool of proteins in the cytoplasm or to enter rough ER of the endomembrane system.

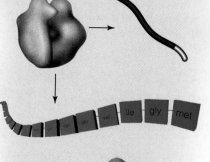

l The two ribosomal subunits separate.

DO MUTATIONS AFFECT PROTEIN SYNTHESIS?

Whenever a cell puts its genetic code into action, it is making precisely those proteins that form its structure and carry out its functions. If something changes a gene's code words, the resulting protein may change, also. If the protein is central to cell architecture or metabolism, we can expect the outcome to be an abnormal cell.

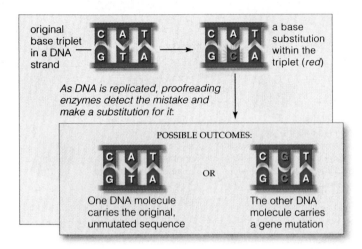

As DNA is replicated, proofreading enzymes detect the mistake and make a substitution for it.

POSSIBLE OUTCOMES:

One DNA molecule carries the original, unmutated sequence

The other DNA molecule carries a gene mutation

a Example of a base-pair substitution

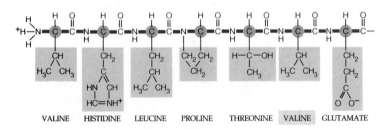

VALINE HISTIDINE LEUCINE PROLINE THREONINE VALINE GLUTAMATE

b Outcome of the base-pair substitution

Figure 14.11 Common types of mutations. **(a)** One example of a base-pair substitution. **(b)** This base-pair substitution is a type of molecular change that caused a single amino acid to be replaced in the beta chains of hemoglobin (valine instead of glutamate). Sickle-cell anemia is the result.

Gene sequences do change. Sometimes one base gets substituted for another in the nucleotide sequence. At other times, an extra base is inserted into the sequence or a base is lost from it. Such small-scale changes in the nucleotide sequence of genes in the DNA molecule are **gene mutations**. There is some leeway here; remember, more than one codon may specify the same amino acid. For instance, if UCU were changed to UCC, it probably wouldn't have dire effects because both codons specify serine. However, many mutations give rise to proteins with altered or lost functions.

Common Gene Mutations and Their Sources

Figure 14.11 shows a common gene mutation. One base (adenine) was wrongly paired with another (cytosine) as DNA was replicated. Specialized DNA polymerases fix such errors in growing DNA strands (Section 13.4). But some keep assembling a new strand right past an error, and such a bypass can establish a mutation in the DNA molecule. This particular mutation is a **base-pair substitution**. Its outcome? One amino acid may replace another during protein synthesis. That is what happens in people who carry Hb^S, the mutant allele that causes sickle-cell anemia (Section 3.8).

Figure 14.12 shows a different mutation. One *extra* base was inserted into a gene region. Remember, DNA polymerases read nucleotide sequences in blocks of three. The insertion shifted the "three-bases-at-a-time" reading frame by one base; hence the term *frameshift* mutation. The altered gene has a different message, so an altered version of the protein will be synthesized. Frameshift mutations fall in broader categories of mutation called **insertions** and **deletions**. In such cases, one to several base pairs are inserted into DNA or deleted from it.

As another example, **transposons**—or transposable elements—may bring about mutation when they jump around in the genome. Barbara McClintock discovered that these segments of DNA move spontaneously from

mRNA transcribed from the DNA

PART OF PARENTAL DNA TEMPLATE

resulting amino acid sequence

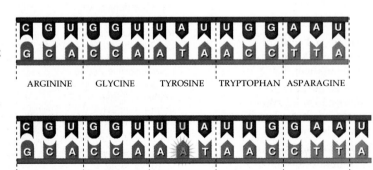

ARGININE GLYCINE TYROSINE TRYPTOPHAN ASPARAGINE

altered message in mRNA

A BASE INSERTION (RED) IN DNA

the altered amino acid sequence

ARGININE GLYCINE LEUCINE LEUCINE GLUTAMATE

Figure 14.12 Example of an insertion, a mutation in which an extra base gets inserted into a gene region of DNA. This insertion has caused a *frameshift*; it has changed the reading frame for base triplets in the DNA and in the mRNA transcript of that region. As a result, the wrong amino acids will be called up when the mRNA transcript becomes translated into protein.

Figure 14.13 Barbara McClintock, who won a Nobel Prize for her research that showed some DNA segments slip into and out of different locations in DNA molecules. The segments are transposons. In her hands is an ear of Indian corn (*Zea mays*). The curiously nonuniform coloration of its kernels sent her on the road to discovery.

Each corn kernel is a seed that can grow into a new corn plant. All of its cells have the same pigment-coding genes. Yet some kernels are colorless or spottily colored. In the ancestor of the plant from which this ear of corn was plucked, a gene in a germ cell left its position in one DNA molecule, invaded a different DNA molecule, and shut down a pigment-encoding gene.

The plant inherited the mutation. As cell divisions proceeded in the growing plant, none of the mutated cell's descendants could synthesize pigment molecules. Wherever they were located, the kernel tissue was colorless. Later, in some cells, the movable DNA slipped out of the pigment-encoding gene. All descendants of *those* cells produced pigment—and colored kernel tissue.

one location to another in the same DNA molecule or to a different one. Often they inactivate genes into which they become inserted. Their unpredictability can cause interesting variations in traits. You can read about two examples in Figure 14.13 and *Critical Thinking* question 3 at this chapter's end.

Causes of Gene Mutations

Many mutations arise spontaneously when DNA is being replicated. This shouldn't be surprising, given the swift pace of replication and the huge pools of free nucleotides concentrated near a growing DNA strand. Specialized DNA polymerases repair most of the mistakes, but they bypass a small number with predictable frequency.

Each gene has a characteristic **mutation rate**. This is the probability that it will mutate spontaneously during a specified interval, such as each DNA replication cycle. (The *rate* isn't the same as the *frequency* of a mutation—the number of times it is found in some population, as in the 500,000 people resulting from 1 million gametes.)

Not all mutations are spontaneous. Many result after exposure to mutagens, or mutation-causing agents in the environment. Two classes of radiation are mutagenic. High-energy wavelengths of **ionizing radiation**, such as x-rays, can damage DNA directly. They also damage it indirectly by the action of free radicals that form when they ionize water and other molecules. Repair enzymes may not restore the altered base sequences. Ionizing radiation that deeply penetrates living tissue leaves a trail of free radicals. X-ray doses used for dental work and internal medicine diagnoses are extremely low to minimize mutation. However, x-rays have a cumulative effect, so repeated exposure even to low levels over the years can cause problems.

Nonionizing radiation simply boosts electrons to a higher energy level. But DNA easily absorbs one form, ultraviolet (UV) light. Two of its nucleotides, cytosine

and thymine, are highly vulnerable to excitation that can alter their base-pairing properties. The next chapter's introduction describes one possible mutation.

Natural and synthetic chemicals accelerate rates of spontaneous mutations. For instance, substances called **alkylating agents** transfer methyl and ethyl groups to reactive sites in DNA's bases or phosphate groups. At the alkylated sites, DNA is more susceptible to base-pair alterations that invite mutation. Many **carcinogens**, or cancer-causing agents, operate by alkylating DNA.

The Proof Is In the Protein

Spontaneous mutations are rare in terms of a human life. The rate for eukaryotes in general ranges between 10^{-4} and 10^{-6} per gene per generation. If one arises in a somatic cell, its good or bad effects won't endure, for it cannot be passed on to offspring. If the mutation arises in a germ cell or gamete, however, it may well enter the evolutionary arena. The same can happen if a mutation arises in an asexually reproducing organism or cell.

In all such cases, nature's test is this: *A protein that is specified by a heritable mutation may have harmful, neutral, or beneficial effects on the individual's ability to function in the prevailing environment.* Also, gene mutations have had powerful evolutionary consequences—and that will be a major theme of the next unit of the book.

A gene mutation is an alteration in one to several bases in the nucleotide sequence of DNA. The most common are base-pair substitutions, base insertions, and base deletions.

Each gene has a spontaneous and characteristic mutation rate, which may be accelerated by exposure to harmful radiation and certain chemicals in the environment.

A protein specified by a mutated gene may have harmful, neutral, or beneficial effects on the ability of an individual to function in the prevailing environment.

SUMMARY

Gold indicates text section

1. Single cells and multicelled organisms cannot stay alive unless they build enzymes and other proteins. A protein consists of one or more polypeptide chains. Each chain is a linear sequence of amino acids. *CI*

   a. The amino acid sequence of a polypeptide chain corresponds to a gene region in one strand of the DNA double helix. That region is a sequence of nucleotide bases. DNA's bases are adenine, thymine, guanine, and cytosine (A, T, G, and C).

   b. Transcription and translation are two steps in the path from genes to proteins (Figure 14.14):

$$\textbf{DNA} \xrightarrow{\textit{transcription}} \textbf{RNA} \xrightarrow{\textit{translation}} \textbf{PROTEIN}$$

2. The path requires three classes of RNA molecules, or ribonucleic acids:

   a. Messenger RNA (mRNA) is the only class of RNA that carries a protein-building message. *14.1, 14.2*

   b. Ribosomal RNA (rRNA) and structural proteins that stabilize it are the components of ribosomes. All polypeptide chains are assembled on ribosomes. *14.2*

   c. Transfer RNA (tRNA) binds free amino acids in the cytoplasm and gives them up at a ribosome, in the sequence dictated by a sequential message in mRNA. Different kinds bind different amino acids. *14.2, 14.3*

3. In transcription, DNA is unwound at a gene region. Exposed bases on one strand function as a template for assembling an RNA strand from the cell's pool of free nucleotides. The RNA-to-DNA rules for base-pairing are that guanine pairs with cytosine, and *uracil*—not thymine—pairs with adenine: *14.1*

   a. Different RNAs are assembled on different genes.

   b. In eukaryotic cells, the mRNA transcripts become modified into final form before being shipped from the nucleus. We call this transcript processing.

4. In translation, mRNA, tRNAs, and rRNA interact to build polypeptide chains. Afterward, the chains twist, fold, and may be additionally modified into a protein's final, three-dimensional shape. *CI, 14.3*

   a. Translation follows a genetic code. The code is a set of sixty-four base triplets; each is a series of three nucleotide bases. *Triplet* refers to the way the bases are "read" three at a time during translation at a ribosome.

   b. A base triplet in mRNA is a codon. An anticodon is a complementary triplet in a tRNA molecule. Some combination of codons specifies what the amino acid sequence of a polypeptide chain will be, start to finish.

5. Translation proceeds through three stages: *14.3*

   a. Initiation. One small ribosomal subunit and one initiator tRNA bind with the mRNA and move along it until they encounter an AUG START codon. The small subunit binds with a large ribosomal subunit.

   b. Chain elongation. tRNAs deliver amino acids to an intact ribosome. Their anticodons base-pair with the mRNA codons. Part of the rRNA of the large ribosomal subunit catalyzes peptide bond formation between every two amino acids, forming a new polypeptide chain.

   c. Chain termination. An mRNA STOP codon moves onto the ribosomal platform, making the polypeptide chain and the mRNA detach from the ribosome.

6. Gene mutations are heritable, small-scale changes in the base sequence of DNA. Many arise spontaneously while DNA is being replicated or after it is exposed to ultraviolet or ionizing radiation, alkylating agents, or some other mutagen in the environment. *14.4*

TRANSCRIPTION *Assembly of RNA on unwound gene regions of DNA molecule*

Pre-mRNA transcript processing

mRNA   **rRNA**   **tRNA**

protein subunits

mature mRNA transcripts

ribosomal subunits

mature tRNA

TRANSLATION

Convergence of RNAs

cytoplasmic pools of amino acids, ribosomal subunits, and tRNAs

At an intact ribosome, synthesis of a polypeptide chain at the binding sites for mRNA and tRNAs

A U G G U G

ile — gly — met

FINAL PROTEIN

For use in cell or for export

Figure 14.14 Summary of the flow of genetic information from DNA to proteins in eukaryotic cells. DNA is transcribed into RNA in the nucleus; RNA is translated in the cytoplasm. Prokaryotic cells don't have a nucleus; both transcription and translation proceed in their cytoplasm.

Review Questions

1. Are the polypeptide chains of proteins assembled on DNA? If so, state how. If not, state how they are assembled, and on which molecules. *CI, 14.3*

2. Name the three classes of RNA and state the function of each class in protein synthesis. *14.1, 14.2*

3. The pre-mRNA transcripts of eukaryotic cells contain both introns and exons. Are the introns or exons snipped out before the transcript leaves the nucleus? *14.1*

4. Distinguish between codon and anticodon. *14.2*

5. Name the three stages of translation. Briefly describe the key events of each stage. *14.3*

6. Define gene mutation. Give three examples of agents that cause mutations. *14.4*

7. Do all mutations arise spontaneously? Are environmental agents always the trigger for mutation? *14.4*

8. Define and then state the possible outcomes of the following types of mutation: base-pair substitution, base insertion, and insertion of a transposon at a new location in the DNA. *14.4*

Self-Quiz   ANSWERS IN APPENDIX III

1. DNA contains many different genes that are transcribed into different _____ .
   - a. proteins
   - b. mRNAs only
   - c. mRNAs, tRNAs, and rRNAs
   - d. all are correct

2. An RNA molecule is _____ .
   - a. a double helix
   - b. usually single-stranded
   - c. always double-stranded
   - d. usually double-stranded

3. An mRNA molecule is produced by _____ .
   - a. replication
   - b. duplication
   - c. transcription
   - d. translation

4. Each codon calls for a specific _____ .
   - a. protein
   - b. polypeptide
   - c. amino acid
   - d. carbohydrate

5. Referring to Figure 14.7, use the genetic code to translate the mRNA sequence UAUCGCACCUCAGGAGACUAG. Notice that the first codon in the frame is UAU. Which amino acid sequence is being specified?

   a. TYR—ARG—THR—SER—GLY—ASP—

   b. TYR—ARG—THR—SER—GLY

   c. TYR—ARG—TYR—SER—GLY—ASP—

6. Anticodons pair with _____ .
   - a. mRNA codons
   - b. DNA codons
   - c. tRNA anticodons
   - d. amino acids

7. Match the terms with the most suitable description.
   - _____ alkylating agent
   - _____ chain elongation
   - _____ exons
   - _____ genetic code
   - _____ anticodon
   - _____ intron
   - _____ codon
   - a. parts of mature mRNA transcript
   - b. base triplet coding for amino acid
   - c. second stage of translation
   - d. base triplet that pairs with codon
   - e. one environmental agent that induces mutation in DNA
   - f. set of sixty-four codons for mRNA
   - g. parts removed from a pre-mRNA transcript

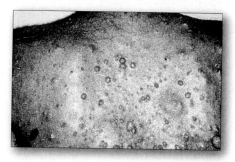

Figure 14.15
Soft skin tumors on a person with neurofibromatosis, an autosomal dominant disorder.

Critical Thinking

1. Sandra discovered a tRNA with a mutation in DNA that encodes the anticodon 3'-AUU instead of 3'-AAU. In cells with the mutated tRNA, what will be the effect on protein synthesis?

2. A DNA polymerase made an error during the replication of an important gene region of DNA. None of the DNA repair enzymes detected or repaired the damage. A portion of the DNA strand with the error is shown here:

```
..AATTCC ACTCCTATGG
..TTAAGG TGAGGATACC
```

After the DNA molecule is replicated and two daughter cells have formed, one cell is carrying a mutation and the other cell is normal. Develop a hypothesis to explain this observation.

3. *Neurofibromatosis* is a human autosomal dominant disorder caused by mutations in the *NF1* gene. It is characterized by soft, fibrous tumors in the peripheral nervous system and skin as well as abnormalities in muscles, bones, and internal organs (Figure 14.15).

   Because the gene is dominant, an affected child usually has an affected parent. Yet in 1991, scientists reported on a boy who had neurofibromatosis yet whose parents did not. When they examined both copies of his *NF1* gene, they found the copy he had inherited from his father contained a transposon. Neither the father nor the mother had a transposon in any of the copies of their own *NF1* genes. Explain the cause of neurofibromatosis in the boy and how it arose.

Selected Key Terms

alkylating agent *14.4*	mRNA (messenger RNA) *14.1*
anticodon *14.2*	mutation rate *14.4*
base sequence *CI*	nonionizing radiation *14.4*
base-pair substitution *14.4*	promoter *14.1*
carcinogen *14.4*	ribonucleic acid (RNA) *CI*
codon *14.2*	RNA polymerase *14.1*
deletion (of base) *14.4*	rRNA (ribosomal RNA) *14.1*
exon *14.1*	transcription *CI*
gene mutation *14.4*	translation *CI*
genetic code *14.2*	transposon *14.4*
insertion (of base) *14.4*	tRNA (transfer RNA) *14.1*
intron *14.1*	uracil *14.1*
ionizing radiation *14.4*	

Readings

Crick, F. 1988. *What Mad Pursuit: A Personal View of Scientific Discovery.* New York: HarperCollins. Crick's autobiography.

Friedberg, E., R. Wagner, and M. Radman. 31 May 2002. "Specialized DNA Polymerases, Cellular Survival, and the Genesis of Mutations." *Science* 296:1627–1630.

15

CONTROLS OVER GENES

When DNA Can't Be Fixed

Not long ago, Laurie Campbell turned eighteen and happened to notice a black mole on her skin. It was an odd, encrusted lump with a ragged border. No fool, she quickly made an appointment with her family doctor, who ordered a biopsy. The mole turned out to be a *malignant melanoma*, the deadliest form of skin cancer.

Moles and other tumors are **neoplasms**, abnormal masses of cells that ignore controls over growth and division. Benign neoplasms stay put. Malignant ones are **cancers**, with cells that can break away, invade other tissues, and give rise to more abnormal masses. Laurie was lucky. She detected a cancer in its earliest stage before it could spread through her body. Now she regularly checks out other skin moles. She is aware of having become a statistic. In 2001, there were more than a million cases of skin cancer in the United States and 7,700 deaths from malignant melanoma.

Laurie is smart. She plotted the position of each mole on her body. Once a month, she uses that body map as a guide for a quick, thorough self-examination. Figure 15.1 shows examples of what she looks for. Laurie also schedules a medical examination every six months.

Ultraviolet wavelengths in the sun's rays, tanning lamps, and other sources of nonionizing radiation can cause skin cancer. For instance, they promote covalent bonding between two neighboring thymine bases in a DNA strand. The result is an abnormal, bulky structure in a DNA molecule —a thymine dimer.

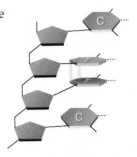

thymine dimer

Normally, at least seven gene products interact as a DNA repair mechanism to remove the bulky lesion. When mutation in one or more of those genes disrupts the mechanism, thymine dimers can accumulate in skin cells. They may trigger cancers by upsetting the normal gene controls over cell growth and division.

You are at risk if you habitually irritate moles, as by shaving or wearing abrasive clothing. You are at risk if your skin is chronically chapped, cracked, or sore. You are at risk if there is a history of cancer in your family or if you have endured radiation therapy. Like Laurie,

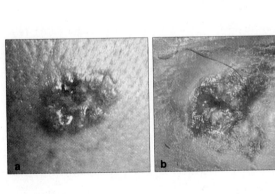

Figure 15.1 Examples of what can happen when repair enzymes can't fix damaged DNA. (**a**) *Basal cell carcinoma*, the most common form of skin cancer. This slow-growing, raised lump may be uncolored, reddish-brown, or black. (**b**) *Squamous cell carcinoma*, the second most common skin cancer. The pink growths, firm to the touch, spread rapidly under the surface of skin exposed to the sun. (**c**) *Malignant melanoma* spreads fastest. The malignant cells form dark, encrusted lumps. They may itch like an insect bite or bleed easily. *Right:* Laurie Campbell avoiding the sun—and melanoma.

you are at risk simply if you burn easily. Tanning or otherwise staying out in the sun without protection is ill advised; damaged DNA is a reality for all of us.

Why start a chapter with such awful prospects? Doing so might help focus your attention on how lucky individuals are when gene controls operate as they should. **Gene controls** are molecular mechanisms that govern when and how fast specific genes will be transcribed and translated, and whether gene products will be switched on or silenced.

In all prokaryotic and eukaryotic cells, many of the controls make transcription rates rise or fall in response to concentrations of nutrients and other substances. For example, genes that specify the enzymes required to metabolize sugars are transcribed in controlled ways. Some controls adjust the rates of transcription according to how much sugar is available. Others activate, inactivate, and degrade the gene products, including the sugar-digesting enzyme molecules that have finished their task.

In eukaryotic cells, still other controls guide mRNA transcript processing, transport of mature RNAs from the nucleus, and how fast RNAs are translated in the cytoplasm. Other controls guide the modification of new polypeptide chains in the endomembrane system.

Controls also guide the contribution that eukaryotic cells will make to a long-term program of growth and development. This is especially so for large, complex, multicelled species. As part of the program, different cell lineages activate and suppress fractions of their genes in different ways. Certain genes are expressed once, some of the time, or not at all. The result? Most cells become specialized in structure, composition, and function. We call this process and its outcome cell differentiation.

Explaining control of gene activity is like trying to explain a full symphony orchestra to someone who has never seen one or heard it perform. You have to identify all of the many separate parts before their interactions start to make sense! Gene controls weave through the story line of many chapters throughout the book. For that reason, take some time now to become acquainted with just a few of the molecular players and their amazing functions.

Key Concepts

1. In cells, a variety of controls govern when, how, and to what extent genes are expressed. The control elements operate in response to preprogrammed schedules of development. They also operate in response to changing chemical conditions and to reception of external signals.

2. Control is exerted by way of regulatory proteins and other molecules that operate before, during, or after gene transcription. The control elements interact with DNA, with RNA that has been transcribed from DNA, or with the resulting polypeptide chains or the final proteins.

3. Prokaryotic cells depend on rapid responses to short-term changes in nutrient availability and other aspects of their surrounding environment. They commonly use regulatory proteins that help make quick adjustments in gene transcription rates, which compensate for the changes.

4. Eukaryotic cells also use controls over short-term shifts in diet and levels of activity. In complex multicelled species, they rely as well on long-term controls over an intricate program of growth and development.

5. Controls over eukaryotic cells come into play when new cells contact one another in developing tissues. They also come into play when these cells start interacting with their neighbors by way of hormones and other signaling molecules.

6. The cells of a multicelled organism inherit the same genes, but different cell types activate or suppress many of the genes in different ways. Their controlled, selective use of their genes leads to the synthesis of the proteins that give each type of cell its distinctive structure, function, and products.

TYPES OF CONTROL MECHANISMS

Different mechanisms control gene expression through interactions with DNA, RNA, and the new polypeptide chains or final proteins. Some mechanisms respond to rising or falling concentrations of a nutrient or some other substance. Others respond to external signaling molecules that call for change.

The control agents include **regulatory proteins** that intervene before, during, or after gene transcription or translation. They also include signaling molecules such as hormones, which initiate change in some cell activity when they dock at suitable receptors.

By **negative control**, regulatory proteins slow down or curtail gene activity. By **positive control**, regulatory proteins promote or enhance gene activities. Control mechanisms for transcription involve noncoding base sequences in DNA that do not specify proteins. For example, **promoters** are base sequences that signal the start of a gene. As another example, the **enhancers** are binding sites for some activator proteins.

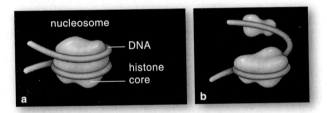

Figure 15.2 How DNA–histone packing in nucleosomes may be loosened to make gene regions available for transcription. Attaching an acetyl group to a histone makes it loosen its grip on DNA wound around it. Enzymes that attach or detach acetyl groups are known to be associated with transcription.

Control also is exerted with chemical modification. For instance, regions of newly replicated DNA can be shut down by **methylation**, the attachment of a methyl group ($-CH_3$) to nucleotide bases. Inactivated genes are often heavily methylated. And certain genes are activated by demethylation. Also, access to genes is partly controlled through **acetylation**: attaching acetyl groups from the histones that structurally organize the DNA (Section 9.1 and Figure 15.2).

When, how, and to what extent a gene is expressed depends on the type of cell and its functions, on the cell's chemical environment, and on signals for change.

Gene expression is controlled by regulatory proteins that interact with one another, with control elements built into the DNA, with RNA, and with newly synthesized proteins.

Control also is exerted through chemical modifications that inactivate or activate specific gene regions or the histone proteins that organize the DNA.

BACTERIAL CONTROL OF TRANSCRIPTION

Think about the dot of the letter "i." About a thousand bacterial cells would stretch side by side across the dot. Just imagine, each of those microscopic specks depends as much on gene controls as you do! When nutrients are plentiful and other environmental conditions also favor growth, the cells swiftly grow and divide. Gene controls promote the rapid synthesis of enzymes that catalyze nutrient digestion and other growth-related activities. Translation starts even before the RNA transcripts are finished. Bacteria, recall, have no nuclear envelope that keeps DNA away from ribosomes in the cytoplasm.

Often, prokaryotic genes for enzymes of a metabolic pathway are clustered together in the same sequence as the reaction steps. All genes in such a sequence may be transcribed as one continuous mRNA transcript.

With this bit of background, consider how one kind of prokaryote adjusts transcription rates downward or upward, depending on the availability of nutrients.

Negative Control of the Lactose Operon

Escherichia coli, an enteric bacterium, lives on sugars and other ingested nutrients in the mammalian gut. While mammals are infants, they live on milk only. Milk does not contain glucose, the sugar that *E. coli* cells prefer. It contains lactose, a different sugar. Once mammals are weaned, their milk intake typically declines or stops.

E. coli cells still take advantage of lactose when it is available. They activate a set of three adjoining genes coding for lactose-metabolizing enzymes. A promoter precedes the genes in the bacterial DNA, and operators are positioned on both sides of it.

An **operator** is a binding site for a repressor protein that can prevent gene transcription. This arrangement, in which a promoter and set of operators control more than one bacterial gene, is an **operon** (Figure 15.3).

In the absence of lactose, the repressor molecule binds to a set of operators. Binding causes the part of the DNA with the promoter to loop outward, as shown in Figure 15.3*c*. When looped this way, the promoter is inaccessible to RNA polymerase. Therefore, the operon genes can't be transcribed when they are not required.

When lactose *is* present, *E. coli* cells convert some of it to allolactose. This sugar binds to the repressor and alters its shape. In altered form, the repressor cannot bind to the operators. The looped DNA unwinds, RNA polymerase can start transcription, and so the lactose-degrading enzymes are produced when required.

Positive Control of the Lactose Operon

E. coli cells pay far more attention to glucose than to lactose. They transcribe genes for its breakdown faster, and continuously. Even when lactose is in the gut, the

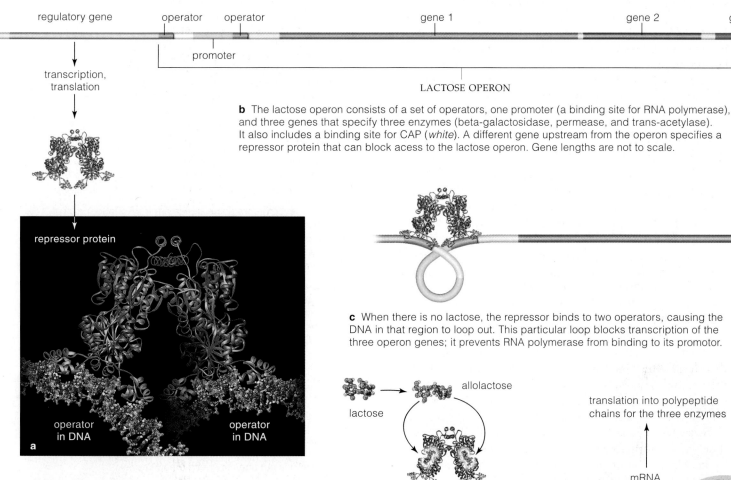

regulatory gene   operator   operator   gene 1   gene 2   gene 3

promoter

transcription,
translation

LACTOSE OPERON

b The lactose operon consists of a set of operators, one promoter (a binding site for RNA polymerase), and three genes that specify three enzymes (beta-galactosidase, permease, and trans-acetylase). It also includes a binding site for CAP (*white*). A different gene upstream from the operon specifies a repressor protein that can block acess to the lactose operon. Gene lengths are not to scale.

repressor protein

operator
in DNA

operator
in DNA

a

c When there is no lactose, the repressor binds to two operators, causing the DNA in that region to loop out. This particular loop blocks transcription of the three operon genes; it prevents RNA polymerase from binding to its promotor.

lactose

allolactose

translation into polypeptide chains for the three enzymes

mRNA

RNA polymerase

operator   promoter   operator   gene 1

Figure 15.3 **(a)** Computer model for the lac repressor protein, bound to two operators of a prokaryotic DNA molecule. **(b–d)** Negative control of the lactose operon. The operon's first gene codes for an enzyme that splits the disaccharide lactose into glucose and galactose. The second enzyme codes for an enzyme that helps transport lactose into cells. The third helps metabolize certain sugars.

d When lactose is present, some is converted to a form that can bind to the repressor protein and alter its shape. The altered repressor cannot bind to operators. RNA polymerase can transcribe the operon genes.

lactose operon is not used much—unless there is no glucose. These conditions call for an **activator** protein called CAP. This activator exerts positive control over the lactose operon by making a promoter more inviting to RNA polymerase. But CAP can't issue the invitation until it is bound to a chemical messenger, cAMP (short for cyclic adenosine monophosphate). When cAMP and the activator are complexed together and bound to the promoter, they make it far easier for RNA polymerase to start transcribing genes.

When glucose is plentiful, ATP forms by glycolysis, but an enzyme needed to synthesize cAMP is inhibited. That enzyme is released from inhibition when glucose is scarce and lactose is available. cAMP accumulates, CAP–cAMP complexes form, and the lactose operon genes are rapidly transcribed. The gene products allow lactose to be used as an alternative energy source.

Humans are born with a gene for lactase, a lactose-digesting enzyme made by intestinal cells. Before three years pass, lactase levels start to decline in genetically predisposed people. *Lactose intolerance* begins, because lactose can't be degraded. It moves into the colon and fans the growth of resident bacterial populations. One gaseous by-product of their metabolism accumulates and distends the colon, causing pain. Short fatty acid chains released by the reactions lead to diarrhea, which is often severe. People can avoid symptoms by drinking milk that has predigested lactose or by taking lactose-digesting enzymes before ingesting dairy products.

Transcription rates of bacterial genes for nutrient-digesting enzymes are quickly adjusted downward and upward by control systems that respond to nutrient availability.

15.3

GENE CONTROLS IN EUKARYOTIC CELLS

Like bacteria, eukaryotic cells control short-term shifts in diet and in levels of activity. If those cells happen to be among hundreds or trillions of cells in a multicelled organism, long-term controls also enter the picture, for gene activities change during development.

Cell Differentiation and Selective Gene Expression

Later in the book, you will be reading about controls over development, particularly in Chapters 32, 43, and 44. For now, tentatively accept this basic premise: All of the cells in your body started out life with the same genes, because every one arose by mitotic cell divisions from one fertilized egg. Many of those inherited genes specify proteins that are essential for the structure and everyday functioning of every cell. That is why those genes are controlled in ways that promote ongoing, low levels of transcription.

Even so, *nearly all of your cells became specialized in composition, structure, and function*. This process of **cell differentiation** proceeds during the development of all multicelled species. It arises as embryonic cells and the cell lineages descended from them activate a fraction of their genes in selective ways.

For example, nearly all of your body cells use genes that specify the enzymes of glycolysis on an ongoing basis. Yet only your immature red blood cells activate genes for hemoglobin. Your liver cells activate genes required to synthesize enzymes that neutralize certain toxins, but they are the only ones that do. When your eyes first formed, only certain cells accessed the genes necessary for synthesis of crystallin. No other cells can activate the genes for this protein, which helped make transparent fibers of the lens in each eye.

Controls Before and After Transcription

In all large, complex eukaryotic organisms, many of the genes that govern housekeeping tasks are continuously transcribed at low levels. In the case of other genes, transcription rates are adjusted up and down. Why? Tissue fluids of such organisms are the body's internal environment. Individual cells of the body continually deliver or secrete substances into that environment, and withdraw substances from it. The ongoing inputs and outputs cause slight shifts in the concentrations of nutrients, signaling molecules, metabolic products, and other solutes. Most often, transcription rates for those genes rise or fall by small degrees in response.

As the examples in Figure 15.4a indicate, some gene sequences are repeatedly duplicated or rearranged in genetically programmed fashion prior to transcription. Besides this, programmed chemical modifications often

a CONTROLS RELATED TO TRANSCRIPTION. At any time, most genes of a multicelled organism are shut down permanently or temporarily. Genes necessary for a cell's everyday tasks are under positive controls that promote ongoing, low levels of transcription. These controls help provide a cell with enough enzymes and other proteins to carry out its most basic functions. Transcription of many other genes shifts. In this case, controls work to assure chemical responsiveness even when concentrations of specific substances rise or fall only slightly.

Also, even before some genes are transcribed, parts of them are amplified, rearranged, or chemically modified in permanent or reversible ways. These changes are not mutations. They are programmed events that affect how the gene will be expressed, if at all.

1. *Gene amplification.* Immature amphibian eggs and glandular cells of some insect larvae copy the same genes again and again when they require enormous numbers of the products. Sometiems, multiple rounds of DNA replication produce hundreds or thousands of copies prior to transcription (Sections 15.4 and 15.5).

2. *DNA rearrangements.* Some genes have many base sequences in the DNA, and they are put together in different ways to make different product molecules. This happens when B lymphocytes, a special class of white blood cells, are forming (Section 39.5). Different cells transcribe and translate the uniquely sequences into different versions of antibodies. These protein weapons act against specific pathogens .

3. *Chemical modification.* Histones and other proteins interact with eukaryotic DNA in organized ways. The DNA–protein packaging, plus chemical modifications to particular sequences, influences gene expression. Normally, just a small fraction of a cell's genes are available for transcription. A dramatic shutdown occurs in female mammals. As Section 15.4 describes, this event is called X chromosome inactivation.

Figure 15.4 Examples of the levels of control over gene expression in eukaryotes.

shut down many genes. So does the orderly packaging of DNA by histones and other chromosomal proteins, as you may realize after reflecting on the organization of eukaryotic chromosomes (Section 9.1).

Controls also come into play after genes have been transcribed. As Figure 15.4b–d indicates, many controls govern RNA transcript processing, transport of mature RNAs from the nucleus, and rates of translation in the cytoplasm. Other controls deal with the modification of new polypeptide chains. Others deal with activating, inhibiting, and breaking down existing proteins.

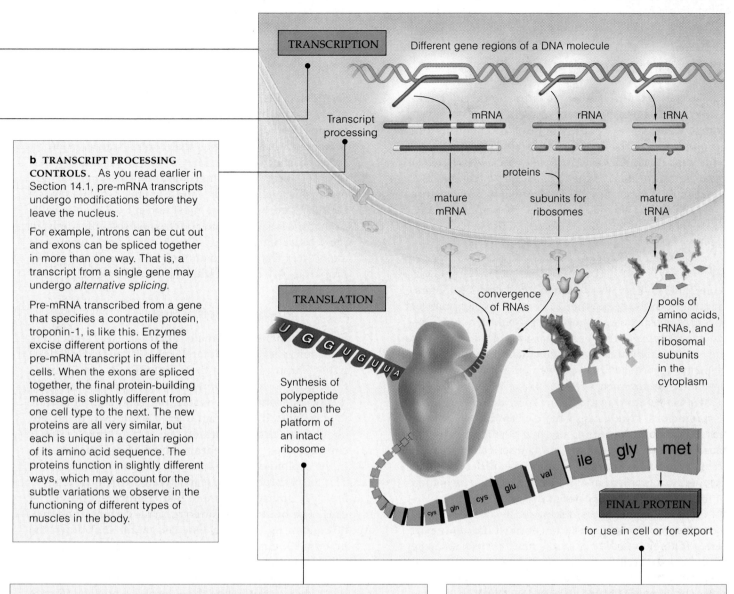

TRANSCRIPTION

Different gene regions of a DNA molecule

Transcript processing

mRNA   rRNA   tRNA

proteins

mature mRNA

subunits for ribosomes

mature tRNA

TRANSLATION

convergence of RNAs

pools of amino acids, tRNAs, and ribosomal subunits in the cytoplasm

Synthesis of polypeptide chain on the platform of an intact ribosome

cys gln cys glu val ile gly met

FINAL PROTEIN

for use in cell or for export

b TRANSCRIPT PROCESSING CONTROLS. As you read earlier in Section 14.1, pre-mRNA transcripts undergo modifications before they leave the nucleus.

For example, introns can be cut out and exons can be spliced together in more than one way. That is, a transcript from a single gene may undergo *alternative splicing*.

Pre-mRNA transcribed from a gene that specifies a contractile protein, troponin-1, is like this. Enzymes excise different portions of the pre-mRNA transcript in different cells. When the exons are spliced together, the final protein-building message is slightly different from one cell type to the next. The new proteins are all very similar, but each is unique in a certain region of its amino acid sequence. The proteins function in slightly different ways, which may account for the subtle variations we observe in the functioning of different types of muscles in the body.

c CONTROLS OVER TRANSLATION. Controls govern when, how fast, and how often an mRNA transcript is translated. Sections 36.2 and 43.3 provide elegant examples.

A transcript's stability affects how many protein molecules can be produced from it. Enzymes digest transcripts from the poly-A tail on up (Section 14.1). The tail's length and its attached proteins affect how fast it is digested. Also, after leaving the nucleus, some transcripts are temporarily or permanently inactivated. Example: In unfertilized eggs, many transcripts are inactivated and stored in the cytoplasm. These "masked messengers" will not be available for translation until after fertilization, when great numbers of protein molecules will be required for the early cell divisions of the new individual.

d CONTROLS FOLLOWING TRANSLATION. Many newly formed polypeptide chains enter the endomembrane system (Section 4.5). They undergo modification, as when enzymes attach specific oligosaccharides or phosphate groups to them.

Diverse control mechanisms govern the activation, inhibition, and stability of enzymes and other molecules used in protein synthesis. Allosteric control of tryptophan synthesis is an example (Section 6.7).

Consider enzymes, the proteins that catalyze nearly all metabolic reactions. Besides selectively transcribing and translating the genes for enzymes, diverse control systems activate and inhibit the molecules of enzymes that have already been synthesized in the cell.

Just imagine the coordination necessary to govern which of the cell's thousands of types of enzymes are to be stockpiled, deployed, or degraded in a specified interval. *That coordination helps govern all short-term and long-term aspects of cell structure and function.*

In multicelled species, a variety of gene controls guide the moment-by-moment activities that maintain cells. Other controls guide intricate, long-term patterns of the body's growth and development.

Cells of complex organisms inherit the same genes, yet most become specialized in composition, structure, and function. This process of cell differentiation arises when different populations of cells activate and suppress their genes in highly selective, unique ways.

TYPES OF CONTROL MECHANISMS

By some estimates, cells of complex organisms rarely use more than 5 to 10 percent of their genes at a given time. One way or another, control mechanisms are keeping most of the genes inactivated. *Which genes are active depends on the type of organism, the stage of growth and development it is passing through, and the controls that are operating at different stages.* You will be coming across some elegant cases of this in later chapters. For now, two examples will make the point.

Homeotic Genes and Body Plans

Some kinds of regulatory proteins bind with promoters, enhancers, and one another to control transcription of specific genes. As an example, most eukaryotic species have **homeotic genes**. These are a class of master genes; they interact with one another and also with control elements to bring about the formation of tissues and organs in accordance with the basic body plan. The master genes are transcribed in specific order in local tissue regions. Gradients in the concentrations of gene products result. Depending on their position relative to these gradients, other genes are transcribed or remain silent. In animal embryos, for instance, different genes respond to a gradient in ways that lead to the formation of the body's anterior–posterior axis (Section 43.5).

Homeotic genes were discovered by way of single mutations in *Drosophila* that transformed one body part into a different one. For example, the *antennapedia* gene is actively transcribed in regions of the embryo that will become the thorax, complete with legs. Its transcription normally is restricted in other regions, such as the one destined to become the head. Figure 15.5a shows what happens when it is wrongly activated in an embryo's head cells. Similarly, a homeotic gene in corn plants controls leaf vein formation. Its mutation gives rise to veins that are twisted instead of organized in flat planes.

Homeotic genes code for homeodomains, which are regulatory proteins that contain a sequence of sixty or so amino acids. That sequence, a "homeobox," can bind to short control sequences in promoters and enhancers (Figure 15.5b). We know of more than a hundred kinds of homeodomain proteins that control transcription by common mechanisms. They occur in all eukaryotes. Many are even interchangeable among distantly related organisms such as yeasts and humans, which suggests they evolved among the most ancient eukaryotic cells. In many cases, the homeobox sequences differ only in *conservative* amino acid substitutions; even when one amino acid has replaced another, it has similar chemical properties. We return to this topic in Chapter 43.

X Chromosome Inactivation

Female humans and female calico cats have something in common at the cellular level. Although both have two X chromosomes in their diploid cells, one is in its threadlike form; the other one is scrunched up even during interphase. The scrunching isn't a chromosome abnormality. It is a programmed shutdown of all but about three dozen genes on *one* of two homologous X chromosomes. This **X chromosome inactivation** occurs in the diploid cells of all female placental mammals.

One X chromosome is inactivated when the females are early embryos, no more than a tiny ball of dividing cells. The outcome is random; *either* chromosome may be condensed this way. One cell might shut down the maternal X chromosome, another cell next to it might shut down the paternal X chromosome or the maternal X chromosome, and so on. The condensed one looks like a dense spot inside the interphase nucleus (Figure 15.6a). Researchers named a condensed X chromosome a **Barr body** after Murray Barr, its discoverer.

Once that first random selection is made in a cell, however, all the descendant cells make the exact same selection as they go on dividing to form tissues. When fully developed, then, *each female mammal bears patches of tissue where genes of the maternal X chromosome are*

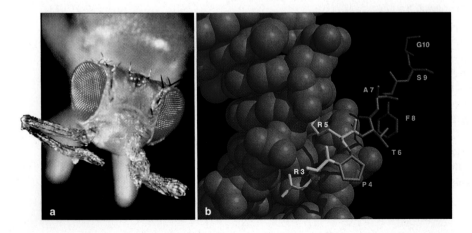

Figure 15.5 (a) Experimental evidence that genes control the development of body parts. In normal *Drosophila* larvae, activation of genes in a certain group of cells gives rise to the head's antennae. A larva with a mutant form of the *antennapedia* gene develops into an adult fly with legs instead of antennae on its head. This is one of the genes controlled by homeodomains, a type of regulatory protein. (b) Model for a homeodomain binding to a transcriptional control sequence in a DNA molecule.

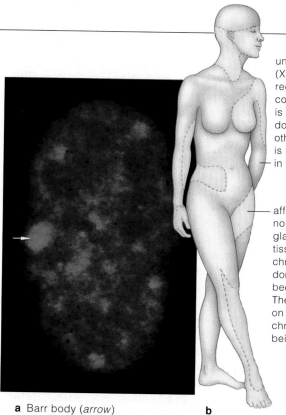

unaffected skin (X chromosome with recessive allele was condensed; its allele is inactivated. The dominant allele on other X chromosome is being expressed in this tissue.)

affected skin with no normal sweat glands (In this tissue, the X chromosome with dominant allele has been condensed. The recessive allele on the other X chromosome is being transcribed.)

a Barr body (*arrow*)   **b**

Figure 15.6 (**a**) Micrograph of an inactivated X chromosome, called a Barr body, as it appears in a human female's somatic cell during interphase. The X chromosome is not condensed this way in human male cells. (**b**) Anhidrotic ectodermal dysplasia, a mosaic pattern of gene expression. The condition arises rarely as a result of random X chromosome inactivation.

Figure 15.7 Why is this female calico cat "calico"? In her cells, one X chromosome carries a dominant allele for the brownish-black pigment melanin. The allele on her other X chromosome specifies yellow fur. At an early stage of the cat's embryonic development, one of the two X chromosomes was inactivated at random in each cell that had formed by then.

In all descendants of those cells, the same chromosome also became inactivated, which left them with only one functional allele for the coat-color trait. We see patches of different colors, depending on which allele was inactivated in cells that formed a given tissue region. (The white patches result from a gene interaction involving the "spotting gene," which blocks melanin synthesis entirely.)

being expressed—and patches where genes of the paternal X chromosome are being expressed. She is a "mosaic" for the X chromosomes! As you know, a pair of alleles on two homologous chromosomes might or might not be identical. When they are not, skin and other tissues of female mammals may have different features from one tissue patch to the next. Mary Lyon, a geneticist, was the discoverer of this **mosaic tissue effect** of random X chromosome inactivation.

We see the mosaic tissue effect in human females who are heterozygous for a rare recessive allele that blocks sweat gland formation. Their skin is a mosaic of tissues with and without sweat glands. This is just one symptom of *anhidrotic ectodermal dysplasia*. Where sweat glands are absent, the mutant allele is on the active X chromosome (Figure 15.6b). You can see the same effect in female calico cats, which are heterozygous for a coat color allele on their X chromosomes (Figure 15.7).

What's the point of the shutdown? Remember, male and female mammals differ in their sex chromosomes (XY versus XX). However, a subset of genes, mainly on the X chromosome's long arm, must be expressed at the *same* levels in males and females. Otherwise, the individual will not develop properly. Inactivating one of the X chromosomes in XX embryos is called **dosage compensation**. It is a control mechanism that balances gene expression between sexes, starting with the early stages of development.

The *XIST* gene governs X chromosome inactivation. Early in the development of a female embryo, the *XIST* gene on each X chromosome is transcribed. Its product, a type of RNA molecule, accumulates like paint on each chromosome to inactivate almost all of the other genes. Next, *XIST* gene transcription ends on one of the two X chromosomes as it becomes the target of methylation. The addition of methyl groups to DNA usually blocks access to genes. Most of the additions take place where there are two adjacent cytosine–guanine pairs. Pairings like this are clustered in control elements in the DNA.

Which genes are transcribed at a given time depends on the type of organism, the stage of its growth and development, and the kinds of controls operating at different stages.

Homeotic gene expression during development is a case in point. Its products map out the basic body plan. Another case is X chromosome inactivation, a control mechanism that balances gene expression between sexes.

EXAMPLES OF SIGNALING MECHANISMS

The story that opened this chapter introduced the idea that a great variety of signals influence gene activity. The following examples from animals and plants will give you a sense of their effects at the molecular level.

Hormonal Signals

Hormones, a major category of signaling molecules, can stimulate or inhibit gene activity in target cells. Any cell with receptors for a given hormone is a target. Animal cells secrete hormones into tissue fluid. Most of these molecules are picked up by the bloodstream, which distributes them to cells some distance away.

Some hormones bind to membrane receptors at the target cell surface. Others enter the cell and promote transcription of genes by binding to specific activator proteins. In turn, base sequences in the DNA molecule called enhancers can bind the activators. One way or another, an activator–enhancer complex ends up right next to the promoter for the targeted gene. Because RNA polymerase can bind avidly with it, the complex facilitates rapid gene transcription.

Consider the effect of **ecdysone**, a hormone with key roles in many insect life cycles. Immature forms called larvae grow rapidly during part of the cycles and continually feed on organic matter, such as leaves. Preparing food for digestion requires copious amounts of saliva. In their salivary gland cells, DNA had been replicated again and again. The multiple copies of DNA molecules stayed together in parallel array, forming a **polytene chromosome**. When the ecdysone binds to its receptor on the salivary gland cells, its signal triggers rapid transcription of the many copies of genes in the DNA. Gene regions that respond to this hormonal signal puff out while being transcribed, as shown in Figure 15.8. After

this, translation of mRNA transcripts made from the genes produces the protein components of saliva.

In vertebrates, certain hormones have widespread effects on gene expression because many types of cells have receptors for them. As one example, the pituitary gland secretes somatotropin, or growth hormone. This hormonal signal stimulates synthesis of all the proteins required for cell division and, ultimately, the body's growth. Most cells have receptors for somatotropin.

Other vertebrate hormones only signal specific cells at specific times. Prolactin, secreted from the pituitary gland, is like this. A few days after a female mammal gives birth, prolactin can be detected in her blood. This hormone activates genes in mammary gland cells that have receptors for it. Those genes have responsibility for milk production. Liver cells and heart cells have the same genes but do not have the receptors necessary to respond to signals from prolactin.

We will return to this topic later on. Particularly in Chapter 32, 36, 43, and 44, you will come across elegant examples of hormonal controls drawn from studies of plant and animal reproduction and development.

Sunlight as a Signal

Plant a few seeds from a corn or bean plant in a pot that holds moist, nutrient-rich soil. Next, let the seeds germinate, but keep them in total darkness. After eight days have passed, they will develop into seedlings that are spindly and pale, because they have no chlorophyll (Figure 15.9). Now expose those seedlings to one burst of dim light from a flashlight. Within ten minutes, the seedlings will start to convert their stockpiles of certain molecules to the activated forms of chlorophylls, the light-trapping pigment molecules that donate electrons for the reactions of photosynthesis.

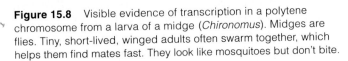

one of the larger chromosome puffs

Figure 15.8 Visible evidence of transcription in a polytene chromosome from a larva of a midge (*Chironomus*). Midges are flies. Tiny, short-lived, winged adults often swarm together, which helps them find mates fast. They look like mosquitoes but don't bite.

Most midge larvae develop in aquatic habitats. To sustain their rapid growth, they feed continually, as on decaying organic material. They require a lot of saliva, and they must continuously transcribe genes for saliva's protein components. Those genes have undergone amplification.

Ecdysone, a hormone, serves as a regulatory protein that helps promote their transcription. Midge chromosomes loosen and puff out in regions where the genes are being transcribed in response to the hormonal signal. Puffs are largest and most diffuse where transcription is most intense. Staining techniques reveal banding patterns in the chromosomes, as in the micrograph.

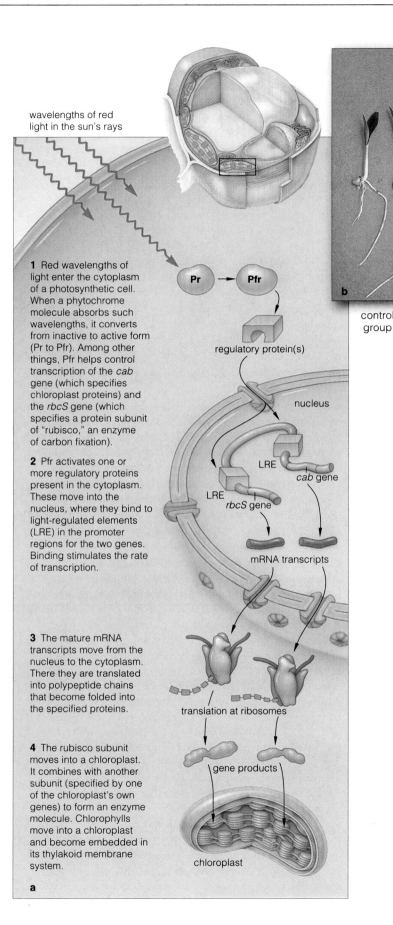

wavelengths of red
light in the sun's rays

Pr → Pfr

regulatory protein(s)

nucleus

LRE

LRE
rbcS gene

cab gene

mRNA transcripts

translation at ribosomes

gene products

chloroplast

1 Red wavelengths of light enter the cytoplasm of a photosynthetic cell. When a phytochrome molecule absorbs such wavelengths, it converts from inactive to active form (Pr to Pfr). Among other things, Pfr helps control transcription of the *cab* gene (which specifies chloroplast proteins) and the *rbcS* gene (which specifies a protein subunit of "rubisco," an enzyme of carbon fixation).

2 Pfr activates one or more regulatory proteins present in the cytoplasm. These move into the nucleus, where they bind to light-regulated elements (LRE) in the promoter regions for the two genes. Binding stimulates the rate of transcription.

3 The mature mRNA transcripts move from the nucleus to the cytoplasm. There they are translated into polypeptide chains that become folded into the specified proteins.

4 The rubisco subunit moves into a chloroplast. It combines with another subunit (specified by one of the chloroplast's own genes) to form an enzyme molecule. Chlorophylls move into a chloroplast and become embedded in its thylakoid membrane system.

a

Figure 15.9 (a) Sunlight as a signal for gene expression. One model for the mechanism by which phytochrome might help to control transcription of genes in plants. Red wavelengths can convert the phytochrome molecule from inactive form (here designated Pr) to active form (Pfr). In this form, the phytochrome can serve as a regulator of transcription.

(b) Experiment showing the effect of an absence of light on corn seedlings. The two seedlings at *left*, the control group, were grown inside a sunlit greenhouse. The other two seedlings, the experimental group, were grown in total darkness for eight days. The dark-grown plants were not able to convert stockpiled precursors of chlorophyll molecules to active form, and they never did green up.

b

control
group

experimental
group

Phytochrome, a blue-green pigment, is a signaling molecule that helps plants adapt over the short term to changes in light conditions. Section 32.4 takes a close look at this molecule. For now, simply be aware that it alternates between active and inactive forms. At sunset, at night, or in shade, far-red wavelengths predominate. At such times, the phytochrome in cells is inactive. It is activated at sunrise, when red wavelengths dominate the sky. Also, the amount of red or far-red wavelengths that a plant intercepts varies from day to night and as seasons change.

Such variations act as controls over phytochrome activity. That activity influences transcription of certain genes at certain times of day and certain times of year. These genes specify a number of enzymes and other proteins that help seeds to germinate, stems to lengthen and then branch, and leaves to grow. The proteins also help flowers, fruits, and seeds to form.

Elaine Tobin and her coworkers at the University of California, Los Angeles, performed experiments that provided evidence in favor of phytochrome control. Using dark-grown seedlings of duckweed (*Lemna*), they discovered a marked increase in the number of certain mRNA transcripts after a one-minute exposure to red light. Exposure had enhanced transcription of the genes for proteins that bind chlorophylls and for rubisco, an enzyme that mediates carbon fixation (Sections 7.6 and 7.7). In the absence of the proteins, chloroplasts do not develop properly and they don't turn green.

Hormones and other signaling molecules, including diverse kinds that respond to environmental stimuli, have profound influence on gene expression.

Lost Controls and Cancer

THE CELL CYCLE REVISITED   Every second, millions of cells in your skin, bone marrow, gut lining, liver, and elsewhere divide and replace worn-out, dead, and dying predecessors. They don't divide willy-nilly. Mechanisms control the expression of genes that specify enzymes and other proteins required for cell growth, DNA replication, chromosome movements, and cytoplasmic division. And they control when the division machinery is put to rest.

The cell cycle has built-in checkpoints, where proteins monitor chromosome structure and other aspects of the preceding phase of the cycle. They also sense whether or not conditions favor division. The proteins are products of **checkpoint genes**, which are the basis of mechanisms that advance, delay, or block the cell cycle.

For example, the protein products called **kinases** are a class of enzymes that phosphorylate molecules. Some signal that DNA replication is finished and call on the cell to make the transition from interphase to mitosis (Section 9.2). The products called **growth factors** invite transcription of genes with roles in the body's growth. Example: one of these, epidermal growth factor (EGF), activates tyrosine kinase when it binds to target cells in epithelial tissues. Binding is a signal to start mitosis. The product of checkpoint gene *p53* can put the brakes on division when chromosomes are damaged.

ONCOGENES   Many cancers are known to arise through mutations in one or more checkpoint genes. A mutated gene that has the potential to induce cancer falls into the broader category of **oncogenes**.

Oncogenes can form following insertion of viral DNA into a cell's DNA. They also can form after mutagens alter the structure of DNA. As you read in Section 14.4, ultraviolet radiation and nonionizing radiation, such as

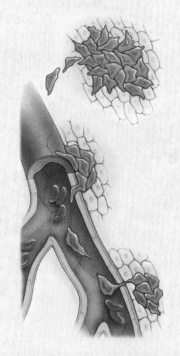

a Cancer cells break away from their home tissue.

b The metastasizing cells become attached to the wall of a blood vessel or lymph vessel. They secrete digestive enzymes onto it. Then they cross the wall at the breach.

c Cancer cells creep or tumble along inside blood vessels, then leave the bloodstream the same way they got in. They start new tumors in new tissues.

Figure 15.11   Steps in metastasis.

x-rays, are mutagens. So are many natural and synthetic compounds, including asbestos and certain substances in tobacco smoke.

And remember those chromosome alterations and gene mutations described in Sections 12.8 and 14.4? Some cancers arise when base substitutions or deletions alter a gene or one of the control elements that deal with its transcription. Other cancers may arise by translocations or by transposons that destabilize gene controls.

CHARACTERISTICS OF CANCER   What happens to a cell when it undergoes cancerous transformation? At the very least, all cancer cells display four characteristics. First, *the plasma membrane and cytoplasm change profoundly.* That membrane becomes more permeable, its proteins become lost or altered, and abnormal proteins form. The cytoskeleton shrinks severely, becomes disorganized, or both (Figure 15.10). Enzyme action shifts, as in amplified reliance on glycolysis. Second, *cancer cells grow and divide abnormally.* Controls against overcrowding in tissues are lost; cell populations reach high densities. The number of small blood vessels that service the growing cell mass increases abnormally. Third, *cancer cells have a weakened capacity for adhesion.* Recognition proteins become lost or altered, so cells cannot stay anchored in proper tissues. They break away and may establish colonies in other, distant tissues. **Metastasis** (meh-TA-stuh-SIS) is the name for the process of abnormal cell migration and tissue invasion (Figure 15.11). Fourth, *cancer cells usually are lethal.* Unless they are eradicated, their uncontrollable divisions put the individual on a painful road to death.

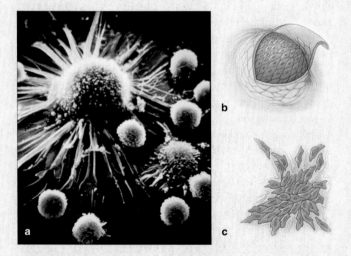

Figure 15.10   (**a**) A patrol of white blood cells surrounding a cancer cell. (**b**) Benign tumor. (**c**) Malignant tumor.

Signal
to die
docks at
receptor.

Signal
causes
activation
of ICE-like
proteases.

Figure 15.12 Artist's representation of weapons of cell death being unleashed inside a cell. Normally, cells self-destruct when they finish their functions or become altered in ways that could threaten the body as a whole. When controls over programmed cell death are lost, an altered cell may start dividing abnormally and give rise to cancer.

Cells of common skin moles and other noncancerous, *benign* neoplasms grow abnormally but slowly, and they retain the surface recognition proteins that hold them in their home tissue. Unless benign neoplasms become too large or irritating, doctors usually leave them alone.

Abnormally growing and dividing cells of a *malignant* neoplasm have destructive physical and metabolic effects on the surrounding tissues. These are grossly disfigured, metastasizing cancer cells (Figure 15.11c). They break loose from home tissues, enter lymph or blood vessels, then slip out and invade other tissues where they do not belong. There, they may start growing as new masses.

Each year in the developed countries alone, 15 to 20 percent of all deaths result from cancer. Cancer isn't just a human problem. Researchers have observed cancers in most of the animal species they have studied to date.

HERE'S TO SUICIDAL CELLS! We conclude with a case study of a gene and its cancerous transformation.

The first cell of a new multicelled individual contains marching orders that will guide its descendants along a program of growth, development, and reproduction, and often death. As part of that program, many cells heed calls to self-destruct when they finish their prescribed function. If they become altered in ways that might pose a threat to the body as a whole, as by infection or cancer, they can execute themselves. **Apoptosis** (app-uh-TOE-sis) is the name for this form of cell death. It starts with molecular signals that activate and unleash lethal weapons of self-destruction, which were stockpiled earlier in the cell.

Protein-cleaving enzymes called **ICE-like proteases** are such weapons. Think of them as sheathed Ninja knives. When popped open, they chop apart structural proteins, including the building blocks of cytoskeletal elements and nucleosomes that organize the DNA (Figure 15.12).

A body cell in the act of suicide shrinks away from its neighbors. Its cytoplasm seems to roil as its surface repeatedly bubbles outward and inward. No longer are its chromosomes extended through the nucleoplasm; they bunch up near the nuclear envelope. The nucleus, then the cell, breaks apart. Suicidal cells or remnants of them are swiftly engulfed by phagocytic white blood cells, which patrol and protect the body's tissues.

The timing of cell death is predictable in some cells, such as keratinocytes. These cells form densely packed sheets of dead cells that are continually sloughed off and replaced at the surface of skin. They have a three-week life span, more or less. Keratinocytes and other body cells, even kinds that are supposed to last for a lifetime, can be induced to die ahead of schedule. All it takes is sensitivity to the signals that can activate ICE-like proteases and other enzymes of death.

Control genes suppress or trigger programmed cell death. One of these, *bcl-2*, helps keep normal body cells from dying before their time. The knives stay sheathed in cancer cells, which are supposed to—but do not—commit suicide on cue. Quite possibly, they lost normal receptors that would permit contact with their signaling neighbors. Maybe they are receiving abnormal signals about when to grow, divide, or cease dividing. In some cancers, for instance, the suppressor gene *bcl-2* has been tampered with or shut down. Apoptosis is no longer in the cards.

Cancer is a multistep process involving more than one oncogene. But researchers have already identified many of the mutated forms of checkpoint genes that contribute to it. Remember how one form of leukemia, described in Chapter 12's introduction, is responding to the molecular targeting of one gene's product? More anticancer drugs and gene therapies should soon follow.

SUMMARY

1. Cells are equipped with controls that govern gene expression; that is, which gene products appear, when, and in what amounts. *When* control mechanisms come into play depends on cell type, on prevailing chemical conditions, and on signals from other cell types that can change a target cell's activities. *CI*

   a. In all prokaryotic and eukaryotic cells, many of the controls adjust transcription rates in response to concentrations of nutrients and other substances.

   b. In eukaryotic cells of complex organisms, controls contribute to a program of growth and development. They allow a fraction of the same genes to be expressed selectively in different lineages of cells. Selective gene expression is the beginning of diverse specializations in cell structure, composition, and function. We call this process cell differentiation.

2. Regulatory proteins (e.g., activators and inhibitors of transcription), hormones, and some DNA sequences are typical control agents. *15.1–15.4*

   a. Diverse control agents interact with one another, with operators and other control elements built into DNA, with RNA, and with gene products.

   b. Gene expression also is controlled with chemical modifications that inactivate and activate gene regions. X chromosome inactivation is an example.

   c. Control elements called promoters are the binding sites in DNA that signal the start of a gene. Enhancers are binding sites in DNA for some activator proteins.

3. In all cell types, two of the most common types of control systems inhibit or enhance gene transcription. In negative control systems, a regulatory protein binds at a specific DNA sequence to *block* transcription of one or more genes. In positive control systems, a regulatory protein binds to DNA to *promote* transcription. *15.1*

4. Most prokaryotic cells do not have great structural complexity and do not undergo complex development. Most of their gene control systems adjust transcription rates in response to nutrient availability. Control of operons (groupings of functionally related prokaryotic genes and control elements) is an example. *15.2*

5. Compared to prokaryotic cells, eukaryotic cells use more complex gene controls. Gene expression changes in response to external conditions and is subject to long-term controls over growth and development. *15.3–15.5*

   a. X chromosome inactivation in XX embryos of all female mammals is an outcome of interactions between a product of a gene (*XIST*) and control elements in one X chromosome. This interaction is an example of dosage compensation, a control mechanism that maintains a crucial balance in gene expression between the sexes.

   b. Hormones and phytochromes are two examples of signaling molecules that have roles in selective gene expression. Both can affect transcription.

6. Checkpoint gene products advance, delay, or block the cell cycle. They include the protein kinases. Some checkpoint gene mutations lead to cancer. A mutated form of such a gene is an oncogene. *15.6*

7. Cancer involves loss of normal controls over the cell cycle and mechanisms of programmed cell death. In all cancer cells, the structure and function of the plasma membrane and cytoplasm are severely compromised. Cancer cells grow and divide abnormally. Mechanisms for self-destruction are suppressed. Cancer cells have a weakened capacity for adhesion to their home tissue. Unless eradicated, cancers typically can be lethal. *15.6*

Review Questions

1. A plant, fungus, and animal consist of diverse cell types. How might the diversity arise, given that body cells in each of these organisms inherit the same set of genetic instructions? As part of your answer, define cell differentiation and the general way that selective gene expression brings it about. *CI, 15.3*

2. In what fundamental way do negative and positive controls of transcription differ? Is the effect of one or the other form of control (or both) reversible? *15.1*

3. Distinguish between: *15.1, 15.2*
   a. repressor protein and activator protein
   b. promoter and operator
   c. methylation and acetylation

4. Describe one type of control of transcription for the lactose operon in *E. coli*, a prokaryotic cell. *15.2*

5. Using the sketch on the facing page, define three types of gene controls and note the levels at which they take effect. Does the sketch work for both prokaryotic and eukaryotic cells? Why or why not? *15.3*

6. What is a Barr body? Does it appear in the cells of human males, human females, or both? Explain your answer. *15.4*

7. Briefly describe the general characteristics of normal cells. Then explain the difference between a benign tumor and a malignant tumor. *15.6*

Self-Quiz
ANSWERS IN APPENDIX III

1. Cell differentiation _____ .
   a. occurs in all complex multicelled organisms
   b. requires unique genes in different cells
   c. involves selective gene expression
   d. both a and c
   e. all of the above

2. The expression of a given gene depends on the _____ .
   a. type of cell and its functions   c. environmental signals
   b. chemical conditions   d. all of the above

3. Regulatory proteins interact with _____ .
   a. DNA   c. gene products
   b. RNA   d. all of the above

4. A base sequence signaling the start of a gene is a(n) _____ .
   a. promoter   c. enhancer
   b. operator   d. activator protein

5. In prokaryotic cells but not eukaryotic cells, a(n) _____ is a type of base sequence that precedes genes of an operon.
   a. lactose molecule   c. operator
   b. promoter   d. both b and c

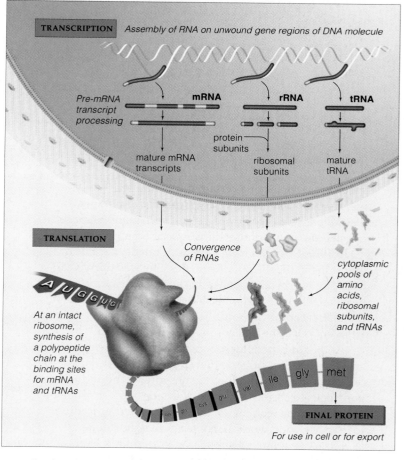

TRANSCRIPTION *Assembly of RNA on unwound gene regions of DNA molecule*

Pre-mRNA
transcript
processing

mRNA **rRNA** **tRNA**

protein
subunits

mature mRNA
transcripts

ribosomal
subunits

mature
tRNA

TRANSLATION

*Convergence
of RNAs*

*cytoplasmic
pools of
amino
acids,
ribosomal
subunits,
and tRNAs*

*At an intact
ribosome,
synthesis of
a polypeptide
chain at the
binding sites
for mRNA
and tRNAs*

A U G G U G

ile gly met

val glu cys gln phe

FINAL PROTEIN

For use in cell or for export

---

transcription of the lactose operon when cells of this bacterial strain are subjected to these conditions:

   a. Lactose and glucose are available.
   b. Lactose is available but glucose is not.
   c. Both lactose and glucose are absent.

2. *Duchenne muscular dystrophy*, a genetic disorder, affects boys almost exclusively. Early in childhood, muscles begin to atrophy (waste away) in affected individuals, who typically die in their teens or early twenties (Section 37.10). Muscle biopsies of a few women who carry an allele that is associated with the disorder reveal some body regions of atrophied muscle tissue. Yet muscles adjacent to a region of atrophy are normal or even larger and more chemically active, as if to compensate for the weakness of the adjoining region. Give a brief explanation of these observations.

3. Unlike most rodents, guinea pigs are well developed at the time of birth. Within a few days, they can eat grass, vegetables, and other plant material. Suppose a breeder decides to separate the baby guinea pigs from their mothers after three weeks. He wants to keep the males and females in different cages. But it is difficult to identify the sex of young guinea pigs. Suggest a test that the breeder can perform to identify their sex.

4. Individuals affected by *pituitary dwarfism* cannot synthesize somatotropin (also called growth hormone). Children with this genetic abnormality will be below the range of normal height. Develop a hypothesis to explain why therapy that is based on somatotropin injections is effective.

5. The closer a mammalian species is to humans in its genetic makeup, the more useful information it may yield in laboratory studies of the mechanisms of cancer. Do you support the use of any mammal for cancer research? Why or why not?

---

6. An operon most typically governs _____ .
   a. bacterial genes   c. genes of all types
   b. a eukaryotic gene   d. DNA replication

7. Eukaryotic genes guide _____ .
   a. fast short-term activities   c. development
   b. overall growth   d. all of the above

8. X chromosome inactivation is an example of _____ .
   a. a chromosome abnormality   c. dysfunctional calico cats
   b. dosage compensation   d. dysfunctional women

9. Hormones may _____ gene transcription in target cells.
   a. promote   c. participate in
   b. inhibit   d. both a and b

10. Apoptosis is _____ .
   a. cell division after severe tissue damage
   b. programmed cell death by suicide
   c. a popping sound in mutated toes

11. ICE-like proteases are _____ .
   a. structural proteins   c. environmental signals
   b. lethal weapons   d. low-temperature enzymes

12. Match the terms with their most suitable descriptions.
   _____ phytochrome   a. helps animal body plan develop
   _____ Barr body   b. mutated form of any gene that
   _____ oncogene   can induce cancer
   _____ homeotic gene   c. hormone with key role in
   _____ ecdysone   insect life cycles
                    d. helps plants adapt to daily
                      and seasonal changes in light
                    e. inactivated X chromosome

Critical Thinking

1. Geraldo isolated a strain of *E. coli* in which a mutation has affected the capacity of CAP to bind to a region of the lactose operon, as it would do normally. State how the mutation affects

Selected Key Terms

acetylation *15.1*	kinase *15.6*
activator *15.2*	metastasis *15.6*
apoptosis *15.6*	methylation *15.1*
Barr body *15.4*	mosaic tissue effect *15.4*
cancer *CI*	negative control system *15.1*
cell differentiation *15.3*	neoplasm *CI*
checkpoint gene *15.6*	oncogene *15.6*
dosage compensation *15.4*	operator *15.2*
ecdysone *15.5*	operon *15.2*
enhancer *15.1*	phytochrome *15.5*
gene control *CI*	polytene chromosome *15.5*
growth factor *15.6*	positive control system *15.1*
homeotic gene *15.4*	promoter *15.1*
hormone *15.5*	regulatory protein *15.1*
ICE-like protease *15.6*	X chromosome inactivation *15.4*

Readings

Duke, R., D. Ojcius, and J. Ding-E Young. December 1996. "Cell Suicide in Health and Disease." *Scientific American*, 80–87.

Murray, A., and M. Kirschner. March 1991. "What Controls the Cell Cycle?" *Scientific American*, 264(3): 56–63.

Tijan, R. February 1995. "Molecular Machines That Control Genes." *Scientific American*, 54–61.

Travis, J. 5 August 2000. "Silence of the Xs." *Science News* 158: 92–94. A look at mechanism of X chromosome inactivation. Raises questions about "junk DNA" (base sequences that have accumulated in DNA yet have no known function).

RECOMBINANT DNA AND GENETIC ENGINEERING

Mom, Dad, and Clogged Arteries

Butter! Bacon! Eggs! Ice cream! Cheesecake! Possibly you think of such foods as enticing, off-limits, or both. After all, who among us doesn't know about animal fats and the dreaded cholesterol?

Soon after you feast on such fatty foods, cholesterol enters the bloodstream. Cholesterol is important. It is a structural component of animal cell membranes, and without membranes, there would be no cells. Cells also remodel cholesterol into a variety of molecules, such as the vitamin D necessary for the development of good bones and teeth. Normally, however, the liver itself synthesizes enough cholesterol for your cells.

Some proteins circulating in blood combine with cholesterol and other substances to form lipoprotein particles. *High*-density lipoproteins, or *HDLs*, transport cholesterol to the liver, where it's metabolized. Usually, *low*-density lipoproteins, or *LDLs*, end up inside cells that store or use cholesterol. But sometimes too many LDLs form. The excess infiltrates the elastic walls of arteries and helps form atherosclerotic plaques (Figure 16.1). These abnormal masses interfere with blood flow. If they clog one of the tiny coronary arteries that deliver blood to the heart, a heart attack may result.

How well you handle dietary cholesterol depends on which alleles you got from your parents. For example, one gene specifies a protein receptor for LDLs. If you inherited two "good" alleles and don't go overboard on the fatty foods, your blood level of cholesterol may stay low to moderate and your arteries may never clog up.

Inherit two copies of a certain mutated allele, and you will develop *familial hypercholesterolemia*, a rare genetic disorder. Cholesterol levels get so high, many of those affected die of heart attacks in childhood or their teens.

In 1992 a woman from Quebec, Canada, became a milestone in the history of genetics. She was thirty years old. Like two brothers felled in their early twenties by heart attacks, she had inherited the mutant allele for the LDL receptor. She survived a heart attack at sixteen. At twenty-six, she had coronary bypass surgery.

At the time, people were hotly debating the risks and the promises of **gene therapy**—the transfer of one or more normal or modified genes into an individual's body cells to correct a genetic defect or boost resistance to disease. The woman opted for an untried procedure to give her body working copies of the good gene.

Surgeons removed about 15 percent of her liver. Researchers put cells from it in a nutrient-rich medium to promote cell growth and division. *And they spliced the functional allele for the LDL receptor into the genetic material of a harmless virus.* The modified virus infected the cultured liver cells, thereby inserting copies of the good gene into them. When the liver cells went on to reproduce, they made copies of the good gene, also.

About a billion modified cells were infused into the woman's portal vein, a blood vessel that leads directly to the liver. Some cells took up residence in the liver and started to make the missing cholesterol receptor. Two years later, a fraction of the woman's liver cells were behaving normally and sponging up cholesterol from blood. Her LDL blood levels declined nearly 20 percent. There was no sign of the arterial clogging that nearly killed her. Her cholesterol levels were still higher than normal. Yet the intervention demonstrated the safety and feasibility of gene therapy for some patients.

As you might gather from this pioneering clinical work, recombinant DNA technology has huge potential for medicine, agriculture, and industry. Think of it! For thousands of years, we humans have been changing

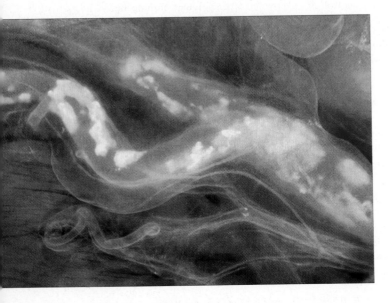

Figure 16.1 Life-threatening plaques (*bright yellow, white*) in one of the coronary arteries. These small arteries deliver blood to the heart. Abnormally high levels of cholesterol contribute to the formation of plaques that may clog them. Gene therapies based on recombinant DNA technology have the potential to counter many such threats to health.

Figure 16.2 One outcome of genetic manipulation by way of artificial selection practices: a large kernel from a modern strain of corn next to the tiny kernels of an ancestral species, which was recovered from a prehistoric cave in Mexico.

genetically based traits. By artificial selection practices, we produced new plants and new breeds of cattle, cats, dogs, and birds from wild ancestral stocks. We were selective agents for meatier turkeys, sweeter oranges, seedless watermelons, spectacular ornamental roses, and big juicy corn kernels (Figure 16.2). We conjured up hybrids such as the mule (horse × donkey) and plants that bear tangelos (tangerine × grapefruit).

Of course, we have to remember we're newcomers on the evolutionary stage. During the 3.8 billion years before we even made our entrance, nature conducted uncountable numbers of genetic experiments. Nature's tools have included mutation, crossing over, and other events that introduce changes in genetic messages. The countless changes gave rise to life's rich diversity.

But the striking thing about human-directed change is that the pace has picked up. Researchers analyze genes with **recombinant DNA technology**. They cut and recombine DNA from different species and insert it into bacterial, yeast, or mammalian cells. The cells replicate their DNA and divide rapidly. They copy the foreign DNA as if it were their own and churn out useful quantities of recombinant DNA molecules. The technology also is the basis of **genetic engineering**. By this process, genes are isolated, modified, and inserted into the same organism or into a different one. Protein products of the modified genes may cover the function of their missing or malfunctioning counterparts.

The new technology does not come without risks. With this chapter, we consider its basic aspects. At the chapter's end, we also address some ecological, social, and ethical questions related to its application.

Key Concepts

1. Genetic experiments have been proceeding for billions of years, through gene mutations, crossing over, genetic recombination, and other natural events.

2. Humans are now purposefully bringing about rapid genetic changes by way of recombinant DNA technology. Such manipulations are called genetic engineering.

3. With this technology, researchers isolate, cut, and splice together gene regions from different species. Then they greatly amplify the number of copies of genes that interest them. The genes, and sometimes the proteins they specify, are produced in quantities that are large enough to use for research and for practical applications.

4. Three activities are at the heart of recombinant DNA technology. First, procedures based on specific types of enzymes are used to cut DNA molecules into fragments. Second, the fragments are inserted into cloning tools, such as plasmids. Third, the fragments containing the genes of interest are identified, then copied rapidly and repeatedly.

5. Genetic engineering involves isolating, modifying, and inserting genes back into the same organism or into a different one. The goal is to beneficially modify traits that the genes influence. Human gene therapy, which focuses on controlling or curing genetic disorders, is an example.

6. The new technology raises social, legal, ecological, and ethical questions regarding its benefits and risks.

A TOOLKIT FOR MAKING RECOMBINANT DNA

Restriction Enzymes

In the 1950s, the scientific community was agog over the discovery of DNA's structure. The excitement gave way to frustration. No one could figure out the sequence of nucleotides, the order of genes and gene regions along a chromosome. Robert Holley and his colleagues did sequence a small tRNA. They used digestive enzymes that broke the molecule into fragments small enough to be chemically characterized. But DNA molecules? These are far longer than RNAs. No one knew how to cut one into fragments long enough to have unique, and thus analyzable, sequences. Digestive enzymes can cut DNA, but not in any particular order. There was no telling how the fragments had been arranged in the molecule.

Then, by accident, Hamilton Smith discovered that *Haemophilus influenzae* chops up foreign DNA inserted into it by a bacteriophage. Extracts from *H. influenzae* cells included an enzyme that restricts itself to a specific kind of site in DNA. It was named a **restriction enzyme**.

In time, several hundred strains of bacteria offered up a toolkit of restriction enzymes that recognize and cut specific sequences of four to eight bases in DNA. Table 16.1 lists a few. They are among many that make *staggered* cuts, which leave single-stranded "tails" on the end of a DNA fragment. Tails made by *TaqI* are two bases long (CG). *EcoRI* makes tails four bases long (AATT).

How many cuts do restriction enzymes make? That depends in part on the molecule. *NotI* only recognizes a rare eight-base sequence in the DNA of mammals, so when it makes its cuts, most of the fragments are tens of thousands of base pairs long. That is long enough for studying the **genome**—all of the DNA in a haploid number of chromosomes for a species. For the human species, the genome is about 3.2 billion base pairs long.

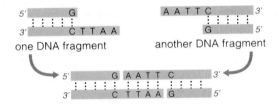

Table 16.1 *Examples of Restriction Enzymes*

Bacterial Source	Enzyme's Abbreviation	Its Specific Cut
Thermus aquaticus	*Taq*I	5' T C G A 3' 3' A G C T 5'
Escherichia coli	*Eco*RI	5' G A A T T C 3' 3' C T T A A G 5'
Nocardia otitidus caviarum	*Not*I	5' G C G G C C G C 3' 3' C G C C G G C G 5'

Modification Enzymes

DNA fragments with staggered cuts have sticky ends. "Sticky" means a restriction fragment's single-stranded tail can base-pair with a complementary tail of any other DNA fragment or molecule cut by the same restriction enzyme. *Mix DNA fragments cut by the same restriction enzyme, and the sticky ends of any two fragments having complementary base sequences will base-pair and form a recombinant DNA molecule:*

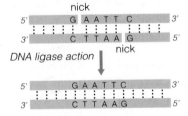

Notice the nicks where such fragments base-pair. **DNA ligase**, a modification enzyme, can seal these nicks:

Cloning Vectors for Amplifying DNA

Restriction and modification enzymes make it possible to insert foreign DNA into bacterial cells. As you know, each bacterial cell has only one chromosome, a circular DNA molecule. Many also inherit plasmids. A **plasmid** is a very small circle of extra DNA that has just a few genes and that gets replicated along with the bacterial chromosome (Figure 16.3). Bacteria usually can survive without plasmids, but some of the genes offer benefits, as when they confer resistance to antibiotics.

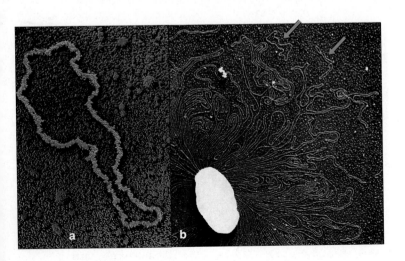

Figure 16.3 **(a)** A single plasmid. **(b)** Plasmids (*arrows*) from a ruptured *Escherichia coli* cell.

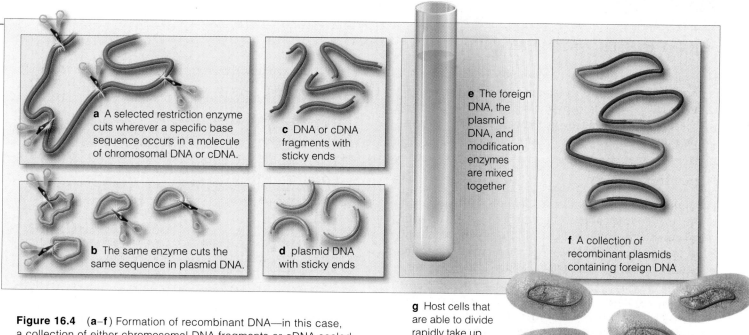

a A selected restriction enzyme cuts wherever a specific base sequence occurs in a molecule of chromosomal DNA or cDNA.

c DNA or cDNA fragments with sticky ends

b The same enzyme cuts the same sequence in plasmid DNA.

d plasmid DNA with sticky ends

e The foreign DNA, the plasmid DNA, and modification enzymes are mixed together

f A collection of recombinant plasmids containing foreign DNA

g Host cells that are able to divide rapidly take up the recombinant plasmids

Figure 16.4 (a–f) Formation of recombinant DNA—in this case, a collection of either chromosomal DNA fragments or cDNA sealed into bacterial plasmids. (g) Recombinant plasmids are inserted into host cells that can rapidly amplify the foreign DNA of interest.

Under favorable conditions, bacteria divide rapidly and often—every thirty minutes, for some species—so that huge populations of genetically identical cells form. Before cell division, replication enzymes duplicate the bacterial chromosome. They also replicate the plasmid, sometimes repeatedly, so a cell can hold many identical copies of foreign DNA. In research laboratories, foreign DNA typically is inserted into a plasmid for replication. The outcome is called a **DNA clone**, because bacterial cells have made many identical, "cloned" copies of it.

A modified plasmid that accepts foreign DNA is a **cloning vector**. It can insert foreign DNA into a host bacterium, yeast, or some other cell that can be the start of a "cloning factory." It may give rise to a population of rapidly dividing descendant cells, all with identical copies of the foreign DNA (Figure 16.4).

Reverse Transcriptase To Make cDNA

Researchers analyze genes to unlock secrets about gene products and how they are put to use. Yet even when a host cell takes up a gene, it may not be able to make the protein. For instance, most eukaryotic genes have introns (noncoding sequences). mRNA transcripts of the genes can't be translated until the introns are snipped out and coding regions (exons) spliced together. Bacterial cells can't snip out the introns, so they often can't translate human genes into proteins.

Researchers get around this problem with **cDNA**, a DNA strand "copied" from a mature mRNA transcript. **Reverse transcriptase** catalyzes transcription in reverse. This enzyme from RNA viruses builds a complementary DNA strand on an mRNA transcript (Figure 16.5). Then

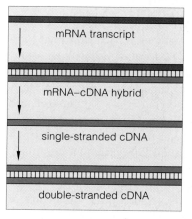

mRNA transcript

mRNA–cDNA hybrid

single-stranded cDNA

double-stranded cDNA

a Reverse transcriptase catalyzes the assembly of a single DNA strand on a mature mRNA transcript. The result is an mRNA–cDNA hybrid molecule.

b Enzymes remove the mRNA and use the cDNA strand as a template to assemble another DNA strand. The result is double-stranded cDNA, "copied" from an mRNA template.

Figure 16.5 Formation of cDNA from an mRNA transcript.

other enzymes remove the RNA from the mRNA–cDNA molecule and substitute a complementary DNA strand. The outcome is double-stranded cDNA.

Double-stranded cDNA can be further modified, as by attaching signals for transcription and translation. The modified cDNA can be inserted into a plasmid for amplification. Bacterial cells exposed to the recombinant plasmids often take them up and then use the cDNA instructions for synthesizing a protein of interest.

Restriction enzymes and modification enzymes cut apart chromosomal DNA or cDNA and splice it into plasmids and other cloning vectors. Recombinant plasmids can be taken up by bacteria or other cells that divide rapidly, thus making multiple, identical quantities of the foreign DNA.

PCR—A FASTER WAY TO AMPLIFY DNA

The polymerase chain reaction, widely known as **PCR**, is another way to amplify fragments of chromosomal DNA or cDNA. These copy-making reactions proceed with astounding speed in test tubes, not in bacterial cloning factories. Primers get them going.

What Are Primers?

Primers are synthetic, short nucleotide sequences, ten to thirty or so nucleotides long, that can base-pair with complementary sequences in DNA. The workhorses of DNA replication—the DNA polymerases—chemically recognize primers as START tags. Following a computer program, machines synthesize a primer one step at a time. How do researchers decide on the order of these nucleotides? First, they must identify short nucleotide sequences located just before and just after the DNA region from a cell that interests them. Then they build primers that have the complementary sequences.

What Are the Reaction Steps?

For PCR, the enzyme of choice is a DNA polymerase extracted from *Thermus aquaticus*, a bacterium that lives in superheated water of hot springs. It has been found in water heaters, also. This enzyme is not destroyed at the elevated temperatures required to unwind a DNA double helix. Most DNA polymerases are denatured and permanently lose their activity at such temperatures.

Researchers mix together primers, the molecules of DNA polymerase and cellular DNA from an organism, and free nucleotides. Next, they expose the mixture to precise cycles of temperature. At the beginning of each temperature cycle, the two strands of all molecules of DNA in the mixture unwind from each other.

Primers become positioned on exposed nucleotides at targeted sites according to base-pairing rules (Figure 16.6). Each round of reactions doubles the number of DNA molecules amplified from the targeted site. Thus, if there are 10 such molecules in a test tube, there soon will be 20, then 40, 80, 160, 320, 640, 1,280, and so on. Any targeted region from a single DNA molecule can be rapidly amplified into billions of molecules.

In short, *PCR amplifies samples that contain even tiny amounts of DNA.* At this writing, it is the amplification procedure of choice in thousands of laboratories all over the world. As you'll see in the next section, such samples can be obtained even from a single hair follicle or drop of blood left at the scene of a crime.

PCR is a method of amplifying chromosomal DNA or cDNA inside test tubes. Compared to cloning methods, PCR is far more rapid.

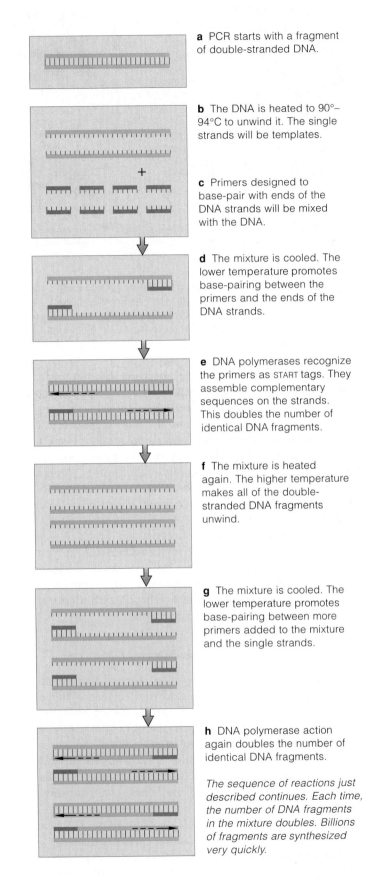

a PCR starts with a fragment of double-stranded DNA.

b The DNA is heated to 90°–94°C to unwind it. The single strands will be templates.

c Primers designed to base-pair with ends of the DNA strands will be mixed with the DNA.

d The mixture is cooled. The lower temperature promotes base-pairing between the primers and the ends of the DNA strands.

e DNA polymerases recognize the primers as START tags. They assemble complementary sequences on the strands. This doubles the number of identical DNA fragments.

f The mixture is heated again. The higher temperature makes all of the double-stranded DNA fragments unwind.

g The mixture is cooled. The lower temperature promotes base-pairing between more primers added to the mixture and the single strands.

h DNA polymerase action again doubles the number of identical DNA fragments.

The sequence of reactions just described continues. Each time, the number of DNA fragments in the mixture doubles. Billions of fragments are synthesized very quickly.

Figure 16.6 The polymerase chain reaction (PCR).

DNA Fingerprints

Except for identical twins, no two people have exactly the same base sequence in their DNA. By detecting the differences in sequences, scientists can distinguish one person from another. As you know, each human has a unique set of fingerprints, a marker of identity. Like all other sexually reproducing species, each human also has a **DNA fingerprint**, a unique array of DNA sequences that were inherited from parents in a Mendelian pattern. DNA fingerprints are so accurate that they can reveal differences even between full siblings.

More than 99 percent of the DNA in all humans is exactly the same. However, DNA fingerprinting focuses only on the portion that tends to be highly variable from one person to the next. Throughout the human genome are **tandem repeats**—short DNA sequences present in many copies, one after the other, in a chromosome.

For example, one individual's DNA might have the five bases TTTTC repeated four times in one location, yet another individual might have them repeated fifteen times in the same location. One might have five repeats of the bases CGG, and another might have fifty of them.

Repetitive DNA sequences have increased or decreased in number over many generations. They slipped into the DNA as spontaneous changes while the DNA was being replicated. In these regions, the mutation rate is relatively high (Section 14.4).

Researchers detect such differences at tandem-repeat sites with **gel electrophoresis**. This laboratory technique uses an electric field to force molecules through a viscous gel. In this case, it separates DNA fragments according to length. Size alone dictates how far each fragment will move through the gel. Tandem repeats of different sizes migrate at different rates.

A gel is immersed in a buffered solution, then DNA fragments from individuals are added to the gel. When an electric current is applied to the solution, one end of the gel takes on a negative charge, and the other end a positive charge. With their negatively charged phosphate groups, DNA fragments migrate through the gel toward the positively charged pole. They do so at different rates and separate into bands according to length. The smaller the fragment is, the faster it migrates. After a set time, researchers identify fragments of different lengths by staining the gel or specifically highlighting fragments that contain tandem repeats.

DNA fingerprints help forensic scientists identify criminals, victims, and innocent suspects. A few drops of blood, semen, or cells from a hair follicle at a crime scene or on a suspect's clothing often yield enough to do the trick. For example, Figure 16.7 shows some tandem-repeat DNA fragments separated by gel electrophoresis. They are DNA fingerprints from seven people and from blood collected at a crime scene. Notice how much all but two of the DNA fingerprints differ. Can you identify

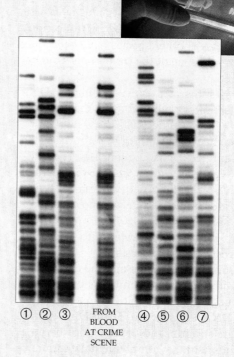

① ② ③ FROM BLOOD AT CRIME SCENE ④ ⑤ ⑥ ⑦

Figure 16.7 *Left:* Comparison of DNA fingerprints from a bloodstain discovered at a crime scene and from blood samples of seven suspects (the circled numbers). Given what you have learned about gel electrophoresis, point out which of the seven is a match. *Above:* The gel shown in this photograph had been stained earlier with a substance that can make DNA fragments fluoresce when placed under ultraviolet light.

which pattern exactly matches the pattern from the blood collected at the crime scene?

DNA fingerprint analysis even confirmed that human bones exhumed from a shallow pit in Siberia belonged to five members of the Russian imperial family, all shot to death in secrecy in 1918.

When DNA fingerprinting was first used as evidence in court, attorneys challenged conclusions based on it. But DNA fingerprinting is now firmly established as an accurate, unambiguous procedure. Today it is routinely submitted as evidence in disputes over paternity. It is widely used to convict the guilty and exonerate innocent suspects. To give an early example, in 1998, an eleven-year-old girl was raped, stabbed, and strangled to death in Elisabethfehn, Germany. In the largest mass screening yet conducted, the police requested blood samples from 16,400 males in and around the vicinity. A single DNA fingerprint from a thirty-year-old mechanic matched the evidence, and he confessed to the crime.

The variation in tandem repeats also can be detected as restriction fragment length polymorphisms, or RFLPs. (These are DNA fragments of different sizes, cleaved by restriction enzymes.) In the case of tandem repeats, a restriction enzyme cleaves the DNA flanking the repeat. Alternatively, researchers might use PCR to amplify the tandem-repeat region. Either way, differences in the size of the fragments, which reveal genetic differences, can be detected with electrophoresis.

HOW IS DNA SEQUENCED?

In 1995, researchers accomplished what was little more than a dream a few decades ago. They determined the full DNA sequence for a species: *Haemophilus influenzae*, a bacterium that causes human respiratory infections. Since then, the genomes of other species have been fully sequenced. A draft sequence of the human genome has now been completed.

The molecular sleuths are using **automated DNA sequencing**. This laboratory method gives the sequence of cloned DNA or PCR-amplified DNA in a few hours.

Researchers use the four standard nucleotides (T, C, A, and G) for automated DNA sequencing. They also use four modified versions, which we can represent as T*, C*, A*, and G*. Each modified version is labeled with a molecule that fluoresces a certain color when it passes through a laser beam. Every time one of these modified nucleotides gets incorporated into a growing DNA strand, it arrests DNA synthesis.

Before the reactions, researchers mix the eight kinds of nucleotides together. They add millions of copies of the DNA to be sequenced along with a type of primer and molecules of DNA polymerase. Then they separate the DNA into single strands, and the reactions begin.

The primer binds with its complementary sequence on one of the strands. DNA polymerase synthesizes a new DNA strand starting at one end of the primer. One by one, it adds nucleotides in the order dictated by the exposed sequence in the template strand. Each time, one of the standard nucleotides *or* one of the modified versions may be attached.

Suppose that DNA polymerase encounters a T in a DNA template strand. It will catalyze the base-pairing of either A or A* to it. If A is added to the new strand, replication will continue. But if A* is added, replication will stop; the modified nucleotide will *block* addition of any more nucleotides to that strand. The same thing happens at each nucleotide in a template strand. When a standard nucleotide is attached, replication proceeds; when a modified version is attached, replication stops.

Remember, the starting mixture contained millions of identical copies of the DNA sequence. Because either a standard or a modified nucleotide could be added at every exposed base, the new strands end at different locations in the sequence. The mixture now contains millions of copies of tagged fragments having different lengths. These can now be separated by length into *sets* of fragments. And each set corresponds to only one of the nucleotides in the entire base sequence.

The automated DNA sequencer is a machine that separates the sets of fragments by gel electrophoresis. The set having the shortest fragments migrates fastest through the gel and reaches the end of it first. The last set to reach the end has the longest fragments. Because the fragments have a modified nucleotide at the 3' end,

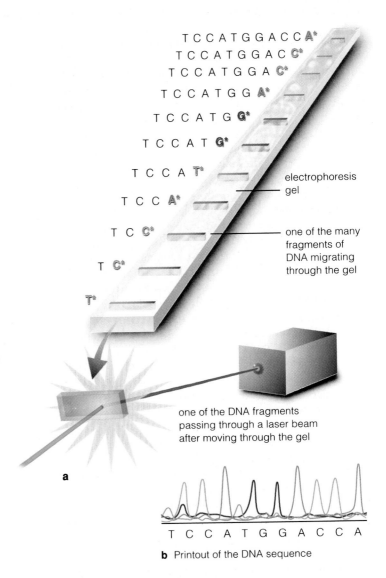

a

b Printout of the DNA sequence

Figure 16.8 Automated DNA sequencing. (**a**) DNA fragments from an organism's genome become labeled at their end with a modified nucleotide that fluoresces a certain color. (**b**) Printout of the DNA sequence used in this example. Each peak indicates absorbance by a particular labeled nucleotide.

each set fluoresces a certain color as it passes through a laser beam (Figure 16.8*a*). The automated sequencer detects the color and indicates which nucleotide is on the end of the fragments in each set. It assembles the information from all nucleotides in the sample, and in this way it reveals the entire DNA sequence.

Figure 16.8*b* is a printout from an automated DNA sequencer. Each peak along that tracing represents the detection of a particular color as the sets of fragments reached the end of the gel.

With automated DNA sequencing, the order of nucleotides in a cloned or amplified DNA fragment can be determined.

FROM HAYSTACKS TO NEEDLES—ISOLATING GENES OF INTEREST

Any genome consists of thousands of genes. *E. coli* has 4,279, for instance, and humans apparently have about 30,000. What if you wanted to learn about or modify the structure of any one of those genes? First, you'd have to isolate that gene from all others in the genome.

There are several ways to do this. If part of the gene sequence is already known, you can design primers to amplify the entire gene or part of it by PCR. Often, though, researchers must isolate and clone a gene. First they make a **gene library**. This is a mixed collection of bacteria that contain different cloned DNA fragments. One is the gene of interest. A *genomic* library contains cloned DNA fragments from an entire genome. A *cDNA* library contains DNA derived from mRNA. Often it is the most useful because it is free of introns. But the gene is still hidden like a needle in a haystack. How can you isolate it? One way is to use a nucleic acid probe.

What Are Probes?

A **probe** is a very short stretch of DNA labeled with a radioisotope so that it can be distinguished from other DNA molecules in a given sample. Part of the probe must be able to base-pair with some portion of the gene of interest. Any base-pairing that takes place between sequences of DNA (or RNA) from different sources is called **nucleic acid hybridization**.

How do you acquire a suitable probe? Sometimes part of the gene or a closely related gene has already been cloned, in which case it can be used as the probe. If the gene's structure is a mystery, you still may be able to work backward from the amino acid sequence of its protein product, assuming that protein is already available. By using the genetic code as a guide (Section 14.2), you could build a DNA probe that is more or less similar to the gene of interest.

Screening for Genes

Once you have a gene library and a suitable probe, you are ready to hunt down the gene. Figure 16.9 shows the steps of one isolation method. The first step is to take bacterial cells of the library and spread them apart on the surface of a gelled growth medium in a petri plate. When spread out sufficiently, individual cells undergo division. Each cell starts a colony of genetically identical cells. The bacterial colonies appear as hundreds of tiny white spots on the surface of a culture medium.

After colonies appear, you lay a nylon filter on top of the colonies. Some cells stick to the filter at locations that mirror the locations of the original colonies. You use solutions to rupture the cells, and the DNA so released sticks to the filter. You denature the DNA to single strands and then add the probes. The probes hybridize

a Bacterial colonies, each derived from a single cell, grow on a culture plate. Each colony is about 1 millimeter across.

b A nitrocellulose or nylon filter is placed on the plate. Some cells of each colony adhere to it. The filter mirrors how the colonies are distributed on the culture plate.

c The filter is lifted off and put into a solution. The cells stuck to it rupture; the cellular DNA sticks to the filter.

d Also, the DNA is denatured to single strands at each site. A radioactively labeled probe is added to the filter. It binds to DNA fragments that have a complementary base sequence.

e The probe's location is identified by exposing the filter to x-ray film. The image that forms on the film reveals the colony that has the gene of interest.

Figure 16.9 Use of a probe to identify bacterial colonies that have taken up a gene library.

only with DNA from the colony that took up the gene of interest. If you expose the probe-hybridized DNA to x-ray film, the pattern formed by the radioactivity will identify that colony. With this information you can now culture cells from that colony alone, knowing that it will be the only one that can replicate the cloned gene.

Probes may be used to identify one particular gene among many in gene libraries. Bacterial colonies that have taken up the library can be cultured to isolate the gene.

USING THE GENETIC SCRIPTS

As researchers decoded the genetic scripts of species, they opened the door to astonishing possibilities. For example, genetically engineered bacteria now produce medically valued proteins. Huge bacterial populations produce useful quantities of the desired gene products in stainless steel vats. *Diabetics* who must receive insulin injections every day are among the beneficiaries. At one time, medical supplies of this pancreatic hormone were extracted only from pigs and cattle. Later on, synthetic genes for human insulin were transferred into *E. coli* cells. The cells were the start of populations that became the first large-scale, cost-effective bacterial factory for proteins. Besides insulin, human somatotropin, blood-clotting factors, hemoglobin, interferon, and a variety of drugs and vaccines are also manufactured with the help of genetically engineered bacteria.

Other kinds of modified bacteria hold potential for industry and for cleaning up environmental messes. For instance, as you know, many microorganisms can break down organic wastes and help cycle nutrients through ecosystems. Certain modified bacteria can break down crude oil into less harmful compounds. When sprayed onto oil spills, as from a shipwrecked supertanker, they might help avert an environmental disaster. Others are genetically engineered to sponge up excess phosphates or heavy metals from the environment.

In addition, bacterial species that contain plasmids offer benefits for basic research, agriculture, and gene therapy. For instance, deciphering the messages encoded in bacterial genes helps us reconstruct the evolutionary history of life. Sections 19.6 and 19.7 will give you an idea of how comparisons of DNA or RNA from different organisms can reveal evolutionary secrets. The next two sections give a few more examples of benefits.

What about the "bad" bunch—the pathogenic fungi, bacteria, and viruses? Natural selection favors mutated genes that improve a pathogen's chances of evading a host organism's natural defenses. Mutation is frequent among pathogens that reproduce rapidly. So designing new antibiotics and other defenses against new gene products is a constant challenge. But if we learn about the genes, we may have advance warning of the "plan of attack." The story of HIV, a rapidly mutating virus that causes AIDS, gives insight into the magnitude of the problem (Section 39.10).

Knowing about genes allows us to genetically engineer beneficial microorganisms for uses in medicine, industry, agriculture, and environmental remediation.

Knowing about genes can help us reconstruct evolutionary histories of species.

Knowing about genes may help us design effective, timely counterattacks against rapidly mutating pathogens.

DESIGNER PLANTS

Regenerating Plants From Cultured Cells

Many years ago, Frederick Steward and his coworkers cultured cells from carrot plants. They induced some of them to develop into small embryos. As you will read in Section 31.7, some embryos grew into whole plants. Today, researchers routinely regenerate crop plants and many other plant species from cultured cells. They use ingenious methods to pinpoint a gene in a culture that may contain even millions of cells. Suppose they add a pathogen's toxin to a culture medium. Suppose a few cells happen to carry a gene that confers resistance to the toxin. Those cells will be the only ones to survive.

Whole plants regenerated from preselected cultured cells can be hybridized with other varieties of plants. In this way, desired genes might be transferred to a plant lineage to improve herbicide resistance, pest resistance, and other traits that can benefit crops and gardens.

Consider this larger aspect of the research: The food supply for most of the human population is vulnerable. Generally, farmers prefer genetically similar varieties of high-yield plants and have discarded the more diverse, older varieties. But genetic uniformity makes food crops vulnerable to many kinds of pathogenic fungi, viruses, and bacteria. That is why botanists comb the world for seeds of the wild ancestors of potatoes, corn, and other crop plants. They send their prizes—seeds with genes of a plant's lineage—to **seed banks**. These safe storage facilities are designed to preserve genetic diversity.

The problem is huge. Example: In 1970 a new fungal strain of *Southern corn leaf blight* destroyed much of the United States corn crop. All of the plants carried the same gene that conferred susceptibility to the disease. Ever since that devastating epidemic, seed companies have been much more attentive to offering genetically diverse corn seeds. And where does the diversity come from? Plant breeders call on the seed banks.

How Are Genes Transferred Into Plants?

The Ti (Tumor-inducing) plasmid from *Agrobacterium tumefaciens* is one vehicle for inserting new or modified genes into plants. *A. tumefaciens* infects many kinds of flowering plants (Figure 16.10). Some Ti plasmid genes

crown gall tumor

Figure 16.10 Crown gall tumor on a woody plant, an abnormal tissue growth triggered by a gene in the Ti plasmid.

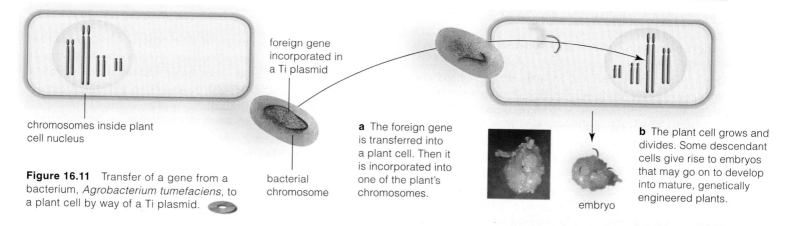

chromosomes inside plant cell nucleus

foreign gene incorporated in a Ti plasmid

Figure 16.11 Transfer of a gene from a bacterium, *Agrobacterium tumefaciens*, to a plant cell by way of a Ti plasmid.

bacterial chromosome

a The foreign gene is transferred into a plant cell. Then it is incorporated into one of the plant's chromosomes.

embryo

b The plant cell grows and divides. Some descendant cells give rise to embryos that may go on to develop into mature, genetically engineered plants.

invade a plant's DNA, then induce the formation of abnormal tissue masses. We call the masses crown gall tumors. Before introducing genes from a plasmid into plant cells, researchers remove the tumor-inducing genes and then insert a desired gene into it. They place plant cells in a culture of the modified bacteria. Some cells may take up the gene, and whole plants may be regenerated from the cellular descendants, as in Figure 16.11. Expression of foreign genes in plants sometimes is dramatic, as in Figure 16.12.

In nature, *A. tumefaciens* infects plants called dicots, which include beans, peas, potatoes and other vital crops. Geneticists have modified the bacterium so that it also delivers genes into monocots that are vital food crops, including wheat, corn, and rice. Some researchers use electric shocks or chemicals to deliver modified genes into plant cells. Some also blast microscopic particles coated with DNA into them.

Despite many obstacles, improved varieties of crop plants have been engineered or are in the works. For example, genetically engineered cotton plants display resistance to a herbicide (Figure 16.13). Farmers spray the herbicide in fields of modified cotton plants to kill weeds. The plants are not affected.

Also on the horizon are engineered plants that can be factories for pharmaceuticals. A few years ago, for example, tobacco plants that were engineered to produce hemoglobin and some other proteins were planted in a test field in North Carolina. Afterward, ecologists found

Figure 16.12 A modified plant that glows in the dark. A firefly gene was inserted into the plant's DNA and is being expressed. (Refer to Section 6.9). The gene's product is luciferase, an enzyme with a role in bioluminescence.

Figure 16.13 (**a**) Control plant (*left*) and three genetically engineered aspen seedlings. Vincent Chiang and coworkers suppressed a regulatory gene involved in a lignin biosynthetic pathway. The modified plants synthesized normal lignin, but not as much. Lignin production decreased by as much as 45 percent—yet cellulose production increased 15 percent. Root, stem, and leaf growth were greatly enhanced. Plant structure did not suffer. Wood harvested from such trees might make it easier to manufacture paper and some clean-burning fuels, such as ethanol. (Lignin, a tough polymer, strengthens the secondary cell walls of plants. Before paper can be made from wood, the lignin must be chemically extracted.)

(**b**) *Left:* Cotton plant, used as the control. *Right:* Genetically engineered cotton plant with a gene for herbicide resistance. Both plants were sprayed with a weedkiller that currently has widespread application in cotton fields.

no trace of foreign genes or proteins in the soil or in other plants or animals in the neighborhood. Another example: In a Stanford University laboratory, mustard plants synthesized biodegradable plastic beads that are suitable for use in manufacturing plastics.

Genetically diverse plant species are essential to protecting our vulnerable food supply. Genetic engineers can design plants with new beneficial traits.

GENE TRANSFERS IN ANIMALS

Supermice and Biotech Barnyards

The first mammals enlisted for experiments in genetic engineering were laboratory mice. Consider an example of this work. R. Hammer, R. Palmiter, and R. Brinster corrected a hormone deficiency that leads to dwarfism in mice. Insufficient levels of somatotropin (or growth hormone) cause the abnormality. The researchers used a microneedle to inject the gene for rat somatotropin into fertilized mouse eggs, which they implanted in an adult female. The gene was successfully integrated into mouse DNA, and the eggs developed into mice. Young mice in which that foreign gene was expressed grew 1–1/2 times larger than their dwarf littermates. In other experiments, researchers transferred the gene for human somatotropin into a mouse embryo. The gene became integrated into mouse DNA, and the modified embryo grew up to be a "supermouse" (Figure 16.14).

Today, human gene transfers are being attempted in research into the molecular basis of genetic disorders. "Biotech barnyards" have animals that compete with bacterial factories as genetically engineered sources of proteins. Goats produce CFTR protein (to treat cystic fibrosis) and TPA (to counter effects of a heart attack). Cattle may soon be producing human collagen, which can help repair cartilage, bone, and skin.

Remember the first case of cloning a mammal from an adult cell? A nucleus extracted from a mammary gland cell of a ewe was inserted into an enucleated egg. Signals from the egg cytoplasm sparked development of an embryo, which was implanted into a surrogate mother. This resulted in *Dolly* (Section 13.5).

For years, researchers had been cloning animals from embryonic tissue, but not adults that already displayed some sought-after trait. Now, however, if they could just refine the steps of their cloning processes, they might be able to maintain some genotype indefinitely.

Not long after Dolly's debut, genetically engineered clones of mice, cattle, and other animals were produced. Example: Steve Stice and his colleagues produced designer cattle. They started with a culture of cells from cattle. They induced the development of six genetically identical calves. They also engineered targeted changes in the cloning cell lineage.

Stice would like to genetically engineer cattle resistant to bovine spongiform encephalopathy (mad cow disease, Section 21.8). His method also has been used to put the human serum albumin gene into the chromosomes of dairy cows. Albumin can be used to control blood pressure. At present, this protein must be separated from large quantities of donated human blood. It would be easier to get quantities of the protein from bountiful supplies of milk.

Given how rapidly the technologies are developing, is the genetic engineering and cloning of humans not far behind? We return to this thought in Section 16.10.

Mapping and Using the Human Genome

In the late 1980s, biologists were in an uproar over a costly proposal before the National Institutes of Health (NIH) to map the entire human genome. Many argued that benefits for medicine and pure research would be incalculable. Others said the mapping couldn't be done and would divert funding from "worthy" endeavors. But then the molecular biologist Leroy Hood invented automated DNA sequencing, and the race was on.

By 1990, the Human Genome Initiative was under way. The international effort came with a projected price tag of 3 billion dollars—about 92 cents for each base pair in the heritable script for human life. By 1991, not even 2,000 genes had been sequenced. The pace picked up after Craig Venter realized even a bit of cDNA could be used as a molecular hook to drag the whole sequence of its parent gene out of a cDNA library. He named the hooks ESTs (for *E*xpressed *S*equence *T*ags).

Later, Venter and Hamilton Smith used a software program, the TIGR Assembler, to decipher the first full genome of an organism—the bacterium *H. influenzae*. In 2000, researchers of the Human Genome Initiative along with Venter's team jointly announced that they had completed a rough draft of the human genome. Its extensive noncoding portions are now being analyzed.

Today, studies of the genome of humans and other organisms are now grouped together as a new research

Figure 16.14 Evidence of a successful gene transfer. Two ten-week-old mouse littermates. *Left:* This one weighs 29 grams. *Right:* This one weighs 44 grams. It grew from a fertilized egg into which a gene for human somatotropin had been inserted.

field: **genomics**. The branch called *structural* genomics is concerned with actual mapping and sequencing of genomes of individuals. The branch called *comparative* genomics is more concerned with possible evolutionary relationships of groups of organisms. The similarities and differences being discovered among the different genomes are analyzed.

Comparative genomics has practical applications as well as potential for research. The starting premise is that the genomes of all existing organisms are derived from common ancestral ones. For example, pathogens share some conserved genes with their human hosts, although their ancestors diverged long ago from the lineage that led to humans. By comparing the shared gene sequences, how those sequences are organized, and where they have come to differ, it may be possible to advance understanding of where immune defenses against pathogens are strongest and where they are the most vulnerable. In such ways, it is helpful to think of genomes as the keys to ancient molecular locks that became more and more complicated over time.

Genomics obviously has potential for human gene therapy. However, now that the human genome is fully sequenced, it still is not easy to manipulate it. Genetic researchers have to insert a modified gene into a host cell of a particular tissue. They have to be sure the gene gets inserted at a suitable site in a given chromosome. They must also make sure that the targeted type of cell will synthesize the specified protein at suitable times, in suitable amounts.

Today, most experimenters employ stripped-down viruses as vectors that put genes into cultured human cells. They know cells are able to incorporate foreign genes into their DNA. But viral genetic material can undergo rearrangements, deletions, and other changes that can shut down or disrupt gene expression. What about developing synthetic, streamlined versions of human chromosomes? Maybe the machinery for DNA replication and protein synthesis will work on them.

Some gene therapies simply put modified cells into a tissue. This may help, even if the cells only make 10 to 15 percent of a required protein. But no one can yet predict where the genes will end up. The danger is that the insertion will disrupt the function of other genes, including those controlling cell growth and division. One-for-one gene swaps by homologous recombination are possible. Oliver Smithies, one of the best at using this process, can put genes right where they should go, but only once every 100,000 or so tries.

Although the technical details are still being worked out, modified genes are being transferred into cells of humans and other mammals during experimental and clinical trials.

SAFETY ISSUES

Many years have passed since foreign DNA was first transferred into a plasmid. That gene transfer ignited a debate that will continue well into the next century. The issue is this: *Do potential benefits of gene modifications and gene transfers outweigh potential dangers?*

Genetically engineered bacteria are "designed" so they cannot survive except in the laboratory. As added precautions, "fail-safe" genes are built into the foreign DNA in case they escape. These genes are silent unless the captives are exposed to environmental conditions— whereupon the genes get activated, with lethal results for their owner. Say the package includes a *hok* gene next to a promoter of the lactose operon (Section 14.2). Sugars are plentiful in the environment. If they were to activate the *hok* gene, the protein product of that gene would destroy membrane function and the wayward cell.

What about a worst-case scenario? Remember how retroviruses are used to insert genes into cultured cells? If they escape from laboratory isolation, what might be the consequences? For instance, check out Section 47.9.

And what about genetically engineered plants and animals released into the environment? For example, Steven Lindlow thought about how frost destroys many crops. Knowing that a surface protein of a bacterium promotes formation of ice crystals, he excised the "ice-forming" gene from bacterial cells. As he hypothesized, spraying "ice-minus bacteria" on strawberry plants in an isolated field prior to a frost would help plants resist freezing. He actually had deleted a *harmful* gene from a species, yet a bitter legal battle ensued. The courts ruled in his favor. His coworkers sprayed a strawberry patch; nothing bad happened.

Then there was one potato plant designed to kill the insects that attack it. It also was too toxic for people to eat. Or think of how crop plants compete poorly with weeds for nutrients. Many have been designed to resist weedkillers so farmers can spray for weeds and not worry about killing their crops. Some of the herbicides do have toxic effects on more than their targets. If crop plants offer herbicide resistance, will farmers be less or more apt to spread them about?

And what if engineered plants or animals transfer modified genes to organisms in the wild? Think of how the advantage in acquiring resources would tilt toward vigorous weeds blessed with herbicide-resistant genes. Such possibilities are why standards for rigorous and extended safety tests are in place *before* the modified organisms enter the environment.

For more on safety issues, turn to *Critical Thinking* question 1 at the chapter's end.

Rigorous safety tests are carried out before genetically modified organisms are released into the environment.

BIOTECHNOLOGY IN A BRAVE NEW WORLD

Before you leave this unit of the book, reflect on what you have learned so far. You started out by examining cell division mechanisms, the means by which parents pass on their DNA to each new generation. You then moved on to the chromosomal and molecular basis of inheritance. You continued with a glimpse into the gene controls that guide the continuation of life from one generation to the next. The sequence you followed parallels the history of genetics research.

And now, with this chapter, you have arrived at a point in time when molecular geneticists hold keys to the kingdom of inheritance. They are now unlocking and changing genomes at a stunning pace. They are making **DNA microarrays**, or gene chips, each with thousands of DNA sequences from a genome stamped onto a glass plate the size of a business card. Within days instead of years, researchers can now find out which genes are silent and which are being expressed in a tissue, pinpoint mutations, track host–microbe interactions, diagnose genetic diseases, and see how drugs or therapies influence gene expression.

Each gene chip is bathed in a solution containing labeled probes. These probes are RNA transcripts that were extracted from the cells being studied—say, cancer cells from a patient. The probes bind only to complementary base sequences on the gene chip and make them glow in fluorescent light. Analysis of the glowing spots on the chip can reveal which of the thousands of genes in the cells are inactive and which are switched on. This kind of molecular knowledge is about to revolutionize how we diagnose, treat, and prevent the many diverse forms of cancer.

And we as a society are working our way through bioethical aspects of such astonishing research even as it is swirling past us.

Who Gets Well?

Think about the potential of gene therapy. The historic case mentioned at the start of this chapter was the first proof that medical applications of genetic engineering might alleviate suffering and save lives. More proof comes from two male infants with a severe combined immune deficiency called *SCID-X1*. With this genetic disorder, having two copies of a certain mutant allele disables the immune system. The infants were living as "bubble boys," in germ-free isolation tents, because they could not fight infections. Then geneticists used a viral vector to insert copies of the nonmutated gene stem cells from the boys' bone marrow. They infused the genetically modified cells into the marrow. Months later, both boys left their isolation tents. For the past year, at least, their immune system is working as it should. The boys are healthy and living at home.

We already have identified more than 15,500 genetic disorders. Should we dismiss them simply because they are rare in the population at large? Whose lives will they touch? In any given year, genetic disorders affect 3 to 5 percent of all newborns and underlie 20 to 30 percent of all infant deaths. They account for 50 percent of all people who are mentally impaired and close to 25 percent of all hospital admissions. And what of the age-related disorders that await all of us? Critics who would like to put a lid on genetic research would do well to put a face on those among us who are badly and painfully bent or broken.

Who Gets Enhanced?

To many of us, human gene therapy to correct genetic disorders seems like a socially acceptable goal. Now take this idea one step further. Is it socially desirable or even acceptable to change some genes of a normal human (or sperm or egg) to alter or enhance traits?

The idea of being able to select desirable human traits is called *eugenic engineering*. Yet who decides which forms of a trait are most "desirable"? Would it be desirable to engineer taller or blue-eyed or fair-skinned boys and girls? Would it be okay to engineer "superhumans" with amazing strength or intelligence?

For a survey conducted not long ago, more than 40 percent of the Americans interviewed said that it would be okay to use gene therapy to make smarter or better looking babies. One poll of British parents found 18 percent willing to use genetic enhancement to prevent children from being aggressive and 10 percent willing to keep them from growing up to be homosexual.

Through gene transfers, J. Z. Tsien and others have produced mice with enhanced memory and learning abilities. Maybe their work heralds help for those who have Alzheimer's disease or dementia, perhaps even for those who just want to have more brain power.

Geneticist Dean Hamer even predicts we will soon be tinkering with genes in order to change forms of behavior held to be socially undesirable. Violent forms of aggression and drug addictions come to mind.

Send in the Clones? Don't Bother, They're Here

Each year, about 75,000 people are on waiting lists for an organ transplant from a human donor, but donors are in short supply. There is talk of harvesting organs from pigs, which function very much like organs from humans do. Transferring an organ from one species to another is called **xenotransplantation**.

But the human immune system battles anything that it recognizes as "nonself," and it would reject an unmodified pig organ at once. At the free surface of cells of a pig organ's blood vessels is a certain sugar component of a glycoprotein. Antibodies circulating in human blood would bind at once to the sugar. Binding would doom the transplant. It would trigger a cascade of reactions that would, within hours, cause massive coagulation inside the vessels (Sections 11.4 and 34.4). Potent drugs can suppress the immune response, but they make the organ recipient vulnerable to infections.

Pig DNA contains two copies of *Ggta1*, the gene for alpha-1,3-galactosyltransferase. This enzyme catalyzes a key step in the biosynthesis of the sugar that human antibodies latch on to. Knowing this, biotechnologists worked to knock the two copies of the gene out of pig DNA. Recently, they did succeed in excising one of the copies. They went on to transfer "knockout cells" (those with the excised gene) into enucleated pig eggs, in the manner described in Section 13.5.

Some eggs developed into embryos, which were implanted in a host sow. The embryos developed into a clone of piglets, each without one copy of the *Ggta1* gene. Two piglets died shortly afterward; another died a few weeks later. The survivors appear to be healthy even though one piglet has a defective eye and small ear flaps. Biotechnologists plan to breed the cloned pigs by conventional methods to produce offspring in which both copies of the *Ggta1* gene are knocked out. (Remember your Mendelian genetics? If both parents lack one copy of the *Ggta1* gene, there is one chance in four that one of their offspring will have no copies.)

A potential problem for xenotransplantation is that pigs carry a virus similar to the one that causes AIDS in humans. For the experiments just described, the researchers used miniature pigs, which apparently cannot transmit the virus to human cells.

Pigs are mammals. So are humans. Is human cloning next? Are there people willing to fund research that might yield a clone of a child, either "as is" or with genetically engineered "enhancements"? Will a human female be able to reproduce copies of herself without involving men? Will men be able to pay to have their DNA inserted into a cell stripped of its nucleus, have the cell implanted in a surrogate mother, and make a little repeat of themselves?

Lee Silver, a biologist at Princeton University, has suggested that anyone who thinks human cloning will move slowly is naive. One couple has already pledged up to 500 million dollars to a controversial company to clone their dead infant. The company says it has lots of potential customers and surrogate mothers lined up.

At this writing, several countries have now banned human cloning, but the bans leave room for cloning technology with uses in basic research. Nonscientists and scientists alike are actively debating whether any form of human cloning should be allowed.

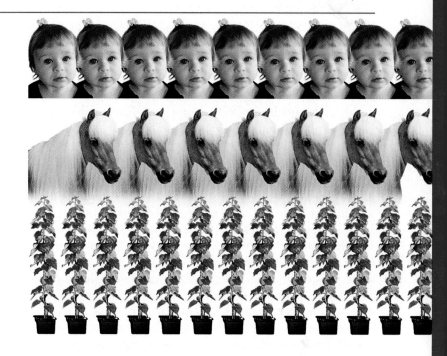

Weighing the Benefits and Risks

Some say that the DNA of any organism must never be altered. Put aside the fact that nature itself has been altering DNA ever since the origin of life. The concern is that we as a species simply do not have the wisdom to bring about beneficial changes without causing irreparable harm to ourselves or to the environment.

To be sure, when it comes to altering human genes, one is reminded of our very human tendency to leap before we look. And yet, when it comes to restricting genetic modifications of any sort, one also is reminded of another old saying: "If God had wanted us to fly, he would have given us wings." Something about the human experience did give us a capacity to imagine wings of our own making—and that capacity carried us to the frontiers of space.

Where are we going from here? To gain perspective on our future, spend some time reading about our past. Ours is a history of survival in the face of challenges, threats, bumblings, and sometimes disasters on a grand scale. It is also a story of our connectedness with the environment and with one another.

The basic questions now confronting you are these: Should we be more cautious, believing the risk takers may go too far? And what do we as a species stand to lose if risks are not taken? There are no simple answers. But weigh the options when you consider arguments for and against genetically engineered food, stem cell research, and other issues in biotechnology, as you will be doing later in this book.

Our ability to manipulate genomes is outpacing attempts to think through some of its bioethical implications.

SUMMARY   *Gold* indicates text section

1. Uncountable numbers of gene mutations and other forms of genetic "experiments" have been proceeding in nature for at least 3 billion years. *CI*

2. Through artificial selection practices, humans have been manipulating the genetic character of a great many species for at least 11,000 years. Recombinant DNA technology enormously expands our capacity to genetically modify organisms. *CI*

3. In genetic engineering, specific genes are modified and inserted into the same organism or a different one. In gene therapy, copies of normal or modified genes are inserted into individuals to correct a genetic defect or boost resistance to disease. *CI, 16.9, 16.10*

4. A genome is all the DNA in the haploid chromosome number for a species. By recombinant DNA technology, a genome can be cut into fragments, then the fragments can be amplified (copied over and over again) to make useful quantities that permit analysis of the nucleotide sequence of the genome or a specific portion of it. *16.1*

5. Researchers work with chromosomal DNA cut by restriction enzymes or with cDNA. (A cDNA strand is transcribed by the enzyme reverse transcriptase from a mature mRNA transcript). Chromosomal DNA is better for questions about DNA regions that control gene expression or that contain introns. cDNA is better for questions about the amino acid sequence of the protein of interest. *16.1*

6. The use of plasmids or other cloning vectors is one way to amplify DNA fragments. Many bacteria contain plasmids: small circles of DNA with a few genes in addition to those of the bacterial chromosome. *16.1*

   a. Certain restriction enzymes make staggered cuts that leave the fragments with single-stranded tails. Such tails base-pair with complementary tails of any other DNA cut by the same enzyme, such as plasmid DNA.

   b. DNA ligase, a modification enzyme, seals base-pairing sites between plasmid DNA and foreign DNA. A plasmid modified to accept foreign DNA is a cloning vector; it can deliver foreign DNA into a bacterium or some other cell that can start a population of rapidly dividing descendant cells. All of the descendants have identical copies of the foreign DNA. Collectively, all of the identical copies are a DNA clone.

7. Currently, the polymerase chain reaction (PCR) is the fastest way to amplify fragments of chromosomal DNA or cDNA. Bacterial factories are not needed; the reactions occur in test tubes. The reactions use a supply of nucleotide building blocks and primers: synthetic, short nucleotide sequences that will base-pair with any complementary DNA sequence and that enzymes (DNA polymerases) recognize to start replication. Each round of replication doubles the number of fragments. *16.2*

8. For sexually reproducing species, no two individuals have exactly the same DNA base sequence (except for identical twins). When restriction enzymes are used to cut an individual's DNA, the result is a unique array of restriction fragments called a DNA fingerprint. *16.3*

   a. The DNA in all humans is more than 99 percent identical except at tandem repeats (short stretches of repeated base sequences, such as TTTC). The number and combination of tandem repeats is unique in each individual and can be detected by gel electrophoresis.

   b. DNA fingerprinting has uses in forensic science, as in resolving crimes and paternity suits.

9. Automated DNA sequencing can rapidly reveal the base sequence of cloned DNA or PCR-amplified DNA fragments. It labels fragments with one of four modified nucleotides, then separates them according to length by gel electrophoresis. Each label fluoresces a certain color under a laser beam. A machine reads each as it peels off the gel and assembles the whole sequence. *16.4*

10. A gene library is a mixed collection of bacterial cells that took up different cloned DNA or cDNA fragments. A gene may be isolated from the library with use of a probe, a very short stretch of radioactively labeled DNA that is known or suspected to be similar to or identical with part of that gene and can base-pair with it. A base pairing between nucleotide sequences from different sources is called nucleic acid hybridization. *16.5*

11. Generally, recombinant DNA technology and genetic engineering have potential for research and applications in medicine and agriculture, in the home and industry. As with any new technology, the potential benefits must be weighed against potential risks, including ecological and social repercussions. *CI, 16.9–16.10*

Review Questions

1. Distinguish between recombinant DNA technology and genetic engineering. *CI*

2. In the following diagram, which restriction enzyme made the cuts (indicated in *red*) in part of a DNA molecule from two different organisms? *16.1*

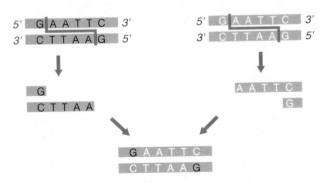

3. Distinguish these terms from one another: *16.1*
   a. chromosomal (genomic) DNA and cDNA
   b. cloning vector and DNA clone

4. Define PCR. Can fragments of chromosomal DNA, cDNA, or both be amplified by PCR? *16.2*

5. Define DNA fingerprinting. Briefly describe which portions of the DNA are used in DNA fingerprinting. *16.3*

6. Outline the steps of automated DNA sequencing. *16.4*

7. Define cDNA library, then briefly explain how a gene can be isolated from it. Define probe and nucleic acid hybridization as part of your answer. *16.5*

8. Give three examples of applications that can be derived from knowledge of an organism's genome. *CI, 16.6–16.8*

9. Name one of the ways in which modified genes have been inserted into mammalian cells. *16.8*

10. Define gene therapy. Once the human genome has been fully sequenced, why will it be difficult to manipulate its genes to advantage? *CI, 16.8*

Self-Quiz ANSWERS IN APPENDIX III

1. _____ is the transfer of normal genes into body cells to correct a genetic defect.
   a. Reverse transcription
   b. Nucleic acid hybridization
   c. Gene mutation
   d. Gene therapy

2. DNA fragments result when _____ cut DNA molecules at specific sites.
   a. DNA polymerases
   b. DNA probes
   c. restriction enzymes
   d. RFLPs

3. Fill in the blank: _____ are small circles of bacterial DNA that are separate from the circular bacterial chromosome.

4. Foreign DNA that was inserted into a plasmid and then replicated many times in a population of bacteria is a _____ .
   a. DNA clone
   b. gene library
   c. DNA probe
   d. gene map

5. By reverse transcription, _____ is assembled on _____ .
   a. mRNA; DNA
   b. cDNA; mRNA
   c. DNA; enzymes
   d. DNA; agar

6. PCR stands for _____ .
   a. polymerase chain reaction
   b. polyploid chromosome restrictions
   c. polygraphed criminal rating
   d. politically correct research

7. By gel electrophoresis, fragments of a gene library can be separated according to _____ .
   a. shape   b. length   c. species

8. Automated DNA sequencing relies on _____ .
   a. supplies of standard and labeled nucleotides
   b. primers and DNA polymerases
   c. gel electrophoresis and a laser beam
   d. all of the above

9. Match the terms with the most suitable description.
   _____ DNA fingerprint
   _____ Ti plasmid
   _____ nature's genetic experiments
   _____ nucleic acid hybridization
   _____ Human Genome Initiative
   _____ eugenic engineering

   a. selecting "desirable" traits
   b. deciphering 3.2 billion base pairs of 23 human chromosomes
   c. used in some gene transfers
   d. unique array of DNA fragments inherited in Mendelian pattern from each of two parents
   e. base pairing of nucleotide sequences from different DNA or RNA sources
   f. mutations, crossovers

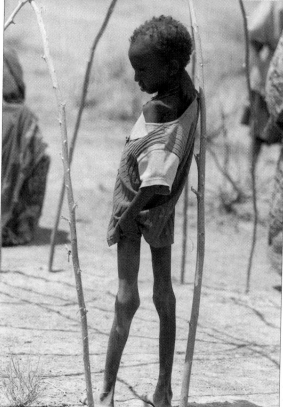

Figure 16.15 *Above:* Activists ripping some genetically modified crop plants from an experimental field in Great Britain. *Below:* Malnourished child in southeastern Ethiopia.

Critical Thinking

1. Avoiding genetically engineered food is probably impossible in the United States. At least 45 percent of cotton, 38 percent of soybean, and 25 percent of corn crops have been engineered to withstand weedkillers or to make their own pesticides. For years, modified corn and soybeans have found their way into breakfast cereals, tofu, soy sauce, vegetable oils, beer, soft drinks, and other food products. They are fed to farm animals.

By contrast, public resistance to genetically engineered food is high in Europe, especially Britain. Many people, including the Prince of Wales, speak out against what the tabloids call "Frankenfood." Protesters routinely vandalize crops (Figure 16.15). Worries abound that such foods may be more toxic, have lower nutritional value, and promote natural selection for antibiotic resistance. Some people worry that designer plants will cross-pollinate with wild plants to produce "superweeds."

Biotechnologists envision a new Green Revolution. They argue that designer plants can hold down food production costs, reduce dependence on pesticides and herbicides, enhance crop yields, and offer improved flavor, nutritional value, even salt tolerance and drought tolerance.

Yet the chorus of critics in Europe may provoke a trade war with the United States. The issue is not small potatoes, so to speak. In 1998, the value of agricultural exports reached about 50 billion dollars. Flattery, threats, and bullying are rampant on both sides of the Atlantic. Restrictions on genetic engineering will have profound impact on United States agriculture, and inevitably the impact will trickle down to what you eat and how much you pay for it.

All of which invites you to read up on scientific research related to this issue and form your own opinions. The alternatives are to be swayed either by media hype (the term Frankenfood, for instance) or by biased reports from groups (such as chemical manufacturers) with their own agendas.

Possibly start with Christopher Bond's article "Politics, Misinformation, and Biotechnology" in the 18 February 2000 *Science* (287:1201). Bond argues that biotechnology attempts to solve real-world problems of sickness, hunger, and dwindling resources; and that "hysteria and unworkable propositions advanced by those who can afford to take their next meal for granted have little currency among those who are hungry." For example, think about the body masses of the individuals in the two photographs in Figure 16.15.

2. Lunardi's Market put out a bin of tomatoes having splendid vine-ripened redness, flavor, and texture. The sign posted above the bin identified them as genetically engineered produce. Most shoppers selected unmodified tomatoes in the adjacent bin even though those tomatoes were pale pink, mealy-textured, and tasteless. Which ones would you pick? Why?

3. Ryan, a forensic scientist, obtained a very small bit of DNA from material at a crime scene. In order to examine the sample by DNA fingerprinting, he must amplify the sample by PCR, the polymerase chain reaction. By his estimate, there are 50,000 copies of the DNA in his sample. Calculate the number of copies Ryan will have after fifteen cycles of PCR.

4. A game warden in Africa confiscated eight ivory tusks from elephants. Some tissue is still attached to the tusks. Now she must determine whether the tusks were taken illegally from northern populations of endangered elephants or from other populations of elephants to the south that can be hunted legally. How can she use DNA fingerprinting to find the answer?

5. The Human Genome Initiative is undergoing completion, and knowledge about a number of the newly discovered genes is already being used to detect genetic disorders. Ask yourself: What will be done with genetic information about individuals? Will insurance companies and potential employers request it? At this writing, many women have already refused to take advantage of genetic screening for a gene associated with the development of breast cancer. Should medical records about people participating in genetic research and genetic clinical services be made available to other individuals? If not, how could such information be protected?

6. Animal rights activists are up in arms over the possibility of raising pigs as organ sources for xenotransplantation. Are you for or against this possibility? Why or why not?

7. What if it were possible to create life in test tubes? This is the question behind attempts to model and eventually create *minimal organisms*, which we define as living cells having the smallest possible set of genes that are necessary to survive and reproduce.

As recent experiments by Craig Venter and Claire Fraser revealed, *Mycoplasma genitalium*, a bacterium with 517 genes (and 2,209 transposons) might be a candidate. By disabling the genes one at a time in the laboratory, they discovered that this bacterium may contain no more than 265–350 essential protein-coding genes.

What if the genes were to be synthesized one at a time and inserted into an engineered cell consisting only of a plasma membrane and cytoplasm? Would the cell come to life?

The possibility that it might prompted Venter and Fraser to seek advice from a panel of bioethicists and theologians. As Arthur Caplan, a bioethicist at the University of Pennsylvania reported, no one on the panel objected to synthetic life research. They felt that much good might come of it, provided scientists didn't claim to have found "the secret of life."

The 10 December 1999 issue of *Science* includes an essay from the panel and an article on *M. genitalium* research. Read both, then write down your thoughts about creating life in a test tube.

Selected Key Terms

automated DNA sequencing *16.4*	genomics *16.8*
cDNA *16.1*	nucleic acid hybridization *16.5*
cloning vector *16.1*	PCR *16.2*
DNA clone *16.1*	plasmid *16.1*
DNA fingerprint *16.3*	primer *16.2*
DNA ligase *16.1*	probe (nucleic acid) *16.5*
DNA microarray *16.10*	recombinant DNA technology *CI*
gel electrophoresis *16.3*	restriction enzyme *16.1*
gene library *16.5*	reverse transcriptase *16.1*
gene therapy *CI*	seed bank *16.7*
genetic engineering *CI*	tandem repeats *16.3*
genome *16.1*	xenotransplantation *16.10*

Readings

Cho, M., et al. 10 December 1999. "Ethical Considerations in Synthesizing a Minimal Genome." *Science* 286: 2087–2090.

"The Human Genome." *Science* 291: 5507. The entire 16 February 2001 issue consolidates current understandings and questions.

Lai, L., et al. Published online 3 January 2002. "Production of 1,3-Galactosyltransferase Knockout Pigs by Nuclear Transfer Cloning," *Science*. For interested and fearless students who want to get right into cutting-edge research.

Pennisi, E. June 2000. "Finally, The Book of Life and Instructions for Navigating It." *Science* 288: 5475.

Watson, J. D., et al. 1992. *Recombinant DNA*. Second edition. New York: Scientific American Books.

On-Line readings at Student Guide for InfoTrac:
www.brookscole.com/biology

APPENDIX I.   CLASSIFICATION SYSTEM

This revised classification scheme is a composite of several that microbiologists, botanists, and zoologists use. The major groupings are agreed upon, more or less. However, there is not always agreement on what to name a particular grouping or where it might fit within the overall hierarchy. There are several reasons why full consensus is not possible at this time.

First, the fossil record varies in its completeness and quality (Section 19.1). As one outcome, the phylogenetic relationship of one group to other groups is sometimes open to interpretation. Today, comparative studies at the molecular level are firming up the picture, but the work is still under way.

Second, ever since the time of Linnaeus, systems of classification have been based on the perceived morphological similarities and differences among organisms. Although some original interpretations are now open to question, we are so used to thinking about organisms in certain ways that reclassification often proceeds slowly.

A few examples: Traditionally, birds and reptiles were grouped in separate classes (Reptilia and Aves); yet there are many compelling arguments for grouping the lizards and snakes in one class and the crocodilians, dinosaurs, and birds in a separate class. Some biologists have favored a six-kingdom system of classification (archaebacteria, eubacteria, protistans, plants, fungi, and animals). Many others now favor a three-domain classification system. The archaebacteria, eubacteria, and eukaryotes (alternatively, the archaea, bacteria, and eukarya) are its major groupings.

Third, researchers in microbiology, mycology, botany, zoology, and other fields of inquiry inherited a wealth of literature, based on classification systems that have been developed over time in each field of inquiry. Many simply do not wish to give up established terminology that offers access to the past.

For example, botanists and microbiologists often use *division*, and zoologists *phylum*, for taxa that actually are equivalent in hierarchies of classification. As another example, opinions are quite polarized with respect to

kingdom Protista, certain members of which could easily be grouped with plants, or fungi, or animals. Indeed, the term "protozoan" is a holdover from an earlier scheme in which some single-celled organisms were ranked as simple animals.

Given the problems, why do systematists work so hard to construct hypotheses regarding the history of life? They do so because classification systems that accurately reflect evolutionary relationships are more useful, with far more predictive power for comparative studies. Such systems also serve as a good framework for studying the living world, which could otherwise be an overwhelming body of knowledge. Importantly, classifications enhance the retrieval of information that relates to living organisms.

Bear in mind, *we include this appendix on classification for your reference purposes only*. Besides being open to revision, it is by no means complete. Names in "quotes" are polyphyletic or paraphyletic groups undergoing revision. The groupings of certain animal phyla reflect common ancestry (Nematoda through Arthropoda, and Nemertea through Annelida), as emerging molecular data suggest. Also, the most recently discovered species, as from the mid-ocean province, are not listed. Many existing and extinct species of the more obscure phyla are not represented. Our strategy is to focus primarily on the organisms mentioned in the text.

PROKARYOTES AND EUKARYOTES COMPARED

As a general frame of reference, note that almost all eubacteria and archaebacteria are microscopic in size. Their DNA is concentrated in a nucleoid (a region of cytoplasm), not in a membrane-bound nucleus. All are single cells or simple associations of cells. They reproduce by prokaryotic fission or budding; they transfer genes by bacterial conjugation.

Table A lists representative types of autotrophic and heterotrophic prokaryotes. The authoritative reference, *Bergey's Manual of Systematic Bacteriology,* has called this a time of taxonomic transition. It references groups mainly by numerical taxonomy (Section 21.3) rather than by phylogeny. Our classification system does reflect evidence of evolutionary relationships for at least some bacterial groups.

The first life forms were prokaryotic. Similarities between Eubacteria and Archaebacteria have more ancient origins relative to the traits of eukaryotes.

Unlike the prokaryotes, all eukaryotic cells start out life with a DNA-enclosing nucleus and other membrane-bound organelles. Their chromosomes have many histones and other proteins attached. They include spectacularly diverse single-celled and multicelled species, which can reproduce by way of meiosis, mitosis, or both.

BACTERIA — EUBACTERIA

ARCHAEA — ARCHAEBACTERIA

EUKARYA — PROTISTA   FUNGI   PLANTAE   ANIMALIA

DOMAIN OF EUBACTERIA (BACTERIA)

KINGDOM EUBACTERIA   Gram-negative and Gram-positive prokaryotic cells. Peptidoglycan in cell wall. Collectively, great metabolic diversity; photosynthetic autotrophs, chemosynthetic autotrophs, and heterotrophs.

PHYLUM FIRMICUTES Typically Gram-positive, thick wall. Heterotrophs. *Bacillus, Staphylococcus, Streptococcus, Clostridium, Actinomycetes.*

PHYLUM GRACILICUTES Typically Gram-negative, thin wall. Autotrophs (photosynthetic and chemosynthetic) and heterotrophs. *Anabaena* and other cyanobacteria. *Escherichia, Pseudomonas, Neisseria, Myxococcus.*

PHYLUM TENERICUTES Gram-negative, wall absent. Heterotrophs (saprobes, pathogens). *Mycoplasma.*

DOMAIN OF ARCHAEBACTERIA (ARCHAEA)

KINGDOM ARCHAEBACTERIA   Methanogens, extreme halophiles, extreme thermophiles. Evolutionarily closer to eukaryotic cells than to eubacteria. All strict anaerobes living in habitats as harsh as those that probably prevailed on the early Earth. Compared with other prokaryotic cells, all archaebacteria have a distinctive cell wall and unique membrane lipids, ribosomes, and RNA sequences. *Methanobacterium, Halobacterium, Sulfolobus.*

Table A   Representative Eubacteria and Archaebacteria Grouped on the Basis of Numerical Taxonomy

Some Major Groups	Main Habitats	Characteristics	Representatives
EUBACTERIA			
Photoautotrophs: Cyanobacteria, green sulfur bacteria, and purple sulfur bacteria	Mostly lakes, ponds; some marine, terrestrial habitats	Photosynthetic; use sunlight energy, carbon dioxide; cyanobacteria use oxygen-producing noncyclic pathway; some also use cyclic route	*Anabaena, Nostoc, Rhodopseudomonas, Chloroflexus*
Photoheterotrophs: Purple nonsulfur and green nonsulfur bacteria	Anaerobic, organically rich muddy soils, and sediments of aquatic habitats	Use sunlight energy; organic compounds as electron donors; some purple nonsulfur may also grow chemotrophically	*Rhodospirillum, Chlorobium*
Chemoautotrophs: Nitrifying, sulfur-oxidizing, and iron-oxidizing bacteria	Soil; freshwater, marine habitats	Use carbon dioxide, inorganic compounds as electron donors; influence crop yields, cycling of nutrients in ecosystems	*Nitrosomonas, Nitrobacter, Thiobacillus*
Chemoheterotrophs: Spirochetes	Aquatic habitats; parasites of animals	Helically coiled, motile; free-living and parasitic species; some major pathogens	*Spirochaeta, Treponema*
Gram-negative aerobic rods and cocci	Soil, aquatic habitats; parasites of animals, plants	Some major pathogens; some fix nitrogen (e.g., *Rhizobium*)	*Pseudomonas, Neisseria, Rhizobium, Agrobacterium*
Gram-negative facultative anaerobic rods	Soil, plants, animal gut	Many major pathogens; one bioluminescent (*Photobacterium*)	*Salmonella, Escherichia, Proteus, Photobacterium*
Rickettsias and chlamydias	Host cells of animals	Intracellular parasites; many pathogens	*Rickettsia, Chlamydia*
Myxobacteria	Decaying organic material; bark of living trees	Gliding, rod-shaped; aggregation and collective migration of cells	*Myxococcus*
Gram-positive cocci	Soil; skin and mucous membranes of animals	Some major pathogens	*Staphylococcus, Streptococcus*
Endospore-forming rods and cocci	Soil; animal gut	Some major pathogens	*Bacillus, Clostridium*
Gram-positive nonsporulating rods	Fermenting plant, animal material; gut, vaginal tract	Some important in dairy industry, others major contaminators of milk, cheese	*Lactobacillus, Listeria*
Actinomycetes	Soil; some aquatic habitats	Include anaerobes and strict aerobes; major producers of antibiotics	*Actinomyces, Streptomyces*
ARCHAEBACTERIA (ARCHAEA)			
Methanogens	Anaerobic sediments of lakes, swamps; animal gut	Chemosynthetic; methane producers; used in sewage treatment facilities	*Methanobacterium*
Extreme halophiles	Brines (extremely salty water)	Heterotrophic; also, unique photosynthetic pigments (bacteriorhodopsin) form in some	*Halobacterium*
Extreme thermophiles	Acidic soil, hot springs, hydrothermal vents	Heterotrophic or chemosynthetic; use inorganic substances as electron donors	*Sulfolobus, Thermoplasma*

DOMAIN OF EUKARYOTES (EUKARYA)

Diverse single-celled, colonial, and multicelled eukaryotic species. Existing types are unlike prokaryotes and most like the earliest forms of eukaryotes. Autotrophs, heterotrophs, or both (Table 22.1). Reproduce sexually and asexually (by meiosis, mitosis, or both). Not a monophyletic group. The kingdom may soon be split into multiple kingdoms, and some of its groups are already being reclassified as plants or fungi.

PHYLUM "MASTIGOPHORA" Flagellated protozoans. Free-living heterotrophs; many are internal parasites. They have one to several flagella. At present, a non-monophyletic grouping of ancient lineages, including the diplomonads, parabasalids, and kinetoplastids. *Trypanosoma, Trichomonas, Giardia.*

PHYLUM EUGLENOPHYTA Euglenoids. Mostly heterotrophs, some photoautotrophs, some both depending on conditions. Most with one short, one long flagellum. Pigmented (red, green) or colorless. Related to kinetoplastids. *Euglena.*

PHYLUM SARCODINA Amoeboid protozoans. Heterotrophs, free-living or endosymbionts, some pathogens. Soft-bodied, with or without shell, pseudopods. Rhizopods (naked amoebas, foraminiferans), actinopods (radiolarians, heliozoans). *Amoeba.*

ALVEOLATES

PHYLUM CILIOPHORA Ciliated protozoans. Heterotrophs, predators or symbionts, some parasitic. All have cilia. Free-living, sessile, or motile. *Paramecium, Didinium,* hypotrichs.

PHYLUM APICOMPLEXA Heterotrophs, sporozoite-forming parasites. Complex structures at head end. Most familiar types known as sporozoans. *Cryptosporidium, Plasmodium, Toxoplasma.*

PHYLUM PYRRHOPHYTA Dinoflagellates. Photosynthetic, mostly, but some heterotrophs. *Pfiesteria, Gymnodinium breve.*

STRAMENOPILES

PHYLUM OOMYCOTA. Water molds. Heterotrophs. Decomposers, some parasites. *Saprolegnia, Phytophthora, Plasmopara.*

PHYLUM CHRYSOPHYTA. Golden algae, yellow-green algae, diatoms. Photosynthetic. Some flagellated. *Mischococcus, Synura, Vaucheria.*

PHYLUM PHAEOPHYTA. Brown algae. Photosynthetic, nearly all endemic to temperate or marine waters. *Macrocystis, Fucus, Sargassum, Ectocarpus, Postelsia.*

GROUPS CLOSELY RELATED TO PLANTS

PHYLUM CHLOROPHYTA Green algae. Mostly photosynthetic, some parasitic. Most freshwater, some marine or terrestrial. *Chlamydomonas, Spirogyra, Ulva, Volvox, Codium, Halimeda.*

PHYLUM RHODOPHYTA Red algae. Mostly photosynthetic, some parasitic. Nearly all marine, some in freshwater habitats. *Porphyra, Bonnemaisonia, Euchema.*

PHYLUM CHAROPHYTA Stoneworts. *Chara.*

GROUPS CLOSELY RELATED TO FUNGI

PHYLUM CHYTRIDIOMYCOTA Chytrids. Heterotrophs; saprobic decomposers or parasites. *Chytridium.*

GROUPS OF SLIME MOLDS

PHYLUM ACRASIOMYCOTA Cellular slime molds. Heterotrophs with free-living, phagocytic amoeboid cells and spore-bearing stages. *Dictyostelium.*

PHYLUM MYXOMYCOTA Plasmodial slime molds. Heterotrophs with free-living, phagocytic amoeboid cells and spore-bearing stages. Aggregate into streaming mass of cells that discard their plasma membrane. *Physarum.*

Multicelled eukaryotes. Nearly all photosynthetic autotrophs with chlorophylls *a* and *b*. Some parasitic. Nonvascular and vascular species, generally with well-developed root and shoot systems. Nearly all adapted to survive dry conditions on land; a few in aquatic habitats. Sexual reproduction predominant with spore-forming chambers and embryos in life cycle; also asexual reproduction by vegetative propagation and other mechanisms.

PHYLUM "BRYOPHYTA" Bryophytes; mosses, liverworts, hornworts. Not a monophyletic group. Seedless, nonvascular, haploid dominance. *Marchantia, Polytrichum, Sphagnum.*

VASCULAR PLANTS

PHYLUM "RHYNIOPHYTA" Earliest known vascular plants; muddy habitats. A polyphyletic group, some are primitive lycophytes. Extinct. *Cooksonia, Rhynia.*

PHYLUM LYCOPHYTA Lycophytes, club mosses. Seedless, vascular. Small leaves, branching rhizomes, vascularized roots and stems. *Lepidodendron* (extinct), *Lycopodium, Selaginella.*

PHYLUM SPHENOPHYTA Horsetails. Seedless, vascular, whorled leaves. Some stems photosynthetic, others nonphotosynthetic, spore-producing. *Calamites* (extinct), *Equisetum.*

PHYLUM CYCADOPHYTA Cycads. Gymnosperm group (vascular, bear "naked" seeds). Tropical, subtropical. Palm-shaped leaves, simple cones on male and female plants. *Zamia.*

PHYLUM PTEROPHYTA. Ferns. Large leaves, usually with sori. Largest group of seedless vascular plants (12,000 species), mainly tropical, temperate habitats. *Pteris, Trichomanes, Cyathea* (tree ferns), *Polystichum.*

PHYLUM PSILOPHYTA Whisk ferns. Seedless, vascular. No obvious roots, leaves on sporophyte, very reduced. *Psilotum.*

PHYLUM "PROGYMNOSPERMOPHYTA" The progymnosperms. Ancestral to early seed-bearing plants; extinct. *Archaeopteris.*

SEED-BEARING PLANTS (A subgroup of vascular plants)

PHYLUM "PTERIDOSPERMOPHYTA" Seed ferns. Fernlike gymnosperms; extinct. *Medullosa.*

PHYLUM CYCADOPHYTA Cycads. Group of gymnosperms (vascular, bear "naked" seeds). Tropical, subtropical. Compound leaves, simple cones on male and female plants. Plants usually palm-like. *Zamia, Cycas.*

PHYLUM GINKGOPHYTA Ginkgo (maidenhair tree). Type of gymnosperm. Seeds with fleshy outer layer. *Ginkgo.*

PHYLUM GNETOPHYTA Gnetophytes. Only gymnosperms with vessels in xylem and double fertilization (but endosperm does not form). *Ephedra, Welwitchia, Gnetum.*

PHYLUM CONIFEROPHYTA Conifers. Most common and familiar gymnosperms. Generally cone-bearing species with needle-like or scale-like leaves.

Family Pinaceae. Pines *(Pinus)*, firs *(Abies)*, spruces *(Picea)*, hemlock *(Tsuga)*, larches *(Larix)*, true cedars *(Cedrus)*.

Family Cupressaceae. Junipers *(Juniperus)*, Cypresses *(Cupressus)*, Bald cypress *(Taxodium)*, redwood *(Sequoia)*, bigtree *(Sequoiadendron)*, dawn redwood *(Metasequoia)*.

Family Taxaceae. Yews. *Taxus.*

PHYLUM ANTHOPHYTA Angiosperms (the flowering plants). Largest, most diverse group of vascular seed-bearing plants. Only organisms that produce flowers, fruits. Some families from several representative orders are listed:

BASAL FAMILIES

Family Amborellaceae. *Amborella.*
Family Nymphaeaceae. Water lilies.
Family Illiciaceae. Star anise.

MAGNOLIIDS

Family Magnoliaceae. Magnolias.
Family Lauraceae. Cinnamon, sassafras, avocados.
Family Piperaceae. Black pepper, white pepper.

EUDICOTS

Family Papaveraceae. Poppies.
Family Cactaceae. Cacti.
Family Euphorbiaceae. Spurges, poinsettia.
Family Salicaceae. Willows, poplars.
Family Fabaceae. Peas, beans, lupines, mesquite.
Family Rosaceae. Roses, apples, almonds, strawberries.
Family Moraceae. Figs, mulberries.
Family Cucurbitaceae. Gourds, melons, cucumbers, squashes.
Family Fagaceae. Oaks, chestnuts, beeches.
Family Brassicaceae. Mustards, cabbages, radishes.
Family Malvaceae. Mallows, okra, cotton, hibiscus, cocoa.
Family Sapindaceae. Soapberry, litchi, maples.
Family Ericaceae. Heaths, blueberries, azaleas.
Family Rubiaceae. Coffee.
Family Lamiaceae. Mints.
Family Solanaceae. Potatoes, eggplant, petunias.
Family Apiaceae. Parsleys, carrots, poison hemlock.
Family Asteraceae. Composites. Chrysanthemums, sunflowers, lettuces, dandelions.

MONOCOTS

Family Araceae. Anthuriums, calla lily, philodendrons.
Family Liliaceae. Lilies, tulips.
Family Alliaceae. Onions, garlic.
Family Iridaceae. Irises, gladioli, crocuses.
Family Orchidaceae. Orchids.
Family Arecaceae. Date palms, coconut palms.
Family Bromeliaceae. Bromeliads, pineapples, Spanish moss.
Family Cyperaceae. Sedges.
Family Poaceae. Grasses, bamboos, corn, wheat, sugarcane.
Family Zingiberaceae. Gingers.

FUNGI   Nearly all multicelled eukaryotic species. Heterotrophs, mostly saprobic decomposers, some parasites. Nutrition based upon extracellular digestion of organic matter and absorption of nutrients by individual cells. Multicelled species form absorptive mycelia within substrates and structures that produce asexual spores (and sometimes sexual spores).

PHYLUM ZYGOMYCOTA   Zygomycetes. Producers of zygospores (zygotes inside thick wall) by way of sexual reproduction. Bread molds, related forms. *Rhizopus, Philobolus.*

PHYLUM ASCOMYCOTA   Ascomycetes. Sac fungi. Sac-shaped cells form sexual spores (ascospores). Most yeasts and molds, morels, truffles. *Saccharomycetes, Morchella, Neurospora, Sarcoscypha, Claviceps, Ophiostoma, Candida, Aspergillus, Penicillium.*

PHYLUM BASIDIOMYCOTA   Basidiomycetes. Club fungi. Most diverse group. Produce basidiospores inside club-shaped structures. Mushrooms, shelf fungi, stinkhorns. *Agaricus, Amanita, Craterellus, Gymnophilus, Puccinia, Ustilago.*

"IMPERFECT FUNGI"   Sexual spores absent or undetected. The group has no formal taxonomic status. If better understood, a given species might be grouped with sac fungi or club fungi. *Arthobotrys, Histoplasma, Microsporum, Verticillium.*

"LICHENS"   Mutualistic interactions between fungal species and a cyanobacterium, green alga, or both. *Lobaria, Usnea, Cladonia.*

KINGDOM ANIMALIA   Multicelled eukaryotes, nearly all with tissues, organs, and organ systems; show motility during at least part of their life cycle; embryos develop through a series of stages. Diverse heterotrophs, predators (herbivores, carnivores, omnivores), parasites, detritivores. Reproduce sexually and, in many species, asexually as well.

PHYLUM PORIFERA   Sponges. No symmetry, tissues. *Euplectella.*

PHYLUM PLACOZOA   Marine. Simplest known animal. Two cell layers, no mouth, no organs. *Trichoplax.*

PHYLUM CNIDARIA   Radial symmetry, tissues, nematocysts.
  Class Hydrozoa. Hydrozoans. *Hydra, Obelia, Physalia, Prya.*
  Class Scyphozoa. Jellyfishes. *Aurelia.*
  Class Anthozoa. Sea anemones, corals. *Telesto.*

PHYLUM MESOZOA   Ciliated, wormlike parasites, about the same level of complexity as *Trichoplax.*

PHYLUM PLATYHELMINTHES   Flatworms. Bilateral, cephalized; simplest animals with organ systems. Saclike gut.
  Class Turbellaria. Triclads (planarians), polyclads. *Dugesia.*
  Class Trematoda. Flukes. *Clonorchis, Schistosoma.*
  Class Cestoda. Tapeworms. *Diphyllobothrium, Taenia.*

PHYLUM ROTIFERA   Rotifers. *Asplancha, Philodina.*

PHYLUM NEMERTEA   Ribbon worms. *Tubulanus.*

PHYLUM MOLLUSCA   Mollusks.
  Class Polyplacophora. Chitons. *Cryptochiton, Tonicella.*
  Class Gastropoda. Snails (periwinkles, whelks, limpets, abalones, cowries, conches, nudibranchs, tree snails, garden snails), sea slugs, land slugs. *Aplysia, Ariolimax, Cypraea, Haliotis, Helix, Liguus, Limax, Littorina, Patella.*
  Class Bivalvia. Clams, mussels, scallops, cockles, oysters, shipworms. *Ensis, Chlamys, Mytelus, Patinopectin.*
  Class Cephalopoda. Squids, octopuses, cuttlefish, nautiluses. *Dosidiscus, Loligo, Nautilus, Octopus, Sepia.*

PHYLUM BRYOZOA   Bryozoans (moss animals).

PHYLUM BRACHIOPODA   Lampshells.

PHYLUM ANNELIDA   Segmented worms.
  Class Polychaeta. Mostly marine worms. *Eunice, Neanthes.*
  Class Oligochaeta. Mostly freshwater and terrestrial worms, many marine. *Lumbricus* (earthworms), *Tubifex.*
  Class Hirudinea. Leeches. *Hirudo, Placobdella.*

PHYLUM NEMATODA   Roundworms. *Ascaris, Caenorhabditis elegans, Necator* (hookworms), *Trichinella.*

PHYLUM TARDIGRADA   Water bears.

PHYLUM ONYCHOPHORA   Onychophorans. *Peripatus.*

PHYLUM ARTHROPODA.
  Subphylum Trilobita. Trilobites; extinct.
  Subphylum Chelicerata. Chelicerates. Horseshoe crabs, spiders, scorpions, ticks, mites.
  Subphylum Crustacea. Shrimps, crayfishes, lobsters, crabs, barnacles, copepods, isopods (sowbugs).
  Subphylum Uniramia.
    Superclass Myriapoda. Centipedes, millipedes.
    Superclass Insecta.
      Order Ephemeroptera. Mayflies.
      Order Odonata. Dragonflies, damselflies.
      Order Orthoptera. Grasshoppers, crickets, katydids.
      Order Dermaptera. Earwigs.

Order Blattodea. Cockroaches.
Order Mantodea. Mantids.
Order Isoptera. Termites.
Order Mallophaga. Biting lice.
Order Anoplura. Sucking lice.
Order Hemiptera. Cicadas, aphids, leafhoppers, spittlebugs, bugs.
Order Coleoptera. Beetles.
Order Diptera. Flies.
Order Mecoptera. Scorpion flies. *Harpobittacus.*
Order Siphonaptera. Fleas.
Order Lepidoptera. Butterflies, moths.
Order Hymenoptera. Wasps, bees, ants.
Order Neuroptera. Lacewings, antlions.

PHYLUM ECHINODERMATA.   Echinoderms.

Class Asteroidea. Sea stars. *Asterias.*
Class Ophiuroidea. Brittle stars.
Class Echinoidea. Sea urchins, heart urchins, sand dollars.
Class Holothuroidea. Sea cucumbers.
Class Crinoidea. Feather stars, sea lilies.
Class Concentricycloidea. Sea daisies.

PHYLUM HEMICHORDATA.   Acorn worms.

PHYLUM CHORDATA.   Chordates.

Subphylum Urochordata. Tunicates, related forms.
Subphylum Cephalochordata. Lancelets.

CRANIATES

Superclass "Agnatha." Jawless fishes, including ostracoderms (extinct).

Class Myxini. Hagfishes.

Class Cephalaspidomorphi. Lampreys.

Subphylum Vertebrata. Jawed vertebrates.

Class "Placodermi." Jawed, heavily armored fishes; extinct.

Class Chondrichthyes. Cartilaginous fishes (sharks, rays, skates, chimaeras).

Class "Osteichthyes." Bony fishes. Not monophyletic.

Subclass Dipnoi. Lungfishes.
Subclass Crossopterygii. Coelacanths, related forms.
Subclass Actinopterygii. Ray-finned fishes.
Order Acipenseriformes. Sturgeons, paddlefishes.
Order Salmoniformes. Salmon, trout.
Order Atheriniformes. Killifishes, guppies.
Order Gasterosteiformes. Seahorses.
Order Perciformes. Perches, wrasses, barracudas, tunas, freshwater bass, mackerels.
Order Lophiiformes. Angler fishes.

TETRAPODS (A subgroup of craniates)

Class Amphibia. Amphibians.
Order Caudata. Salamanders.
Order Anura. Frogs, toads.
Order Apoda. Apodans (caecilians).

AMNIOTES (A subgroup of tetrapods)

Class "Reptilia." Skin with scales, embryo protected and nutritionally supported by extraembryonic membranes.
Subclass Anapsida. Turtles, tortoises.
Subclass Lepidosaura. *Sphenodon*, lizards, snakes.
Subclass Archosaura. Dinosaurs (extinct), crocodiles, alligators.

Class Aves. Birds. In some of the more recent classification systems, dinosaurs, crocodilians, and birds are grouped in the same category, the archosaurs.
Order Struthioniformes. Ostriches.
Order Sphenisciformes. Penguins.
Order Procellariiformes. Albatrosses, petrels.
Order Ciconiiformes. Herons, bitterns, storks, flamingoes.
Order Anseriformes. Swans, geese, ducks.
Order Falconiformes. Eagles, hawks, vultures, falcons.
Order Galliformes. Ptarmigan, turkeys, domestic fowl.
Order Columbiformes. Pigeons, doves.
Order Strigiformes. Owls.
Order Apodiformes. Swifts, hummingbirds.
Order Passeriformes. Sparrows, jays, finches, crows, robins, starlings, wrens.
Order Piciformes. Woodpeckers, toucans.
Order Psittaciformes. Parrots, cockatoos, macaws.

Class Mammalia. Skin with hair; young nourished by milk-secreting glands of adult.

Subclass Prototheria. Egg-laying mammals (monotremes; duckbilled platypus, spiny anteaters).
Subclass Metatheria. Pouched mammals or marsupials (opossums, kangaroos, wombats, Tasmanian devil).
Subclass Eutheria. Placental mammals.
Order Edentata. Anteaters, tree sloths, armadillos.
Order Insectivora. Tree shrews, moles, hedgehogs.
Order Chiroptera. Bats.
Order Scandentia. Insectivorous tree shrews.
Order Primates.
Suborder Strepsirhini (prosimians). Lemurs, lorises.
Suborder Haplorhini (tarsioids and anthropoids).
Infraorder Tarsiiformes. Tarsiers.
Infraorder Platyrrhini (New World monkeys).
Family Cebidae. Spider monkeys, howler monkeys, capuchin.
Infraorder Catarrhini (Old World monkeys and hominoids).
Superfamily Cercopithecoidea. Baboons, macaques, langurs.
Superfamily Hominoidea. Apes and humans.
Family Hylobatidae. Gibbon.
Family "Pongidae." Chimpanzees, gorillas, orangutans.
Family Hominidae. Existing and extinct human species (*Homo*) and humanlike species, including the australopiths.
Order Lagomorpha. Rabbits, hares, pikas.
Order Rodentia. Most gnawing animals (squirrels, rats, mice, guinea pigs, porcupines, beavers, etc.).
Order Carnivora. Carnivores.
Suborder Feloidea. Cats, mongooses, hyenas.
Suborder Canoidea. Dogs, weasels, skunks, otters, raccoons, pandas, bears.
Order Pinnipedia. Seals, walruses, sea lions.
Order Proboscidea. Elephants; mammoths (extinct).
Order Sirenia. Sea cows (manatees, dugongs).
Order Perissodactyla. Odd-toed ungulates (horses, tapirs, rhinos).
Order Tubulidentata. African aardvarks.
Order Artiodactyla. Even-toed ungulates (camels, deer, bison, sheep, goats, antelopes, giraffes, etc.).
Order Cetacea. Whales, porpoises.

APPENDIX II. UNITS OF MEASURE

Metric-English Conversions

Length

English		Metric
inch	=	2.54 centimeters
foot	=	0.30 meter
yard	=	0.91 meter
mile (5,280 feet)	=	1.61 kilometer

To convert	multiply by	to obtain
inches	2.54	centimeters
feet	30.00	centimeters
centimeters	0.39	inches
millimeters	0.039	inches

Weight

English		Metric
grain	=	64.80 milligrams
ounce	=	28.35 grams
pound	=	453.60 grams
ton (short) (2,000 pounds)	=	0.91 metric ton

To convert	multiply by	to obtain
ounces	28.3	grams
pounds	453.6	grams
pounds	0.45	kilograms
grams	0.035	ounces
kilograms	2.2	pounds

Volume

English		Metric
cubic inch	=	16.39 cubic centimeters
cubic foot	=	0.03 cubic meter
cubic yard	=	0.765 cubic meters
ounce	=	0.03 liter
pint	=	0.47 liter
quart	=	0.95 liter
gallon	=	3.79 liters

To convert	multiply by	to obtain
fluid ounces	30.00	milliliters
quart	0.95	liters
milliliters	0.03	fluid ounces
liters	1.06	quarts

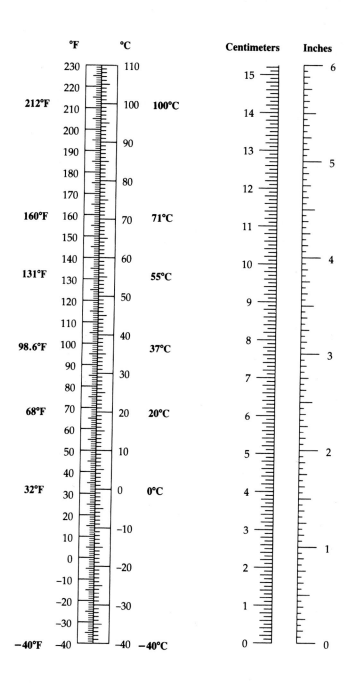

APPENDIX III. ANSWERS TO SELF-QUIZZES

CHAPTER 1

1. Metabolism — 5
2. Homeostasis — 5
3. cell — 6
4. adaptive — 10
5. mutation — 10
6. d — 4
7. d — 4, 10
8. d — 10–11
9. d — 13
10. c — 13
11. a — 13
12. c — 10
    e — 11
    d — 13
    b — 12
    a — 12

CHAPTER 2

1. b — 22
2. c — 27
3. e — 28
4. e — 29
5. f — 30–31
6. acid, base — 30
7. c — 20
   a — 31
   b — 25
   d — 28–29

CHAPTER 3

1. d — 36
2. e — 39
3. f — 40
4. b — 42
5. d — 44, 50
6. d — 47
7. d — 50
8. c — 44
   e — 50
   b — 43
   d — 50
   a — 40

CHAPTER 4

1. c — 56
2. d — 60
3. d — 60, 74
4. False; many cells have a wall — 60, 74
5. c — 56, 76
6. e — 60, 66
   d — 67
   a — 56, 64
   b — 64
   c — 64

CHAPTER 5

1. c — 82
2. c — 82, 83
3. a — 81, 84
4. d — 84
5. b — 91
6. d — 86–87
7. b — 89
8. d — 92–93

CHAPTER 6

1. c — 99
2. d — 100
3. b — 102
4. a — 102–103
5. d — 104
6. e — 105, 108–109
7. a — 106, 107
8. a — 104
9. c — 102, 106
   g — 102
   a — 102
   d — 102
   e — 102
   b — 102, 105
   f — 102

CHAPTER 7

1. carbon dioxide; light energy from sun — 114
2. d — 116
3. b — 122
4. c — 116, 125
5. d — 122
6. e — 116, 124
7. c — 125
8. c — 125
9. d — 120
   e — 122
   a — 125
   b — 125
   c — 122

CHAPTER 8

1. d — 134
2. c — 138, 139
3. b — 134, 140
4. c — 134, 140
5. d — 142
6. c — 142, 143i
7. b — 142
8. d — 144, 145i
9. b — 134, 136
   c — 142
   a — 138–139
   d — 140, 140i

CHAPTER 9

1. d — 152, 152t
2. b — 152
3. c — 152
4. a — 153
5. c — 154
6. a — 155, 157
7. b — 154
8. d — 156
   b — 156
   c — 157
   a — 157

CHAPTER 10

1. c — 163, 164
2. a — 164
3. d — 164
4. b — 164–165
5. d — 165
6. c — 165
7. c — 166
8. d — 167
9. d — 164
   a — 164
   c — 166
   b — 165

CHAPTER 11

1. a — 179
2. b — 179
3. a — 179
4. b — 179
5. c — 180
6. a — 181
7. d — 182
8. b — 182
   d — 180
   a — 179
   c — 179

CHAPTER 12

1. c — 196
2. c — 201
3. d — 203, 206–208
4. a — 203
5. e — 206–207
6. c — 208
7. d — 208
8. c — 201
   e — 207
   d — 208
   b — 206
   a — 194
   f — 200

CHAPTER 13

1. c — 220
2. d — 221
3. c — 221
4. a — 223
5. d — 222–223
6. c — 224
7. d — 223
   b — 221, 222
   c — 222
   a — 221, 221i

CHAPTER 14

1. c — 228
2. b — 228
3. c — 228
4. c — 230
5. a — 230i
6. a — 230
7. e — 235
   c — 232
   a — 229
   f — 230
   d — 230
   g — 229
   b — 230

CHAPTER 15

1. d — 242
2. d — 239, 240
3. d — 240
4. a — 240
5. c — 240, 240i
6. a — 240
7. d — 242–243
8. b — 245
9. d — 246
10. b — 249
11. b — 249
12. d — 247
    e — 244
    b — 248
    a — 244
    c — 246

CHAPTER 16

1. d — 252
2. c — 254
3. Plasmids — 254
4. a — 255
5. b — 255
6. a — 256
7. b — 257
8. d — 258
9. d — 257
   c — 260–261
   f — 253
   e — 259
   b — 262
   a — 264

APPENDIX IV. ANSWERS TO GENETICS PROBLEMS

CHAPTER 11

1. a. *AB*

    b. *AB, aB*

    c. *Ab, ab*

    d. *AB, Ab, aB, ab*

2. a. All of the offspring will be *AaBB*.

    b. 1/4 *AABB* (25% each genotype)
    1/4 *AABb*
    1/4 *AaBB*
    1/4 *AaBb*

    c. 1/4 *AaBb* (25% each genotype)
    1/4 *Aabb*
    1/4 *aaBb*
    1/4 *aabb*

    d. 1/16 *AABB* (6.25%)
    1/8 *AaBB* (12.5%)
    1/16 *aaBB* (6.25%)
    1/8 *AABb* (12.5%)
    1/4 *AaBb* (25%)
    1/8 *aaBb* (12.5%)
    1/16 *AAbb* (6.25%)
    1/8 *Aabb* (12.5%)
    1/16 *aabb* (6.25%)

3. Yellow is recessive. Because F_1 plants have a green phenotype and must be heterozygous, green must be dominant over the recessive yellow.

4. a. *ABC*

    b. *ABc, aBc*

    c. *ABC, aBc, ABc, aBc*

    d. *ABC, aBC, AbC, abC, ABc, aBc, Abc, abc*

5. Because all F_1 plants of this dihybrid cross had to be heterozygous for both genes, then 1/4 (25%) of the F_2 plants will be heterozygous for both genes.

6. a. The mother must be heterozygous $I^A i$. The male with type B blood could have fathered the child if he were heterozygous $I^B i$.

    b. Genotype alone cannot prove the accused male is the father. Even if he happens to be heterozygous, *any* male who carries the *i* allele could be the father, including those heterozygous for type A blood ($I^A i$) or type B blood ($I^B i$) and those with type O blood (*ii*).

7. A mating between a mouse from a true-breeding, white-furred strain and a mouse from a true-breeding, brown-furred strain would provide you with the most direct evidence.
    Because true-breeding strains of organisms typically are homozygous for a trait being studied, all F_1 offspring from this mating should be heterozygous. Record the phenotype of each F_1 mouse, then let them mate with one another. Assuming only one gene locus is involved, these are possible outcomes for the F_2 offspring:

    a. All F_1 mice are brown, and their F_2 offspring segregate 3 brown : 1 white. *Conclusion*: Brown is dominant to white.

    b. All F_1 mice are white, and their F_2 offspring segregate 3 white : 1 brown. *Conclusion*: White is dominant to brown.

    c. All F_1 mice are tan, and the F_2 offspring segregate 1 brown : 2 tan : 1 white. *Conclusion*: The alleles at this locus show incomplete dominance.

8. You cannot guarantee that the puppies will not develop the disorder without more information about Dandelion's genotype. You could do so only if she is a heterozygous carrier, if the male is free of the alleles, and if the alleles are recessive.

9. Fred could use a testcross to find out if his pet's genotype is *WW* or *Ww*. He can let his black guinea pig mate with a white guinea pig having the genotype *ww*.
    If any F_1 offspring are white, then the genotype of his pet is *Ww*. If the two guinea pig parents are allowed to mate repeatedly and all the offspring of the matings are black, then there is a high probability that his pet guinea pig is *WW*.
    (If, say, ten offspring are all black, then the probability that the male is *WW* is about 99.9 percent. The greater the number of offspring, the more confident Fred can be of his conclusion.)

10. a. 1/2 red, 1/2 pink

    b. All pink

    c. 1/4 red, 1/2 pink, 1/4 white

    d. 1/2 pink, 1/2 white

11.   9/16 walnut comb

        3/16 rose comb

        3/16 pea comb

        1/16 single comb

12. Because both parents are heterozygotes ($Hb^A Hb^S$), the following are the probabilities for each child:

   a. $1/4$ $Hb^S Hb^S$

   b. $1/4$ $Hb^A Hb^A$

   c. $1/2$ $Hb^A Hb^S$

13. A mating of two $M^L M$ cats yields:

   $1/4$ homozygous dominant (MM)

   $1/2$ heterozygous ($M^L M$)

   $1/4$ homozygous recessive ($M^L M^L$)

Because $M^L M^L$ is lethal, the probability that any one kitten among the survivors will be heterozygous is $2/3$.

14. a. Both parents must be heterozygotes (Aa). Their children may be albino (aa) or unaffected (AA or Aa).

   b. All are aa.

   c. The albino father must be aa. They have an albino child, so the mother must be Aa. (If she were AA, they could not have an albino child.) The albino child is aa. The three unaffected children are Aa. There is a 50% chance that any child of theirs will be albino. The observed 3:1 ratio is not surprising, given the small number of offspring.

15. a. All of the offspring will have medium-red color corresponding to the genotype $A^1 A^2 B^1 B^2$.

   b. All possible genotypes could appear in the following proportions:

$1/16$	$A^1 A^2 B^1 B^1$	dark red
$1/8$	$A^1 A^1 B^1 B^2$	medium-dark red
$1/16$	$A^1 A^1 B^2 B^2$	medium red
$1/8$	$A^1 A^2 B^1 B^1$	medium-dark red
$1/4$	$A^1 A^2 B^1 B^2$	medium red
$1/8$	$A^1 A^2 B^2 B^2$	light red
$1/16$	$A^2 A^2 B^1 B^1$	medium red
$1/8$	$A^2 A^2 B^1 B^2$	light red
$1/16$	$A^2 A^2 B^2 B^2$	white

CHAPTER 12

1. a. Human males (XY) inherit their X chromosome only from their mother.

   b. In males, an X-linked allele will be found only on his one X chromosome. Males can produce two kinds of gametes: one kind with a Y chromosome free of the gene, and the other kind with an X chromosome bearing the X-linked allele.

   c. One. Each gamete of a woman who is homozygous for an X-linked allele will have an X chromosome that carries the allele.

   d. Two. If a female is heterozygous for an X linked allele, half of the gametes that she produces will contain one of the alleles and the other half will contain the other allele.

2. All of the offspring should be heterozygous for the gene and all should have long wings. However, because some have vestigial wings, the dominant allele might have mutated because of radiation.

3. Because Marfan syndrome is a case of autosomal dominant inheritance and because one parent bears the allele, the probability of any child inheriting the mutant allele is 50%.

4. Because the phenotype appeared in every generation shown in the diagram, this must be a pattern of autosomal dominant inheritance.

5. A daughter could develop this type of muscular dystrophy only if she were to inherit two X-linked recessive alleles—one from her father and one from her mother.

   However, if a son bears the allele for the disorder on his X chromosome, then it will be expressed. He will develop the disorder, and most likely he will not father children because of his early death.

6. If no crossover occurs between the two genes, then half the chromosomes will carry alleles AB and half will carry alleles ab.

7. a. Nondisjunction could occur in anaphase I or anaphase II of meiosis.

   b. As a result of a translocation, chromosome 21 (which is small) may become attached to the end of chromosome 14. Even though the chromosome number of the new individual would be 46, its somatic cells would contain the translocated chromosome 21, in addition to two normal chromosomes 21.

8. In the mother, a crossover between the two genes at meiosis generates an X chromosome that carries neither mutant allele.

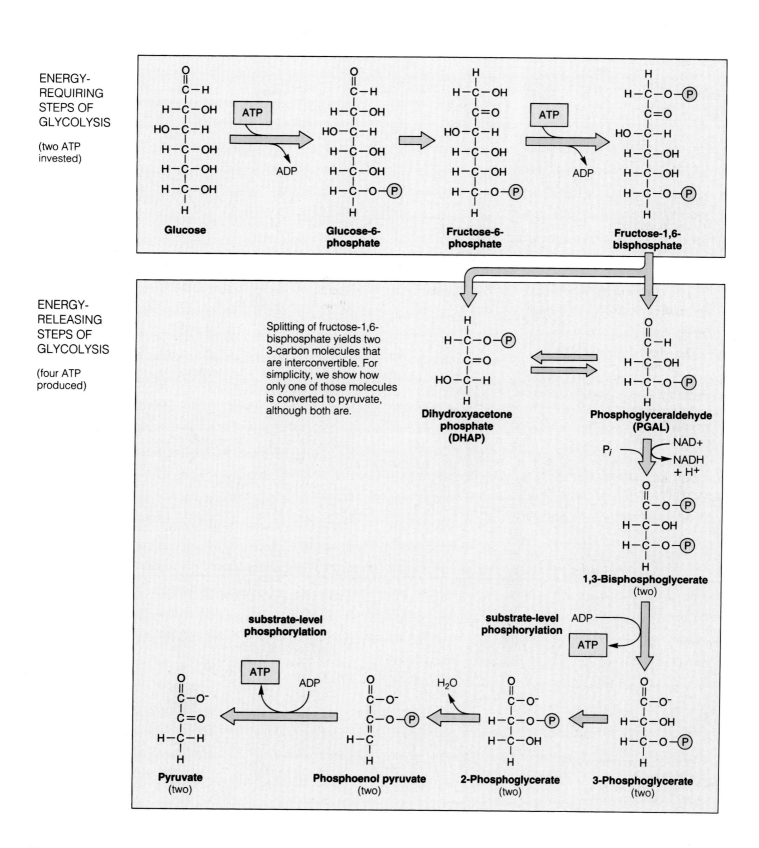

Figure A   Glycolysis, ending with two 3-carbon pyruvate molecules for each 6-carbon glucose molecule entering the reactions. The *net* energy yield is two ATP molecules (two invested, four produced).

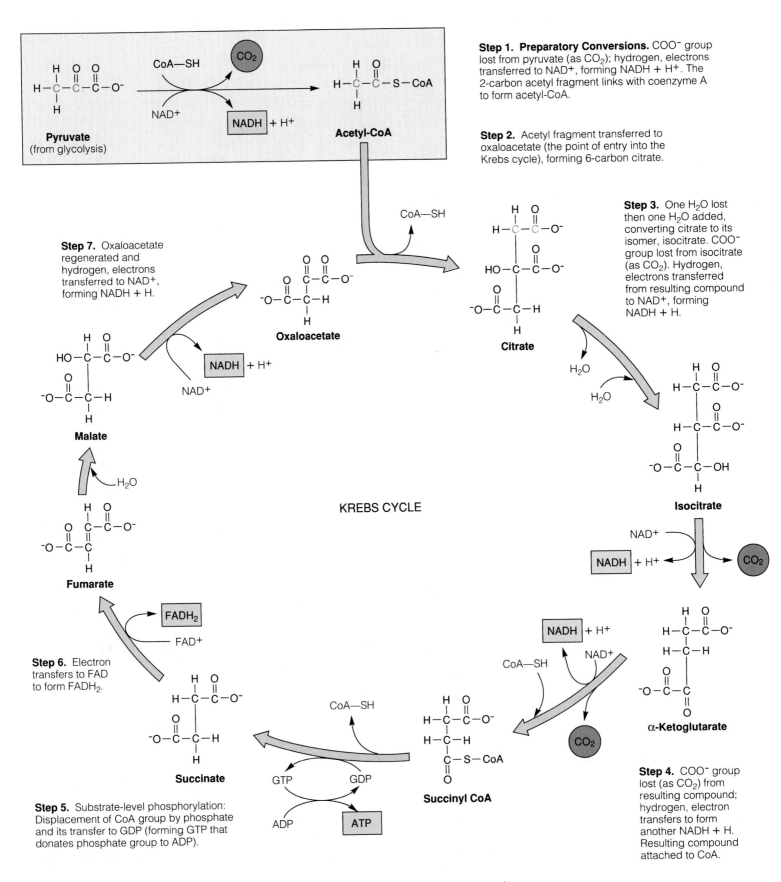

Step 1. Preparatory Conversions. COO^- group lost from pyruvate (as CO_2); hydrogen, electrons transferred to NAD^+, forming $NADH + H^+$. The 2-carbon acetyl fragment links with coenzyme A to form acetyl-CoA.

Step 2. Acetyl fragment transferred to oxaloacetate (the point of entry into the Krebs cycle), forming 6-carbon citrate.

Step 3. One H_2O lost then one H_2O added, converting citrate to its isomer, isocitrate. COO^- group lost from isocitrate (as CO_2). Hydrogen, electrons transferred from resulting compound to NAD^+, forming $NADH + H$.

Step 7. Oxaloacetate regenerated and hydrogen, electrons transferred to NAD^+, forming $NADH + H$.

Step 6. Electron transfers to FAD to form $FADH_2$.

Step 5. Substrate-level phosphorylation: Displacement of CoA group by phosphate and its transfer to GDP (forming GTP that donates phosphate group to ADP).

Step 4. COO^- group lost (as CO_2) from resulting compound; hydrogen, electron transfers to form another $NADH + H$. Resulting compound attached to CoA.

KREBS CYCLE

Pyruvate (from glycolysis)

Acetyl-CoA

Oxaloacetate

Citrate

Isocitrate

Malate

α-Ketoglutarate

Fumarate

Succinate

Succinyl CoA

Figure B Krebs cycle, also known as the citric acid cycle. *Red* identifies carbon atoms entering the cyclic pathway (by way of acetyl-CoA) and leaving (by way of carbon dioxide). These cyclic reactions run twice for each glucose molecule that has been degraded to two pyruvate molecules.

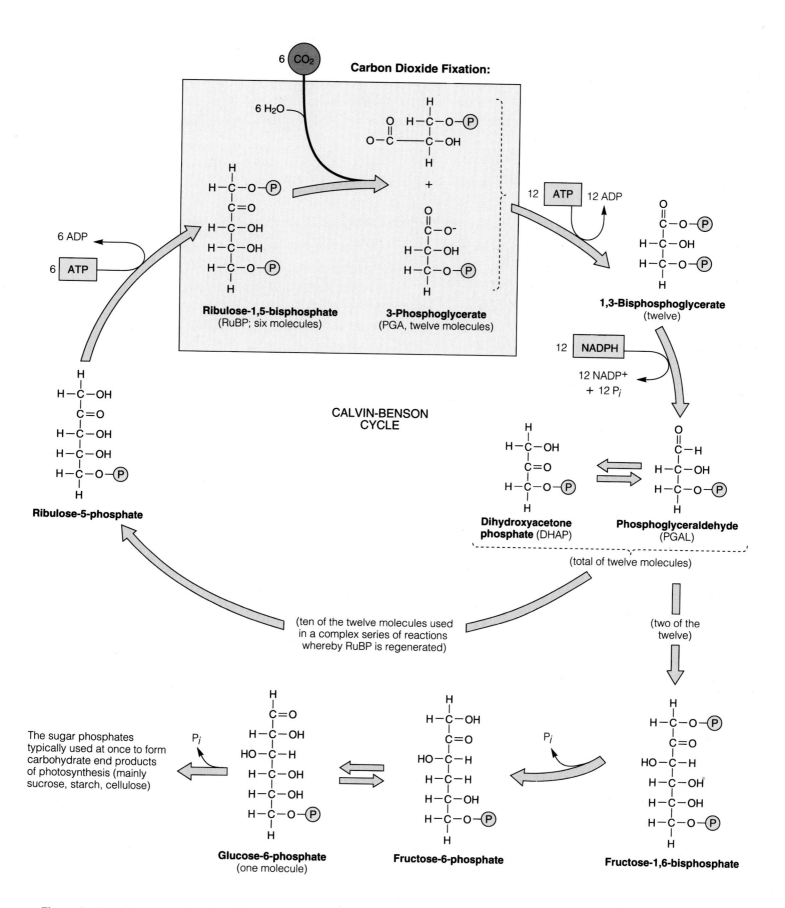

Figure C Calvin–Benson cycle of the light-independent reactions of photosynthesis.

Neutral, nonpolar side group

glycine (gly)

alanine (ala)

valine (val)

isoleucine (ile)

leucine (leu)

phenylalanine (phe)

proline (pro)

methionine (met)

Neutral, polar side group

serine (ser)

threonine (thr)

tyrosine (tyr)

tryptophan (trp)

asparagine (asn)

glutamine (gln)

cysteine (cys)

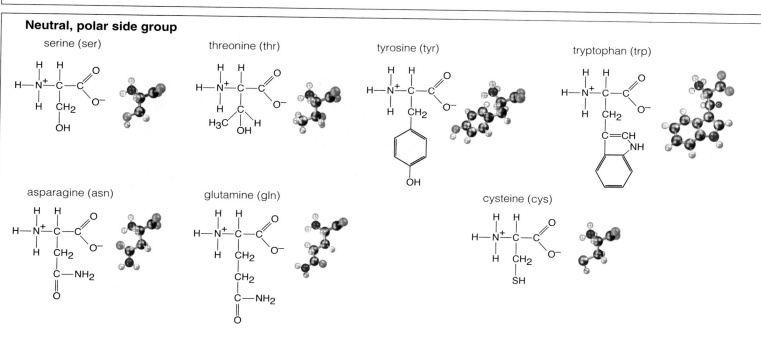

Acidic side group

aspartic acid (asp)

glutamic acid (glu)

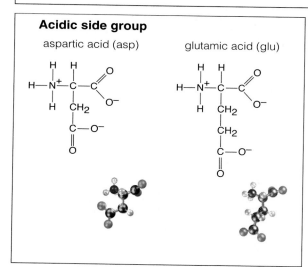

Basic side group

lysine (lys)

arginine (arg)

histidine (his)

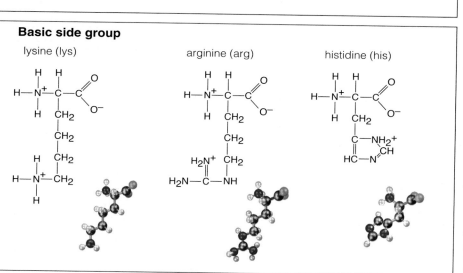

APPENDIX VII.   PERIODIC TABLE OF THE ELEMENTS

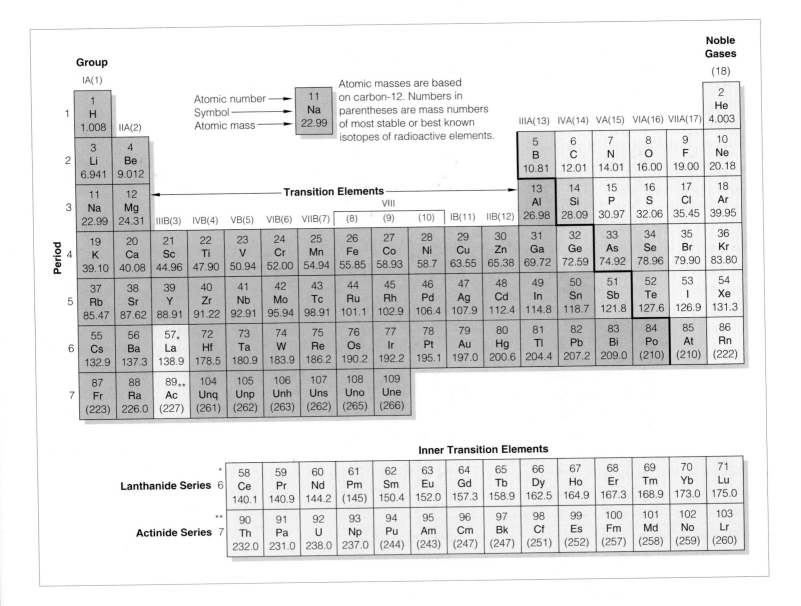

GLOSSARY

ABC transporter One of a class of ATP-driven membrane pumps, each for a specific substrate (e.g., ions, sugars, amino acids).

abortion Premature, spontaneous or induced expulsion of the embryo or fetus from uterus.

absorption spectrum Range of wavelengths that one or more specified pigments can absorb.

acetylation Attachment of an acetyl group to a compound, such as DNA.

acetyl-CoA (uh-SEET-ul) Coenzyme A attached to a two-carbon fragment from pyruvate, which it transfers to oxaloacetate for the Krebs cycle.

acid [L. *acidus*, sour] Any substance that, when dissolved in water, donates hydrogen ions.

activation energy A specific minimum amount of energy required to get a specific reaction going, with or without the help of an enzyme. Reactions differ in the amount required.

activator Regulatory protein that enhances a cell activity (e.g., gene transcription).

active site Crevice in an enzyme molecule where a specific reaction is catalyzed.

active transport Pumping of a specific solute across a cell membrane against its concentration gradient, through a transport protein's interior. Requires an energy boost, typically from ATP.

adaptive trait Any aspect of form, function, or behavior that helps the individual survive and reproduce under prevailing conditions.

adenine (AH-de-neen) A purine; a nitrogen-containing base in certain nucleotides.

adhesion protein Plasma membrane protein that helps cells locate tissue mates and stick together.

ADP Adenosine diphosphate (ah-DEN-uh-seen die-FOSS-fate). A nucleotide coenzyme.

aerobic respiration (air-OH-bik) [Gk. *aer*, air, + *bios*, life] Oxygen-requiring pathway of ATP formation; from glycolysis, to Krebs cycle, to electron transport phosphorylation. Typical net energy yield: 36 ATP per glucose molecule.

alcohol Organic compound containing one or more hydroxyl groups (—OH); it dissolves readily in water. Sugars are examples.

alcoholic fermentation Anaerobic ATP-forming pathway. Pyruvate from glycolysis is degraded to acetaldehyde, which accepts electrons from NADH to form ethanol; NAD+ needed for the reactions is regenerated. Net yield: 2 ATP.

alkylating agent Any substance that transfers methyl or ethyl groups to a compound.

allele (uh-LEEL) One of two or more molecular forms of a gene that arise by mutation and code for different versions of the same trait.

amino acid (uh-MEE-no) Organic compound with an H atom, amino group, acid group, and R group, all covalently bonded to a carbon atom. Subunit of polypeptide chains.

anaerobic electron transfer (an-uh-ROW-bik) [Gk. *an*, without, + *aer*, air] Flow of electrons through transfer chains in plasma membrane that drives ATP formation. A compound other than oxygen is the final electron acceptor.

anaphase (AN-uh-faze) Of mitosis, stage when sister chromatids of each chromosome move to opposite spindle poles. Of anaphase I (meiosis), each duplicated chromosome and its homologue move to opposite poles. Of anaphase II (meiosis), sister chromatids of each chromosome move to opposite poles.

aneuploidy (AN-yoo-ploy-dee) Having one extra or one less chromosome relative to the parental chromosome number.

animal Multicelled, aerobic, motile predator or parasite; usually with tissues, organ systems; developed through embryonic stages.

anthocyanin Blue or red accessory pigment.

antibiotic Metabolic product of soil microbes that kills bacterial competitors for nutrients.

anticodon Series of three nucleotide bases in tRNA; can base-pair with an mRNA codon.

apoptosis (APP-oh-TOE-sis) Programmed cell death of body cells finished with prescribed functions or altered, as by infection or cancer.

Archaebacteria Prokaryotic domain; closer to eukaryotic cells than to eubacteria; includes methanogens, halophiles, and thermophiles; also called a kingdom.

artificial selection Selection of traits among a population under contrived conditions.

asexual reproduction Any reproductive mode by which offspring arise from a single parent and inherit the genes of that parent only.

atom Smallest unit of an element that retains the element's properties.

ATP Adenosine triphosphate (ah-DEN-uh-seen try-FOSS-fate). Nucleotide with adenine, ribose, and three phosphate groups; delivers energy to most energy-requiring metabolic reactions.

ATP/ADP cycle ADP forms from ATP, then ATP is regenerated from ADP, through the transfer of inorganic phosphate or a phosphate group.

automated DNA sequencing Fast method of sequencing cloned or PCR-amplified DNA.

autosome Any chromosome of a type that is the same in males and females of the species.

autotroph (AH-toe-trofe) [Gk. *autos*, self, + *trophos*, feeder] Any organism that makes its own food with an environmental energy source (e.g., sunlight) and CO_2 as its carbon source.

bacteriophage (bak-TEER-ee-oh-fahj) Category of viruses that infect bacterial cells.

Barr body Randomly condensed one of two X chromosomes in cells of female mammals.

basal body Centriole which, after giving rise to microtubules of a flagellum or cilium, stays attached to its base in cytoplasm.

base Any substance that accepts hydrogen ions (H+) when dissolved in water; OH− forms after this. Also, an organic compound with a single or double ring structure containing nitrogen.

base-pair substitution One amino acid has replaced another during protein synthesis.

base sequence Sequential order of bases in a DNA or RNA strand.

binding energy Energy released when weak bonds that form between a substrate, enzyme, and any cofactor break during the transition state; the partial payment on energy cost of a reaction (the activation energy).

biofilm Huge microbial populations anchored to surfaces (e.g., lung epithelia) by their own sticky, stiff polysaccharide secretions.

biology The scientific study of life.

bioluminescence Fluorescent light produced by an organism through an ATP-driven reaction involving enzymes (luciferases).

biosphere [Gk. *bios*, life, + *sphaira*, globe] All regions of the Earth's waters, crust, and atmosphere in which organisms live.

buffer system A weak acid and the base that forms when it dissolves in water. The two work as a pair to counter slight shifts in pH.

bulk flow In response to a pressure gradient, movement of more than one kind of molecule in the same direction in the same medium.

C3 plant Plant that uses three-carbon PGA as the first intermediate for carbon fixation.

C4 plant Plant that uses oxaloacetate (a four-carbon compound) as the first intermediate for carbon fixation. CO_2 is fixed twice, in two cell types; helps counter photorespiration.

calcium pump Active transporter protein specific for calcium ions.

Calvin–Benson cycle Light-independent cyclic reactions of photosynthesis that form sugars, using ATP energy, NADPH, and CO_2.

CAM plant Type of plant that conserves water by opening stomata only at night, when it fixes carbon dioxide by means of a C4 pathway.

cancer Malignant tumor; mass of altered cells that divide abnormally. Potentially lethal.

carbohydrate Molecule of carbon, hydrogen, and oxygen mostly in a 1:2:1 ratio. Main kinds are monosaccharides, oligosaccharides, and polysaccharides. They are structural materials, energy stores, and transportable energy forms.

carbon fixation First of the light-independent reactions. Rubisco, an enzyme, affixes carbon (from CO_2) to RuBP or to another compound for entry into the Calvin–Benson cycle.

carcinogen (kar-SIN-uh-jen) Any substance or agent that can trigger cancer.

carotenoid (kare-OTT-en-oyd) An accessory pigment of photosynthesis; e.g., fucoxanthin.

cDNA DNA molecule copied from a mature mRNA transcript by reverse transcription.

cell [L. *cella*, small room] Smallest living unit, it can survive and reproduce on its own, given its DNA, raw materials, and an energy source.

cell communication Mechanisms by which free-living cells of a species or the cells of a multicelled organism coordinate their activities; involves sending and receiving, transducing, and responding to signaling molecules.

cell cortex A dynamic, crosslinked mesh of cytoskeletal elements under plasma membrane.

cell cycle Events by which a cell increases in mass, roughly doubles its cytoplasm, duplicates its DNA, and divides in two. Extends from the time a cell forms until it completes division.

cell differentiation Key development process. Cell lineages become specialized in structure, composition, and function by activating and suppressing part of the genome selectively.

cell junction Site where cells are interacting physically, functionally, or both.

cell plate formation Process of cytoplasmic division in plant cells; after nuclear division, a cross-wall with plasma membrane on both surfaces forms and divides the cytoplasm.

cell theory All organisms consist of one or more cells, the smallest units with a capacity for independent life that no longer spontaneously arise under existing conditions on Earth.

cell wall A semirigid, permeable structure external to the plasma membrane; helps many cells retain their shape and resist rupturing.

central vacuole Fluid-filled storage organelle of plant cell; its growth enhances cell surface area.

centriole (SEN-tree-ohl) Structure that gives rise to microtubules of cilia and flagella.

centromere (SEN-troh-meer) A constricted area of a chromosome that has attachment sites for spindle microtubules during nuclear division.

checkpoint gene Its protein product can help delay, advance, or block cell cycle.

chemical bond A union between the electron structures of two or more atoms or ions.

chemical energy Potential energy of molecules.

chemical equilibrium Time when a reversible reaction runs both ways at about same rate.

chemoautotroph (KEE-moe-AH-toe-trofe) Any prokaryotic cell that synthesizes its own food using carbon dioxide as the carbon source and an inorganic substance as the energy source.

chlorophyll (KLOR-uh-fill) [Gk. *chloros*, green, + *phyllon*, leaf] Main photosynthetic pigment. Chlorophylls absorb all wavelengths of visible light but not much of green and yellow ones.

chloroplast (KLOR-uh-plast) The organelle of photosynthesis in plants and many protistans.

chromatin A cell's collection of DNA and all of the proteins associated with it.

chromosome (CROW-moe-some) [Gk. *chrōma*, color, + *soma*, body] Of eukaryotic cells, a DNA molecule, duplicated or unduplicated, with many associated proteins. Of prokaryotic cells, a circular DNA molecule.

chromosome number All chromosomes in a given type of cell. *See* haploidy; diploidy.

cilium (SILL-ee-um), plural **cilia** Short motile or sensory structure of certain eukaryotic cells; its core is a 9 + 2 array of microtubules.

cleavage furrow Ringlike depression defining the cutting plane for a dividing animal cell.

cloning Making a genetically identical copy of DNA or of an organism.

cloning vector Any plasmid, viral DNA, or some other piece of DNA that researchers use to isolate and amplify DNA of interest.

ccodominance In heterozygotes, simultaneous expression of a pair of nonidentical alleles that specify different phenotypes.

codon One of 64 possible base triplets in an mRNA strand. A code word for an amino acid in a polypeptide chain; a few codons also act as START or STOP signals for translation.

coenzyme Enzyme helper; a nucleotide that transfers electrons and H atoms stripped from substrates to a different reaction site.

cofactor Metal ion, coenzyme, or other organic compound that assists an enzyme in catalysis or that transfers atoms, functional groups, or electrons to a different reaction site.

cohesion Capacity to resist rupturing when placed under tension (stretched).

communication protein Part of a protein complex that forms an open channel between cytoplasm of adjoining cells.

community All populations in a habitat. Also, a group of organisms with similar life-styles.

compound Molecule consisting of two or more elements in unvarying proportions (e.g., H_2O).

concentration gradient A difference in the number per unit volume of molecules or ions of a substance between two regions. Molecules collide constantly and careen outward to a region where they are less concentrated. All substances tend to diffuse down such gradients.

condensation reaction Covalent bonding of two molecules into a larger one; water often forms as a by-product.

conservation of mass, law of The total mass of all substances entering a reaction equals the total mass of all products.

consumer [L. *consumere*, to take completely] A heterotroph that feeds on cells or tissues of other organisms (e.g., herbivores, carnivores).

continuous variation Of a population, a more or less continuous range of small differences in a given trait among its individuals.

control group Group used as a standard for comparison with an experimental group and, ideally, identical with it in all respects except for the one variable being studied.

covalent bond (koe-VAY-lunt) [L. *con*, together, + *valere*, to be strong] Sharing of one or more electrons between atoms or groups of atoms. Nonpolar bonds share them equally. Polar bonds (slightly positive at one end, negative at the other) share them unequally.

crossing over At prophase I of meiosis, an interaction in which nonsister chromatids of a pair of homologous chromosomes break at corresponding sites and exchange segments; genetic recombination is the result.

cytoplasm (SIGH-toe-plaz-um) All cell parts, particles, and semifluid substances between the plasma membrane and nucleus or nucleoid.

cytoplasmic division After nuclear division, a splitting of the parent cell cytoplasm that completes the formation of daughter cells.

cytosine (SIGH-toe-seen) Pyrimidine; one of the nitrogen-containing bases in nucleotides.

cytoskeleton Inner, interconnected system of protein elements that reinforce, organize, and move the eukaryotic cell and its structures.

decomposer [L. *dis–*, to pieces] Prokaryotic or fungal heterotroph; gets carbon and energy from products or remains of organisms. Helps cycle nutrients to producers in ecosystems.

deductive logic Pattern of thinking; making inferences about specific consequences or predictions that must follow from a hypothesis.

deletion At cytological level, loss of a segment from a chromosome. At molecular level, loss of one to a few base pairs from a DNA molecule.

denaturation (deh-NAY-chur-AY-shun) Loss of a molecule's three-dimensional shape as weak bonds (e.g., hydrogen bonds) are disrupted.

development The series of genetically guided embryonic and post-embryonic stages by which morphologically distinct, specialized body parts emerge in a new multicelled individual.

diffusion Net movement of like molecules or ions down their concentration gradient.

dihybrid cross An intercross between two F_1 heterozygotes that are identical for two gene loci; the dihybrids are offspring of parents that bred true for different versions of two traits.

diploidy (DIP-loyd-ee) Presence of two of each type of chromosome (i.e., pairs of homologs) in a cell nucleus at interphase. *Compare* haploidy.

disease Outcome of infection when defenses aren't mobilized fast enough and a pathogen's activities interfere with normal body functions.

DNA Deoxyribonucleic acid (dee-OX-ee-RYE-bow-new-CLAY-ik) Of cells and many viruses, the molecule of inheritance. H bonds join DNA's two helically twisted nucleotide strands, one of which has instructions (in its base sequence) for synthesizing all of the enzymes and other proteins required to build and maintain cells.

DNA clone Many identical copies of DNA that was inserted into plasmids and later amplified.

DNA fingerprint Unique array of DNA sequences inherited in a Mendelian pattern.

DNA ligase (LYE-gaze) Enzyme that seals new base-pairings during DNA replication.

DNA microarray Gene chip stamped with thousands of DNA sequences from a genome.

DNA polymerase (poe-LIM-uh-raze) Enzyme of replication and repair that assembles a new strand of DNA on a parent DNA template.

DNA repair Enzyme-mediated process that fixes small-scale alterations in a DNA strand by restoring the original base sequence.

DNA replication Any process by which a cell duplicates its DNA molecules before dividing.

domain Part or all of a polypeptide chain that forms a structurally stable, functional unit.

dosage compensation Any mechanism that balances gene expression between the sexes during critical early stages of development.

double-blind study Different investigators independently collect, then compare data.

duplication Gene sequence repeated several to many hundreds or thousands of times. Even normal chromosomes have such sequences.

ecdysone Hormone of many insect life cycles; roles in metamorphosis, molting.

ecosystem Array of organisms, together with their environment, interacting through a flow of energy and a cycling of materials.

egg Mature female gamete; an ovum.

electric gradient Difference in electric charge between adjoining regions.

electromagnetic spectrum All wavelengths from radiant energy less than 10^{-5} nm long to radio waves more than 10 km long.

electron Negatively charged unit of matter, with particulate and wavelike properties, that occupies one of the orbitals around the atomic nucleus. Atoms gain, lose, or share electrons.

electron transfer chain Organized array of membrane-bound enzymes and cofactors that accept and donate electrons in series. It sets up an electrochemical gradient that makes H^+ flow across the membrane. The flow energy drives ATP formation at ATP synthases.

electron transfer phosphorylation Last stage of ATP formation by aerobic respiration; uses the hydrogen and electrons that many reduced coenzymes deliver to electron transfer chains.

element Fundamental form of matter that has mass, occupies space, and cannot be broken apart into a different form of matter, at least by ordinary physical or chemical means.

endocytosis (EN-doe-sigh-TOE-sis) Cell uptake of substances via vesicle formation. Receptor-mediated endocytosis, phagocytosis, and the bulk transport of extracellular fluid are three modes of endocytosis.

endomembrane system Series of organelles in which new polypeptide chains are modified and lipids are synthesized.

endoplasmic reticulum or **ER** (EN-doe-PLAZ-mik reh-TIK-yoo-lum) Organelle that starts at nucleus and curves through cytoplasm. New polypeptides get side chains in rough ER (has ribosomes on its cytoplasmic side); smooth ER (has no ribosomes) is a site of lipid synthesis.

energy Capacity to do work.

energy carrier Molecule that delivers energy from one reaction site to another; mainly ATP.

enhancer A short DNA base sequence that is a binding site for an activator protein.

entropy (EN-trow-pee) Measure of the degree of disorder in a system (how much energy has become disorganized, usually as dispersing heat, and is no longer available to do work). All systems need energy inputs to counter entropy.

enzyme (EN-zime) A type of protein or one of the few RNAs that catalyze reactions between substances, most often at functional groups.

epistasis (eh-PISS-tah-sis) Interaction among the products of two or more gene pairs.

Eubacteria Domain of all prokaryotic cells except archaebacteria; also called a kingdom.

eukaryotic cell (yoo-CARE-EE-oh-tic) [Gk. *eu*, good, + *karyon*, kernel] Cell having a nucleus and other membrane-bound organelles.

evaporation [L. *e-*, out, + *vapor*, steam] The conversion of a substance from a liquid state to a gaseous state by an input of heat energy.

evolution, biological [L. *evolutio*, unrolling] Genetic change in a line of descent. Outcome of microevolutionary events: gene mutation, natural selection, genetic drift, and gene flow.

exocytosis (EK-so-sigh-TOE-sis) Release of a vesicle's contents at cell surface as it fuses with and becomes part of the plasma membrane.

exon One of the base sequences of an mRNA transcript that will become translated.

experiment, scientific Test that simplifies observation in nature or in the laboratory by manipulating and controlling the conditions under which the observations are made.

F_1, F_2 In genetics, first and second generations.

FAD Flavin adenine dinucleotide, a type of nucleotide coenzyme; it transfers electrons and unbound protons (H^+) between reaction sites. At such times it is abbreviated $FADH_2$.

fat Lipid with a glycerol head and one, two, or three fatty acid tails. Unsaturated tails have single covalent bonds in the carbon backbone; saturated tails have one or more double bonds.

fatty acid Molecule with a backbone of up to 36 carbon atoms, a carboxyl group ($—COO^-$ or $—COOH$) at one end, and hydrogen atoms at most or all of the other bonding sites.

feedback inhibition Mechanism by which a cellular change resulting from some activity shuts down the activity that brought it about.

fertilization [L. *fertilis*, to carry, to bear] The fusion of a sperm nucleus with the nucleus of an egg, thus forming a zygote.

first law of thermodynamics The total amount of energy in the universe is constant. Energy cannot be created from nothing and existing energy cannot be destroyed.

flagellum (fluh-JELL-um), plural **flagella** A motile structure of many free-living eukaryotic cells. Its core has a 9 + 2 array of microtubules. Typically longer than cilia.

fluid mosaic model A cell membrane is a mix of lipids (organized as a bilayer) and proteins. Structural lipids make it largely impermeable to water-soluble molecules yet impart fluidity by packing variations and motions. Diverse proteins perform most membrane functions (e.g., transport, signal reception).

fluorescence A destabilized molecule emits light when reverting to more stable form.

free radical Any unbound molecular fragment with an unpaired electron.

functional group An atom or a group of atoms that is covalently bonded to the carbon backbone of an organic compound and that influences its chemical behavior.

fungus Eukaryotic heterotroph; it secretes enzymes that digest food to bits that its cells absorb (extracellular digestion, absorption). Saprobic types use nonliving organic matter; parasitic types feed on living organisms.

gamete (GAM-eet) [Gk. *gametēs*, husband, and *gametē*, wife] Haploid cell, formed by meiotic cell division of a germ cell; required for sexual reproduction. Eggs and sperm are examples.

gel electrophoresis Laboratory technique used to distinguish among molecules. Applied electric field forces them to migrate through a viscous gel and distance themselves from one another by length, size, or electric charge.

gene [German *pangan*, after Gk. *pan*, all; *genes*, to be born] Unit of information for a heritable trait, passed from parents to offspring.

gene control One of the molecular mechanisms that govern when and how fast specific genes will be transcribed and translated, and whether gene products will be activated or inactivated.

gene library Mixed collection of bacteria that house many different cloned DNA fragments.

gene mutation A small-scale change in the nucleotide sequence of a DNA molecule.

gene therapy Generally, a transfer of one or more normal genes into an organism to correct or lessen adverse effects of a genetic disorder.

genetic abnormality A rare or less common version of a heritable trait.

genetic code [After L. *genesis*, to be born] The correspondence between nucleotide triplets in DNA (then mRNA) and specific sequences of amino acids in a polypeptide chain; the basic language of protein synthesis in cells.

genetic disorder Any inherited condition that causes mild to severe medical problems.

genetic engineering Deliberately altering the information content of DNA molecules.

genetic recombination Result of any process that puts new genetic information into a DNA molecule (e.g., by crossing over).

genome All the DNA in a haploid number of chromosomes for a given species.

genomics Study of the genome of humans and other organisms.

genotype (JEEN-oh-type) Genetic constitution of an individual; a single gene pair or the sum total of an individual's genes.

genus, plural **genera** (JEEN-US, JEN-er-ah) [L. *genus*, race or origin] A grouping of species that are more closely related to one another in their morphology, ecology, and history than to other species at the same taxonomic level.

germ cell Animal cell of a lineage set aside for sexual reproduction; gives rise to gametes.

global warming A long-term increase in the temperature of the Earth's lower atmosphere.

glycolysis (gly-CALL-ih-sis) [Gk. *glykys*, sweet, + *lysis*, breaking apart] Breakdown of glucose or another organic compound to two pyruvate molecules. First stage of aerobic respiration, fermentation, and anaerobic electron transfer.

Oxygen has no role in glycolysis, which occurs in the cytoplasm of all cells. Two NADH form. Net yield: 2 ATP per glucose molecule.

Golgi body (GOHL-gee) Organelle of lipid assembly, polypeptide chain modification, and packaging of both in vesicles for export or for transport to locations in cytoplasm.

growth factor Protein that plays a role in the body's growth (e.g., by promoting mitosis).

guanine Nitrogen-containing base in one of four nucleotide monomers of DNA or RNA.

haploidy (HAP-loyd-ee) Presence of only half of the parental number of chromosomes in a spore or gamete, as brought about by meiosis.

heat Thermal energy; a form of kinetic energy.

HeLa cell Cancer cell of a lineage established for research; now used in many laboratories.

heterotroph (HET-er-oh-trofe) [Gk. *heteros*, other, + *trophos*, feeder] Organism unable to make its own organic compounds; feeds on autotrophs, other heterotrophs, organic wastes.

heterozygous condition (HET-er-oh-ZYE-guss) [Gk. *zygoun*, join together] Having a pair of nonidentical alleles at a gene locus (that is, on a pair of homologous chromosomes).

histone Type of protein intimately associated with eukaryotic DNA and largely responsible for organization of eukaryotic chromosomes.

HLA One of a class of recognition proteins important in immune responses.

homeostasis (HOE-me-oh-STAY-sis) [Gk. *homo*, same, + *stasis*, standing] State in which physical and chemical aspects of internal environment (blood, interstitial fluid) are being maintained within ranges suitable for cell activities.

homeotic gene In all major animal groups, one of the master genes, the products of which interact with one another and with control elements to map out the overall body plan.

homologous chromosome (huh-MOLL-uh-gus) [Gk. *homologia*, correspondence] Of cells with a diploid chromosome number, one of a pair of chromosomes identical in size, shape, and gene sequence, and that interact at meiosis. Nonidentical sex chromosomes (e.g., X and Y) also interact as homologues during meiosis.

homozygous dominant condition Having a pair of dominant alleles (*AA*) at a gene locus (on a pair of homologous chromosomes).

homozygous recessive condition Having a pair of recessive alleles (*Aa*) at a gene locus (on a pair of homologous chromosomes).

hormone [Gk. *hormon*, stir up, set in motion] Signaling molecule secreted by one cell that stimulates or inhibits activities of any cell with receptors for it. Animal hormones are picked up and transported by the bloodstream.

hybrid offspring Of a genetic cross, offspring having a pair of nonidentical alleles for a trait.

hydrocarbon An organic compound that has only hydrogen bonded to a carbon backbone.

hydrogen bond Weak interaction between a small, highly electronegative atom and an H atom already taking part in a polar covalent bond in the same molecule or a different one.

hydrogen ion Free (or unbound) proton; one hydrogen atom that lost its electron and now bears a positive charge (H^+).

hydrolysis (high-DRAWL-ih-sis) [L. *hydro*, water, + Gk. *lysis*, loosening] Cleavage reaction that breaks covalent bonds and splits a molecule into two or more parts. H^+ and OH^- (from a water molecule) are often attached to the newly exposed bonding sites.

hydrophilic substance [Gk. *philos*, loving] A polar molecule or molecular region that easily dissolves in water (e.g., sugars).

hydrophobic substance [Gk. *phobos*, dreading] A nonpolar molecule or molecular region that strongly resists dissolving in water (e.g., oils).

hydrostatic pressure Pressure exerted by a volume of fluid against a wall, membrane, or some other structure that encloses the fluid.

hydrothermal vent A steaming fissure in the deep ocean floor; has unique ecosystems.

hypertonic solution A fluid having a greater solute concentration relative to another fluid.

hypothesis In science, a possible explanation of a phenomenon, one that has the potential to be proved false by experimental tests.

hypotonic solution A fluid that has a lower solute concentration relative to another fluid.

ICE-like protease One enzyme of apoptosis.

in vitro fertilization Conception outside the body ("in glass" petri dishes or test tubes).

incomplete dominance Condition in which one allele of a pair is not fully dominant; a heterozygous phenotype somewhere between both homozygous phenotypes emerges.

independent assortment theory Mendelian theory that, by the end of meiosis, each pair of homologous chromosomes (and linked genes on each one) are sorted before shipment to gametes independently of how the other pairs were sorted. Later modified to account for the disruptive effect of crossing over on linkages.

induced-fit model When bound to an active site, a substrate promotes reactivity by altering an enzyme's shape; it brings about a more precise molecular fit between the two.

inductive logic Pattern of thinking; deriving a general statement from specific observations.

inheritance The transmission, from parents to offspring, of genes that specify structures and functions characteristic of the species.

insertion Insertion of one to a few bases into a DNA strand. Also, a movable attachment of muscle to bone.

intermediate Substance that forms between the start and end of a metabolic pathway.

intermediate filament Cytoskeletal element; mechanically strengthens some animal cells.

interphase Of a cell cycle, interval between nuclear divisions when a cell increases in mass and roughly doubles the number of its cytoplasmic components. It also duplicates its chromosomes (replicates its DNA) during interphase, but *not* between meiosis I and II.

intron A noncoding portion of a pre-mRNA transcript; excised before translation.

inversion Part of a chromosome that became oriented in reverse, with no molecular loss.

ion, negatively charged (EYE-on) An atom or a molecule that acquired an overall negative charge by gaining one or more electrons.

ion, positively charged Atom or molecule that acquired an overall positive charge by losing one or more electrons.

ionic bond Two ions being held together by the attraction of their opposite charge.

ionizing radiation High-energy wavelengths.

isotonic solution A fluid with the same solute concentration as another fluid being studied.

isotope (EYE-so-tope) One of two or more forms of an element's atoms that differ in the number of neutrons.

karyotype (CARE-ee-oh-type) Preparation of metaphase chromosomes sorted by length, centromere location, other defining features.

kinase Enzyme (e.g., a protein kinase) that catalyzes a phosphate-group transfer.

kinetic energy Energy of motion.

Krebs cycle A stage of aerobic respiration, in mitochondria only; it (and a few preparatory steps) breaks down pyruvate to CO_2 and H_2O. Many coenzymes accept protons (H^+) and electrons from intermediates and deliver them to the next stage of reactions; 2 ATP form.

lactate fermentation An anaerobic pathway of ATP formation. Pyruvate from glycolysis is converted to three-carbon lactate, and NAD^+ is regenerated. Net energy yield: 2 ATP.

light-dependent reactions The first stage of photosynthesis. Sunlight energy is trapped and converted to chemical energy of ATP, NADPH, or both, depending on the pathway.

light-independent reactions Second stage of photosynthesis; sugar-building reactions that require phosphate-group transfers from ATP, electrons and H atoms from NADPH, and carbon from CO_2. The phosphorylated sugars from these reactions are then converted to end products (e.g., sucrose, cellulose, starch).

linkage group All genes on a chromosome.

lipid A mostly greasy or oily hydrocarbon; resists dissolving in water but dissolves in nonpolar substances. All cells use lipids as storage forms of energy, structural materials as in membranes, and cell products.

lipid bilayer Phospholipids, mostly, arranged in two layers; the structural basis of all cell membranes. Hydrophobic tails are sandwiched between the hydrophilic heads; the heads are dissolved in intracellular or extracellular fluid.

lysosome (LYE-so-sohm) Important organelle of intracellular digestion.

meiosis (my-OH-sis) [Gk. *meioun*, to diminish] Two-stage nuclear division process that halves the chromosome number of a parental germ cell nucleus, to the haploid number. Basis of gamete formation (and meiospore formation).

messenger RNA (mRNA) A single strand of ribonucleotides transcribed from DNA, then translated into a polypeptide chain. The only RNA encoding protein-building instructions.

metabolic pathway (MEH-tuh-BALL-ik) Orderly sequence of enzyme-mediated reactions by which cells maintain, increase, or decrease the concentrations of particular substances.

metabolism (meh-TAB-oh-lizm) [Gk. *meta*, change] All the controlled, enzyme-mediated chemical reactions by which cells acquire and use energy to synthesize, store, degrade, and eliminate substances in ways that contribute to growth, survival, and reproduction.

metaphase Of meiosis I, stage when all pairs of homologous chromosomes have become positioned at the spindle equator. Of mitosis or meiosis II, all the duplicated chromosomes are positioned at the spindle equator.

metastasis Abnormal migration of cancer cells, which may establish colonies in other tissues.

methylation Attachment of a methyl group to an organic compound; a common gene control.

microfilament Cytoskeletal element of two thin, helically twisted polypeptide chains; functions in movement, cell shape.

micrograph Photograph of an image brought into view with the aid of a microscope.

microtubule (my-crow-TUBE-yool) Cylinder of tubulin subunits; cytoskeletal element with roles in most eukaryotic cell movements.

mitochondrion (MY-toe-KON-dree-on) Double-membrane organelle of ATP formation. Only site of aerobic respiration's second and third stages. May have endosymbiotic origins.

mitosis (my-TOE-sis) [Gk. *mitos*, thread] Type of nuclear division that maintains the parental chromosome number for daughter cells. The basis of growth in size, tissue repair, and often asexual reproduction for eukaryotes.

mixture Two or more elements intermingled in proportions that can and usually do vary.

model Theoretical, detailed description or analogy that helps people visualize something that has not yet been directly observed.

molecule Two or more atoms of the same or different elements joined by chemical bonds.

monohybrid cross Intercross between two F_1 heterozygotes that are identical for one gene locus; offspring of two parents that breed true for different forms of a trait.

monomer Small molecule used as a subunit of polymers, such as sugar monomers of starch.

monosaccharide (MON-oh-SAK-ah-ride) [Gk. *monos*, alone, single, + *sakcharon*, sugar] One of the simple carbohydrates (e.g., glucose).

mosaicism, mosaic tissue effect Cells of same type express genes differently, so phenotypic differences emerge in same type of tissue. E.g., occurs by X chromosome inactivation in female mammals; also by nondisjunction in any cell after fertilization (only descendants of altered cell inherit the abnormal chromosome number).

motor protein Type of protein (e.g., myosin) attached to microfilaments and microtubules; used in cell movements (e.g., contraction).

multicelled organism Organism composed of many cells with coordinated metabolic activity; most show extensive cell differentiation into tissues, organs, and organ systems.

multiple allele system Three or more slightly different molecular forms of a gene that occur among individuals of a population.

mutation [L. *mutatus*, a change, + *-ion*, act, result, or process] Heritable change in DNA's molecular structure. Original source of all new alleles and, ultimately, the diversity of life.

mutation rate Of a gene locus, the probability that a spontaneous mutation will occur during or between DNA replication cycles.

NAD⁺ Nicotinamide adenine dinucleotide. A nucleotide coenzyme; abbreviated NADH when carrying electrons and H^+ to a reaction site.

natural selection Microevolutionary process; the outcome of differences in survival and reproduction among individuals that differ in details of heritable traits.

negative control Use of regulatory proteins to trigger a slowdown in gene expression.

neoplasm Abnormally grown mass of cells; if benign, remains in tissue. If malignant, it metastasizes; it is a cancer.

neutron Unit in the atomic nucleus that has mass but no electric charge.

nondisjunction Failure of sister chromatids or a pair of homologous chromosomes to separate during meiosis or mitosis. Daughter cells end up with too many or too few chromosomes.

non-ionizing radiation Wavelengths that boost electrons to a higher energy level. DNA easily absorbs one form, ultraviolet light.

nuclear envelope Outermost portion of the cell nucleus; consists of a double membrane (two lipid bilayers and associated proteins).

nucleic acid (new-CLAY-ik) Single- or double-stranded chain of four kinds of nucleotides joined at their phosphate groups. Nucleic acids (e.g., DNA, RNA) differ in base sequences.

nucleic acid hybridization Any base-pairing between DNA or RNA from different sources.

nucleoid (NEW-KLEE-oid) Portion of bacterial cell interior in which the DNA is physically organized but not enclosed by a membrane.

nucleolus (new-KLEE-oh-lus) [L. *nucleolus*, tiny kernel] In a nondividing cell nucleus, a site for assembling protein and RNA subunits that will later join up as ribosomes in the cytoplasm.

nucleosome (NEW-klee-oh-sohm) A stretch of eukaryotic DNA looped twice around a spool of histone molecules; one of many units that give condensed chromosomes their structure.

nucleotide (NEW-klee-oh-tide) Small organic compound with deoxyribose (a five-carbon sugar), a nitrogenous base, and a phosphate group. Monomer for adenosine phosphates, nucleotide coenzymes, and nucleic acids.

nucleus (NEW-klee-us) [L. *nucleus*, a kernel] Of atoms, a central core of one or more protons and (in all but hydrogen atoms) neutrons. In a eukaryotic cell, the organelle that physically separates DNA from cytoplasmic machinery.

oligosaccharide (oh-LIG-oh-SAC-uh-rid) Short-chain carbohydrate of two or more covalently bonded sugar monomers (e.g., disaccharides).

oncogene (ON-koe-jeen) Any gene having the potential to induce cancerous transformation.

oocyte Type of immature egg.

operator Very short base sequence between a promoter and bacterial genes; a binding site for a repressor that can block transcription.

operon Promoter–operator sequence serving more than one bacterial gene; part of a control that adjusts transcription rates up or down.

organelle (or-guh-NELL) Membrane-bound sac or compartment in the cytoplasm having one or more specialized metabolic functions. Most eukaryotic cells have a profusion of them.

organic compound A molecule containing carbon and at least one hydrogen atom.

osmosis (oss-MOE-sis) [Gk. *osmos*, pushing] In response to a water concentration gradient, the diffusion of water between two regions that a selectively permeable membrane separates.

osmotic pressure Pressure that operates after hydrostatic pressure develops in an enclosed region (e.g., a cell); it counters water's inward diffusion (stops further rises in fluid volume).

oxaloacetate (ox-AL-oh-ASS-ih-tate) A four-carbon compound with roles in metabolism (e.g., the point of entry into the Krebs cycle).

oxidation–reduction reaction A transfer of an electron (and often an unbound proton, or H^+) between atoms or molecules.

passive transport Event in which a transport protein that spans a cell membrane passively permits a solute to diffuse through its interior. Also called facilitated diffusion.

PCR Polymerase chain reaction. A method of enormously amplifying the quantity of DNA fragments cut by restriction enzymes.

pedigree Diagram of the genetic connections among related individuals through successive generations; uses standardized symbols.

peroxisome A vesicle; its enzymes digest fatty acids and amino acids to hydrogen peroxide, which is then converted to harmless products.

PGA Phosphoglycerate (FOSS-foe-GLISS-er-ate) Important intermediate of glycolysis and of the Calvin–Benson cycle.

PGAL Phosphoglyceraldehyde. Intermediate of glycolysis and of the Calvin–Benson cycle.

pH scale Measure of the concentration of free hydrogen ions (H^+) in blood, water, and other solutions. pH 0 is the most acidic, 14 the most basic, and 7, neutral.

phagocytosis Of some cells, engulfment of an extracellular target by way of pseudopod formation and endocytosis.

phenotype (FEE-no-type) [Gk. *phainein*, to show + *typos*, image] Observable trait or traits of an individual that arise from gene interactions and gene–environment interactions.

phospholipid Organic compound that has a glycerol backbone, two fatty acid tails, and a hydrophilic head of two polar groups (one being phosphate). Phospholipids are the main structural component of cell membranes.

phosphorylation (FOSS-for-ih-LAY-shun) An enzyme activates a molecule by attaching inorganic phosphate to it or by mediating a phosphate-group transfer to it, as from ATP.

photoautotroph Photosynthetic autotroph; any organism that synthesizes its own organic compounds using CO_2 for carbon atoms and sunlight for energy. Nearly all plants, some protistans, and a few bacteria do this.

photolysis (foe-TALL-ih-sis) [Gk. *photos*, light, +-*lysis*, breaking apart] Reactions split water molecules using photon energy. The released electrons, hydrogen used in noncyclic pathway of photosynthesis; oxygen is a by-product.

photon One unit of energy of visible light.

photosynthesis Sunlight energy trapped, converted to chemical energy (ATP, NADPH, or both); then synthesis of sugar phosphates that are converted to sucrose, cellulose, starch, and other end products. The main pathway by which energy and carbon enter the web of life.

photosystem Cluster of many light-trapping pigments in a photosynthetic membrane.

phycobilin (FIE-koe-BY-lin) Type of accessory pigment; notably abundant in red algae and in cyanobacteria.

phytochrome A light-sensitive pigment. Its controlled activation and inactivation affect plant hormone activities that govern leaf expansion, stem branching, stem lengthening, and often seed germination and flowering.

pigment Any light-absorbing molecule.

plant Generally, a multicelled photoautotroph with well-developed root and shoot systems; photosynthetic cells that include starch grains as well as chlorophylls *a* and *b*; and cellulose, pectin, and other polysaccharides in cell walls.

plasma membrane Outermost cell membrane; structural and functional boundary between cytoplasm and the fluid outside the cell.

plasmid A small, circular molecule of extra bacterial DNA that carries a few genes and is replicated independently of the chromosome.

pleiotropy (PLEE-oh-troe-pee) [Gk. *pleon*, more, + *trope*, direction] Positive or negative effects on two or more traits owing to expression of alleles at a single gene locus. Effects may or may not emerge at the same time.

polymer Large molecule of three to millions of monomers of the same or different kinds.

polypeptide chain An organic compound of three or more amino acids joined by peptide bonds; atoms of its backbone have this pattern: —N—C—C—N—C—C— . All proteins are composed of one or more polypeptide chains.

polyploidy (POL-ee-PLOYD-ee) Having three or more of each type of chromosome in the nucleus of a eukaryotic cell at interphase.

polysaccharide [Gk. *polus*, many, + *sakcharon*, sugar] Straight or branched chain of many covalently linked sugar units of the same or different kinds. In nature, the most common types are cellulose, starch, and glycogen.

polytene chromosome Insect chromosome consisting of many in-parallel copies of DNA.

population All individuals of the same species that are occupying a specified area.

positive control Use of regulatory proteins to promote gene expression.

potential energy A stationary object's capacity to do work owing to its position in space or to the arrangement of its parts.

prediction Statement about what you should observe in nature if you were to go looking for a particular phenomenon; the if–then process.

pressure gradient A difference in pressure between two adjoining regions.

primary wall A thin, flexible plant cell wall of cellulose, polysaccharides, and glycoproteins; allows growing cells to divide or change shape.

primer Short nucleotide sequence designed to base-pair with any complementary DNA sequence; acts as a START tag for replication.

probability The chance that each outcome of a given event will occur is proportional to the number of ways the outcome can be reached.

probe Very short stretch of DNA labeled with a radioisotope; it is designed to base-pair with part of a gene in a DNA sample being studied.

producer Autotroph (self-feeder); it nourishes itself using sources of energy and carbon from the physical environment. Photoautotrophs and chemoautotrophs are examples.

product Substances left at end of a reaction.

prokaryotic cell (pro-CARE-EE-oh-tic) [L. *pro*, before, + Gk. *karyon*, kernel] Archaebacterium or eubacterium; single-celled organism, most often walled; lacks the profusion of membrane-bound organelles observed in eukaryotic cells.

promoter Short stretch of DNA to which RNA polymerase can bind and start transcription.

prophase Of mitosis, a stage when duplicated chromosomes start to condense, microtubules form a spindle, and the nuclear envelope starts to break up. Duplicated pairs of centrioles (if present) are moved to opposite spindle poles.

protein Organic compound of one or more polypeptide chains folded and twisted into a globular or fibrous shape, overall.

protistan (pro-TISS-tun) [Gk. *prōtistos*, primal] Photoautotroph or heterotroph (or both) unlike bacteria; some like earliest eukaryotic cells. Has a nucleus, larger ribosomes, mitochondria, ER, Golgi bodies, chromosomes with numerous proteins, and cytoskeletal microtubules. Range in size from microscopic algae to giant kelps.

proton Positively charged particle; one or more in the nucleus of each atom. An unbound (free) proton is called a hydrogen ion (H^+).

pseudopod A dynamically extending lobe of cytoplasm used for motility or engulfment.

Punnett-square method Construction of a simple diagram as a way to predict probable outcomes of a genetic cross.

pyruvate (PIE-roo-vate) Organic compound with a backbone of three carbon atoms. Two molecules form as end products of glycolysis.

radioisotope Unstable atom (uneven number of protons and neutrons). It spontaneously emits particles and energy, and so decays into a different atom over a predictable time span.

reactant Substance that enters a reaction.

reaction center The only molecule (a special chlorophyll *a*) that can pass electrons out of a photosystem to a nearby acceptor molecule.

receptor, molecular Membrane protein (or cytoplasmic protein) that triggers a change in cell activities after it binds signaling molecule.

reciprocal cross A paired cross. In the first cross, one parent displays the trait of interest. In the second, the other parent displays it.

recognition protein One of a class of plasma membrane proteins that distinguish *nonself* (foreign) from *self* (belonging to a body tissue).

recombinant DNA technology Procedures by which DNA molecules from different species are isolated, cut up, spliced together, and then enormously amplified to useful quantities.

regulatory protein Component of mechanisms that control transcription, translation, and gene products by interacting with DNA, RNA, new polypeptide chains, or proteins (e.g., enzymes).

reproduction Any process by which a parental cell or organism produces offspring. Among eukaryotes, asexual modes (e.g., binary fission, budding, vegetative propagation) and sexual modes. Prokaryotes use prokaryotic fission only. Viruses cannot reproduce themselves; host organisms execute their replication cycle.

rrestriction enzyme One of a class of bacterial enzymes that can cut apart foreign DNA that infects a cell, as by viral attack. Important tool of recombinant DNA technology.

reverse transcription Synthesis of DNA on an RNA template by reverse transcriptase, a viral enzyme. Basis of RNA virus replication cycle and of cDNA synthesis in the laboratory.

ribosomal RNA (rRNA) Type of RNA that combines with proteins to form ribosomes, on which polypeptide chains are assembled.

ribosome Two subunits of rRNA and proteins briefly joined together as a structure on which mRNA is translated into polypeptide chains.

RNA Ribonucleic acid. Any of a class of single-stranded nucleic acids that function in transcribing and translating the genetic instructions encoded in DNA into proteins.

RNA polymerase Enzyme that catalyzes the assembly of RNA strands on DNA templates.

rubisco RuBP carboxylase; an enzyme that catalyzes attachment of the carbon atom from CO_2 to RuBP and so starts the Calvin–Benson cycle of the light-independent reactions.

RuBP Ribulose bisphosphate; five-carbon organic compound; used for carbon fixation in the Calvin–Benson cycle, which regenerates it.

salt Compound that releases ions other than H^+ and OH^- in solution.

sampling error Use of a sample or subset of a population, an event, or some other aspect of nature for an experimental group that is not large enough to be representative of the whole.

second law of thermodynamics A law of nature stating that the spontaneous direction of energy flow is from organized forms to less organized forms; with each conversion, some energy is randomly dispersed in a form (usually heat) not as useful for doing work.

secondary wall Of older plant cells no longer growing but in need of structural support, a wall on the inner surface of the primary wall. Contains lignin in older cells of woody plants.

seed bank A safe storage facility where genes of diverse plant lineages are being preserved.

segregation, theory of [L. *se-*, apart, + *grex*, herd] Mendelian theory. Sexually reproducing organisms inherit pairs of genes (on pairs of homologous chromosomes), the two genes of each pair are separated from each other at meiosis, and they end up in separate gametes.

selective permeability Of a cell membrane, a capacity to let some substances but not others cross at certain sites, at certain times owing to its bilayer structure and its transport proteins.

sex chromosome A chromosome with genes that affect sexual traits. Depending on the species, somatic cells have one or two sex chromosomes of the same or different type (e.g., in mammals, XX females, XY males).

sexual reproduction Production of offspring by meiosis, gamete formation, and fertilization.

shell model Model of electron distribution in which all orbitals available to electrons of atoms occupy a nested series of shells.

sister chromatid (CROW-mah-tid) Of a duplicated chromosome, one of two DNA molecules (and associated proteins) attached at the centromere until they are separated from each other at mitosis or meiosis; each is then designated a separate chromosome.

sodium–potassium pump Type of membrane transport protein that, when activated by ATP, selectively transports potassium ions across a membrane against its concentration gradient, and passively allows sodium ions to cross in the opposite direction.

solute (SOL-yoot) [L. *solvere*, to loosen] Any substance dissolved in a solution.

somatic cell (so-MAT-ik) [Gk. *sōmā*, body] Any body cell that is not a germ cell. (Germ cells are the forerunners of gametes.)

species (SPEE-sheez) [L. *species*, a kind] One kind of organism. Of sexually reproducing organisms, one or more natural populations in which individuals are interbreeding and are reproductively isolated from other such groups.

sperm [Gk. *sperma*, seed] Mature male gamete.

spindle apparatus Dynamic, temporary array of microtubules that moves chromosomes in precise directions during mitosis or meiosis.

spore A reproductive or resting structure of one or a few cells, often walled or coated; used for resisting harsh conditions, dispersal, or both. May be nonsexual or sexual, as formed by way of meiosis. Some bacteria as well as sporozoans, fungi, and plants form spores.

sterol (STAIR-all) Lipid with a rigid backbone of four fused carbon rings (e.g., cholesterol). Sterols differ in the number, position, and type of their functional groups.

stimulus [L. *stimulus*, goad] A specific form of energy (e.g., pressure, light, and heat) that activates a sensory receptor able to detect it.

stoma (STOW-muh), plural **stomata** [Gk. *stoma*, mouth] A gap between two guard cells in leaf or stem epidermis. Opens or closes to control CO_2 movement into a plant and H_2O and O_2 out of it. Stomata help plants conserve water.

stroma [Gk. *strōma*, bed] A semifluid matrix between the thylakoid membrane system and two outer membranes of a chloroplast; a zone where sucrose, starch, cellulose, and other end products of photosynthesis are assembled.

substrate Reactant or precursor for a specific enzyme-mediated metabolic reaction.

substrate-level phosphorylation The direct, enzyme-mediated transfer of a phosphate group from a substrate to a molecule, as when an intermediate of glycolysis is made to give up a phosphate group to ADP, forming ATP.

surface-to-volume ratio Mathematical relation in which the volume of an object expands in three dimensions (e.g., length, width, depth) but its surface area expands in only two dimensions; a constraint on cell size and shape.

syndrome A set of symptoms that may not individually be a telling clue but collectively characterize a genetic disorder or disease.

tandem repeat One of many short sequences of DNA, occurring one after the other, in a chromosome. Used in DNA fingerprinting.

telophase (TEE-low-faze) Of meiosis I, a stage when one member of each pair of homologous chromosomes has arrived at a spindle pole. Of mitosis and of meiosis II, the stage when chromosomes decondense into threadlike structures and two daughter nuclei form.

temperature A measure of the kinetic energy of ions or molecules in a specified region.

test A means to determine the accuracy of a prediction, as by conducting experimental or observational tests and by developing models. Scientific tests are conducted under controlled conditions in nature or the laboratory.

testcross Experimental cross to determine whether an individual of unknown genotype that shows dominance for a trait is either homozygous dominant or heterozygous.

theory, scientific An explanation of the cause or causes of range of related phenomena. It has been rigorously tested but is still open to tests, revision, and tentative acceptance or rejection.

thylakoid Inner chloroplast membrane often folded as interconnected flattened sacs; forms a single compartment for hydrogen ions. Light-trapping pigments, and enzymes used to form ATP, NADPH, or both, are embedded in it.

thymine A nitrogen-containing base; one of the nucleotides in DNA (not in RNA).

tracer Substance with an attached radioisotope that researchers can track after delivering it into a cell, body, ecosystem, or some other system. Its emissions are detected as it moves through a pathway or reaches a destination.

transcription [L. *trans*, across, + *scribere*, to write] First stage of protein synthesis. An RNA strand is assembled on exposed bases of an unwound strand of a DNA double helix. The transcript has a complementary base sequence.

transfer RNA (tRNA) An RNA that binds with and delivers amino acids to a ribosome and that pairs with an mRNA codon during the translation stage of protein synthesis.

transition state The point when a reaction can run either to product or back to reactant.

translation Stage of protein synthesis when an mRNA's base sequence becomes converted to the amino acid sequence of a new polypeptide chain by rRNA, tRNA, and mRNA interactions.

translocation Of cells, movement of a stretch of DNA to a new chromosomal location with no molecular loss. Of vascular plants, distribution of organic compounds by way of phloem.

transport protein Membrane protein that passively or actively assists specific molecules or ions across a membrane's lipid bilayer. The solutes pass through a channel in its interior.

transposon DNA segment that can randomly move to different locations in a genome; may inactivate genes into which it inserts itself and cause changes in phenotype.

triglyceride (neutral fat) A neutral fat; a lipid with three fatty acid tails attached to a glycerol unit. Triglycerides are the animal body's most abundant lipid and its richest energy source.

true breeding lineage Of sexually reproducing species, a lineage in which only one version of a trait appears over the generations in all parents and their offspring.

uracil (YUR-uh-sill) Nitrogen-containing base of a nucleotide in RNA but not DNA. Like thymine, uracil can base-pair with adenine.

variable Of an experimental test, a specific aspect of an object or event that may differ over time and among individuals. A single variable is directly manipulated in an attempt to support or disprove a prediction.

vesicle (VESS-ih-kul) [L. *vesicula*, little bladder] One of various small, membrane-bound sacs in cytoplasm that function in the transport, storage, or digestion of substances.

wavelength A wavelike form of energy in motion. The horizontal distance between the crests of every two successive waves.

wax Organic compound; long-chain fatty acids packed together and attached to long-chain alcohols or carbon rings. Waxes have a firm consistency and repel water.

X chromosome A type of sex chromosome. An XX mammalian embryo becomes female; an XY pairing causes it to develop into a male.

X chromosome inactivation One of the two X chromosomes in somatic cells of mammalian females condenses, which inactivates most of its genes. A dosage compensation mechanism.

X-ray diffraction image Pattern that forms on film exposed to x-rays that have been directed at a molecule; reveals positions of atoms, not the molecular structure.

xenotransplantation The transfer of an organ from one species to another.

Y chromosome Distinctive chromosome in males or females of many species, but not both (e.g., human males XY; human females, XX).

ART CREDITS AND ACKNOWLEDGMENTS

This page is an extension of the copyright page. We made every effort to trace ownership of all copyrighted material and to secure permission from copyright holders. Should any question arise concerning the use of any material, we will make necessary corrections in future printings. Many thanks to the illustrators who rendered our art and to authors, publishers, and agents for granting permission to use their material.

TABLE OF CONTENTS
Page iv Above left, © Gary Head; above right, model by Dr. David B. Goodin, The Scripps Research Institute; below left, DigitalVision/ Picture Quest. **Page v** Mary Osborn, Max Planck Institute for Biophysical Chemistry, Goettingen, FRG. **Page vi** Larry West/FPG. **Page vii** Above, Lisa Starr; below, © 2002 PhotoDisc, computer enhanced by Lisa Starr. **Page ix** Left to right, Lisa Starr with John McNamara, www.paleodirect.com; © Alfred Kamajian; Jean Paul Tibbles; Christopher Ralling. **Page x** Left to right, K. G. Murti/Visuals Unlimited; © Oliver Meckes/Photo Researchers, Inc.; Robert C. Simpson/Nature Stock; © Jim Christensen, Fine Art Digital Photographic Images. **Page xi** Above left to right, © 2002 PhotoDisc/Getty Images; Stanley Sessions, Hartwick College; Pieter Johnson; below left, Jane Burton/Bruce Coleman. **Page xii** Left to right, © David Parker/ SPL/Photo Researchers, Inc.; Ed Reschke; David Cavagnaro/Peter Arnold, Inc.; Juergen Berger, Max Planck Institute for Developmental Biology, Tuebingen, Germany. **Page xiii** Left, © Cory Gray; right, © David Scharf/Peter Arnold, Inc. **Page xiv** Left to right, Eric A. Newman; © 2002 Ken Usami/PhotoDisc /Getty Images; Kenneth Garrett/National Geographic Image Collection. **Page xiv** Left to right, Bone Clones®, www.boneclones.com; Dr. John D. Cunningham/Visuals Unlimited; NSIBC/SPL/Photo Researchers, Inc. **Page xvi** Left, © micrograph Ed Reschke; right, Gunter Ziesler/Bruce Coleman. **Page xvii** Left to right, Dr. Maria Leptin, Institute of Genetics, University of Koln, Germany; © Minden Pictures; Nigel Cook /Daytona Beach News Journal/Corbis Sygma. **Page xviii** Left to right, © W. Perry Conway/CORBIS; Australian Broadcasting Company; Robert Vrijenhoek, MBARI. **Page xix** Gerry Ellis/The Wildlife Collection

INTRODUCTION NASA Space Flight Center

CHAPTER 1
1.1 John McColgan, Bureau of Land Management. **1.2** Lisa Starr. **1.3** All, Jack de Coningh. **1.4** © Y. Arthrus-Bertrand/Peter Arnold, Inc. **1.5** Left, art, Gary Head; photographs, left, Paul DeGreve/ FPG; right, Norman Meyers/Bruce Coleman. **1.6** Art, Lisa Starr; photographs, (a) Walt Anderson /Visuals Unlimited; (b) Gregory Dimijian/Photo Researchers, Inc.; (c) Alan Weaving/Ardea, London. **1.7** Page 8, clockwise from above, © Lewis Trusty/Animals Animals; *Emiliania huxleyi* photograph, Vita Pariente, scanning electron micrograph taken on a Jeol T330A instrument at Texas A&M University Electron Microscopy Center; Carolina Biological Supply Company; R. Robinson/Visuals Unlimited, Inc.; © Oliver Meckes/Photo Researchers, Inc.; © James Evarts; art, Gary Head; page 9, clockwise from above left, Ed Reschke; Edward S. Ross; Edward S. Ross; Robert C. Simpson/Nature Stock; © Stephen Dalton/Photo Researchers, Inc.; CNRI/SPLPhoto Researchers, Inc.; © P. Hawtin, University of Southampton/SPL/Photo Researchers, Inc.; Gary Head. **1.8** All, J. A. Bishop and L. M. Cook. **1.9** (All) Courtesy Derrell Fowler, Tecumseh, Oklahoma. **1.10** Gary Head. **1.11** All, © Gary Head. **1.12** Art, Gary Head and Preface, Inc.; photographs, Dr. Douglas Coleman, The Jackson Laboratory. **1.13** Courtesy Dr. Michael S. Donnenberg. **1.14** James Carmichael Jr./NHPA. **Page 18** © Digital Vision/PictureQuest

Page 19 UNIT I © Cabisco/Visuals Unlimited

CHAPTER 2
2.1 (a) Art, Lisa Starr; (b) Jack Carey. **2.2** Lisa Starr. **2.3** Photograph, Gary Head; art, Lisa Starr. **2.4** (a) © John Griffin/MediChrome; (b, c) Art, Raychel Ciemma; (d) Dr. Harry T. Chugani, M.D., UCLA School of Medicine. **2.5** Micrograph, Maris and Cramer. **2.6, 2.7** Lisa Starr. **2.8** Gary Head and Lisa Starr. **2.9** (a,b) Art, Lisa Starr; micrograph, © Bruce Iverson. **2.10** Vandystadt/Photo Researchers, Inc. **2.11** Lisa Starr. **2.12** (a–c) Art, Lisa Starr; (b, right) photographs © Steve Lissau/Rainbow; (c, right) © Kennan Ward /CORBIS. **2.13** (a) H. Eisenbeiss/Frank Lane Picture Agency; (b) art, Lisa Starr. **2.14** Lisa Starr. **2.15** Art, Lisa Starr; photographs, © 2000 PhotoDisc, Inc. **2.16** Michael Grecco/Picture Group. **Page 33** Left, Lisa Starr; right, Raychel Ciemma

CHAPTER 3
3.1 Above, Dave Schiefelbein; below, © Ron Sanford/Michael Agliolo/Photo Researchers. **Page 35** NASA. **Page 36**, Lisa Starr. **3.2, 3.3, 3.4**, Lisa Starr. **3.5** Art, Gary Head. **3.6** Photograph, Tim Davis/Photo Researchers, Inc.; art, Lisa Starr. **3.7, 3.8, 3.9** Lisa Starr. **3.10** Art and photograph, Lisa Starr. **3.11** © David Scharf/Peter Arnold, Inc. **3.12** (a) Art, Precision Graphics; (b) art, Lisa Starr. **3.13** (a) art, Precision Graphics; (b) Clem Haagner /Ardea, London. **3.14** (a) Art, Lisa Starr ; (b) Art, Precision Graphics. **3.15** (a) Art, Precision Graphics; (b) Larry Lefever/Grant Heilman Photography, Inc.; (c) Kenneth Lorenzen. **Page 43** Lisa Starr. **3.16** All photographs © 2002 PhotoDisc/Getty Images; art, Gary Head. **3.17, 3.18, 3.19** Lisa Starr. **3.20** (a) Photo, Al Giddings/Images Unlimited; art, Lisa Starr; (b) art, Lisa Starr. **3.21** (a,b) Lisa Starr ; (c,d) Gary Head and Lisa Starr. **3.22** (a) art, Preface, Inc.; (b,c) Stanley Flegler/Visuals Unlimited. **3.23** Gary Head and Lisa Starr. **3.24** Gary Head. **3.25** Lisa Starr. **3.26** © 2002 Charlie Waite/Stone/Getty Images. **Page 53** Lisa Starr

CHAPTER 4
4.1 (a) © Bettmann/Corbis; (b) Armed Forces Institute of Pathology; (c) National Library of Medicine; (d) Francis A. Countway Library of Medicine. **4.2** (a) Manfred Kage/Bruce Coleman; (b) George J. Wilder/Visuals Unlimited. **4.3** (a) Precision Graphics; (b) Raychel Ciemma; (c) Lisa Starr. **4.4** Lisa Starr. **4.5** Photograph, Hans Pfletschinger; art, Raychel Ciemma. **4.6** (a) Leica Microsystems, Inc., Deerfield, IL; (b) Art, Gary Head. **4.7** (a) Photograph, George Musil/Visuals Unlimited; art, Gary Head. **4.8** Art, Raychel Ciemma; micrograph, Driscoll, Youngquist, Baldeschwieler/CalTech/Science Source/Photo Researchers, Inc. **4.9** Jeremy Pickett-Heaps, School of Botany, University of Melbourne. **4.10** Raychel Ciemma and Precision Graphics. **4.11** Micrographs, Stephen Wolfe; art, Raychel Ciemma. **4.12** (a, left) Don W. Fawcett/Visuals Unlimited; (a, right) A. C. Faberge, *Cell and Tissue Research*, 151:403–415, 1974; Art, Raychel Ciemma; (b) art, Lisa Starr. **4.13** Art, Raychel Ciemma and Precision Graphics. **4.14** Micrographs (a,b) Don W. Fawcett /Visuals Unlimited; art, Raychel Ciemma. **4.15** Art left, Raychel Ciemma; right, Robert Demarest after a model by J. Kephart; micrograph, Gary Grimes. **4.16** Micrograph, Keith R. Porter; art, Raychel Ciemma. **4.17** Art above, Lisa Starr; micrograph, L. K. Shumway. **4.18** Art, Raychel Ciemma and Preface, Inc.; micrograph, M. C. Ledbetter, Brookhaven National Laboratory. **4.19** Micrograph, G. L. Decker; art, Raychel Ciemma and Preface, Inc. **4.20** (a) © J. W. Shuler/Photo Researchers, Inc.; (b) Courtesy Dr. Vincenzo Cirulli, Laboratory of Developmental Biology, The Whittier Institute for Diabetes, University of California, San Diego, California; (c) Courtesy Mary Osborn, Max Planck Institute for Biophysical Chemistry, Goettingen, FRG. **4.21** Lisa Starr. **4.22** (a) John Lonsdale, www.johnlonsdale.net; (b) David C. Martin, PhD. **4.23, 4.24** Lisa Starr. **4.25** Photographs (a) © Lennart Nilsson; (b) CNRI/ SPL/Photo Researchers, Inc.; art, Precision Graphics after Stephen Wolfe, *Molecular and Cellular Biology*, Wadsworth, 1993; (c) © Andrew Syred/SPL/Photo Researchers, Inc.; (d) art, Precision Graphics after Stephen L. Wolfe, *Molecular and Cellular Biology*, Wadsworth, 1993. **4.26** Ronald Hoham, Dept. of Biology, Colgate University. **4.27** Photographs, (a) Walter Hodge/ Peter Arnold, Inc.; (c) courtesy of Burlington Mills; micrograph, Biophoto Associates/Photo Researchers; art, Raychel Ciemma. **4.28** (a) George S. Ellmore; (b) Ed Reschke. **4.29** Raychel Ciemma and Lisa Starr. **4.30** (a) Art, Lisa Starr; (b) courtesy Dr. G. Cohen–Bazire; (c) K. G. Murti/Visuals Unlimited; (d) R. Calentine /Visuals Unlimited; (e) Gary Gaard, Arthur Kelman. **Page 78** Left, Raychel Ciemma; right, Lisa Starr. **Page 79** Lisa Starr

CHAPTER 5
5.1 (a) © Abraham Menashe; (b) Lisa Starr. **5.2** (a,b) Art, Precision Graphics; (c) art, Raychel Ciemma. **5.3, 5.4, 5.5** Lisa Starr. **Page 86** Raychel Ciemma after Pinto daSilva, D. Branton, *Journal of Cell Biology*, 45:98, by permission of The Rockefeller University Press. **5.6** Gary Head and Precision Graphics. **5.7** Lisa Starr. **5.8** Raychel Ciemma and Gary Head. **5.9** Lisa Starr and Gary Head. **5.10** Lisa Starr. **5.11** Lisa Starr. **5.12** Precision Graphics. **5.13** (a) Art, Raychel Ciemma; (b) M. Sheetz, R. Painter, and S. Singer, *Journal of Cell Biology*, 70:193 (1976) by permission, The Rockefeller University Press. **5.14** Lisa Starr. **5.15** Raychel Ciemma. **5.16** Lisa Starr. **5.17** Micrographs (a–d) M. M. Perry and A. M. Gilbert; (e) Courtesy John Heuser/www.cellbio.wustl.edu. **5.18** (a) © Juergen Berger/Max Planck Institute/ SPL/ Photo Researchers, Inc.; (b) Art, Raychel Ciemma. **5.19** Raychel Ciemma. **5.20** Lisa Starr. **5.21** © Prof. Marcel Bessis/SPL/Photo Researchers, Inc. **5.22** Frieder Sauer/Bruce Coleman

CHAPTER 6
6.1 (a,b) Both models by Dr. David B. Goodin, The Scripps Research Institute. **6.2** (a,b) Photographs by Gary Head; (c) art, Raychel Ciemma. **6.3** Evan Cerasoli. **6.4** Above, NASA; below, Manfred Kage /Peter Arnold, Inc. **6.5** Lisa Starr, using photographs © 1997, 1998 from the Jet Propulsion Laboratory/California Institute of Technology and NASA, 1972. **6.6** Gary Head. **6.7** Lisa Starr. **6.8** Precision Graphics. **6.9** Lisa Starr and Gary Head. **6.10** Gary Head. **6.11** (a,b) Lisa Starr, from B. Alberts et al, *Molecular Biology of the Cell*, 1983, Garland Publishing; (c) Lisa Starr. **6.12** (a,b) Lisa Starr using photographs © 1997, 1998 from Jet Propulsion Laboratory, California Institute of Technology and NASA, 1972. **6.13** (a,b) Thomas A. Steitz. **6.14** Lisa Starr. **6.15** Lisa Starr. **6.16** Lisa Starr. **6.17** (a) Douglas Faulkner/Sally Faulkner Collection; (b,c) art, Gary Head. **6.18** Left, courtesy Dr. Edward C. Klatt; right, courtesy of Downstate Medical Center, Department of Pathology, Brooklyn, NY. **6.19** (a) © Frank Borges Llosa/www.frankley.com. (b) Sara Lewis, Tufts University; (c) Art, Lisa Starr. **6.20** Professor J. Woodland Hastings, Harvard University.

6.21 (a,b) C. Contag, *Molecular Microbiology*, November, 1985, 18 (4):593. "Photonic Detection of Bacterial Pathogens in Living Hosts." Reprinted by permission of Blackwell Science. **Page 113** Model by Dr. David B. Goodin, The Scripps Research Institute

CHAPTER 7

7.1 Art, Raychel Ciemma; micrograph, Carolina Biological Supply Company. **7.2** Art, Lisa Starr; photograph, Wernhner Krutein/PhotoVault. **Page 115** Lisa Starr. **7.3** Lisa Starr with Preface, Inc. **Page 117** Gary Head. **7.4** Gary Head and Lisa Starr with photograph from David Neal Parks. **7.5** (a) Precision Graphics; (b, above) © 2002 PhotoDisc; (b, below) Bobby Mantoni. **Page 118** Gary Head. **7.6** (a,b) Lisa Starr after Stephen L. Wolfe, *Molecular and Cellular Biology*, Wadsworth; (c) Precision Graphics and Gary Head after Govindjee. **7.7** Left, Precision Graphics; (a,b) Lisa Starr. **7.8** (a) Douglas Faulkner/Sally Faulkner Collection; (b) Herve Chaumeton/Agence Nature. **7.9** Larry West/FPG. **7.10** Lisa Starr. **7.11** Lisa Starr. **Page 122** Gary Head. **7.12** Lisa Starr. **7.13** Lisa Starr. **7.14** E. R. Degginger. **7.15** Lisa Starr. **7.16** Lisa Starr. **Page 125** Gary Head. **7.17** (a, above) © Bill Beatty/Visuals Unlimited; (a,below) micrograph, Bruce Iverson, computer-enhanced by Lisa Starr; (b, above) © 2001 PhotoDisc; (b, below) micrograph by Ken Wagner/Visuals Unlimited, computer-enhanced by Lisa Starr. **Page 126** Gary Head. **7.18** Gary Head. **7.19** © 2002/Jeremy Woodhouse/PhotoDisc/Getty Images; Gary Head. **7.20** (a,b) NASA. **7.21** Lisa Starr. **Page 130** Lisa Starr. **7.22** Steve Chamberlain, Syracuse University. **7.23** Clockwise from above left, Gary Head; © 2002/PhotoDisc/Getty Images; C. B. Frith and D. W. Frith/Bruce Coleman, Ltd.; © 2002/PhotoDisc/Getty Images; © Rudiger Lhenen /SPL/Photo Researchers, Inc.

CHAPTER 8

8.1 Stephen Dalton/Photo Researchers, Inc., computer-enhanced by Lisa Starr. **8.2** Raychel Ciemma and Gary Head. **8.3** (a, left) Gary Head; (a, right) Ed Reschke; (b) Paolo Fioratti; (c) Lisa Starr with Gary Head. **Page 136** Lisa Starr. **8.4** page 136, Raychel Ciemma and Lisa Starr; page 137, Lisa Starr and Gary Head, after Ralph Taggart. **Page 137** Lisa Starr. **8.5** (a) Art, Raychel Ciemma; micrograph, Keith R. Porter; (b,c) art, Lisa Starr. **8.6** Photo, © 2001 PhotoDisc, Inc.; art left, Raychel Ciemma; art right, Lisa Starr. **8.7** Above, Raychel Ciemma; below, Lisa Starr. **8.8** Lisa Starr with Preface, Inc. **8.9** Lisa Starr. **Page 142** Lisa Starr. **8.10** (a) Lisa Starr with Gary Head; (b) Adrian Warren/Ardea, London; (c) David M. Phillips/Visuals Unlimited. **8.11** William Grenfell/Visuals Unlimited. **8.12** Art, Lisa Starr; photograph, Gary Head. **Page 146** Art, Gary Head; photograph R. Llewellyn/SuperStock. **Page 147** Lisa Starr. **Page 148** Thomas D. Mangelsen/Images of Nature

Page 149 UNIT II © Francis Leroy, Biocosmos/Science Photo Library/Photo Researchers

CHAPTER 9

9.1 Left and right, center, Chris Huss; right above and below, Tony Dawson. **9.2** Gary Head. **9.3** Art, Lisa Starr and Raychel Ciemma; micrographs (a) C. J. Harrison et al, *Cytogenetics and Cell Genetics*, 35: 21–27 © 1983 S. Karger, A. G. Basel; (c) B. Hamkalo; (d) O. L. Miller, Jr., Steve L. McKnight. **9.4** Raychel Ciemma and Gary Head. **9.5** All micrographs courtesy Andrew S. Bajer, University of Oregon. **9.6** Gary Head. **9.7** Art, Raychel Ciemma and Gary Head; micrographs, © Ed Reschke. **9.8** Art, Lisa Starr; micrograph, © R. Calentine/Visuals Unlimited. **9.9** Art, Raychel Ciemma; (d) micrograph, © D. M. Phillips/Visuals Unlimited. **9.10** (a–c, e) Lennart Nilsson from

A Child Is Born © 1966, 1977 Dell Publishing Company, Inc.; (d) Lennart Nilsson from *Behold Man*, © 1974 by Albert Bonniers Förlag and Little, Brown & Company, Boston. **9.11** Photograph right, Courtesy of the Family of Henrietta Lacks; micrograph, Dr. Pascal Madaule, France. **Page 161** Raychel Ciemma

CHAPTER 10

10.1 (a) Jane Burton/Bruce Coleman; (b) Dan Kline /Visuals Unlimited. **10.2** Raychel Ciemma. **10.3** © CNRI/SPL/Photo Researchers, Inc. **Page 165** Raychel Ciemma. **10.4** Micrographs, with thanks to the John Innes Foundation Trustees, computer-enhanced by Gary Head; art, Raychel Ciemma; **10.5, 10.6** Raychel Ciemma. **10.7** Art, Preface, Inc.; photograph, David Maitland/Seaphot Limited. **10.8** Lisa Starr. **10.9** Micrograph, David M. Phillips/Visuals Unlimited; art, Lisa Starr. **10.10** Raychel Ciemma. **10.11** Precision Graphics. **Page 175** Raychel Ciemma. **10.12** © Richard Corman/Corbis Outline

CHAPTER 11

11.1 Tom Cruise photo, ©AFP/CORBIS; Charles Barkley photo, Focus on Sports; Gregor Mendel painting, The Moravian Museum, Brno; Joan Chen photo, Fabian/Corbis Sygma. **11.2** Art, Jennifer Wardrip; photograph, Jean M. Labat/Ardea, London. **11.3** Lisa Starr. **11.4** Precision Graphics. **11.5** Raychel Ciemma and Precision Graphics. **11.6** Precision Graphics. **11.7** Raychel Ciemma with Precision Graphics. **Page 182** Raychel Ciemma and Precision Graphics. **11.8** Raychel Ciemma. **11.9** Raychel Ciemma and Precision Graphics. **11.10** Photographs, William E. Ferguson; art, Raychel Ciemma. **11.11** Art, Precision Graphics. **11.12** (a) © Bettmann/CORBIS; (b) art, Lisa Starr. **11.13** Art, Preface, Inc.; photographs, (a,b) Michael Stuckey/Comstock, Inc.; (c) Bosco Broyer; photograph, Gary Head. **11.14** David Hosking. **11.15** Photographs, Ted Somes. **11.16** Above and below, Frank Cezus/FPG; Frank Cezus/FPG; © 2001 PhotoDisc, Inc.; Ted Beaudin /FPG; Stan Sholik/FPG. **11.17** (a) Photographs courtesy Ray Carson, University of Florida News and Public Affairs; art, Gary Head. **11.18** Photograph, Jane Burton/Bruce Coleman; art, D. Hennings and V. Hennings. **11.19** Photograph, © Pamela Harper/Harper Horticultural Slide Library; art, Lisa Starr, referencing Professor Otto Wilhelm Thomé, Flora von Deutschland Österreich und der Schweiz, 1885, Gera, Germany. **11.20** Left, Eric Crichton/Bruce Coleman; right, William E. Ferguson. **11.21** Evan Cerasoli. **Page 192** Gary Head. **11.22** Leslie Faltheisek. Clacritter Manx. **11.23** © Joe McDonald/Visuals Unlimited

CHAPTER 12

12.1 From "Multicolor Spectral Karyotyping of Human Chromosomes," by E. Schrock, T. Ried et al, *Science*, 26, July 1966, 273:495. Used by permission of E. Schrock and T. Reid and the American Association for the Advancement of Science, computer-enhanced by Lisa Starr. **12.2** From P. Maslak, Blast Crisis of Chronic Myelogenous Leukemia. Posted online December 5, 2001. ASH Image Bank. Copyright American Society of Hematology, used with permission. **12.3** Raychel Ciemma. **12.4** Photograph, Charles D. Winters/Photo Researchers; micrograph, Omikron/Photo Researchers; art, Raychel Ciemma. **12.5** Art, Precision Graphics and Gary Head; photographs left, © 2001 EyeWire; right, © 2001 PhotoDisc, Inc. **12.6** (a) Photograph, from Lennart Nilsson, *A Child Is Born*, © 1966, 1977 Dell Publishing Company, Inc.; (b) Robert Demarest after Patten, Carlson & others; (c) Robert Demarest with the permission of M. Cummings, *Human Heredity: Principles and Issues*, p. 126 third edition, © 1994 Brooks/Cole. All rights reserved. **12.7** Art,

Raychel Ciemma and Preface, Inc.; photograph, Carolina Biological Supply Company. **12.8** Raychel Ciemma. **12.9** (a,b) Precision Graphics; (b) Dr. Victor A. McKusick; (c) Steve Uzzell. **Page 204** Precision Graphics. **12.10** Lisa Starr. **12.11** Giraudon/Art Resource, New York. **12.12** (a) Lisa Starr; (b) photograph © Bettmann/Corbis, art after V. A. McKusick, *Human Genetics*, second edition, © 1969, reprinted by permission, Prentice-Hall, Inc., Englewood Cliffs, N.J. **12.13** C. J. Harrison. **12.14** Eddie Adams/AP/Wide World Photos. **Page 206** Precision Graphics. **12.15** (a,b) Courtesy G. H. Valentine. **12.16** From "Multicolor Spectral Karyotyping of Human Chromosomes," by E. Schrock, T. Ried, et al, *Science*, 26, July 1966, 273:496. Used by permission of E. Schrock and T. Reid and the American Association for the Advancement of Science. **Page 207** Precision Graphics. **12.17** Raychel Ciemma. **12.18** Left, permission of Carole Lafrate; (center) courtesy of Peninsula Association for Retarded Children and Adults, San Mateo Special Olympics, Burlingame, CA; right, Courtesy Special Olympics; karyotype, © 1997, Hironao Numabe, M.D., Tokyo Medical University. **12.19** Preface, Inc. **12.20** Left, UNC Medical Illustration and Photography; right, © 1997, Hironao Numabe, M.D., Tokyo Medical University. **12.21** (a) © 1997, Hironao Numabe, M.D., Tokyo Medical University; (b, inset) Stefan Schwarz. **12.22** Raychel Ciemma. **12.23** Art, Lisa Starr; photographs, courtesy Lennart Nilsson from *A Child Is Born*, © 1966, 1977 Dell Publishing Company, Inc. **12.24** Photograph, (a) Fran Heyl Associates © Jacques Cohen, computer-enhanced by © Pix Elation; (b) Raychel Ciemma. **12.25** Carolina Biological Supply Company. **12.26** Precision Graphics. **12.27** Raychel Ciemma. **Page 215** © Mitchell Gerber/CORBIS

CHAPTER 13

13.1 A. C. Barrington Brown, © 1968 J. D. Watson. **13.2** Lisa Starr. **13.3** Raychel Ciemma. **13.4** (a,b) Raychel Ciemma; (c) Lee D. Simon/Science Source /Photo Researchers, Inc. **13.5** Raychel Ciemma. **13.6** Art, above, Raychel Ciemma; below, Gary Head; micrograph, Biophoto Associates/SPL/ Photo Researchers, Inc. **Page 221** Preface, Inc. **13.7** Lisa Starr. **13.9, 13.10** Precision Graphics. **13.11** (a) Daniel Fairbanks; (b) PA News Photo Library. **13.12** Courtesy of Advanced Cell Technology, Inc., Worcester, Massachusetts

CHAPTER 14

14.1 Above, Dennis Hallinan/FPG; below, © Bob Evan/Peter Arnold, Inc. **14.2, 14.3** Precision Graphics. **14.4** Lisa Starr. **14.5** Gary Head. **14.6, 14.7** Precision Graphics. **14.8** (a) model by Dr. David B. Goodin, The Scripps Research Institute; (b,c) Lisa Starr. **14.9** (a) Courtesy of Thomas A. Steitz from Science; (b) Lisa Starr. **14.10** Lisa Starr. **14.11** (a) Lisa Starr; (b) Precision Graphics. **14.12** Lisa Starr. **14.13** Left, Nik Kleinberg; right, Peter Starlinger. **14.14** Lisa Starr. **14.15** Courtesy of the National Neurofibromatosis Foundation

CHAPTER 15

15.1 Photos (a) Ken Greer/Visuals Unlimited; (b) Biophoto Associates/Science Source/Photo Researchers; (c) James Stevenson/SPL/Photo Researchers; right, Gary Head. **Page 238** Lisa Starr. **15.2** Raychel Ciemma and Gary Head. **15.3** Lisa Starr. **15.4** Raychel Ciemma. **15.5** (a) Carolina Biological Supply Company; (b) UCSF Computer Graphics Laboratory, National Institutes, NCRR Grant 01081. **15.6** (a) Dr. Karen Dyer Montomery; (b) Raychel Ciemma. **15.7** Jack Carey. **15.8** Above, Raychel Ciemma; below, W. Beerman. **15.9** (a) Raychel Ciemma and Gary Head; (b) Frank B. Salisbury. **15.10** (a) Lennart Nilsson © Boehringer Ingelheim International GmbH; (b, c) Betsy Palay/Artemis. **15.11** Betsy

Index of Applications

More Applications in the Text

A quick reference to chapter opening topics and Focus Essays on medical, environmental, and social issues

For additional big-picture applications, see the list of CONNECTIONS ESSAYS on page xxiii of the Preface

Student CD Animations and Interactions

Red algae *interaction*
Porphyra life cycle *animation*
Cellular slime mold *animation*

CHAPTER 23
Evolutionary tree for plants *animation*
Haploid to diploid dominance *interaction*
Milestones in plant evolution *interaction*
Liverwort *animation*
Moss life-cycle *animation*
Seedless vascular plants *interaction*
Fern life-cycle *animation*
Pinus cones *interaction*
Life cycle of pine *animation*
Life cycle of a lily *animation*

CHAPTER 24
Mycelium *animation*
Life cycle of a club fungus *animation*
Basidiocarp diversity *interaction*
Rhizopus life cycle *animation*
Sac fungi *interaction*
Predatory fungus *video*
Lichen *animation*
Mycorrhiza *interaction*

CHAPTER 25
Body symmetry *interaction*
Body cavities *interaction*
Cleavage patterns *interaction*
Coelom formation *interaction*
Sponge body plan *animation*
Nematocyst *animation*
Two common cnidarian body plans *animation*
Life cycle of *Obelia* *animation*
Organ systems of a planarian *interaction*
Life cycle of a beef tapeworm *animation*
Blood fluke life cycle *interaction*
Body plan of a roundworm *animation*
Rotifer *video*
Molluscan classes *interaction*
Torsion in gastropods *animation*
Snail body-plan *interaction*
Clam body-plan graphic *animation*
Cuttlefish body plan graphic *animation*
Marine polychaete photos *interaction*
Feeding leech *interaction*
Earthworm body plan *interaction*
Chelicerates *interaction*
Crustacean *interaction*
Insect headparts *interaction*
Insect development *interaction*
Body plan of a sea star *interaction*
Water–vascular system *video*

CHAPTER 26
Tunicate *animation*
Lancelet *animation*
Jawless fishes *interaction*
Gill-supporting structures *interaction*
Cartilaginous fishes *interaction*
Bony fish body *animation*
Lobed-fin fish and early amphibian *interaction*

Salamander and fish locomotion *animation*
Body plan of a crocodile *animation*
Tortoise *interaction*
Feather development *animation*
Bird's egg *animation*
Avian bone and muscle *interaction*
Mammalian origins and radiations *interaction*
Mammalian dentition *interaction*
Primate skeletons *interaction*
Skulls of extinct primates *interaction*
Fossils of australopiths *animation*
Homo skulls *interaction*
Genetic distance between human groups *animation*
Summary of human evolution *interaction*

CHAPTER 27
Five major extinctions *interaction*
Group-specific diversity patterns *interaction*
Habitat loss and fragmentation *interaction*
Whales caught, graphic *animation*
Biodiversity hot spots *interaction*

CHAPTER 28
Morphology of a tomato plant *animation*
Structures in human respiration *interaction*
Adaptation questions *animation*
Extracellular fluid *animation*
Feedback components *interaction*
Feedback control of temperature *animation*
Compartmentalization *interaction*
Leaf movement *movie animation*
ABC model for flowering *animation*
Communication among neurons *animation*

CHAPTER 29
Plant tissue systems *interaction*
Cutting tissue specimens *interaction*
Apical meristems *interaction*
Ground tissues *interaction*
Vascular tissues *interaction*
Monocots and dicots graphic *interaction*
Shoot differentiation *interaction*
Monocot and dicot stems *interaction*
Monocot and dicot leaves *interaction*
Simple and compound leaf *interaction*
Leaf structure *interaction*
Root types *interaction*
Root cross-section *interaction*
Walnut tree, twig *interaction*
Secondary growth *animation*
Secondary growth in a root *interaction*
Layers in a woody stem *interaction*
Annual rings *interaction*

CHAPTER 30
Plant nutrient *interaction*
Soil profile *interaction*
Water absorption *animation*
Water transport *animation*
Stomata *interaction*
Interdependent processes *animation*
Pressure flow *animation*

CHAPTER 31
Pollination *animation*
Bee-attracting flower pattern *interaction*
Flowering plant life cycle diagram *animation*
Flower parts *interaction*
Microspores to pollen *animation*
Megaspores to eggs *animation*
Double fertilization *animation*
Capsella seed development *animation*
Apple fruit structure *interaction*

CHAPTER 32
Dicot and monocot development *interaction*
Cellular basis for plant growth *interaction*
Cell shapes *interaction*
Coleoptile tip transplant *animation*
Phototropism *interaction*
Gravitropism *video*
Statolith movement *interaction*
Phytochrome interconversion *animation*
Flowering and daylength *interaction*
Daylength and dormancy *interaction*
Vernalization *interaction*

CHAPTER 33
General epithelial structure *animation*
Types of simple epithelium *interaction*
Cell junctions *interaction*
Soft connective-tissue *interaction*
Specialized connective-tissue *interaction*
Muscle tissue *interaction*
Organ systems *interaction*
Germ layers *interaction*
Directional terms and planes of symmetry *interaction*
Body cavities *interaction*

CHAPTER 34
Functional zones of a motor neuron *interaction*
Membrane potential defined *interaction*
Ion concentrations *interaction*
Ion concentrations *animation*
Measuring membrane potential *animation*
Threshold level *interaction*
Action potential step-by-step *interaction*
Action potential *animation*
Neuromuscular junction *interaction*
Synapse function *animation*
Synaptic integration *interaction*
Structure of a nerve *animation*
Saltatory conduction *animation*
Direction of information flow *interaction*
Stretch reflex *animation*
Nerve net *interaction*
Bilateral nervous systems *interaction*
Regions of the vertebrate brain *interaction*
Nervous system divisions *animation*
Autonomic nerves *animation*
Autonomic effects *interaction*
Spinal cord *interaction*
Human brain development *interaction*

Internet Exercises/Media Labs